AF871151

Molecular Insights into Proteinopathies

Molecular Insights into Proteinopathies

Increased Scope From Neurodegeneration to Other Pathologies of the Brain

Edited by

EVA ŽEROVNIK

Department of Biochemistry and Molecular and Structural Biology, Jožef Stefan Institute, Ljubljana, Slovenia; Jožef Stefan International Postgraduate School (IPS), Ljubljana, Slovenia

ROBERT LAYFIELD

Department of Protein Biochemistry, The University of Nottingham Medical School Queen's Medical Centre Nottingham, Nottingham, United Kingdom

ELSEVIER

ACADEMIC PRESS

An imprint of Elsevier

Academic Press is an imprint of Elsevier
125 London Wall, London EC2Y 5AS, United Kingdom
50 Hampshire Street, 5th Floor, Cambridge, MA 02139, United States

Copyright © 2027 Elsevier Inc. All rights are reserved, including those for text and data mining, AI training, and similar technologies.

For accessibility purposes, images in electronic versions of this book are accompanied by alt text descriptions independently provided by Elsevier, generated with the aid of AI. For more information, see https://www.elsevier.com/about/accessibility or Radarweg 29, 1043 NX Amsterdam, Netherlands.

Books and Journals published by Elsevier comply with applicable product safety requirements. For any product safety concerns or queries, please contact our authorised representative, Elsevier B.V., at productsafety@elsevier.com.

Publisher's note: Elsevier takes a neutral position with respect to territorial disputes or jurisdictional claims in its published content, including in maps and institutional affiliations.

No part of this publication may be reproduced or transmitted in any form or by any means, electronic or mechanical, including photocopying, recording, or any information storage and retrieval system, without permission in writing from the publisher. Details on how to seek permission, further information about the Publisher's permissions policies and our arrangements with organizations such as the Copyright Clearance Center and the Copyright Licensing Agency, can be found at our website: www.elsevier.com/permissions.

This book and the individual contributions contained in it are protected under copyright by the Publisher (other than as may be noted herein).

Notices
Knowledge and best practice in this field are constantly changing. As new research and experience broaden our understanding, changes in research methods, professional practices, or medical treatment may become necessary.

Practitioners and researchers must always rely on their own experience and knowledge in evaluating and using any information, methods, compounds, or experiments described herein. In using such information or methods they should be mindful of their own safety and the safety of others, including parties for whom they have a professional responsibility.

To the fullest extent of the law, neither the Publisher nor the authors, contributors, or editors, assume any liability for any injury and/or damage to persons or property as a matter of products liability, negligence or otherwise, or from any use or operation of any methods, products, instructions, or ideas contained in the material herein.

ISBN: 978-0-443-34176-2

For Information on all Academic Press publications
visit our website at https://www.elsevier.com/books-and-journals

Publisher: Megan Ball
Acquisitions Editor: Michelle Fisher
Editorial Project Manager: Prachi Mittal
Production Project Manager: Vijayaraj Purushothaman
Cover Designer: Christian Bilbow

Working together to grow libraries in developing countries
www.elsevier.com • www.bookaid.org

Typeset by MPS Limited, Chennai, India

Contents

Part 4 Contemporary biophysical methods and machine learning

List of contributors

Hannah Bailey
Queen's Medical Centre, School of Life Sciences, University of Nottingham Medical School, Nottingham, United Kingdom

Nicholas J. Bradshaw
Faculty of Biotechnology & Drug Development, University of Rijeka, Rijeka, Croatia

Eva De Lago
Departamento de Bioquímica y Biología Molecular, Facultad de Medicina, Instituto Universitario de Investigación en Neuroquímica, Universidad Complutense, Madrid, Spain; Instituto Ramón y Cajal de Investigación Sanitaria (IRYCIS), Madrid, Spain; Centro de Investigación Biomédica en Red de Enfermedades Neurodegenerativas (CIBERNED), Madrid, Spain

Meredith Edwards
John Curtin School of Medical Research, College of Science and Medicine, The Australian National University, Canberra, ACT, Australia

Javier Fernández-Ruiz
Departamento de Bioquímica y Biología Molecular, Facultad de Medicina, Instituto Universitario de Investigación en Neuroquímica, Universidad Complutense, Madrid, Spain; Instituto Ramón y Cajal de Investigación Sanitaria (IRYCIS), Madrid, Spain; Centro de Investigación Biomédica en Red de Enfermedades Neurodegenerativas (CIBERNED), Madrid, Spain

Altijana Hromić-Jahjefendić
Genetics and Bioengineering Study Program, Faculty of Engineering and Natural Sciences, International University of Sarajevo, Sarajevo, Bosnia and Herzegovina

Maja Juković
Faculty of Biotechnology & Drug Development, University of Rijeka, Rijeka, Croatia

Madeleine King
Division of Physiology Pharmacology and Neuroscience, School of Life Sciences, Queen's Medical Centre, University of Nottingham Medical School, Nottingham, United Kingdom

Kaja Kolmancic
Department of Nuclear Medicine, Institute of Oncology, Ljubljana, Slovenia

Lesia Kolomiiets
Institute of Biochemistry and Biophysics, Polish Academy of Sciences, Warsaw, Poland; Institute of Molecular Biology and Genetics, National Academy of Sciences of Ukraine, Kyiv, Ukraine

Matea Kršanac
Faculty of Biotechnology & Drug Development, University of Rijeka, Rijeka, Croatia

Robert Layfield
Department of Protein Biochemistry, The University of Nottingham Medical School Queen's Medical Centre Nottingham, Nottingham, United Kingdom

Weidong Le
Key Laboratory of Liaoning Province for Research on the Pathogenic Mechanisms of Neurological Diseases, First Affiliated Hospital of Dalian Medical University, Dalian, Liaoning, P.R. China

Aristea Pavešić Radonja
Psychiatry Clinic, Clinical Hospital Centre Rijeka, Rijeka, Croatia; Faculty of Medicine, University of Rijeka, Rijeka, Croatia

Matej Perovnik
Department of Neurology, University Medical Centre Ljubljana, Ljubljana, Slovenia

Lea Pogačnik da Silva
Department of Food Science and Technology, Biotechnical Faculty, University of Ljubljana, Ljubljana, Slovenia

Nataša Poklar Ulrih
Department of Food Science and Technology, Biotechnical Faculty, University of Ljubljana, Ljubljana, Slovenia

Carmen Rodríguez-Cueto
Departamento de Bioquímica y Biología Molecular, Facultad de Medicina, Instituto Universitario de Investigación en Neuroquímica, Universidad Complutense, Madrid, Spain; Instituto Ramón y Cajal de Investigación Sanitaria (IRYCIS), Madrid, Spain; Centro de Investigación Biomédica en Red de Enfermedades Neurodegenerativas (CIBERNED), Madrid, Spain

Abas Sezer
Genetics and Bioengineering Study Program, Faculty of Engineering and Natural Sciences, International University of Sarajevo, Sarajevo, Bosnia and Herzegovina

Jereme G Spiers
John Curtin School of Medical Research, College of Science and Medicine, The Australian National University, Canberra, ACT, Australia

Ilijana Stanivuk
Psychiatry Clinic, Clinical Hospital Centre Rijeka, Rijeka, Croatia; Faculty of Medicine, University of Rijeka, Rijeka, Croatia

Joern R Steinert
School of Life Sciences, Faculty of Medicine and Health Sciences, University of Nottingham, Nottingham, United Kingdom

Jasmin Šutković
Genetics and Bioengineering Study Program, Faculty of Engineering and Natural Sciences, International University of Sarajevo, Sarajevo, Bosnia and Herzegovina

Vladimir N. Uversky
Department of Molecular Medicine, University of South Florida, Tampa, FL, United States

Natasha Vassileff
John Curtin School of Medical Research, College of Science and Medicine, The Australian National University, Canberra, ACT, Australia

Katja Venko
Theory Department, National institute of Chemistry, Ljubljana, Slovenia

Panpan Wang
Key Laboratory of Liaoning Province for Research on the Pathogenic Mechanisms of Neurological Diseases, First Affiliated Hospital of Dalian Medical University, Dalian, Liaoning, P.R. China

Hui Xu
Key Laboratory of Liaoning Province for Research on the Pathogenic Mechanisms of Neurological Diseases, First Affiliated Hospital of Dalian Medical University, Dalian, Liaoning, P.R. China

Beti Zaharija
Faculty of Biotechnology & Drug Development, University of Rijeka, Rijeka, Croatia

Eva Žerovnik
Department of Biochemistry and Molecular and Structural Biology, Jožef Stefan Institute, Ljubljana, Slovenia; Jožef Stefan International Postgraduate School (IPS), Ljubljana, Slovenia

Igor Zhukov
Institute of Biochemistry and Biophysics, Polish Academy of Sciences, Warsaw, Poland

Preface

Proteins are unusual molecules, built from simple chains of polymerized amino acids. As scientists, we started to study them from a physical chemistry point of view, understanding their stability and conformations using various spectroscopy methods, including circular dichroism and fluorescence. Then, we developed a theoretical understanding of their complex energy landscapes and the problem of finding the right path to a correctly folded state. We uncovered intermediate states on these folding pathways, most notably the so-called molten globule, believed to form by hydrophobic amino acids collapsing into a molten core, while the more hydrophilic and charged amino acids would be surface exposed and in contact with the aqueous environment.

Much later, the problem of protein folding was resolved, and even the process of protein folding in the cellular environment was understood, identifying a role for chaperone proteins, the helpers of proper protein folding, and where needed, disaggregation, in the crowded environment of the cell. When some proteins misfold and become aggregated (noting there are many different forms of protein aggregates from amorphous to regular amyloid fibrils, from loose condensates in droplets to irreversible solid and fibrillar aggregates) and form clumps inside the cell, the two endogenous protein degradation systems can clear these aggregates. However, some protein aggregates are physiologically relevant and exert incompletely understood functions, regulating molecular interactions and even building memories.

Today, provided with an amino acid sequence, one can use sophisticated sofware that predict a protein fold with high accuracy (AlphaFold 2), and also complexes with other proteins or small molecules (AlphaFold 3 and similar). By machine learning and neural networking, properties of the protein as a whole and its constituent independently folded parts (domains) can be predicted.

This brings us to our book "*Molecular Insights Into Proteinopathies: Increased Scope From Neurodegeneration to Other Pathologies of the Brain,*" related to the problem of proteinopathies: the common pathologies where protein misfolding and aggregation play a major role. In this book, we reveal some of the proteinopathies relevant to brain pathology. Not only Alzheimer disease, but also other diseases of the brain and the central nervous system, such as amyotrophic lateral sclerosis (ALS), as well as major psychiatric illnesses and epilepsy, may have protein aggregation as a complication, driving disease symptoms. By describing some of the underlying mechanisms operating in these conditions, we hope to provide a basis for rational development of better therapies for proteinopathies. Such therapies may be oriented specifically to specific disease-associated proteins, or more broadly, small molecule modulators may offer opportunties to broadly target proteinopathies in a protein-agnostic manner.

We thank all the authors for contributing their chapters and sharing their knowledge and perspectives.

Acknowledgments

I would like to acknowledge the opportunity to work with many good professors in the field. To name just a few: Vito Turk (proteolysis, cathepsins, and their inhibitors), Franc Gubenšek and Savo Lapanje (physical chemistry and protein denaturation) at Jožef Stefan Institute and University of Ljubljana, Slovenia, respectively, and Roger H. Pain (protein conformation and protein folding, especially the role of molten globule intermediates) at the University of Newcastle upon Tyne, United Kingdom. Collaborations with other leaders were also fruitful. One of the important collaborations was with Jonathan P. Waltho (Biomolecular NMR) at the University of Sheffield, United Kingdom. For funding my research, apart from ARIS projects, the program P1-0140 (led by Boris Turk) is acknowledged.

For continuous support in writing this book I would like to thank my close family and many colleagues and friends at Jožef Stefan Institute, especially Vida Puizdar.

—Eva Žerovnik

I would like to acknowledge in particular the late Prof. R. John Mayer who first sparked my interest in protein degradation and neurodegeneration at the University of Nottingham along with Dr. Mike Landon and Prof. Jim Lowe, as well as Prof. Maria Spillantini in Cambridge. I also would like to acknowledge many more collaborators and funders—past and present (too many to name in person, but including the MND Association and StandAgainstMND)—who supported our research. Finally, I also would like to recognize the courage of the millions of people worldwide who live, or lived with, the devastating conditions that the editors and authors have the privilege to research. Many of these people touched my own life and helped shape me as a person, especially the late Sam Perkins, who it was a privilege to know, sadly only for a short time.

—Robert Layfield

PART 1

Protein aggregation in different proteinopathies

CHAPTER 1

Amyloid-pore hypothesis: prefibrillar oligomers interactions with membranes; a generic cause of neurodegeneration?

Eva Žerovnik[1,2]
[1]Department of Biochemistry and Molecular and Structural Biology, Jožef Stefan Institute, Ljubljana, Slovenia,
[2]Jožef Stefan International Postgraduate School (IPS), Ljubljana, Slovenia

1.1 Introduction

The amyloid cascade hypothesis was formulated by Hardy and Higgins (1992). In general terms, it proposes that amyloid-beta (Aβ) protein aggregation into amyloid fibrils is an initial event upstream of other molecular aberrations that promote neurodegeneration. The original amyloid cascade hypothesis was later modified by Selkoe and Hardy (2016). The latest evidence indicates that the most toxic species are not in fact mature amyloid plaques, but rather soluble prefibrillar oligomers of Aβ that are precursors to insoluble fibrillar aggregates. Hashimoto et al. (2003) discussed possible mechanisms of toxicity associated with these soluble oligomers, highlighting evidence that Aβ and α-synuclein oligomers (the latter relevant to Parkinson's disease [PD] and other synucleinopathies) directly interact with mitochondrial and plasma membranes, thereby promoting damage and ensuing toxicity. Accumulation of Aβ and α-synuclein oligomers in the mitochondrial membrane may result in the release of cytochrome c with the subsequent activation of the apoptosis cascade. Conversely, oxidative stress and mitochondrial dysfunction associated with AD and PD, leading to increased membrane permeability and cytochrome c release, may further promote Aβ and α-synuclein oligomerization, creating a kind of vicious circle. The modulated amyloid cascade hypothesis of AD can be referred to as the "amyloid-pore" or the "ion channel" hypothesis (Kagan et al., 2002) (Fig. 1.1).

Naturally occurring pore-forming proteins (PFPs) exist in virtually all domains of life (Zerovnik, 2021) and, by disrupting cellular membranes, depending on the pore size, they can cause ion and small-molecule imbalance, or even inappropriate protein efflux/influx, influencing cell-signaling pathways and ultimately cell fate. Such PFPs exist from bacteria to viruses and also shape mammalian host defense systems (e.g., antimicrobial peptides[AMPs]), as part of the innate immune response. There is strong and growing evidence in neurodegenerative disorders that amyloid toxicity could, using parallel

Molecular Insights into Proteinopathies
DOI: https://doi.org/10.1016/B978-0-443-34176-2.00001-8
© 2027 Elsevier Inc. All rights are reserved, including those for text and data mining, AI training, and similar technologies.

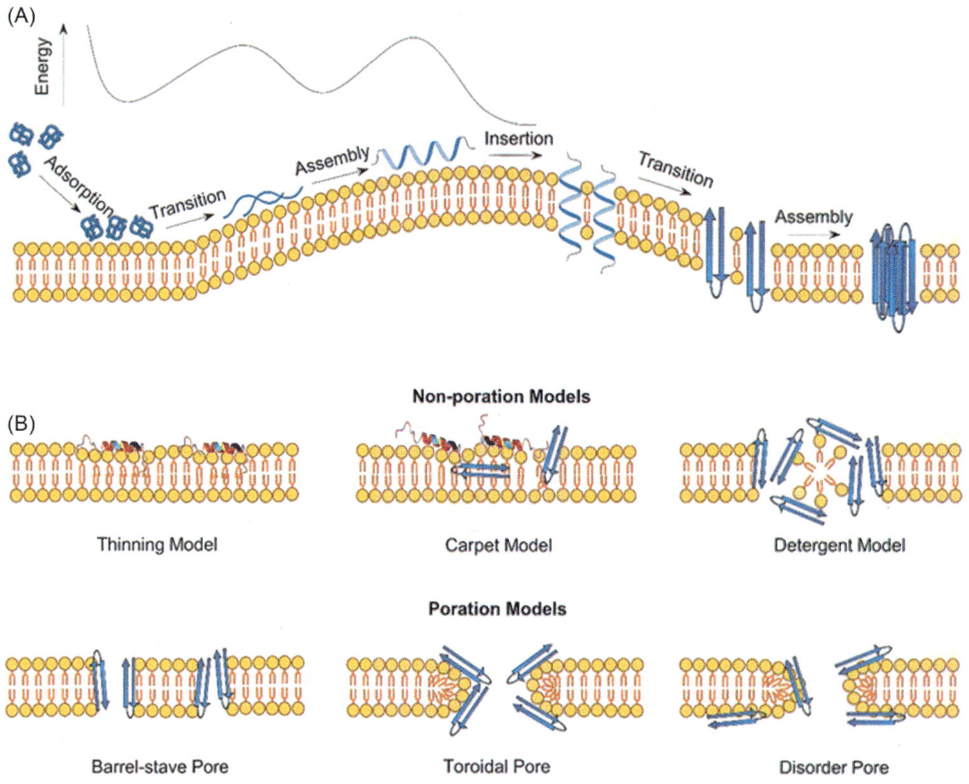

Figure 1.1 (A) A general peptide–membrane interaction process in a free-energy landscape. It involves initial peptide adsorption, structural transition, self-assembly, and insertion pathways. (B) Illustrates distinct nonporation (carpet, detergent, and membrane thinning models) and poration (barrel-stave, toroidal, and disorder pores) membrane-disruption mechanisms of antimicrobial peptides (AMPs) and amyloid peptides (AMYs). *Adapted from Tang, Y., Zhang, Y., Zhang, D., Liu, Y., Nussinov, R., Zheng, J., 2024. Exploring pathological link between antimicrobial and amyloid peptides. Chemical Society Reviews. 53 (17), 8713–8763. https://doi.org/10.1039/d3cs00878a. License Creative Commons 3.0.*

mechanisms to those mediated by PFPs and AMPs (Tang et al., 2024), result from prefibrillar oligomers forming amyloid pores (channels) in cellular membranes (Kayed et al., 2003). If not by perforating membranes directly, toxicity might also arise from oligomers promoting membrane voltage imbalance or from the direct interaction of oligomers with transmembrane channels (see Fig. 1.1). For example, PD pathomechanisms may include interference of α-synuclein oligomers with voltage-dependent anion channels proteins in the outer mitochondrial membrane (Risiglione et al., 2021).

The aim of this chapter is to review what is known about amyloid pores and which areas remain open for further research. Towards the end of the chapter, insights on amyloid pores from a theoretical perspective are given. Finally, the chapter considers whether we could, in the near future, »see« or visualize such oligomer-induced small pores/channels in living systems using advanced microscopy and single-molecule spectroscopy.

1.2 The mechanism of protein aggregation into amyloid

In general terms, the mechanism of amyloidogenesis of proteins consists of the nucleation phase, the growth phase, and the stationary (plateau) phase. The kinetics usually follow a sigmoidal curve. During the lag phase, oligomeric seeds are generated, and in the growth phase, partially unfolded monomers add to the seeds, forming protofibrils, which consolidate into mature fibrils at the end of the reaction. The growth phase varies from protein to protein and can be either linear or exponential (Dovidchenko et al., 2014). Theoretical and experimental aspects of the reactions with a nucleation stage have been studied for a wide range of molecules of different sizes, not only proteins (Chayen, 2005), including, for example, in the crystallization of other macromolecules.

In a study by Dovidchenko et al. (2014), the authors state: "To make a complete description of the mechanism of amyloidogenesis, it is necessary to find all possible conformational states and oligomers of the protein during its aggregation path." Furthermore, they claim that "the description would not be complete without a corresponding determination of kinetic and thermodynamic parameters for all potential forms and states of the protein in the course of aggregation."

Our group has studied the mechanisms of protein folding and aggregation of a model amyloidogenic protein—human stefin B (cystatin B). One should be aware that any protein, under appropriate conditions (not necessarily physiological), can undergo a transition into the amyloid state (Chiti and Dobson, 2006). Not surprisingly, the process of human stefin B amyloid fibril formation (Škerget et al., 2009; Taler-Verčič et al., 2013) was found to be similar to that of some other amyloidogenic proteins. In contrast, in the process of amyloid fibril formation of human stefin B, we observed conformational changes and domain swapping (Žerovnik et al., 2011), leading to the formation of off-pathway tetramers (Jenko Kokalj et al., 2007; Taler-Verčič et al., 2013), which reduced the velocity of fibril formation at higher protein concentrations. On the amyloidogenic pathway (see Fig. 1.2), we have observed partially unfolded (open) monomers, which undergo seeded oligomerization and grow into protofibrils, finally forming mature fibrils. We suspect that oligomers are the species interacting with biological membranes and forming amyloid pores—to be explained in the next sections of this chapter. Whether oligomers are on-pathway to fibril formation, or off-pathway, remains somewhat unclear (see Fig. 1.2). The morphology of the species was observed by several microscopy approaches, in addition to dynamic light scattering (not shown) to determine their approximate size (see Fig. 1.3).

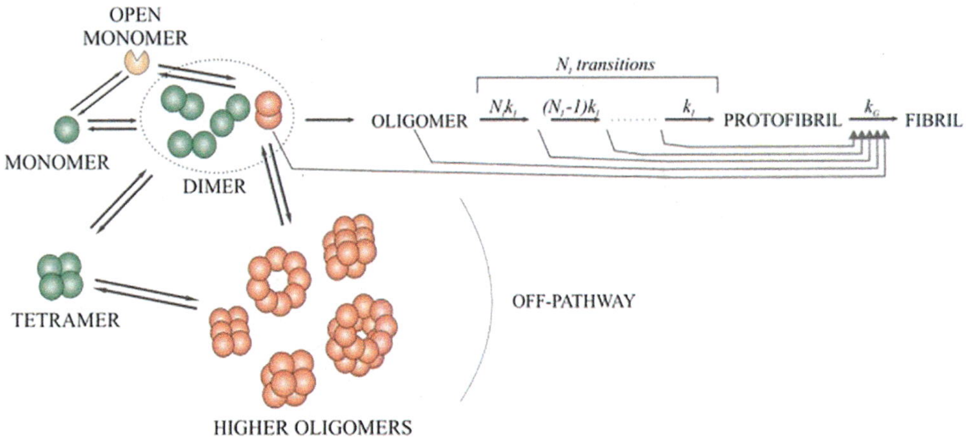

Figure 1.2 The kinetic scheme of amyloid formation by human stefin B, based on previous studies (Jenko Kokalj et al., 2007; Škerget et al., 2009; Žerovnik et al., 2007) and the results presented by Taler-Verčič et al. (2013). The off-pathway explanation has been confirmed and improved experimentally. Nontoxic species are colored green, toxic species are colored red, and potentially toxic species (open monomer) are colored orange. *Reproduced from Taler-Verčič, A., Kirsipuu, T., Friedemann, M., Noormägi, A., Polajnar, M., Smirnova, J., et al., 2013. The role of initial oligomers in amyloid fibril formation by human stefin B. International Journal of Molecular Sciences. 14 (9), 18362–18384. https://doi.org/10.3390/ijms140918362. Open access: Creative Commons CC BY license.*

To simulate the mechanism of sigmoidal fibril growth after the primary nucleation, more kinetic data were evaluated to uncover a more complex picture. In addition, to primary nucleation, secondary nucleation (Cohen et al., 2013) and fragmentation (Kundel et al., 2018) of fibrils gain importance under different circumstances. The situation in vivo may be somewhat different, as chaperones play a role in mediating folding and regulating fibril growth (Sinnige, 2022). Also, *in cellulo*, two endogenous protein degradation pathways, the ubiquitin-proteasome system and autophagy, play various roles in removing protein oligomers and aggregates (Ciechanover and Kwon, 2015).

We contributed to understanding the role of proline residues as a switch (Taler-Vercic et al., 2017) in the processes of oligomerization (Jenko Kokalj et al., 2007) and amyloid fibril formation (Hasanbašić et al., 2019). In the stefin B tetramer, the structure of which was determined both by NMR and X-ray diffraction, Pro74 was found to be in a *cis* isomeric state. The results of our first study (Smajlović et al., 2009) on the role of Pro74 in protein aggregation demonstrated that peptidyl-prolyl *cis/trans* isomerism of this residue, regulated by a peptidyl-prolyl isomerase (PPI), cyclophilin A, is rate-limiting. In the second study performed nearly 10 years later (Hasanbašić et al., 2019), we mutated all five prolines of stefin B (Fig. 1.4) and their combinations. The results showed that Pro74 *cis/trans* isomerization is a key event, acting as a switch in the

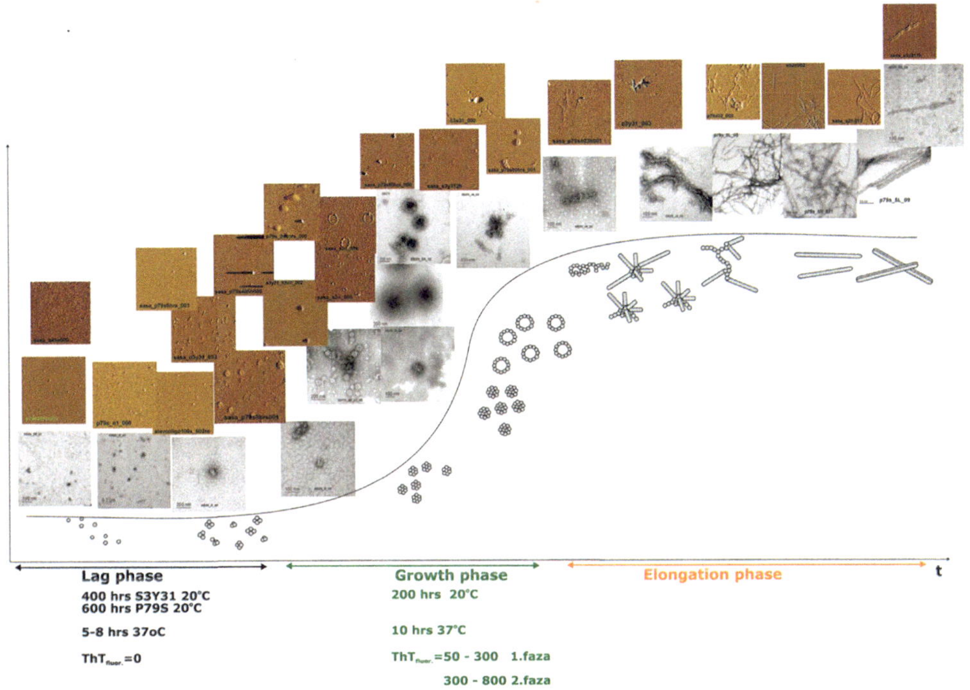

Figure 1.3 Morphology of different oligomers formed by human stefin B at different time points in the growth phase of amyloid formation, as observed by atomic force microscopy and transmission electron microscopy; note how rings and hollow balls appear at different phases. *Credit to S. Jenko Kokalj, E. Žerovnik and D. Turk, unpublished. TEM was performed by Magda Tušek Žnidarič (NIB, Ljubljana) and AFM contributed by co-worker from Biophysics Miha Škarabot (JSI, Ljubljana).*

transition to amyloid fibrils (Fig. 1.3). We simulated the kinetics of all the mutants and, unexpectedly, the nucleation phase, rather than the fibril elongation phase, was the step affected by Pro *cis/trans* isomerization.

Low et al. (2022) similarly report that the PPI-ase domain of CyP40 interacts with proline residues in the β turns of amyloid fibrils and isomerizes them (from *cis* to *trans* or *vice versa*), which disrupts the β-turn structure and thus disaggregates the amyloid fibrils. Others have also discussed theoretically that certain residues with specific properties, such as glycines and prolines, modulate amyloid fibril structure and growth (Taylor and Staniforth, 2022).

1.3 Toxic oligomers of amyloid-forming proteins make pores in biological membranes

Aβ peptide derives from a much larger transmembrane protein, amyloid precursor protein, by proteolysis. Aβ binds to the plasma membrane either in a carpeting manner,

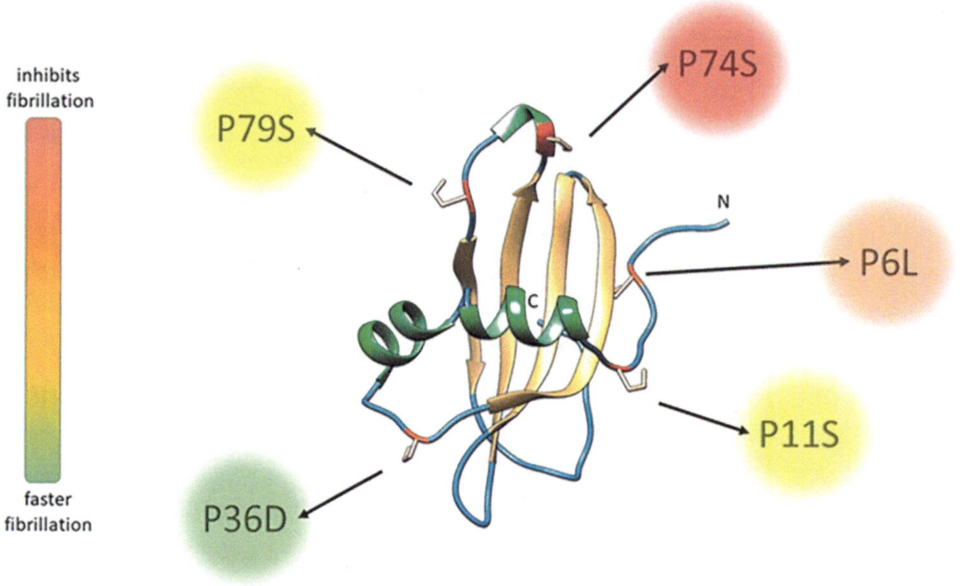

Figure 1.4 Regulation of amyloid formation by key proline residues in human stefin B. The proline mutants influenced the distribution of oligomers between monomers, dimers, and tetramers. The P74S single mutant, two double mutants (P6L/P74S and P74S/P79S), and the triple mutant P6L/P11S/P74S did not fibrillate. Pro74 *cis/trans* isomerization proved to be the key event, acting as a switch toward an amyloid transition. *Reproduced from Hasanbašić, S., Taler-Verčič, A., Puizdar, V., Stoka, V., Tušek Žnidarič, M., Vilfan, A., et al., 2019. Prolines affect the nucleation phase of amyloid fibrillation reaction; mutational analysis of human stefin B. ACS Chemical Neuroscience. 10 (6), 2730–2740. https://doi.org/10.1021/acschemneuro.8b00621. The article is licensed under the Open Access Creative Commons CC-BY license.*

in a detergent-like manner, or by amyloid-pore formation (Wiatrak et al., 2021) (see Fig. 1.1). The pores (i.e., transmembrane channels) facilitate the migration of ions and water through the membrane, leading to intracellular influx of Ca^{2+}, which can lead to excitotoxicity and neuronal death. The cation-permeable pores likely form in the plasma membrane (Shah et al., 2002) but also in the membranes of intracellular organelles, such as mitochondria (Vassallo, 2021). Of interest, the more amyloid-prone Aβ-42 is also much more prone to forming pores in membranes (Bode et al., 2017) than is Aβ-40. However, synergy between Aβ-40 and Aβ-42 in perforating membranes has been reported (Chang et al., 2018).

Di Scala et al. (2016) described the molecular mechanism of amyloid-pore formation by Aβ and α-synuclein. A panel of amyloid-forming fragments of the two proteins was compared. The results showed a two-step mechanism, where gangliosides and cholesterol, components of lipid membranes each interact with specific structural motifs of Aβ and α-synuclein. Fusco et al. (2016) characterized membrane-bound α-synuclein. In order to improve the study, bearing in mind that the α-synuclein pore could be contained in a "soluble" fraction, the team (Fusco et al., 2017) used both solid-state and solution NMR to determine the structural constraints of α-synuclein membrane interactions. The structured region of α-synuclein strongly inserted into lipid bilayers and disrupted their integrity, leading ultimately to cell death. Mutations that prevented membrane interaction also prevented toxicity.

In another study, the early stage of Aβ peptide interaction with synaptic plasma membranes was examined by using solid-state NMR (Kenyaga et al., 2022). The authors showed that different parts of the hydrophobic sequence of Aβ peptides initiated membrane binding and β-strand assembly. Lipid dynamics were slowed down by reduced lateral diffusion. The early events influenced the structural polymorphism of membrane-associated Aβ fibrils. Based on these results, they proposed a mechanism by which Aβ-lipid interactions drive the membrane-associated nucleation process (Kenyaga et al., 2022).

Many reports can be found on the interaction of Aβ with cellular membranes (Bode et al., 2017) and with membranes in animal models (Julien et al., 2018), confirming the hypothesis of "live" membrane perforation by the so-called amyloid toxins. For example, in *C. elegans*, the membrane repair response was turned on when Aβ was fed to the animals (Julien et al., 2018). In addition, a protective effect of oleuropein aglycone in an animal model of transgenic *C. elegans* strains expressing Aβ-42 was observed, with reduced amyloid plaques and less motor deficit (Diomede et al., 2013). Cellular models may lead us to a deeper understanding of the detrimental effects of soluble oligomers of Aβ on neurones and glial cells, eventually leading to the development of innovative therapies for AD (Fontana et al., 2020).

Most oligomers of amyloid-forming proteins other than Aβ are also prone to form such pores (Lashuel et al., 2002; Glabe and Kayed, 2006). A review of possible structures of oligomers of Aβ, α-synuclein, tau, and prion protein as obtained experimentally and by molecular dynamics (MD) appeared recently (Cehlar et al., 2024). However, detailed structures of amyloid pores remain elusive, as the dimensions are smaller than those that confocal microscopy can resolve. Structures of Aβ oligomers in interaction with prepared lipid bilayers can be resolved using cryo-EM and atomic force microscopy (AFM) (Viles, 2023), whereas imaging proteins and pores in living systems is more challenging (see Figs. 1.5 and 1.6).

1.4 Oligomers induce ionic currents in lipid membranes

Using the patch-clamp technique and fluorescence imaging, it was concluded that Aβ does not form a single type of ion channel, but instead contributes to the formation of an array of conducting pores in biological membranes (Sepúlveda et al., 2014). The same authors (Sepúlveda et al., 2014) found that Aβ (in a comparative study with

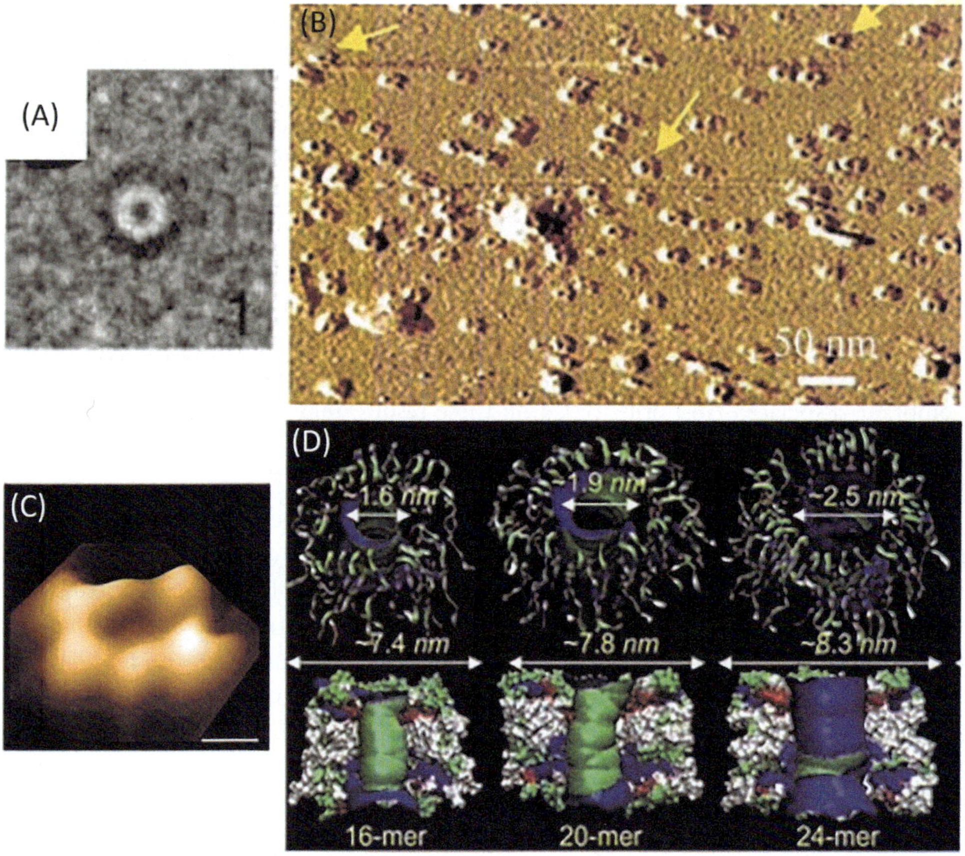

Figure 1.5 (A) Negatively-stained transmission electron microscopy 2D class averages of Aβ-42 annular oligomers (Arctic mutant). (B) Many annular oligomers of Aβ-42 embedded within a supported lipid bilayer, imaged by atomic force microscopy (AFM). (C) Single annular oligomers of Aβ-42 imaged by AFM. (D) Molecular dynamics simulation showing a possible β-barrel conformation of Aβ molecules, with a series of structures (1.6–2.5 nm internal diameter, from 16 to 24 Aβ molecules). *References are given in the original paper. Modified (chopped) from: Viles, J.H., 2023. Imaging amyloid-β membrane interactions: ion-channel pores and lipid-bilayer permeability in Alzheimer's dDisease. Angewandte Chemie-International Edition. 62 (25), https://doi.org/10.1002/anie.202215785. License: CC BY 4.0.*

gramicidin) perforated lipid membranes, producing small (chloride-selective) pores as well as larger, less selective pores.

Bode et al. (2017) first characterized Aβ-40 and Aβ-42 preparations with transmission electron microscopy and thioflavin T (ThT) fluorescence. They also used patch clamping to monitor transmembrane currents associated with the different preparations. Ion channel conductance results suggested that Aβ-42 oligomers, in

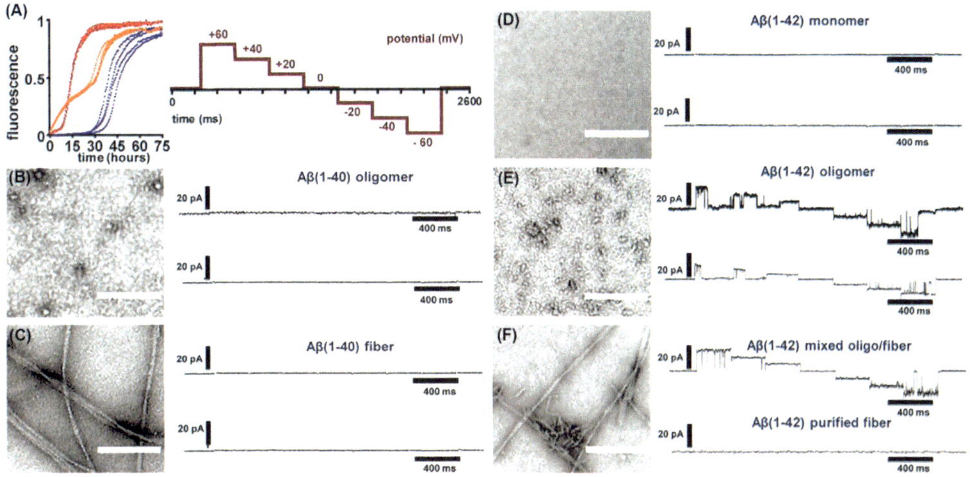

Figure 1.6 Structurally distinct Aβ-40 and Aβ-42 preparations have markedly different abilities to induce ion channel currents in biological membranes. (A) Left, thioflavin T fiber growth kinetics were monitored for solubilized Aβ-40 (*blue*), solubilized Aβ-42 (*orange*), and size-exclusion chromatography-purified Aβ-42 (red) at 10 μM Aβ concentrations (four wells per preparation). Right, a voltage-ramp protocol was applied to clamp the cell membrane at six membrane step potentials. Current was recorded at each potential, and two representative traces recorded from separate membrane patches are shown with corresponding transmission electron microscopy (TEM) images for Aβ-40 oligomers (B) and fibers (C), Aβ-42, monomers (D) and oligomers (E), and a single representative trace for both oligomer-containing fiber and purified fiber preparations (F). All Aβ preparations were at 5 μM (monomer equivalent) at pH 7.4. TEM scale bar, 50 nm. *Reproduced from Bode, D.C., Baker, M.D., Viles, J.H., 2017. Ion channel formation by amyloid-β42 oligomers but not amyloid-β40 in cellular membranes. Journal of Biological Chemistry. 292 (4), 144–1413. https://doi.org/10.1074/jbc.M116.762526. Open access license: CC BY 4.0*

contrast to Aβ-40 oligomers, can form distinct pores with diameters of 1.7, 2.1, and 2.4 nm. Neither Aβ monomers nor fibrils were able to do the same (see Fig. 1.6).

Using electrophysiological methods, Numaguchi et al. (2024) evaluated the channel-forming activity of Aβ-42 in lipid membranes of different compositions, aimed at mimicking neuronal cell membranes. They showed that oligomeric Aβ-42 indeed formed channels in the membranes. Current measurements were compared to ThT fluorescence and AFM images of the corresponding samples. It was shown that membrane cholesterol was one of the key components that promoted Aβ-42 oligomerization, resulting in the formation of the channel structure. Current measurements and western blots showed that Aβ-42 oligomers of the order of 22-mers formed channels with a diameter of ~1.3 nm. Furthermore, current measurements showed that the addition of the oligomerization inhibitor epicatechin-3-gallate (EGCG) (epigallocatechin gallate) decreased the channel-forming activity of the Aβ-42 oligomers.

Similar to other amyloidogenic proteins described above, in the case of our model protein, human stefin B, we observed cytotoxicity when annular protofibrils (obtained at pH 3, high salt, but also stable at pH 7) were added to cultured cells (Anderluh et al., 2005). The same preparations interacted with biological membranes (Anderluh et al., 2005). Later, we embarked on more studies of the protein–lipid bilayers interaction and were able to confirm that the prefibrillar oligomers were more toxic to cells than the lower-order oligomers and monomers. Anderluh and Žerovnik (2012), Anderluh et al. (2005), Čeru et al. (2008) showed that the hexamers, octamers, 12- and 16-mers were the species interacting most strongly with acidic phospholipid bilayers.

Rabzelj et al. (2008) measured pore formation by the wild-type stefin B (stB-wt) and the naturally occurring stB-Y31 (where Tyr replaces glutamic acid) variant in planar lipid membranes using electrophysiology. The ionic current flowing through the membrane increased stepwise after the addition of stB-wt protein. After the opening of the first two or three pores, some rapid closures or flickering of small channels were noted, corresponding to conductance levels 1 and 2. The amplitude of each step was used to calculate the characteristic pore conductance. In contrast, the current flowing through the membrane induced by the addition of the stB-Y31 variant in the prefibrillar state produced the conductance characteristic of larger single pores. In accordance with the different pore sizes, selective and larger pores were detected for stB-wt and its prefibrillar form, respectively (Rabzelj et al., 2008). Even less selective current flow, resembling total membrane disruption (breakage), was seen for the stB-G4R pathogenic mutant, found in some patients with progressive myoclonus epilepsy of type 1.

1.5 Theoretical studies of protein oligomer: membrane interactions

In order to obtain insight into oligomerization and oligomer-membrane interactions of Aβ-42, Brown and Bevan (2016) used united-atom MD simulations on the microsecond scale. They simulated tetramer formation and also the interaction of the Aβ tetramer with lipid bilayers. The conclusion was that the Aβ-42 tetramer interaction with 1-palmitoyl-2-oleoyl-sn-glycero-3-phosphocholine (POPC) (a major phospholipid in eukaryotic membranes and raft-like membranes) obeyed a carpeting model.

Voelker et al. (2017) structurally characterized Aβ-40 and Aβ-42 monomers through pentamers, which were obtained from previously derived coarse-grained (DMD4B-HYDRA) simulations to all-atom MD simulations with explicit solvent. The results indicated that Aβ-40 and Aβ-42 trimers, tetramers, and pentamers formed water-permeable pores. The propensity for pore formation correlated with oligomer size and was highest for Aβ-42 pentamers. These findings confirm the extraordinary ability of Aβ oligomers to form pores in pure water prior to their insertion into a membrane and thus provide support for the ion channel hypothesis of AD.

Using Quencher Induced Step Length Increase in Photobleaching (QSLIP), Dey et al. (2002) measured membrane insertion of Aβ oligomers: Aβ-40, Aβ-42 and AβCha (a synthetic isoform with F19 replaced by cyclohexylalanine) (Korn et al., 2018) at the single-molecule level. As noted earlier, it has been shown by others that Aβ-42 oligomers are more toxic than Aβ-40 (Krishtal et al., 2015), whereas AβCha (with the crucial F19-L34 contact disturbed) is the least toxic (Korn et al., 2018). Using this method, the authors were able to determine the depth of penetration of the amino-terminal of the Aβ oligomers into the lipid bilayer. Under similar conditions, the N-terminus of Aβ-42 was inserted closer to the center of the lipid bilayer, significantly deeper than that of Aβ-40. This also correlated with the increase in reactive oxygen species in neuronal cells, where the highest increase was observed for Aβ-42, which again correlated with toxicity.

Aβ, α-synuclein, as well as Tau and TDP-43 proteins are intrinsically disordered proteins (IDPs) and are prone to form condensates and amyloid aggregates. However, globular proteins can also form amyloid structures, including protofibrils, which can perforate membranes (Lashuel et al., 2002; Glabe and Kayed, 2006). For example, Žganec et al. (2015), using discrete MD and the DMD4B-HYDRA force field, confirmed that other nonpathological amyloidogenic proteins, even non-IDPs, can form oligomers with morphologies that could insert into lipid membranes. In this work, they were able to demonstrate annular, branched, and elongated morphologies of the pentamers and, in particular, decamers of stefin B-var2 (E31Y variant). The higher-order oligomers displayed a partially unfolded structure upon oligomerization and formed specific quaternary structures, which resemble pore-like conformations, in accordance with in vitro studies of membrane interaction and toxicity exerted by the protofibrils, and higher than that of hexamer oligomers (Anderluh et al., 2005; Čeru et al., 2008; Rabzelj et al., 2008).

In addition, to Aβ-40 and Aβ-42, other C-terminally truncated or elongated Aβ peptides, including Aβ-38 and Aβ-43, have been detected in AD patient brains. Pandey and Urbanc (2024) examined oligomer formation by all of these species using DMD4B-HYDRA simulations, which were previously applied to the two predominant alloforms, Aβ-40 and Aβ-42. The results indicated that disorder and solvent exposure of the N-termini in Aβ oligomers (Aβ-42), as well as the size of the elongated oligomer (Aβ-43), both contribute to increased toxicity.

Similarly, Ciudad et al. (2020) studied the insertion of Aβ-42 tetramers and octamers into lipid bilayers using MD simulations. They proposed that the architecture of the Aβ tetramer (PDB: 6RHY) could enable pore formation. They suggested that the α-helical parts of the tetramer could undergo a secondary-structure transition from α to β, especially in lipid rafts rich in gangliosides. Further biophysical experiments confirmed the MD simulations (Viles, 2023).

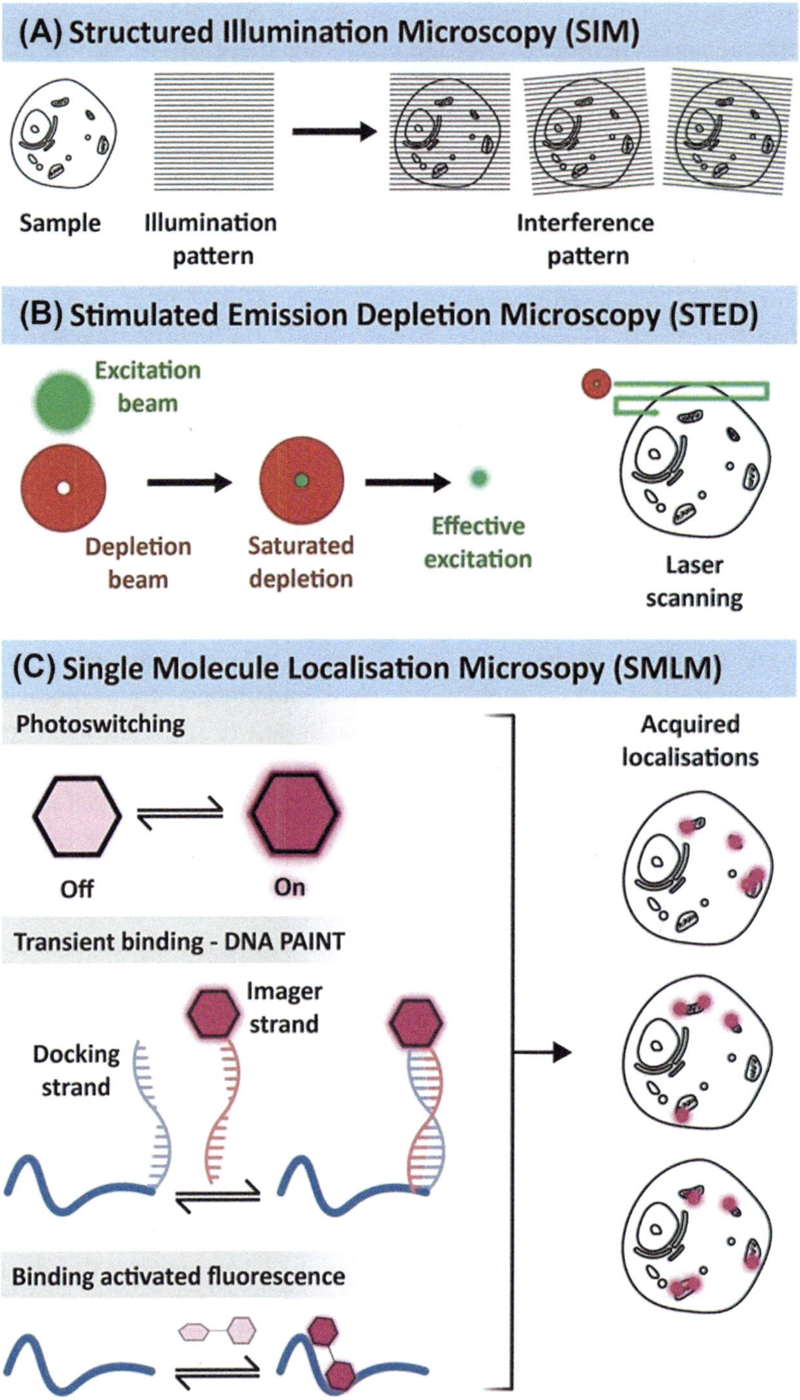
(A) Structured Illumination Microscopy (SIM)
Sample
Illumination pattern
Interference pattern
(B) Stimulated Emission Depletion Microscopy (STED)
Excitation beam
Depletion beam
Saturated depletion
Effective excitation
Laser scanning
(C) Single Molecule Localisation Microsopy (SMLM)
Photoswitching
Off
On
Transient binding - DNA PAINT
Imager strand
Docking strand
Binding activated fluorescence
Acquired localisations

In order to prevent amyloid toxicity, compounds that prevent Aβ-42 oligomerization and pore formation are needed. Catechins are polyphenolic flavonoids with strong antioxidant effects. They are comprised of epigallocatechin, epicatechin, and EGCG, found for example, in tea and different berries. In their study, Firouzi and Sowlati-Hashjin (2023) used an ensemble docking strategy to identify catechin binding sites in monomeric Aβ-42. It has been shown that hydrophobic aromatic interactions and hydrogen bonding are crucial for the binding of catechins to Aβ-42. By blocking the central hydrophobic region of Aβ, catechins, especially EGCG, exert potent inhibition against amyloid aggregation. These findings were confirmed with more detailed multimicrosecond MD simulations.

A recent review summarized and evaluates MD studies on small-molecule inhibitors of amyloid and amyloid-pore formation by Aβ (Atanasova, 2025). The authors evaluated studies on the inhibitory action of endogenous compounds, such as dopamine and norepinephrine. Other compounds, such as serotonin, melatonin, and ATP, were included in the review. They also considered MD studies of repurposed drugs inhibiting Aβ amyloid formation, among them, propafenone (Atanasova, 2025). The authors suggested that to gain a deeper understanding of the anti-amyloidogenic compound inhibitory mechanism against Aβ oligomerization and aggregation, a further comprehensive analysis should be performed, considering key stabilizing interactions on protofibrillar and fibrillar forms (Atanasova, 2025).

1.6 Visualizing pores in living membranes

Confocal microscopy lacks the resolution to visualize amyloid pores in live cells. The small nm size and heterogeneous morphology of the prefibrillar oligomers, protofibrils, and mature amyloid fibrils can be challenging to detect in this context. In a recent review, Bhuskute et al. (2024) provide an overview of different superresolution imaging techniques. They discuss the strengths and limitations of each technique, offering insights into their applicability for observing different amyloid systems. In order to visualize various amyloid morphologies (including amyloid pores) with superresolution imaging, different labeling methods, ranging from small molecules, organic dyes, and chemical probes, to larger fluorescent proteins, are likely to be of utmost importance (Bhuskute et al., 2024) (Fig. 1.7).

Figure 1.7 Principles of superresolution microscopy techniques used in imaging of amyloids: (A) structured illumination microscopy (SIM); (B) stimulated emission depletion (STED) microscopy; (C) single-molecule localization microscopy (SMLM), including photo-switching, transient binding, and binding-activated fluorescence techniques. *Reproduced from Bhuskute, K.R., Kikuchi, K., Luo, Z., Kaur, A., 2024. Visualizing amyloid assembly at the nanoscale: insights from super-resolution imaging. Analysis and Sensing. 4 (4), e202400001. https://doi.org/10.1002/anse.202400001 Open access creative commons licence: CC BY-NC-ND 4.0*

A fluorescent amyloid sensor, AmyBlink-1 has been developed for superresolution imaging of amyloid structures (Kaur et al., 2022). AmyBlink-1 exhibits a 5-fold increase in the ratio of the green (ThT) to red (Alexa Fluor 647) emission intensities upon interaction with amyloid fibrils (Kaur et al., 2022). Using AmyBlink-1, a resolution of of ≈30 nm can be achieved, making it possible to distinguish the length and skew of individual amyloid morphologies (Kaur et al., 2022). Live-cell superresolution imaging across a range of protein targets in mammalian cells has been developed and is termed mLIVE-PAINT (Bhaskar, 2007). In the PAINT method, a protein of interest is tagged with one partner of a coiled-coil (CC) peptide. This tagged protein can be localized *in cellulo*, as another CC peptide fused to a fluorescent protein transiently binds to it.

1.7 Summary

In this chapter, I gathered information supporting the modified amyloid cascade hypothesis, namely, that oligomers of amyloid-forming proteins perforate membranes and, in such a way, cause damage to cells. As noted, the cation-permeable pores may form in the plasma membrane (Shah et al., 2002) but also in membranes of intracellular organelles, including mitochondria (Vassallo, 2021) and lysosomal membranes (Zaretsky et al., 2022). Ca^{2+} influx into neurons then leads to excitotoxicity and neuronal death. After briefly introducing the concepts of protein aggregation pathways to amyloid fibrils with intermediate stages of oligomers and protofibrils, in Section 1.2, evidence from morphological and electrophysiological experiments confirming indirectly pore formation are described in Sections 1.3 and 1.4, while in Section 1.5, theoretical backgrounds to pore formation are overviewed, mostly performed by MD calculations of the oligomerization and lipid-membrane interactions of amyloidogenic peptides, such as Aβ-42—often in comparison with Aβ-40. In my view, the membrane damage could be a primary common cause of neurodegenerative diseases. There is growing evidence that amyloid toxicity could, using similar mechanisms to those mediated by PFPs and AMPs, result from prefibrillar oligomers forming amyloid pores (channels) in cellular membranes (Tang et al., 2024).

Abbreviations

EGCG	epicatechin-3-gallate
POPC	1-palmitoyl-2-oleoyl-sn-glycero-3-phosphocholine
QSLIP	Quencher Induced Step Length Increase in Photobleaching
ROS	reactive oxygen species
IDP	intrinsically disordered protein
MD	molecular dynamics
DMD4B-HYDRA force field	the four-bead protein model in combination with discrete molecular dynamics (DMD)

mLIVE-PAINT	the mammalian cell version of LIVE-PAINT method
PD	Parkinson's disease
AD	Alzheimer's disease
EPM1	progressive myoclonus epilepsy of type 1
SIM	structured illumination microscopy
STED	stimulated emission depletion microscopy
SMLM	single molecule localisation microscopy
ThT	thioflavin T
AMPs	antimicrobial peptides
PFPs	pore-forming proteins
PPI	peptidyl-prolyl isomerase
AFM	atomic force microscopy
TEM	transmission electron microscopy
SEC	size-exclusion chromatography

Acknowledgement

My work was supported throughout by several projects (J7-4050, J1-8742) and the program P1-0140 by the Slovenian Research and Innovation agency (ARIS).

References

Anderluh, G., Žerovnik, E., 2012. Pore formation by human stefin B in its native and oligomeric states and the consequent amyloid induced toxicity. Frontiers in Molecular Neuroscience 5 (2012). Available at https://doi.org/10.3389/fnmol.2012.00085.

Anderluh, G., Gutirrez-Aguirre, I., Rabzelj, S., Čeru, S., Kopitar-Jerala, N., Maček, P., et al., 2005. Interaction of human stefin B in the prefibrillar oligomeric form with membranes. The FEBS Journal 272 (12), 3042–3051. Available at https://doi.org/10.1111/j.1742-4658.2005.04717.x.

Atanasova, M., 2025. Small-molecule inhibitors of amyloid beta: insights from molecular dynamics-part a: endogenous compounds and repurposed drugs. Pharmaceuticals (Basel) 18 (3), 306. Available at https://doi.org/10.3390/ph18030306.

Bhaskar, H., 2007. Super-resolution imaging of proteins inside live mammalian cells with mLIVE-PAINT. bioRxiv: the preprint server for biology 604574.

Bhuskute, K.R., Kikuchi, K., Luo, Z., Kaur, A., 2024. Visualizing amyloid assembly at the nanoscale: insights from super-resolution imaging. Analysis and Sensing 4 (4). Available at https://doi.org/10.1002/anse.202400001.

Bode, D.C., Baker, M.D., Viles, J.H., 2017. Ion channel formation by amyloid-β42 oligomers but not amyloid-β40 in cellular membranes. Journal of Biological Chemistry 292 (4), 144–1413. Available at https://doi.org/10.1074/jbc.M116.762526.

Brown, A.M., Bevan, D.R., 2016. Molecular dynamics simulations of amyloid β -peptide (1-42): tetramer formation and membrane interactions. Biophysical Journal 111 (5), 937–949. Available at https://doi.org/10.1016/j.bpj.2016.08.001.

Cehlar, O., Njemoga, S., Horvath, M., Cizmazia, E., Bednarikova, Z., Barrera, E.E., 2024. Structures of oligomeric states of tau protein, amyloid-β, α-synuclein and prion protein implicated in Alzheimer's disease, Parkinson's disease and prionopathies. International Journal of Molecular Sciences 25 (23), 13049. Available at https://doi.org/10.3390/ijms252313049.

Čeru, S., Kokalj, S.J., Rabzelj, S., Škarabot, M., Gutierrez-Aguirre, I., Kopitar-Jerala, N., et al., 2008. Size and morphology of toxic oligomers of amyloidogenic proteins: a case study of human stefin B. Amyloid: The International Journal of Experimental and Clinical Investigation: The Official Journal of the International Society of Amyloidosis 15 (3), 147–159. Available at https://doi.org/10.1080/13506120802193555.

Chang, C.C., Edwald, E., Veatch, S., Steel, D.G., Gafni, A., 2018. Interactions of amyloid-β peptides on lipid bilayer studied by single molecule imaging and tracking. Biochimica et Biophysica Acta - Biomembranes 1860 (9), 1616–1624. Available at https://doi.org/10.1016/j.bbamem.2018.03.017.

Chayen, N.E., 2005. Methods for separating nucleation and growth in protein crystallisation. Progress in Biophysics and Molecular Biology 88 (3), 329–337. Available at https://doi.org/10.1016/j.pbiomolbio.2004.07.007.

Chiti, F., Dobson, C.M., 2006. Protein misfolding, functional amyloid, and human disease. Annual Review of Biochemistry 75, 333–366. Available at https://doi.org/10.1146/annurev.biochem.75.101304.123901.

Ciechanover, A., Kwon, Y.T., 2015. Degradation of misfolded proteins in neurodegenerative diseases: therapeutic targets and strategies. Experimental and Molecular Medicine 47, e147. Available at https://doi.org/10.1038/EMM.2014.117.

Ciudad, S., Puig, E., Botzanowski, T., Meigooni, M., Arango, A.S., Do, J., et al., 2020. Aβ(1-42) tetramer and octamer structures reveal edge conductivity pores as a mechanism for membrane damage. Nature Communications 11, 3014. Available at https://doi.org/10.1038/s41467-020-16566-1.

Cohen, S.I.A., Linse, S., Luheshi, L.M., Hellstrand, E., White, D.A., Rajah, L., et al., 2013. Proliferation of amyloid-β42 aggregates occurs through a secondary nucleation mechanism. Proceedings of the National Academy of Sciences of the United States of America 110 (24), 9758–9763. Available at http://doi.org/10.1073/pnas.1218402110.

Dey, A., Abhsihek Patil, A., Arumugam, S., Maiti, S., 2024. Single-molecule mapping of amyloid-β oligomer insertion into lipid bilayers. bioRxiv. Available at https://doi.org/10.1101/2024.02.21.581487.

Di Scala, C., Yahi, N., Boutemeur, S., Flores, A., Rodriguez, L., Chahinian, H., et al., 2016. Common molecular mechanism of amyloid pore formation by Alzheimer's β-amyloid peptide and α-synuclein. Scientific Reports 6, 28781. Available at https://doi.org/10.1038/srep28781.

Diomede, L., Rigacci, S., Romeo, M., Stefani, M., Salmona, M., 2013. Oleuropein aglycone protects transgenic *C. elegans* strains expressing Aβ42 by reducing plaque load and motor deficit. PLoS One 8 (3), e58893. Available at https://doi.org/10.1371/journal.pone.0058893.

Dovidchenko, N.V., Finkelstein, A.V., Galzitskaya, O.V., 2014. How to determine the size of folding nuclei of protofibrils from the concentration dependence of the rate and lag-time of aggregation. I. Modeling the amyloid protofibril formation. Journal of Physical Chemistry B 118 (5), 1189–1197. Available at https://doi.org/10.1021/jp4083294.

Dovidchenko, N.V., Leonova, E.I., Galzitskaya, O.V., 2014. Mechanisms of amyloid fibril formation. Biochemistry (Moscow) 79 (13), 1515–1527. Available at https://doi.org/10.1134/s0006297914130057.

Firouzi, Sowlati-Hashjin, 2023. Identification of catechins' binding sites in monomeric Aβ(42) through ensemble docking and MD simulations. International Journal of Molecular Sciences 24.

Fontana, I.C., Zimmer, A.R., Rocha, A.S., Gosmann, G., Souza, D.O., Lourenco, M.V., et al., 2020. Amyloid-β oligomers in cellular models of Alzheimer's disease. Journal of Neurochemistry 155 (4), 348–369. Available at https://doi.org/10.1111/jnc.15030.

Fusco, G., De Simone, A., Arosio, P., Vendruscolo, M., Veglia, G., Dobson, C.M., 2016. Structural ensembles of membrane-bound α-synuclein reveal the molecular determinants of synaptic vesicle affinity. Scientific Reports 6 (1). Available at https://doi.org/10.1038/srep27125.

Fusco, G., Chen, S.W., Williamson, P.T.F., Cascella, R., Perni, M., Jarvis, J.A., et al., 2017. Structural basis of membrane disruption and cellular toxicity by α-synuclein oligomers. Science (New York, N.Y.) 358 (6369), 1440–1443. Available at https://doi.org/10.1126/science.aan6160.

Glabe, C.G., Kayed, R., 2006. Common structure and toxic function of amyloid oligomers implies a common mechanism of pathogenesis. Neurology 66 (1_suppl_1), S74–S78. Available at https://doi.org/10.1212/01.wnl.0000192103.24796.42.

Hardy, J.A., Higgins, G.A., 1992. Alzheimer's disease: the amyloid cascade hypothesis. Science (New York, N.Y.) 256 (5054), 184–185. Available at https://doi.org/10.1126/science.1566067.

Hasanbašić, S., Taler-Verčič, A., Puizdar, V., Stoka, V., Tušek Žnidarič, M., Vilfan, A., et al., 2019. Prolines affect the nucleation phase of amyloid fibrillation reaction; mutational analysis of human stefin B. ACS Chemical Neuroscience 10 (6), 2730–2740. Available at https://doi.org/10.1021/acschemneuro.8b00621.

Hashimoto, M., Rockenstein, E., Crews, L., Masliah, E., 2003. Role of protein aggregation in mitochondrial dysfunction and neurodegeneration in Alzheimer's and Parkinson's diseases. Neuromolecular Medicine 4 (1-2), 21–36. Available at https://doi.org/10.1385/NMM:4:1-2:21.

Jenko Kokalj, S., Gunčar, G., Štern, I., Morgan, G., Rabzelj, S., Kenig, M., et al., 2007. Essential role of proline isomerization in stefin B tetramer formation. Journal of Molecular Biology 366 (5), 1569–1579. Available at https://doi.org/10.1016/j.jmb.2006.12.025.

Julien, C., Tomberlin, C., Roberts, C.M., Akram, A., Stein, G.H., Silverman, M.A., et al., 2018. In vivo induction of membrane damage by β-amyloid peptide oligomers. Acta neuropathologica communications 6 (1), 131. Available at https://doi.org/10.1186/s40478-018-0634-x.

Kagan, B.L., Hirakura, Y., Azimov, R., Azimova, R., Lin, M.C., 2002. The channel hypothesis of Alzheimer's disease: current status. Peptides 23 (7), 1311–1315. Available at https://doi.org/10.1016/S0196-9781(02)00067-0.

Kaur, A., Adair, L.D., Ball, S.R., New, E.J., Sunde, M., 2022. A fluorescent sensor for quantitative super-resolution imaging of amyloid fibril assembly** Angewandte Chemie - International Edition 61 (10). Available at https://doi.org/10.1002/anie.202112832.

Kayed, R., Head, E., Thompson, J.L., McIntire, T.M., Milton, S.C., Cotman, C.W., et al., 2003. Common structure of soluble amyloid oligomers implies common mechanism of pathogenesis. Science (New York, N.Y.) 300 (5618), 486–489. Available at https://doi.org/10.1126/science.1079469.

Kenyaga, J.M., Cheng, Q., Qiang, W., 2022. Early stage β-amyloid-membrane interactions modulate lipid dynamics and influence structural interfaces and fibrillation. Journal of Biological Chemistry 298.

Korn, A., Surendran, D., Krueger, M., Maiti, S., Huster, D., 2018. Ring structure modifications of phenylalanine 19 increase fibrillation kinetics and reduce toxicity of amyloid β (1–40). Chemical Communications 54 (43), 5430–5433. Available at https://doi.org/10.1039/c8cc01733f.

Krishtal, J., Bragina, O., Metsla, K., Palumaa, P., Tõugu, V., 2015. Toxicity of amyloid beta 1-40 and 1-42 on SH-SY5Y cell line. SpringerPlus 4 (S1). Available at https://doi.org/10.1186/2193-1801-4-s1-p19.

Kundel, F., Hong, L., Falcon, B., McEwan, W.A., Michaels, T.C.T., Meisl, G., et al., 2018. Measurement of Tau filament fragmentation provides insights into prion-like spreading. ACS Chemical Neuroscience 9 (6), 1276–1282. Available at https://doi.org/10.1021/acschemneuro.8b00094.

Lashuel, H.A., Hartley, D., Petre, B.M., Walz, T., Lansbury, P.T., 2002. Amyloid pores from pathogenic mutations. Nature 418 (6895). Available at https://doi.org/10.1038/418291a 291–291.

Low, K.J.Y., Venkatraman, A., Mehta, J.S., Pervushin, K., 2022. Molecular mechanisms of amyloid disaggregation. Journal of Advanced Research 36, 113–132. Available at https://doi.org/10.1016/j.jare.2021.05.007.

Numaguchi, Y., Tsukakoshi, K., Takeuchi, N., Suzuki, Y., Ikebukuro, K., Kawano, R., et al., 2024. Real-time monitoring of the amyloid β1–42 monomer-to-oligomer channel transition using a lipid bilayer system. PNAS Nexus 3 (1), 2752–6542. Available at https://doi.org/10.1093/pnasnexus/pgad437.

Pandey, R., Urbanc, B., 2024. Oligomer formation by physiologically relevant C-terminal isoforms of amyloid β-protein. Biomolecules 14 (7), 774. Available at https://doi.org/10.3390/biom14070774.

Rabzelj, S., Viero, G., Gutiérrez-Aguirre, I., Turk, V., Dalla Serra, M., Anderluh, G., et al., 2008. Interaction with model membranes and pore formation by human stefin B – studying the native and prefibrillar states. The FEBS Journal 275 (10), 2455–2466. Available at https://doi.org/10.1111/j.1742-4658.2008.06390.x.

Risiglione, P., Zinghirino, F., Di Rosa, M.C., Magrì, A., Messina, A., 2021. Alpha-synuclein and mitochondrial dysfunction in Parkinson's disease: the emerging role of VDAC. Biomolecules 11 (5), 718. Available at https://doi.org/10.3390/biom11050718.

Selkoe, D.J., Hardy, J., 2016. The amyloid hypothesis of Alzheimer's disease at 25 years. EMBO Molecular Medicine 8 (6), 595–608. Available at https://doi.org/10.15252/emmm.201606210.

Sepúlveda, F.J., Fierro, H., Fernandez, E., Castillo, C., Peoples, R.W., Opazo, C., et al., 2014. Nature of the neurotoxic membrane actions of amyloid-β on hippocampal neurons in Alzheimer's disease. Neurobiology of Aging 35 (3), 472–481. Available at https://doi.org/10.1016/j.neurobiolaging.2013.08.035.

Shah, S.I., Demuro, A., Ullah, G., 2022. Modeling the kinetics of amyloid beta pores and long-term evolution of their Ca2+ toxicity. bioRxiv: the preprint server for biology. Available at https://doi.org/10.1101/2022.05.02.490365.

Sinnige, T., 2022. Molecular mechanisms of amyloid formation in living systems. Chemical Science 13 (24), 7080–7097. Available at https://doi.org/10.1039/d2sc01278b.

Škerget, K., Vilfan, A., Pompe-Novak, M., Turk, V., Waltho, J.P., Turk, D., et al., 2009. The mechanism of amyloid-fibril formation by stefin B: temperature and protein concentration dependence of the rates. Proteins: Structure, Function and Bioinformatics 74 (2), 425–436. Available at https://doi.org/10.1002/prot.22156.

Smajlović, A., Berbić, S., Schiene-Fischer, C., Tušek-Žnidarič, M., Taler, A., Jenko-Kokalj, S., et al., 2009. Essential role of Pro 74 in stefin B amyloid-fibril formation: dual action of cyclophilin A on the process. FEBS Letters 583 (7), 1114–1120. Available at https://doi.org/10.1016/j.febslet.2009.02.037.

Taler-Vercic, A., Hasanbašić, S., Berbić, S., Stoka, V., Turk, D., Žerovnik, E., 2017. Proline residues as switches in conformational changes leading to amyloid fibril formation. International Journal of Molecular Sciences 18 (3), 549. Available at 10.3390/ijms18030549.

Taler-Verčič, A., Kirsipuu, T., Friedemann, M., Noormägi, A., Polajnar, M., Smirnova, J., et al., 2013. The role of initial oligomers in amyloid fibril formation by human stefin B. International Journal of Molecular Sciences 14 (9), 18362–18384. Available at https://doi.org/10.3390/ijms140918362.

Tang, Y., Zhang, Y., Zhang, D., Liu, Y., Nussinov, R., Zheng, J., 2024. Exploring pathological link between antimicrobial and amyloid peptides. Chemical Society Reviews 53 (17), 8713–8763. Available at https://doi.org/10.1039/d3cs00878a.

Taylor, A.I.P., Staniforth, R.A., 2022. General principles underpinning amyloid structure. Frontiers in Neuroscience 16. Available at https://doi.org/10.3389/fnins.2022.878869.

Vassallo, N., 2021. Amyloid pores in mitochondrial membranes. Neural Regeneration Research 16 (11), 2225–2226. Available at https://doi.org/10.4103/1673-5374.310682.

Viles, J.H., 2023. Imaging amyloid-β membrane interactions: ion-channel pores and lipid-bilayer permeability in Alzheimer's disease. Angewandte Chemie - International Edition 62 (25). Available at https://doi.org/10.1002/anie.202215785.

Voelker, M.J., Barz, B., Urbanc, B., 2017. Fully atomistic Aβ40 and Aβ42 oligomers in water: observation of porelike conformations. Journal of Chemical Theory and Computation 13 (9), 4567–4583. Available at https://doi.org/10.1021/acs.jctc.7b00495.

Wiatrak, B., Piasny, J., Kuźniarski, A., Gąsiorowski, K., 2021. Interactions of amyloid-β with membrane proteins. International Journal of Molecular Sciences 22 (11), 6075. Available at https://doi.org/10.3390/ijms22116075.

Zaretsky, D.V., Zaretskaia, M.V., Molkov, Y.I., 2022. Membrane channel hypothesis of lysosomal permeabilization by beta-amyloid. Neuroscience Letters 770, 136338. Available at https://doi.org/10.1016/j.neulet.2021.136338.

Zerovnik, E., 2021. Viroporins vs. other pore-forming proteins: what lessons can we take? Frontiers in chemistry 9, 626059. Available at 10.3389/fchem.2021.626059.

Žerovnik, E., Škarabot, M., Škerget, K., Giannini, S., Stoka, V., Jenko-Kokalj, S., et al., 2007. Amyloid fibril formation by human stefin B: influence of pH and TFE on fibril growth and morphology. Amyloid 14 (3), 237–247. Available at https://doi.org/10.1080/13506120701461137.

Žerovnik, E., Stoka, V., Mirtič, A., Gunčar, G., Grdadolnik, J., Staniforth, R.A., et al., 2011. Mechanisms of amyloid fibril formation – focus on domain-swapping. The FEBS Journal 278 (13), 2263–2282. Available at https://doi.org/10.1111/j.1742-4658.2011.08149.x.

Žganec, M., Žerovnik, E., Urbanc, B., 2015. Assembly of stefin B into polymorphic oligomers probed by discrete molecular dynamics. Journal of Chemical Theory and Computation 11 (5), 2355–2366. Available at https://doi.org/10.1021/acs.jctc.5b00067.

CHAPTER 2

In vivo molecular brain imaging of metabolic dysfunction and protein accumulation in neurodegenerative diseases

Matej Perovnik[1] and Kaja Kolmancic[2]
[1]Department of Neurology, University Medical Centre Ljubljana, Ljubljana, Slovenia,
[2]Department of Nuclear Medicine, Institute of Oncology, Ljubljana, Slovenia

2.1 Introduction

Neurodegenerative brain disorders (NDs) are progressive, chronic diseases characterized by the accumulation of misfolded proteins, abnormal brain function, and gradual cell death. The disease process begins years or even decades before the manifestation of clinical symptoms. This chapter focuses on the use of functional nuclear imaging, which is used to indirectly characterize these pathological changes years before morphological changes are seen on structural computed tomography (CT) or MRI brain scans. The most commonly used clinical investigations are 2-[F]fluoro-2-deoxy-D-glucose positron emission tomography/computer tomography ([^{18}F]FDG PET/CT) and dopamine transporter (DAT) imaging with [^{123}I] ioflupane single-photon emission computed tomography (SPECT). Clinically, [^{18}F]FDG PET)/CT is employed to differentiate between different types of atypical parkinsonisms and dementia syndromes based on changes in synaptic brain activity, and DAT SPECT is used to confirm the loss of striatal dopaminergic neurons, for example, in Parkinson's disease. Additionally, new radiotracers are being developed that can bind directly to abnormally accumulated amyloid β, tau, and α-synuclein (αSyn). These radiotracers show pathological processes in vivo and are used to follow disease progression, monitor treatment response, and study pathophysiology mechanisms.

2.2 Functional nuclear/molecular imaging

2.2.1 Physics of positron emission tomography and single-photon emission computed tomography imaging

Functional nuclear imaging, such as PET and SPECT, involves the use of radiolabeled molecules, known as radiopharmaceuticals or radiotracers, to assess physiological and pathophysiological processes within the body. These compounds typically consist of a

Molecular Insights into Proteinopathies
DOI: https://doi.org/10.1016/B978-0-443-34176-2.00013-4
© 2027 Elsevier Inc. All rights are reserved, including those for text and data mining, AI training, and similar technologies.

radioactive isotope bound to a biologically active molecule or a synthetic substrate. The radionuclides employed in diagnostic imaging are primarily gamma or positron emitters. Among positron-emitting isotopes, fluorine-18 (^{18}F) is the most commonly utilized due to its relatively long half-life (~110 minutes), which offers logistical advantages over other positron emitters such as gallium-68 (^{68}Ga), carbon-11 (^{11}C), nitrogen-13 (^{13}N), and oxygen-15 (^{15}O).

In PET, β^+ decay (positron emission) occurs when a proton in the nucleus is transformed into a neutron, releasing a positron. This positron travels a short distance within tissues before annihilating with an electron, resulting in the emission of two 511 keV photons that travel in opposite directions. PET/CT scanners detect these coincident photons and reconstruct the spatial distribution of the radiotracer. Concurrent CT or MRI acquisition enables attenuation correction and precise anatomical coregistration of the PET signal. The most widely used radiotracer is [^{18}F]FDG, a glucose analog that serves as a surrogate marker of cellular glucose metabolism. [^{18}F]FDG is transported into cells via glucose transporters and subsequently phosphorylated by hexokinase II to form FDG-6-phosphate. Due to its structural modification and negative charge, FDG-6-phosphate is metabolically trapped within the cell, allowing its accumulation in metabolically active tissues. Elevated uptake of [^{18}F]FDG is commonly observed in malignant tumors, as well as in sites of inflammation or infection. In NDs, [^{18}F]FDG PET is used to evaluate regional cerebral glucose metabolism and thus indirectly assess synaptic activity. Distinct metabolic patterns facilitate the differentiation of various forms of dementia and atypical parkinsonian syndromes APS). Furthermore, ^{18}F (and less frequently^{11}C) can be chemically conjugated to ligands targeting pathological proteins, enabling the in vivo imaging of amyloid-β (Aβ), tau, and αSyn.

SPECT is an imaging method in which three-dimensional (3D) functional images of brain regions are obtained after an injection of a gamma photon-emitting radiopharmaceutical. SPECT uses rotating gamma cameras that collect photons from multiple angles to reconstruct a 3D image. The most common gamma emitters are ^{99m}Tc and ^{123}I. They are essential in the diagnostic evaluation of infectious and inflammatory processes, bone lesions, suspected pulmonary embolism, renal obstruction and function, and more. DAT SPECT is most commonly used in the field of neurodegenerative diseases (Lin and Lamsback, 2009; O'Malley et al., 2020).

In the diagnostic workflow of NDs, clinicians also employ CT and MRI, especially to evaluate vascular comorbidities and rule out structural causes, for example, tumors. Certain radiological signs seen on MRI, such as medial temporal atrophy in Alzheimer's disease (AD), or midbrain atrophy in progressive supranuclear palsy (PSP), can aid in the accurate diagnosis. However, structural changes are observable later in the disease course than functional brain changes, and imaging with functional MRI is still confined to pure research applications in NDs. Since it is beyond the scope of this chapter to discuss this further, we refer the interested reader elsewhere (Young et al., 2020; Risacher and Saykin, 2019).

2.3 Metabolic brain changes in neurodegenerative disorders

2.3.1 Image analysis

NDs are characterized by specific changes in regional glucose metabolism. [^{18}F]FDG PET images can be assessed in different ways (Fig. 2.1).

In clinical practice, they are initially assessed qualitatively by an expert nuclear medicine physician or a radiologist with or without the assistance of a neurologist (Nobili et al., 2018). However, visual analysis can be subjective and is dependent on the reader's expertise (Ng et al., 2007; Yamane et al., 2014). Inter-reader variability and the difficulty of detecting subtle diffuse changes prompted the development of additional analytical tools. The assessment is thus most often supplemented, as suggested by the joint recommendation of the Society of Nuclear Medicine and Molecular Imaging and the European Association of Nuclear Medicine, by various analytical approaches, either univariate or multivariate (Arbizu et al., 2025). More recently, various machine learning and deep learning analytical approaches are being developed (Deatsch et al., 2022; Etminani et al., 2022; Wu et al., 2022). While the role of artificial intelligence is expected to become more prominent in the future, the new models need to be subjected to the same scientific rigor as traditional approaches before clinical translation (Saboury et al., 2023; Myszczynska et al., 2020).

2.3.2 Univariate image analysis

Mass univariate testing, based on statistical parametric mapping (SPM) (Friston, 2007), software provided by camera manufacturers (e.g., syngo.via or Siemens) or in-house

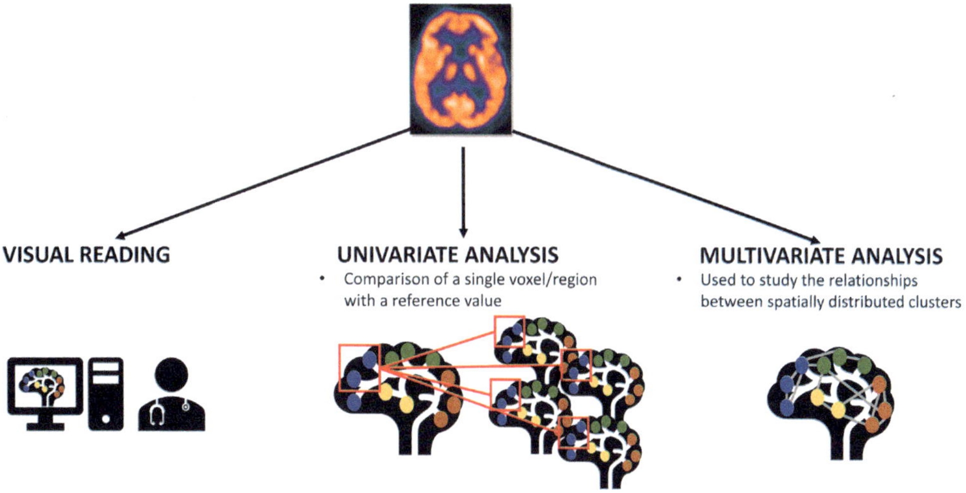

Figure 2.1 Schematic representations of approaches to F-fluorodeoxyglucose positron emission tomography/computer tomography image analysis.

pipelines, is employed to compare voxel or regional changes in the patient's brain with those of a reference population. Statistically significant differences are then plotted on a template and subjected to visual interpretation. These analyses increase sensitivity to subtle changes and reduce observer bias. SPM-maps have been shown to have higher sensitivity and specificity in dementia differential diagnosis compared to standard [^{18}F] FDG PET/CT readings (Perani et al., 2014). In research settings, these methods are invaluable as they offer a reliable, quantifiable approach to assessing metabolic differences between groups, within individuals, before and after intervention, etc.

2.3.3 Multivariate image analysis

Beyond mass univariate comparisons, multivariate techniques examine spatial patterns to capture network-level changes. Commonly applied approaches are principal component analysis (PCA), independent component analysis, seed-based correlation, sparse inverse covariance estimation, and others (Perovnik et al., 2023; Sala et al., 2023).

Among those, the most commonly used is a scaled subprofile model based on PCA that has been developed to identify metabolic covariance networks associated with numerous NDs (Spetsieris and Eidelberg, 2011). The network is initially identified using [^{18}F]FDG PET brain imaging data from a representative group of diseased individuals and a comparative group of healthy controls. The extent of the network changes can then be prospectively quantified on a single-subject basis using a topographic profile rating algorithm to obtain the so-called subject score (Perovnik et al., 2023; Spetsieris and Eidelberg, 2011). Using this approach, reproducible metabolic covariance networks have been identified for the major NDs (Fig. 2.2). A prime example is the so-called Parkinson's disease-related pattern (PDRP), characterized by bilateral changes in basal ganglia, premotor, and parietal association cortices (Ma et al., 2007; Tomše et al., 2017). This network has been identified in various populations worldwide, and the subject scores have been consistently shown to correlate with motor impairment (Schindlbeck and Eidelberg, 2018). Furthermore, elevated PDRP levels have been noted in prodromal PD patients, that is, patients with rapid eye movement (REM) sleep behavior disorder, and have been used to predict disease progression (Carli et al., 2025; Tang et al., 2024).

2.3.4 Metabolic changes in Parkinsonian syndromes

PD and APS such as multiple system atrophy (MSA), PSP, and corticobasal syndrome (CBS) often present with overlapping Parkinsonian features, and a third or more of cases are initially misdiagnosed (Rizzo et al., 2016). [F][^{18}F]FDG PET can improve early diagnostic accuracy. Visual inspection and mass univariate testing reveal characteristic changes that can aid in differential diagnosis (Meyer et al., 2017). The metabolic hallmark in MSA is marked

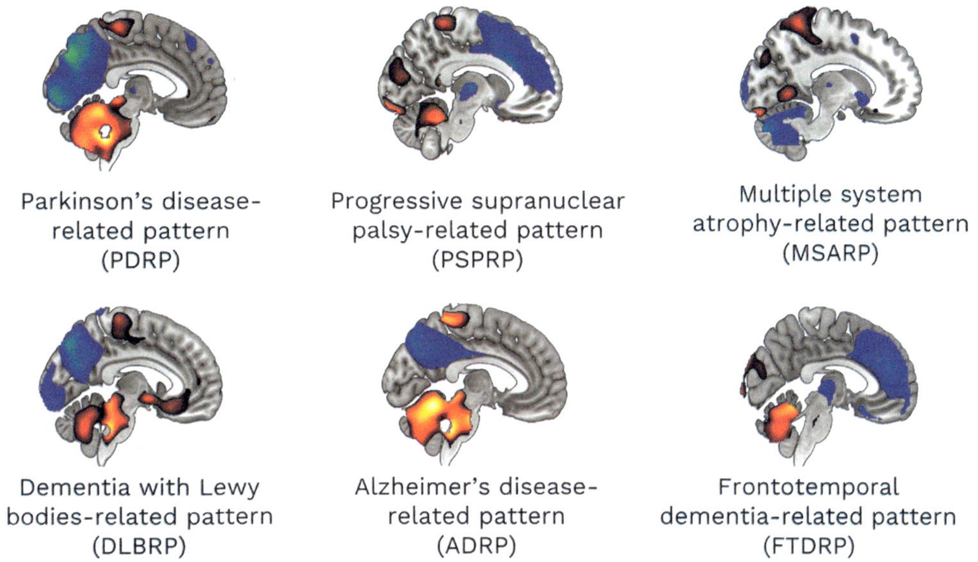

Figure 2.2 Specific disease-related metabolic brain patterns of common neurodegenerative disorders.

hypometabolism in the posterior putamen and the cerebellum. In PSP, hypometabolism is observed in the medial and dorsolateral frontal cortex, caudate, thalamus, and upper brainstem. Whereas in CBS, asymmetric hypometabolic changes are seen in the frontoparietal cortex and the striatum. Interestingly, scans of patients with PD often do not show significant metabolic changes; however, elevated PDRP scores are consistently observed even early in the disease course (Schindlbeck et al., 2020). Multivariate analytical approaches have been employed to identify and validate characteristic metabolic brain networks for MSA (i.e., MSA-related pattern (Eckert et al., 2008; Tomše et al., 2022)), PSP (i.e., PSP-related pattern (Eckert et al., 2008; Tomše et al., 2022)), and pathologically confirmed patients with corticobasal degeneration (i.e., CBD-related pattern (Niethammer et al., 2014)).

2.3.5 Metabolic changes in neurodegenerative cognitive syndromes

Neurodegenerative dementias present primarily with cognitive decline; the most common forms are AD, dementia with Lewy bodies (DLB), and frontotemporal dementia (FTD). Again, misdiagnosis is not uncommon (Rizzo et al., 2018; Spetsieris and Eidelberg, 2011) [^{18}F]FDG PET imaging in AD shows hypometabolism in the precuneus, posterior cingulate gyri, parietal, and temporal lobes. In DLB, we observe reduced glucose metabolism in the precuneus with relative sparing of the posterior cingulate cortex, bilateral parietal and temporal regions, and additional involvement of the occipital lobe. In FTD, frontal and temporal hypometabolism is seen (Brown et al., 2014). Metabolic brain networks for these disorders have been described. AD-related pattern (ADRP) has been identified and validated across different sites and patient populations (Perovnik et al., 2022; Mattis et al., 2016; Perovnik et al., 2022). ADRP is characterized by hypometabolic changes predominantly observed in the medial temporal lobes and precuneus, with covarying preserved metabolism in the cerebellum. Elevated ADRP subject scores are already observable in prodromal AD and can be used to track disease progression (Meles et al., 2017; Perovnik et al., 2023). The two other main causes of neurodegenerative dementias, that is, DLB and FTD, are also characterized by the appearance of characteristic metabolic brain networks, DLBRP and FTDRP, respectively (Nazem et al., 2018; Perovnik et al., 2022; Perovnik et al., 2025).

2.3.6 Dopamine transporter imaging

Dopamine, an intermediate in the biosynthetic pathway from the amino acid tyrosine to adrenaline, is primarily released by neurons in the nigrostriatal pathway. The DAT, located on the presynaptic terminals of dopaminergic neurons, mediates the reuptake of dopamine from the synaptic cleft. Imaging of DAT is made possible with the use of [^{123}I]FP-CIT (N-ω-fluoropropyl-2β-carbomethoxy-3β-nortropane). This compound binds with high affinity to presynaptic DAT in the striatum—specifically the caudate nucleus and putamen—and is visualized using SPECT imaging. Reduced or absent striatal uptake is indicative of dopaminergic neurodegeneration. An analogous compound labeled with ^{18}F has also been developed for use as a PET tracer (Oh et al., 2023). Other parts of the nigrostriatal pathway can also be assessed with PET imaging (Ibrahim and et al., 2016; Nag et al., 2021).

DAT SPECT is most commonly used to differentiate neurodegenerative Parkinsonian syndromes from other, similar disorders in which the dopaminergic neurons are preserved, such as essential tremor, drug-induced Parkinsonism, or vascular Parkinsonism (Morbelli et al., 2020). It can also be used for differential diagnosis of dementia. In DLB, 90% of patients have a positive DAT SPECT to aid in the differential diagnosis with AD. However, in the prodromal phases of the disease (i.e., mild cognitive impairment with Lewy bodies), 20% of patients show preserved integrity of the dopaminergic neurons, thus reducing the accuracy of this biomarker (Nobili et al., 2016). Even though DAT SPECT is not used to differentiate between Parkinsonian syndromes, research has shown that typically an asymmetrical pattern of impairment is observed in PD, while in MSA, especially in the cerebellar variant of the disease, dopaminergic activity may be preserved (Koros et al., 2020; Muñoz et al., 2011). In PSP, striatal dopaminergic uptake is usually symmetrically and substantially reduced (Antonini et al., 2003).

2.4 Imaging of misfolded proteins

2.4.1 Amyloid positron emission tomography

A defining pathological feature of AD is the accumulation of Aβ plaques and tau-containing neurofibrillary tangles (NFTs), which begin to develop approximately 10–20 years before the onset of clinical symptoms. Aβ peptides, consisting of 40 to 42 amino acids, exhibit a propensity to aggregate into soluble oligomers, which subsequently coalesce into extracellular diffuse plaques (primarily composed of Aβ42) and eventually into dense-core neuritic plaques that also contain tau-immunoreactive neurites. Aβ40, by contrast, preferentially deposits in the walls of cerebral blood vessels, resulting in cerebral amyloid angiopathy (CAA) (Johnson et al., 2016). In CAA, Aβ deposition within the walls of cerebral blood vessels leads to vascular damage, increasing the susceptibility to hemorrhage. This pathology is commonly associated with lobar intracerebral hemorrhages or cerebral microbleeds. CAA can manifest in both sporadic and familial forms. One rare hereditary variant is hereditary cystatin C CAA (HCCAA), a rare Icelandic disorder characterized by recurrent cerebral hemorrhages in otherwise healthy young individuals. HCCAA follows an autosomal dominant inheritance pattern and is caused by a point mutation in the *CST3* gene, which encodes human cystatin C (hCC). This mutation results in a leucine-to-glutamine substitution at position 68 (L68Q-CST3), promoting the formation of amyloidogenic variants of hCC (Snorradottir et al., 2025).

A thioflavin-T analog, ^{11}C-labeled Pittsburgh compound B ([^{11}C]PiB), was the first radiotracer that enabled in vivo imaging of amyloid plaques (Klunk et al., 2004). The short half-life (~20 minutes) of the ^{11}C isotope restricts its use to research facilities equipped with an on-site cyclotron. To overcome this limitation, several second-generation [^{18}F] labeled radiotracers have been developed, namely, [^{18}F]-florbetapir, [^{18}F]-florbetaben, [^{18}F]-flutemetamol, and, most recently, third-generation tracers, such as [^{18}F]-flutafuranol (Villemagne et al., 2018). Different Aβ PET tracers have high concordance (Hatashita et al., 2014; Su et al., 2019).

The distribution pattern of amyloid uptake in AD observed on PET imaging aligns with postmortem findings, predominantly involving the frontal, cingulate, precuneus, striatum, parietal, and lateral temporal cortices, while the occipital, sensorimotor, and medial temporal cortices are less affected (Villemagne, 2016). Early and maximal radiotracer uptake is typically localized to the posterior cingulate cortex/precuneus and medial prefrontal regions (Villeneuve et al., 2015). In CAA, tracer retention is frequently observed in the occipital lobes (Johnson et al., 2007). Amyloid PET interpretation is generally dichotomized into positive or negative scans based on visual assessment. A negative scan exhibits preserved gray-to-white matter contrast, with uptake confined to white matter, whereas a positive scan shows cortical gray matter binding equal to or exceeding that of white matter (Fig. 2.3).

The diagnostic validity of amyloid PET has been established through PET–autopsy correlation studies, in which antemortem imaging findings were compared to postmortem Aβ pathology distribution and burden. Visual reads of [^{18}F]-labeled tracer scans achieved sensitivity of 88%–98% and specificity of 80%–95% for identifying moderate-to-frequent neuritic plaques as defined by the CERAD criteria (Johnson et al., 2007; Curtis et al., 2015; Sabri et al., 2015). Additionally, amyloid PET classification shows strong concordance with the CSF Aβ42/40 ratio (Campbell et al., 2021).

Amyloid PET imaging is crucial for the development of new therapeutic approaches. Aducanumab (FDA-approved since 2021 and later taken from the market) is a monoclonal antibody targeting aggregated Aβ for the treatment of mild cognitive impairment and mild AD dementia. The approval was based on aducanumab's ability to reduce amyloid burden detected by PET in a dose-dependent manner (Rabinovici, 2021). Amyloid PET has also served as a key efficacy endpoint in clinical trials of other anti-Aβ monoclonal antibodies, including donanemab (Mintun et al., 2021) and lecanemab (Swanson et al., 2021), which are as of 2026 used in clinical practice worldwide.

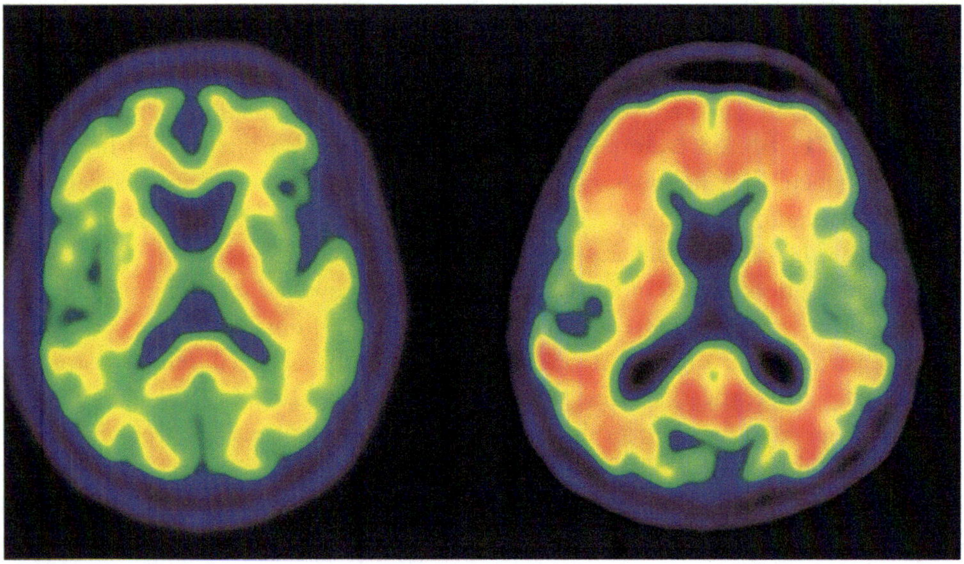

Figure 2.3 An example of a negative (*left*) and positive (*right*) amyloid positron emission tomography scan.

2.5 Tau positron emission tomography

Tau protein is a microtubule-associated protein essential for neuronal stability and intracellular transport of organelles, secretory vesicles, and neurotransmitters (Goedert et al., 1991; Buée et al., 2000). Abnormal hyperphosphorylation of tau leads to the aggregation of tau into straight filaments, twisted ribbons, or paired helical filaments, collectively known as NFTs (Buée et al., 2000). Such abnormal tau aggregation is a hallmark of AD and other tauopathies, including CBD, FTD, and PSP. In humans, six tau isoforms are expressed. These differ by the presence of either three (3R) or four (4R) microtubule-binding repeats. Both 3R and 4R isoforms are present in AD. In contrast, PSP and CBD are predominantly associated with 4R isoforms (Goedert et al., 2017).

The first generation of tau PET tracers includes [^{18}F]-flortaucipir, [^{11}C]PBB3, and the [^{18}F]-THK family. Postmortem pathology studies have shown that [^{18}F]-flortaucipir binding correlates with postmortem NFT Braak staging in AD brains (Marquie et al., 2017), with the greatest uptake observed in the inferior and lateral temporal lobes. The uptake also significantly correlated with the clinical severity of dementia (Pontecorvo et al., 2017). Despite these strengths, [^{18}F]-flortaucipir displays off-target binding. Autoradiographic studies have identified nonspecific uptake in regions such as the medial temporal cortex (possibly age-related), substantia nigra, choroid plexus calcifications, and leptomeningeal melanin (Lowe et al., 2016). Moreover, this tracer does not reliably detect non-AD tauopathies such as PSP and CBD (Malpetti et al., 2022; Fleisher et al., 2020). Tau PET is incorporated in the latest AD diagnostic criteria for disease staging (Jack et al., 2024).

In 2020, Schroter et al., demonstrated that the tracer [^{11}C]PBB3 could distinguish between tauopathies based on distinct regional binding patterns. Uptake was elevated in the dorsolateral frontal and motor cortices in corticobasal syndrome, and in the frontal and temporal cortices in AD, compared to PSP (Schröter et al., 2020). PSP patients exhibited mild but significant tracer uptake in the midbrain relative to AD. However, [^{11}C]PBB3 has several limitations, such as low dynamic range, metabolic instability, and off-target binding in the basal ganglia, the longitudinal sinus, and the choroid plexus (Okamura et al., 2018). Novel, second-generation tracers such as [^{18}F]-MK6240, [^{18}F]-RO948, [^{18}F]-PI2620, [^{18}F]-GTP1, and [^{18}F]-JNJ-64326067 are being developed to minimize the off-target binding.

2.6 Alpha-synuclein positron emission tomography

PD, DLB, and MSA are characterized by the abnormal accumulation of α-αSyn protein aggregates. αSyn is widely expressed neuronal protein, constituting up to 1% of the total cytosolic protein content in primary neurons, with predominant localization at presynaptic terminals (Spillantini et al., 1997). Although its exact physiological role remains incompletely understood, evidence from in vitro and in vivo models suggests that αSyn plays a role in synaptic vesicle trafficking, regulation of neurotransmitter release, and is involved in stabilizing cellular membranes (Larsen et al., 2006; Nemani et al., 2010; Calabresi et al., 2023).

αSyn undergoes several post-translational modifications, although the underlying molecular mechanisms and their biological significance are not fully elucidated (Wang et al., 2016). In PD and DLB, αSyn aggregates into intraneuronal inclusion bodies, known as Lewy bodies. In early-stage PD, these inclusions predominantly appear in the substantia nigra pars compacta and later extend to cortical regions and the limbic system. Conversely, in DLB, cortical involvement occurs at earlier stages. In MSA, αSyn accumulates primarily in the cytoplasm of neurons and oligodendrocytes (Malfertheiner et al., 2021).

The correlation between the structural diversity of αSyn aggregates and their functional consequences remains poorly understood. The development of diagnostic imaging agents is further complicated by the heterogeneity of αSyn conformations and the typically low abundance of aggregates. Moreover, the β-sheet-rich fibrillar structures of αSyn resemble those of Aβ and tau, necessitating high specificity in tracer design. Given the intracellular localization of αSyn aggregates, any effective PET tracer must be capable of penetrating both cellular membranes and the blood–brain barrier.

Since 2010, numerous αSyn-targeting PET tracers have been under development. Initial candidates such as [^{11}C] or [^{18}F]-labeled BF-227 produced inconsistent results and lacked selectivity for αSyn over Aβ and tau fibrils (Xiang et al., 2025). The small molecule Anle138b, in contrast, selectively binds to misfolded prion proteins and αSyn aggregates in vivo, without affinity for native monomeric αSyn. In preclinical PD models, anle138b reduced αSyn oligomer formation by approximately 77% and attenuated disease progression (Challis et al., 2020). The radiotracer form, [^{11}C] anle253b, is currently in development and used in rodent studies (Maurer et al., 2020)

The radiotracer [^{18}F]ACI-12589 demonstrated specific binding to αSyn fibrils in the cerebellar white matter of MSA patients, while showing no significant affinity for Aβ or tau. However, this tracer failed to generate detectable PET signals in individuals with PD or DLB (Smith et al., 2023).

Another candidate, [^{18}F]-F0502B, has shown selective binding to αSyn aggregates in postmortem brain tissue, particularly within the striatum and substantia nigra of PD patients. Notably, it did not exhibit significant binding to Aβ plaques or NFTs associated with AD, indicating a degree of specificity. Furthermore, it showed higher binding in two nonhuman primate models of PD compared to healthy controls (Xiang, 2023). The utility of [^{18}F]-F0502B for in vivo imaging in patients with αSyn pathology remains to be validated.

2.7 Conclusions

Brain imaging techniques, especially functional imaging methods such as PET and SPECT (and fMRI - not in the scope of this review), are critical tools for diagnosing and monitoring NDs. These techniques can detect pathological processes at early and even presymptomatic disease stages, significantly improving early diagnosis and differential diagnosis. Furthermore, novel PET radiotracers enable the monitoring of disease progression and the effectiveness of novel treatments by directly visualizing pathological proteins like Aβ, tau, and αSyn.

While peripheral blood biomarkers are emerging as accessible and less invasive diagnostic tools, they currently lack the spatial resolution and specificity provided by brain imaging (Hansson, 2021). However, combining brain imaging with peripheral biomarkers could improve diagnostic accuracy, offer complementary insights into disease pathology, and assist in tracking treatment responses. Despite their diagnostic value, it is essential to minimize the use of radiotracers due to radiation exposure. Clinicians must weigh diagnostic benefits against the risks and adhere strictly to recommended guidelines to limit unnecessary exposure.

Declaration of AI and AI-assisted technologies in the writing process

Statement: during the preparation of this work, the author(s) used ChatGPT to improve the readability of the text. After using this tool/service, the author(s) reviewed and edited the content as needed and take(s) full responsibility for the content of the publication.

Acknowledgements

M.P. was supported by the Slovenian Research and Innovation Agency (ARIS) through grants P1- 0389 and J3- 60059.

References

Antonini, A., Benti, R., De Notaris, R., Tesei, S., Zecchinelli, A., Sacilotto, G., et al., 2003. 123I-Ioflupane/SPECT binding to striatal dopamine transporter (DAT) uptake in patients with Parkinson's disease, multiple system atrophy, and progressive supranuclear palsy. Neurological Sciences 24 (3), 149–150. Available at https://doi.org/10.1007/s10072-003-0103-5.

Arbizu, J., Morbelli, S., Minoshima, S., Barthel, H., Kuo, P., Van Weehaeghe, D., et al., 2025. SNMMI Procedure standard/EANM practice guideline for brain [18 F]FDG PET imaging, version 2.0. Journal of Nuclear Medicine 66 (Supplement 2), S45. Available at https://doi.org/10.2967/jnumed.124.268754.

Brown, R.K.J., Bohnen, N.I., Wong, K.K., Minoshima, S., Frey, K.A., 2014. Brain PET in suspected dementia: patterns of altered FDG metabolism. Radiographics 34 (3), 684–701. Available at https://doi.org/10.1148/rg.343135065.

Buée, L., Bussière, T., Buée-Scherrer, V., Delacourte, A., Hof, P.R., 2000. Tau protein isoforms, phosphorylation and role in neurodegenerative disorders. Brain Research Reviews 33 (1), 95–130. Available at https://doi.org/10.1016/S0165-0173(00)00019-9.

Calabresi, P., Mechelli, A., Natale, G., Volpicelli-Daley, L., Di Lazzaro, G., Ghiglieri, V., 2023. Alpha-synuclein in Parkinson's disease and other synucleinopathies: from overt neurodegeneration back to early synaptic dysfunction. Cell Death and Disease 14 (3). Available at https://doi.org/10.1038/s41419-023-05672-9.

Campbell, MR, Ashrafzadeh-Kian, S, Petersen, RC, Mielke, MM, Syrjanen, JA, van Harten, AC, Lowe, VJ, Jack CR, Jr, Bornhorst, JA, lgeciras-Schimnich A, A, et al., 2021. P-tau/Aβ42 and Aβ42/40 ratios in CSF are equally predictive of amyloid PET status. Alzheimers Dement (Amst) 13 (1), e12190. Available at 10.1002/dad2.12190.

Carli, G., Dortmond, A., Janzen, A., Sittig, E., De Meyer, E., Leenders, K.L., et al., 2025. Parkinson's Disease-related pattern in isolated REM sleep behaviour disorder as a prodromal progression marker: 8-Year follow-up changes assessed at three time points. Journal of Nuclear Medicine and Molecular Imaging 52 (10), 3550–3556. Available at https://doi.org/10.1007/s00259-025-07260-9.

Challis, C., Hori, A., Sampson, T.R., Yoo, B.B., Challis, R.C., Hamilton, A.M., et al., 2020. Gut-seeded α-synuclein fibrils promote gut dysfunction and brain pathology specifically in aged mice. Nature Neuroscience 23 (3), 327–336. Available at https://doi.org/10.1038/s41593-020-0589-7.

Curtis, C., Gamez, J.E., Singh, U., Sadowsky, C.H., Villena, T., Sabbagh, M.N., et al., 2015. Phase 3 trial of flutemetamol labeled with radioactive fluorine 18 imaging and neuritic plaque density. JAMA Neurology 72 (3), 287–294. Available at https://doi.org/10.1001/jamaneurol.2014.4144.

Deatsch, A., Perovnik, M., Namías, M., Trošt, M., Jeraj, R., 2022. Development of a deep learning network for Alzheimer's disease classification with evaluation of imaging modality and longitudinal data. Physics in Medicine & Biology 67 (19), 195014. Available at https://doi.org/10.1088/1361-6560/ac8f10.

Eckert, T., Tang, C., Ma, Y., Brown, N., Lin, T., Frucht, S., et al., 2008. Abnormal metabolic networks in atypical parkinsonism. Movement Disorders 23 (5), 727–733. Available at https://doi.org/10.1002/mds.21933.

Etminani, K., Soliman, A., Davidsson, A., Chang, J.R., Martínez-Sanchis, B., Byttner, S., et al., 2022. A 3D deep learning model to predict the diagnosis of dementia with Lewy bodies, Alzheimer's disease, and mild cognitive impairment using brain 18F-FDG PET. Journal of Nuclear Medicine and Molecular Imaging 49 (2), 563–584. Available at https://doi.org/10.1007/s00259-021-05483-0.

Fleisher, A.S., Pontecorvo, M.J., Devous, M.D., Lu, M., Arora, A.K., Truocchio, S.P., et al., 2020. Positron emission tomography imaging with [18F]flortaucipir and postmortem assessment of Alzheimer disease neuropathologic changes. JAMA Neurology 77 (7), 829–839. Available at https://doi.org/10.1001/jamaneurol.2020.0528.

Friston, K., 2007. CHAPTER 2 - Statistical parametric mapping. In: Friston, K., Ashburner, J., Kiebel, S., Nichols, T., Penny, W. (Eds.), Statistical Parametric Mapping: The Analysis of Functional Brain Images. Academic Press, pp. 10–31. Available at 10.1016/b978-012372560-8/50002-4.

Goedert, M., Crowther, R.A., Garner, C.C., 1991. Molecular characterization of microtubule-associated proteins tau and map2. Trends in Neurosciences 14 (5), 193–199. Available at https://doi.org/10.1016/0166-2236(91)90105-4.

Goedert, M., Eisenberg, D.S., Crowther, R.A., 2017. Propagation of tau aggregates and neurodegeneration. Annual Review of Neuroscience 40, 189–210. Available at https://doi.org/10.1146/annurev-neuro-072116-031153.

Hansson, O., 2021. Biomarkers for neurodegenerative diseases. Nature Medicine 27 (6), 954–963. Available at https://doi.org/10.1038/s41591-021-01382-x.

Hatashita, S., Yamasaki, H., Suzuki, Y., Tanaka, K., Wakebe, D., Hayakawa, H., 2014. 18F]Flutemetamol amyloid-beta PET imaging compared with [11C]PIB across the spectrum of Alzheimer's disease. Journal of Nuclear Medicine and Molecular Imaging 41 (2), 290–300. Available at https://doi.org/10.1007/s00259-013-2564-y.

Ibrahim, N., Kusmirek, J., Struck, A.F., Floberg, J.M., Perlman, S.B., Gallagher, C., et al., 2016. The sensitivity and specificity of F-DOPA PET in a movement disorder clinic. Journal of Nuclear Medicine and Molecular Imaging 6, 102–109.

Jack, C.R., Andrews, J.S., Beach, T.G., Buracchio, T., Dunn, B., Graf, A., et al., 2024. Revised criteria for diagnosis and staging of Alzheimer's disease: Alzheimer's association workgroup. Alzheimer's & Dementia 20 (8), 5143–5169. Available at https://doi.org/10.1002/alz.13859.

Johnson, K.A., Gregas, M., Becker, J.A., Kinnecom, C., Salat, D.H., Moran, E.K., et al., 2007. Imaging of amyloid burden and distribution in cerebral amyloid angiopathy. Annals of Neurology 62 (3), 229–234. Available at https://doi.org/10.1002/ana.21164.

Johnson, K.A., Schultz, A., Betensky, R.A., Becker, J.A., Sepulcre, J., Rentz, D., et al., 2016. Tau positron emission tomographic imaging in aging and early Alzheimer disease. Annals of Neurology 79 (1), 110–119. Available at https://doi.org/10.1002/ana.24546.

Klunk, W.E., Engler, H., Nordberg, A., Wang, Y., Blomqvist, G., Holt, D.P., et al., 2004. Imaging Brain amyloid in alzheimer's disease with pittsburgh compound-B. Annals of Neurology 55 (3), 306–319. Available at https://doi.org/10.1002/ana.20009.

Koros, C., Simitsi, A.-M., Prentakis, A., Papagiannakis, N., Bougea, A., Pachi, I., et al., 2020. DaTSCAN (123I-FP-CIT SPECT) imaging in early versus mid and late onset Parkinson's disease: longitudinal data from the PPMI study. Parkinsonism & Related Disorders 77, 36–42. Available at https://doi.org/10.1016/j.parkreldis.2020.06.019.

Larsen, K.E., Schmitz, Y., Troyer, M.D., Mosharov, E., Dietrich, P., Quazi, A.Z., et al., 2006. Synuclein overexpression in PC12 and chromaffin cells impairs catecholamine release by interfering with a late step in exocytosis. Journal of Neuroscience 26 (46), 11915–11922. Available at http://doi.org/10.1523/JNEUROSCI.3821-06.2006.

Lin, E.C., Lamsback, A., 2009. PET and PET/CT: A Clinical Guide: Second Edition. Thieme, New York. Available at 10.1055/b-002-74271.

Lowe, V.J., Curran, G., Fang, P., Liesinger, A.M., Josephs, K.A., Parisi, J.E., et al., 2016. An autoradiographic evaluation of AV-1451 Tau PET in dementia. Acta Neuropathologica Communications 4, 58. Available at https://doi.org/10.1186/s40478-016-0315-6.

Ma, Y., Tang, C., Spetsieris, P.G., Dhawan, V., Eidelberg, D., 2007. Abnormal metabolic network activity in Parkinson's disease: test-retest reproducibility. Journal of Cerebral Blood Flow and Metabolism 27 (3), 597–605. Available at https://doi.org/10.1038/sj.jcbfm.9600358.

Malfertheiner, K., Stefanova, N., Heras-Garvin, A., 2021. The concept of alpha-synuclein strains and how different conformations may explain distinct neurodegenerative disorders. Frontiers Neurology 12, 737195. Available at 10.3389/fneur.2021.737195.

Malpetti, M., Kaalund, S.S., Tsvetanov, K.A., Rittman, T., Briggs, M., Allinson, K.S.J., et al., 2022. In Vivo 18 F-Flortaucipir PET does not accurately support the staging of progressive supranuclear palsy. Journal of Nuclear Medicine 63 (7), 1052–1057. Available at https://doi.org/10.2967/jnumed.121.262985.

Marquie, M., Siao Tick Chong, M., Antón-Fernández, A., Verwer, E.E., Sáez-Calveras, N., Meltzer, A.C., et al., 2017. [F-18]-AV-1451 binding correlates with postmortem neurofibrillary tangle Braak staging. Acta Neuropathologica 134, 619–628. Available at https://doi.org/10.1007/s00401-017-1740-8.

Mattis, P.J., Niethammer, M., Sako, W., Tang, C.C., Nazem, A., Gordon, M.L., et al., 2016. Distinct brain networks underlie cognitive dysfunction in Parkinson and Alzheimer diseases. Neurology 87 (18), 1925–1933. Available at https://doi.org/10.1212/WNL.0000000000003285.

Maurer, A., Leonov, A., Ryazanov, S., Herfert, K., Kuebler, L., Buss, S., et al., 2020. Christian griesinger, 11 C radiolabeling of anle253b: a putative PET tracer for Parkinson's disease that binds to α-synuclein fibrils in vitro and crosses the blood-brain barrier. ChemMedChem 15 (5), 411–415. Available at https://doi.org/10.1002/cmdc.201900689.

Meles, S.K., Pagani, M., Arnaldi, D., De Carli, F., Dessi, B., Morbelli, S., et al., 2017. The Alzheimer's disease metabolic brain pattern in mild cognitive impairment. Journal of Cerebral Blood Flow and Metabolism 37 (12), 3643–3648. Available at https://doi.org/10.1177/0271678X17732508.

Meyer, P.T., Frings, L., Rücker, G., Hellwig, S., 2017. 18F-FDG PET in Parkinsonism: differential diagnosis and evaluation of cognitive impairment. Journal of Nuclear Medicine 58 (12), 1888–1898. Available at https://doi.org/10.2967/jnumed.116.186403.

Mintun, M.A., Lo, A.C., Evans, C.D., Wessels, A.M., Ardayfio, P.A., Andersen, S.W., et al., 2021. Donanemab in early Alzheimer's disease. Journal of Medicine 384 (18), 1691–1704. Available at https://doi.org/10.1056/NEJMoa2100708.

Morbelli, S., Esposito, G., Arbizu, J., Barthel, H., Boellaard, R., Bohnen, N.I., et al., 2020. EANM practice guideline/SNMMI procedure standard for dopaminergic imaging in Parkinsonian syndromes 1.0. Journal of Nuclear Medicine and Molecular Imaging 47 (8), 1885–1912. Available at https://doi.org/10.1007/s00259-020-04817-8.

Muñoz, E., Iranzo, A., Rauek, S., Lomeña, F., Gallego, J., Ros, D., et al., 2011. Subclinical nigrostriatal dopaminergic denervation in the cerebellar subtype of multiple system atrophy (MSA-C). Journal of Neurology 258 (12), 2248–2253. Available at https://doi.org/10.1007/s00415-011-6108-8.

Myszczynska, M.A., Ojamies, P.N., Lacoste, A.M.B., Neil, D., Saffari, A., Mead, R., et al., 2020. Applications of machine learning to diagnosis and treatment of neurodegenerative diseases. Nature Reviews Neurology 16 (8), 440–456. Available at https://doi.org/10.1038/s41582-020-0377-8.

Nag, S., Jahan, M., Tóth, M., Nakao, R., Varrone, A., Halldin, C., 2021. PET imaging of VMAT2 with the novel radioligand [18 F]FE-DTBZ-d4 in nonhuman primates: comparison with [11 C]DTBZ and [18 F]FE-DTBZ. ACS Chemical Neuroscience 12 (24), 4580–4586. Available at https://doi.org/10.1021/acschemneuro.1c00651.

Nazem, A., Tang, C.C., Spetsieris, P., Dresel, C., Gordon, M.L., Diehl-Schmid, J., et al., 2018. A multivariate metabolic imaging marker for behavioral variant frontotemporal dementia. Alzheimer's & Dementia: Diagnosis, Assessment & Disease Monitoring 10 (1), 583–594. Available at https://doi.org/10.1016/j.dadm.2018.07.009.

Nemani, V.M., Lu, W., Berge, V., Nakamura, K., Onoa, B., Lee, M.K., et al., 2010. Increased expression of α-synuclein reduces neurotransmitter release by inhibiting synaptic vesicle reclustering after endocytosis. Neuron 65 (1), 66–79. Available at https://doi.org/10.1016/j.neuron.2009.12.023.

Ng, S., Villemagne, V.L., Berlangieri, S., Lee, S.-T., Cherk, M., Gong, S.J., et al., 2007. Visual assessment versus quantitative assessment of 11C-PIB PET and 18F-FDG PET for detection of Alzheimer's disease. Journal of Nuclear Medicine 48 (4), 547–552. Available at https://doi.org/10.2967/jnumed.106.037762.

Niethammer, M., Tang, C.C., Feigin, A., Allen, P.J., Heinen, L., Hellwig, S., et al., 2014. A disease-specific metabolic brain network associated with corticobasal degeneration. Brain 137 (11), 3036–3046. Available at https://doi.org/10.1093/brain/awu256.

Nobili, F., Arnaldi, D., Morbelli, S., 2016. Is dopamine transporter invariably impaired at the time of diagnosis in dementia with Lewy bodies? Journal of Nuclear Medicine and Molecular Imaging 43 (6), 1056–1059. Available at https://doi.org/10.1007/s00259-016-3323-7.

Nobili, F., Festari, C., Altomare, D., Agosta, F., Orini, S., Van Laere, K., et al., 2018. Automated assessment of FDG-PET for differential diagnosis in patients with neurodegenerative disorders. Journal of Nuclear Medicine and Molecular Imaging 45 (9), 1557–1566. Available at https://doi.org/10.1007/s00259-018-4030-3.

O'Malley, J.P., Ziessman, H.A., Thrall, J.H., 2020. Nuclear Medicine and Molecular Imaging: The Requisites, fifth ed. Elsevier, Amsterdam.

Oh, M., Oh, S.J., Lee, S.J., Oh, J.S., Chung, S.J., Kim, J.S., 2023. Diagnostic accuracy of (18)F-FP-CIT PET for clinically uncertain Parkinsonian syndrome. Scientific Reports 13 (1), 15069. Available at 10.1038/s41598-023-42135-9.

Okamura, N., Harada, R., Ishiki, A., Kikuchi, A., Nakamura, T., Kudo, Y., 2018. The development and validation of tau PET tracers: current status and future directions. Clinical and Translational Imaging 6 (4), 305–316. Available at https://doi.org/10.1007/s40336-018-0290-y.

Perani, D., Della Rosa, P.A., Cerami, C., Gallivanone, F., Fallanca, F., Vanoli, E.G., et al., 2014. Validation of an optimized SPM procedure for FDG-PET in dementia diagnosis in a clinical setting. NeuroImage: Clinical 6, 445–454. Available at https://doi.org/10.1016/j.nicl.2014.10.009.

Perovnik, M., Rus, T., Schindlbeck, K.A., Eidelberg, D., 2023. Functional brain networks in the evaluation of patients with neurodegenerative disorders. Nature Reviews Neurology 19 (2), 73–90. Available at https://doi.org/10.1038/s41582-022-00753-3.

Perovnik, M., Simončič, U., Jamšek, J., Gregorič Kramberger, M., Brumberg, J., Meyer, P.T., et al., 2025. Metabolic brain networks in dementia with Lewy bodies: from prodromal to manifest disease stages. Journal of Neurology, Neurosurgery & Psychiatry 97, 316–324. Available at https://doi.org/10.1136/jnnp-2025-336935.

Perovnik, M., Tang, C.C., Namías, M., Eidelberg, D., 2023. Longitudinal changes in metabolic network activity in early Alzheimer's disease. Alzheimer's & Dementia 19 (9), 4061–4072. Available at https://doi.org/10.1002/alz.13137.

Perovnik, M., Tomše, P., Jamšek, J., Tang, C., Eidelberg, D., Trošt, M., 2022. Metabolic brain pattern in dementia with Lewy bodies: relationship to Alzheimer's disease topography. NeuroImage: Clinical 35, 103080. Available at https://doi.org/10.1016/j.nicl.2022.103080.

Perovnik, M, Tomše, P, amšek J, J, Emeršič, A, Tang, C, Eidelberg, D, Trošt, M, et al., 2022. Identification and validation of Alzheimer's disease-related metabolic brain pattern in biomarker confirmed Alzheimer's dementia patients. Sci Rep 12 (1), 11752. Available at https://doi.org/10.1038/s41598-022-15667-9.

Pontecorvo, M.J., Devous, M.D., Navitsky, M., Lu, M., Salloway, S., Schaerf, F.W., et al., 2017. Relationships between flortaucipir PET tau binding and amyloid burden, clinical diagnosis, age and cognition. Brain 140 (3), 748–763. Available at https://doi.org/10.1093/brain/aww334.

Rabinovici, G.D., 2021. Controversy and progress in Alzheimer's disease - FDA approval of aducanumab. Journal of Medicine 385 (9), 771–774. Available at https://doi.org/10.1056/NEJMp2111320.

Risacher, S.L., Saykin, A.J., 2019. Neuroimaging in aging and neurologic diseases. Handbook of Clinical Neurology 167, 191–227. Available at https://doi.org/10.1016/B978-0-12-804766-8.00012-1.

Rizzo, G., Copetti, M., Arcuti, S., Martino, D., Fontana, A., Logroscino, G., 2016. Accuracy of clinical diagnosis of Parkinson disease. Neurology 86 (6), 566–576. Available at https://doi.org/10.1212/WNL.0000000000002350.

Rizzo, G., Arcuti, S., Copetti, M., Alessandria, M., Savica, R., Fontana, A., et al., 2018. Accuracy of clinical diagnosis of dementia with Lewy bodies: a systematic review and meta-analysis. Journal of Neurology, Neurosurgery & Psychiatry 89 (4), 358–366. Available at https://doi.org/10.1136/jnnp-2017-316844.

Saboury, B., Bradshaw, T., Boellaard, R., Buvat, I., Dutta, J., Hatt, M., et al., 2023. Artificial intelligence in nuclear medicine: opportunities, challenges, and responsibilities toward a trustworthy ecosystem. Journal of Nuclear Medicine 64 (2), 188–196. Available at https://doi.org/10.2967/jnumed.121.263703.

Sabri, O., Sabbagh, M.N., Seibyl, J., Barthel, H., Akatsu, H., Ouchi, Y., et al., 2015. Florbetaben PET imaging to detect amyloid beta plaques in Alzheimer's disease: phase 3 study. Alzheimer's & Dementia 11 (8), 964–974. Available at https://doi.org/10.1016/j.jalz.2015.02.004.

Sala, A., Lizarraga, A., Caminiti, S.P., Calhoun, V.D., Eickhoff, S.B., Habeck, C., et al., 2023. Brain connectomics: time for a molecular imaging perspective? Trends in Cognitive Sciences 27 (4), 353–366. Available at https://doi.org/10.1016/j.tics.2022.11.015.

Schindlbeck, K.A., Eidelberg, D., 2018. Network imaging biomarkers: insights and clinical applications in Parkinson's disease. The Lancet Neurology 17 (7), 629–640. Available at https://doi.org/10.1016/S1474-4422(18)30169-8.

Schindlbeck, K.A., Lucas-Jiménez, O., Tang, C.C., Morbelli, S., Arnaldi, D., Pardini, M., et al., 2020. Metabolic network abnormalities in drug-naïve Parkinson's disease. Movement Disorders 35 (4), 587–594. Available at https://doi.org/10.1002/mds.27960.

Schröter, N., Blazhenets, G., Frings, L., Barkhausen, C., Jost, W.H., Weiller, C., et al., 2020. Tau imaging in the 4-repeat-tauopathies progressive supranuclear palsy and corticobasal syndrome: A 11C-pyridinyl-butadienyl-benzothiazole 3 PET pilot study. Clinical Nuclear Medicine 45 (4), 283–287. Available at https://doi.org/10.1097/RLU.0000000000002949.

Smith, R., Capotosti, F., Schain, M., Ohlsson, T., Vokali, E., Molette, J., et al., 2023. The alpha-synuclein PET tracer [18F] ACI-12589 distinguishes multiple system atrophy from other neurodegenerative diseases. Nature Communication 14, 6750. Available at https://doi.org/10.1038/s41467-023-42305-3.

Snorradottir, A.O., Gutierrez-Uzquiza, A., Bragado, P., March, M.E., Kao, C., Arkink, E.B., et al., 2025. N -Acetylcysteine for hereditary cystatin C amyloid angiopathy. JAMA Neurology 82 (5), 486. Available at https://doi.org/10.1001/jamaneurol.2025.0326.

Spetsieris, P.G., Eidelberg, D., 2011. Scaled subprofile modeling of resting state imaging data in Parkinson's disease: methodological issues. Neuroimage 54 (4), 2899–2914. Available at https://doi.org/10.1016/j.neuroimage.2010.10.025.

Spillantini, M.G., Schmidt, M.L., Lee, V.M.Y., Trojanowski, J.Q., Jakes, R., Goedert, M., 1997. α-synuclein in Lewy bodies. Nature 388 (6645), 839–840. Available at https://doi.org/10.1038/42166.

Su, Y., Flores, S., Wang, G., Hornbeck, R.C., Speidel, B., Joseph-Mathurin, N., et al., 2019. Comparison of Pittsburgh compound B and florbetapir in cross-sectional and longitudinal studies. Alzheimer's & Dementia: Diagnosis, Assessment & Disease Monitoring 11 (1), 180–190. Available at https://doi.org/10.1016/j.dadm.2018.12.008.

Swanson, C.J., Zhang, Y., Dhadda, S., Wang, J., Kaplow, J., Lai, R.Y.K., et al., 2021. A randomized, double-blind, phase 2b proof-of-concept clinical trial in early Alzheimer's disease with lecanemab, an anti-Aβ protofibril antibody. Alzheimer's Research and Therapy 13 (1). Available at https://doi.org/10.1186/s13195-021-00813-8.

Tang, C.C., Nakano, Y., Vo, A., Nguyen, N., Schindlbeck, K.A., Mattis, P.J., et al., 2024. Longitudinal network changes and phenoconversion risk in isolated REM sleep behavior disorder. Nature Communications 15, 10797. Available at https://doi.org/10.1038/s41467-024-54695-z.

Tomše, P., Jensterle, L., Grmek, M., Zaletel, K., Pirtošek, Z., Dhawan, V., et al., 2017. Abnormal metabolic brain network associated with Parkinson's disease: replication on a new European sample. Neuroradiology 59 (5), 507–515. Available at https://doi.org/10.1007/s00234-017-1821-3.

Tomše, P., Rebec, E., Studen, A., Perovnik, M., Rus, T., Ležaić, L., et al., 2022. Abnormal metabolic covariance patterns associated with multiple system atrophy and progressive supranuclear palsy. Physical Medical 98, 131–138. Available at https://doi.org/10.1016/j.ejmp.2022.04.016.

Villemagne, V.L., 2016. Amyloid imaging: past, present and future perspectives. Ageing Research Reviews 30, 95–106. Available at https://doi.org/10.1016/j.arr.2016.01.005.

Villemagne, V.L., Doré, V., Burnham, S.C., Masters, C.L., Rowe, C.C., 2018. Imaging tau and amyloid-? Proteinopathies in Alzheimer disease and other conditions. Nature Reviews Neurology 14 (4), 225–236. Available at https://doi.org/10.1038/nrneurol.2018.9.

Villeneuve, S., Rabinovici, G.D., Cohn-Sheehy, B.I., Madison, C., Ayakta, N., Ghosh, P.M., et al., 2015. Existing Pittsburgh compound-B positron emission tomography thresholds are too high: statistical and pathological evaluation. Brain 138 (7), 2020–2033. Available at https://doi.org/10.1093/brain/awv112.

Wang, C., Zhao, C., Li, D., Tian, Z., Lai, Y., Diao, J., et al., 2016. Versatile structures of α-Synuclein. Frontiers in Molecular Neuroscience 9, 48. Available at https://doi.org/10.3389/fnmol.2016.00048.

Wu, P., Zhao, Y., Wu, J., Brendel, M., Lu, J., Ge, J., et al., 2022. Differential diagnosis of parkinsonism based on deep metabolic imaging indices. Journal of Nuclear Medicine 63 (11), 1741–1747. Available at https://doi.org/10.2967/jnumed.121.263029.

Xiang, Jie, Tao, Youqi, Xia, Yiyuan, Luo, Shilin, Zhao, Qinyue, Li, Bowei, Zhang, Xiaoqian, Sun, Yunpeng, Xia, Wencheng, Zhang, Mingming, Kang, Seong Su, Ahn, Eun-Hee, Liu, Xia, Xie, Fang, Guan, Yihui, Yang, Jenny J., Bu, Lihong, Wu, Shengxi, Wang, Xiaochuan, Cao, Xuebing, Liu, Cong, Zhang, Zhentao, Li, Dan, Ye, Keqiang, et al., 2023. Development of an α-synuclein positron emission tomography tracer for imaging synucleinopathies. Cell 186 (1610), 3367.e19. Available at 10.1016/j.cell.2023.06.004.

Xiang, J., Zhang, Z., Wu, S., Ye, K., 2025. Positron emission tomography tracers for synucleinopathies. Molecular Neurodegeneration 20 (1). Available at https://doi.org/10.1186/s13024-024-00787-9.

Yamane, T., Ikari, Y., Nishio, T., Ishii, K., Ishii, K., Kato, T., et al., 2014. Visual-statistical interpretation of 18F-FDG-PET images for characteristic Alzheimer patterns in a multicenter study: inter-rater concordance and relationship to automated quantitative evaluation. Journal of Neuroradiology 35 (2), 244–249. Available at https://doi.org/10.3174/ajnr.A3665.

Young, P.N.E., Estarellas, M., Coomans, E., Srikrishna, M., Beaumont, H., Maass, A., et al., 2020. Imaging biomarkers in neurodegeneration: current and future practices. Alzheimer's Research and Therapy 12 (1). Available at https://doi.org/10.1186/s13195-020-00612-7.

CHAPTER 3

Proteins that show aggregation in the brains of patients with schizophrenia and other chronic mental illnesses

Beti Zaharija[1], Ilijana Stanivuk[2,3], Matea Kršanac[1], Maja Juković[1], Aristea Pavešić Radonja[2,3], and Nicholas J. Bradshaw[1]
[1]Faculty of Biotechnology & Drug Development, University of Rijeka, Rijeka, Croatia,
[2]Psychiatry Clinic, Clinical Hospital Centre Rijeka, Rijeka, Croatia,
[3]Faculty of Medicine, University of Rijeka, Rijeka, Croatia

3.1 The problem of mental illness

Mental illnesses are complex conditions that can be chronic, impair quality of life, and require significant healthcare resources. Due to insufficient knowledge of their biological background, a final diagnosis can only be made based on criteria described in the Diagnostic and Statistical Manual of Mental Disorders (DSM-5) (APA, 2013) or the International Statistical Classification of Diseases (ICD-11) (WHO, 2025). Mental illnesses affect an individual's overall functioning, resulting in emotional distress, diminished quality of life, social isolation, stigma, and prejudice. One in four people worldwide will suffer from a mental or neurological disorder in their lifetime. In addition, people with mental health problems are at increased risk of developing physical illnesses. This is due to a weakened immune system, harmful lifestyle and behavioral patterns, and nonadherence to medical and social treatment guidelines (Sartorius, 2007). Mental disorders have a fairly strong heritability; however, various environmental variables can influence the onset of these diseases (Robinson et al., 2023). We can assume that these stressful environmental conditions lower the necessary threshold for the occurrence of mental illness. Prominent examples of chronic mental illnesses include schizophrenia (SCZ), bipolar disorder (BPD), and major depressive disorder (MDD).

SCZ is a chronic mental illness that manifests itself in episodes. During periods of remission, affected individuals usually show no signs of illness, provided they receive appropriate treatment between episodes. Relapses can be mitigated with continued treatment. SCZ is a disorder of brain function that results in the patient no longer being able to distinguish between real and imaginary sensations and is characterized by delusions, hallucinations, disorganized thinking, abnormal motor behavior, and negative

Molecular Insights into Proteinopathies
DOI: https://doi.org/10.1016/B978-0-443-34176-2.00009-2
© 2027 Elsevier Inc. All rights are reserved, including those for text and data mining, AI training, and similar technologies.

symptoms (APA, 2013; WHO, 2025) (Box 3.1). Worldwide, SCZ has a reported prevalence of approximately 0.3% in the population, with higher figures also reported depending on the methodology used (Solmi et al., 2023). Men and women are affected similarly by SCZ, with a male/female ratio of 1.1 (Solmi et al., 2023).

MDD has a range of symptoms and manifestations. Individuals may have their initial episode in their 20s, with a subsequent peak occurring in their 50s. Depression affects significantly more women than men (Albert, 2015). Divorce or separation, prior

BOX 3.1 Symptoms of schizophrenia (SCZ).

Examples of positive psychotic symptoms include atypical behavior, cognitive deficits, delusions, and hallucinations. Delusions are erroneous convictions devoid of any foundation in reality. The sufferer may perceive themselves as another individual, perhaps a renowned historical or contemporary figure. Patients sometimes believe that individuals are telepathically accessing their thoughts, exerting influence over them, orchestrating conspiracies, surveying them, or intercepting their communications. Auditory hallucinations are the most prevalent form of hallucinations, serving as a significant indicator of SCZ.

Apathy, emotional blunting, and impaired verbal communication exemplify negative symptoms. Muted affect, emotional withdrawal, inadequate interpersonal relationships, lethargy, apathetic social disengagement, rigid mental processes, absence of spontaneity, and impaired abstract reasoning are some of the adverse symptoms. These are often seen as obstacles to daily living and are linked to extended hospitalizations and worse social functioning. Affective symptoms indicate alterations in the emotional aspect of personality. The loss of emotional warmth is a common symptom in the first stages of the disease, resulting in emotional detachment in patients. Patients frequently recognize this alteration and express dissatisfaction with their sense of disconnection from others and the absence of emotional intimacy. Their involvement in daily activities, occupational engagement, and overall life satisfaction are all adversely affected.

Cognitive symptoms encompass difficulties in goal planning and attainment, maintaining concentration and attention, distributing attentional resources, assessing functions, and monitoring performance. These symptoms render the patient incapable of managing daily activities.

Public perceptions of mental illness frequently highlight aggressive characteristics, rendering psychiatric patients menacing. This detrimental stigma is particularly associated with SCZ, which Torrey asserts is one of the foremost barriers to accessing mental healthcare in contemporary society (Torrey, 2011). Numerous studies indicate that individuals with SCZ exhibit a greater propensity for aggressive behavior compared to the general population (Cho et al., 2019). Aggressive behavior in SCZ stems from present clinical psychotic symptoms. A heightened risk of aggression is associated with positive symptoms of SCZ. Violence is more likely to manifest when an individual has imperative hallucinations (Bo et al., 2011). The presence of substance use disorders markedly elevates the likelihood of violence (Ward et al., 2023).

depressive episodes, stress, trauma, and first-degree relatives with a diagnosis of depressive disorder are additional risk factors for MDD (Di Manno et al., 2015).

BPD, meanwhile, is a severe, recurrent psychiatric mood disorder characterized by episodes of mood alteration, including mania, hypomania, and depression, with intervals of relative euthymia between those episodes. Lifetime occurrence is estimated at approximately 1%–3% of the population, is more common in women, and represents 10%–20% of all mood disorders (Merikangas et al., 2007). Despite all this data, BPD is often misdiagnosed, which delays proper treatment and has a negative impact on the prognosis and consequently, leads to poorer long-term outcomes. Depressive episodes often predominate, which means that many patients are initially misdiagnosed with MDD. There is partial overlap between the symptoms of MDD (Box 3.2) and BPD (Box 3.3).

SCZ, BPD, and MDD illustrate the complexity of chronic psychiatric illnesses. Although traditionally classified as distinct, they share overlapping symptoms: mood

BOX 3.2 Symptoms of major depressive disorder (MDD).

For diagnosis with a major depressive episode, an individual must have five or more of these symptoms during a 2-week timeframe, for most of each day, with at least one symptom being either a depressed mood or a reduced interest or pleasure: experiencing apathy or despair as their primary emotion; a significant reduction in pleasure or interest in most activities is experienced nearly everyday; daily observable changes in appetite and weight changes; subjective sensations of restlessness and also a decrease in physical activity and cognitive functioning; constant fatigue or exhaustion is experienced; constant feelings of remorse or not being good enough; not being able to make decisions or having less practical reasoning when doing everyday tasks; suicidal ideas without a clear plan, thoughts of death that come up over and over, or past suicide attempts or methods.

A depressive episode is characterized by depressive, low mood, fatigue, hopelessness, cognitive symptoms, loss of interest and pleasure, reduced energy, lack of motivation, and suicidality. Sleep is often disturbed, with a variety of insomnia to hypersomnia. Appetite changes are common and can be associated with weight loss or gain. Cognitive symptoms are common and include poor concentration, indecisiveness, and negative ruminations. Feelings of worthlessness, excessive guilt, or hopelessness are prominent. Psychomotor changes may be present, either in the form of retardation (slowed movements, speech, and reduced facial expressiveness) or agitation (restlessness, inability to sit still). Suicidal ideation is a critical feature, ranging from passive wishes not to live to active plans or attempts. Overall, depressive episodes are marked by significant distress and functional impairment, affecting social, occupational, and personal domains and/or interests. Approximately 25% of patients have a relapse after 6 months, 58% within 5 years, and 85% within 15 years of recovery (Schlechter et al., 2024). The incidence of prior relapses therefore heightens the probability of a later relapse.

BOX 3.3 Symptoms of bipolar disorder (BPD).

The first episode of BPD is most commonly a depressive one, followed by an alternation of depressive and manic episodes in the later course of the disease. Isolated manic episodes occur in 10%–20% of patients (Nierenberg et al., 2023). Pharmacological management of BPD is complex and often requires long-term polypharmacy. Lithium remains used as the standard, particularly for its antisuicidal properties (Geddes and Miklowitz, 2013). However, its narrow therapeutic window, side effects, and the need for regular monitoring reduce adherence and use. Other mood stabilizers, including valproate and lamotrigine, are being widely used, as well as atypical antipsychotics, which have proven efficacy in mania and bipolar depression. The use of antidepressants remains controversial, as they may induce emergent mania or rapid cycling, especially when not combined with mood stabilizers. This uncertainty marks the need for greater precision in tailoring and fitting therapy to each individual patient. Pharmacogenomic approaches hold well in predicting treatment response, particularly for lithium, but they remain in the early stages of clinical translation (Hou et al., 2016).

Manic episodes are characterized by an elevated or irritated mood, increased level of energy, reduced need for sleep, fast speech difficult to interrupt, grandiosity, and risk-taking behaviors, lasting at least 1 week and present most of the day. The appearance of these patients can change in terms of dressing colorfully, or in a mismatched manner. Heavy make-up is also common, and grooming is often neglected. Their psychomotor activity is characterized by restlessness, overactivity, and disinhibition with frequent gesturing and exaggerated facial expressions. Higher libido and general disinhibition can lead to changed, risky behavior that is potentially harmful to themselves. A manic episode can reach a psychotic level with delusions of grandiosity, wealth, and power, often accompanied by hallucinations. Hypomania presents with similar but less severe symptoms, often still slowing and disturbing daily functioning. Depressive episodes show the same symptomatology as MDD. In some cases, mixed states may occur, in which manic and depressive symptoms coexist at the same time and are particularly difficult to treat. Similarly, the rapid cycling variant of the disease, defined as four or more mood episodes in one year, complicates treatment and is associated with a worse prognosis. BPD is not a static illness but a dynamic and heterogeneous condition that, together with SCZ and depression, challenges the boundaries and uses of traditional psychiatric classification.

dysregulation, cognitive deficits, and psychotic symptoms, which suggest continuous symptoms rather than strictly separate biological categories (Craddock and Owen, 2010). Despite extensive scientific research, their biological basis still remains insufficiently clarified. Current diagnostic strategies are based on clinical assessment according to the guidelines of DSM-5 and ICD-11 (APA, 2013; WHO, 2025), which are descriptive and, like therapeutic strategies, are also limited in precision, objectivity, and efficacy. There are still no satisfactory biological markers available for the detection of any major mental illness or to predict reaction to any specific treatment. There is therefore a great need for a better understanding of the underlying biology of major mental illness in order to improve diagnosis, increase the range of treatment options, and ultimately work towards a more personalized medicine approach.

Over the last several decades, the principal means of identifying molecular targets for these aims has been genetics, initially using a "candidate gene" approach arising from linkage studies and follow-up association studies, and later through the use of genome-wide association studies. These efforts represent a major undertaking and an impressive level of collaboration between many research groups worldwide, revealing hundreds of loci implicated variously in SCZ, BPD, and MDD (Koromina et al., 2025; McIntosh et al., 2024; Trubetskoy et al., 2022). This information is valuable for understanding general pathological approaches, but ultimately leads to very few specific molecular targets, given the low effect size of individual findings.

3.2 Proteinopathy in mental illness

An alternative approach to identifying target molecules, that is complementary to genetic analyzes, has emerged in recent years. This is built around the idea that chronic cases of mental illness, most likely in a subset of patients, may arise from specific proteins misfolding and forming aggregates in the brains of patients (Korth, 2012). This idea was inspired in large part by the fact that similar protein aggregates are an established and central pathophysiological concept in neurodegenerative diseases. Several review papers have laid out the arguments that led to the hypothesis of proteinopathy occurring in major mental illness. We therefore refer the reader to these reviews for more detailed insight into the theoretical basis for investigating protein aggregation in mental illness (Bradshaw and Korth, 2019; Hui et al., 2022; Ochneva et al., 2022), and we will focus here instead on experimental evidence relating to this hypothesis. In summary, however, this represents the idea that after synthesis of proteins as linear peptide chains, they fold to form their correct three-dimensional final configuration, either as a stand-alone monomer or part of a larger protein complex. If a protein is folded incorrectly, in order to maintain correct protein homeostasis (proteostasis), the cell uses various clearance pathways, such as the ubiquitin–proteasome system (UPS) and macroautophagy, to degrade the misfolded proteins. In some instances, however, either due to ineffective protein clearance, cellular stress, and/or mutation, this process is insufficient, and the misfolded proteins can accumulate as large aggregates. These aggregates consist of many copies of the misfolded protein (or in some cases, multiple misfolded proteins) and take the form of β-sheet-rich amyloid, unfolded aggregates, or liquid-like condensates (Juković et al., 2024) and may either aggregate alone or coaggregate with other proteins (Fig. 3.1). Such aggregates are well established to exist in neurodegenerative diseases, where the aggregating proteins, especially smaller oligomeric and protofibrillar states, are often toxic and lead to neuronal loss. Such protein aggregates are now known to also exist in at least some patients with major mental

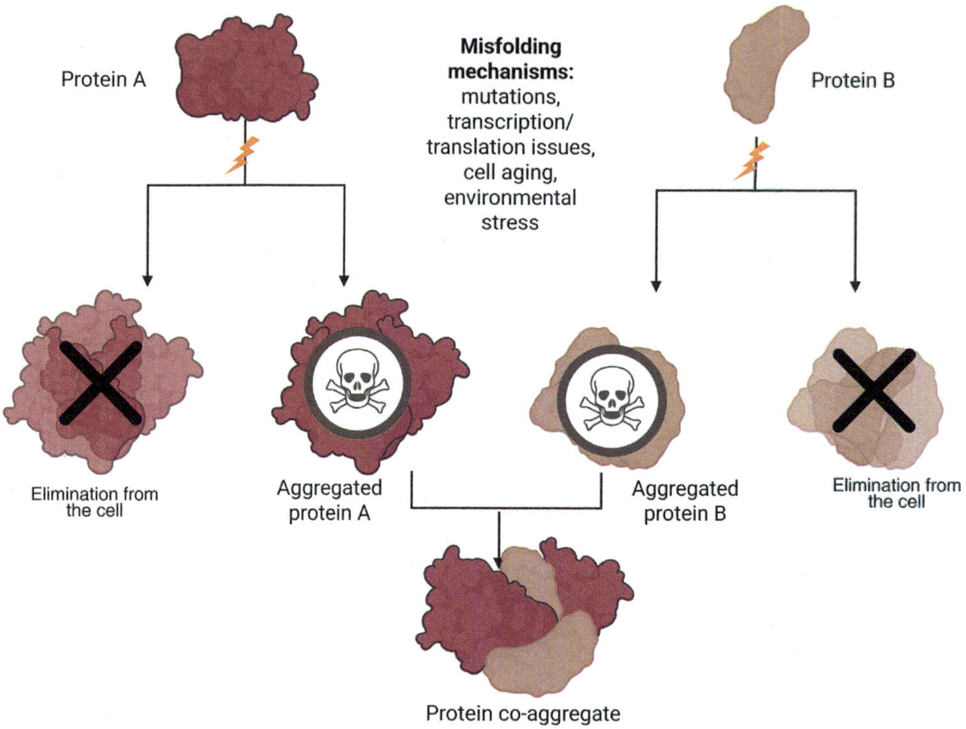

Figure 3.1 Aggregation and coaggregation. Protein aggregation can be induced by a variety of different mechanisms. While some proteins can aggregate by themselves, spontaneously or in response to stress, others only do so as a result of physical interaction with another protein, in a process of "co-aggregation." *Created in BioRender. Bradshaw, N. (2025) https://BioRender.com/gia3ybn.*

illness, although they seemingly do not possess the same degree of toxicity, given the absence of widespread cell death in these conditions (Bradshaw and Korth, 2019).

In the study of proteins that may aggregate in major mental illness, a major approach has been the isolation of insoluble proteins by ultracentrifuge-based fractionation from homogenized brain samples, on the grounds that protein aggregates tend to be larger and less soluble than their correctly folded, functional counterparts (Bradshaw and Korth, 2019; Juković et al., 2024). The group of Nucifora et al. (2019) used this approach in three sets of brain samples from SCZ patients and matched control individuals and consistently found that subgroups of SCZ patients had higher total levels of insoluble protein than the control individuals or patients from outside this subgroup. Crucially, the presence of insoluble protein in the brain correlated with the presence of protein ubiquitination in the brain (Nucifora et al., 2019). Ubiquitination is a key part of the UPS, by which ubiquitin groups are added to proteins in order to target them for degradation by the proteasome. The presence of high ubiquitin levels is therefore consistent with the presence of high levels of misfolded protein that the cell has targeted for degradation. The cooccurrence of high levels of ubiquitination and insoluble protein in the same subset of SCZ patient brain samples therefore strongly implies the presence of disturbed proteostasis and misfolded protein aggregates forming in this patient group.

Several other lines of evidence also suggest disruption of the UPS in major mental illness. Early findings from several microarray experiments in the brains of patients with SCZ have found downregulation of a variety of genes encoding components of the UPS compared to control individuals (Vawter et al., 2001, 2002; Middleton et al., 2002; Konradi et al., 2004). A more recent study looked at this in multiple SCZ patient cohorts, finding consistent downregulation of multiple proteasome subunits in patients and controls, while also noting that these came from a specific subgroup of SCZ patients (Hertzberg et al., 2021), much as Nucifora et al. (2019) had seen a subset of patients show ubiquitinated and misfolded protein.

A study in the blood of patients with SCZ or BPD found the UPS to be one of the most altered pathways in both patients with BPD and psychosis, although not for SCZ diagnosis (Bousman et al., 2010). Notably, the genes they found associated with these diagnoses were not the same ones seen for SCZ in the brain. The same research group also looked at the effect of the expression of several UPS genes on symptom severity in psychosis, based on the blood of patients with SCZ or BPD. Several ubiquitin-related genes were indeed found to have expression levels that correlated (either positively or negatively) with positive symptoms of psychosis (Bousman et al., 2010).

At a protein level, several subunits of the proteasome, along with other UPS proteins, have been seen to be dysregulated in the superior temporal cortex or middle frontal gyrus of patients with SCZ (Scott et al., 2016; Andrews et al., 2017). More recently, two ubiquitin-related proteins were also found to be downregulated in the blood of patients with BPD (Aydınoğlu et al., 2024). Other evidence of disrupted UPS in SCZ patients comes from two studies by the Meador-Woodruff group. In one, they showed that while total free ubiquitin and ubiquitination were reduced in brain samples of patients with SCZ, relative to controls, levels of K63 ubiquitination, which is associated with lysosomal degradation and endosomal trafficking, increased (Rubio et al., 2013). In the other, they managed to show anomalies in the proteasome of patients with SCZ in a compartment-specific manner (Scott and Meador-Woodruff, 2020).

Beyond the UPS, apoptosis is also used for maintenance of proteostasis in many cell types, and research into the brains of individuals with SCZ, BPD, and MDD reveals distinct alterations in the balance of pro and antiapoptotic factors, particularly in the temporal and frontal cortices. Jarskog et al. (2004) examined *postmortem* temporal cortex (Brodmann area 21) tissue from 15 patients with SCZ, BPD, MDD, and matched controls. They found that the Bax/Bcl-2 protein ratio was approximately 50% higher in SCZ patients compared to controls, indicating increased vulnerability of cortical cells to apoptosis. However, caspase-3 levels—both inactive and activated forms—did not differ significantly, suggesting that while apoptotic susceptibility is elevated, active apoptosis may not be occurring in this illness phase. Notably, the Bax and Bcl-2 individual protein levels did not differ significantly, but their ratio was altered, distinguishing SCZ's apoptotic profile from classic neurodegenerative diseases where caspase-3 activation is more prominent. In contrast, patients with BPD or MDD showed Bax/Bcl-2 ratios similar to controls, highlighting a potential specificity of this apoptotic imbalance in SCZ. Earlier work by Jarskog et al. (2000) also reported reduced Bcl-2 protein levels in the temporal cortex of SCZ patients, confirmed by ELISA and Western blot, reinforcing the notion of diminished antiapoptotic protection in this disorder (Jarskog et al., 2000). These findings align with other studies suggesting that altered expression of Bcl-2 family proteins contributes to the pathophysiology of SCZ, potentially underlying the cortical gray matter volume reductions observed in neuroimaging studies. However, the cause of death was not linked to these protein alterations in the analyzed samples.

Further complexity arises from studies of the extrinsic apoptotic pathway in MDD and SCZ. García-Fuster et al. (2024) investigated fas-associated death domain (FADD) protein and its regulator PEA-15 in the prefrontal cortex (Brodmann area 9). They found no difference in total FADD protein between MDD and SCZ subjects and controls, but increased phosphorylated FADD (p-FADD) in MDD, which may modulate apoptotic signaling. Moreover, total PEA-15 was elevated in MDD, potentially inhibiting pro-survival ERK1/2 signaling, which was reduced in antidepressant-free MDD subjects. These findings suggest a complex interplay between intrinsic (Bax/Bcl-2) and extrinsic (FADD/PEA-15/ERK1/2) apoptotic regulators that may contribute to cell death and neuroplasticity deficits in mood disorders and SCZ.

In summary, SCZ is characterized by an increased Bax/Bcl-2 ratio in the temporal cortex, indicating heightened apoptotic vulnerability without evidence of active caspase-3-mediated apoptosis, distinguishing it from classical neurodegenerative disorders. BPD shows a broader activation of apoptotic pathways, including caspase-3, along with synaptic protein loss. MDD and SCZ also involve dysregulation of extrinsic apoptotic pathway components, reflecting complex apoptotic and neuroplasticity alterations. These apoptotic protein imbalances likely contribute to the neuropathology underlying these psychiatric disorders and may represent targets for future therapeutic interventions.

There is therefore a strengthening case that proteostasis is disrupted in the brains of at least a subset of patients with SCZ, in particular, as well as seemingly BPD and MDD. In addition to general increases in protein aggregation, multiple specific proteins have also been implicated as aggregating in the brains of patients with these conditions. It should be noted that proteins that aggregate lose their normal function, and some may also gain a toxic function. Interplay between the two sides results in different specific and general phenomena of proteinopathies. We will therefore discuss these aggregating proteins in this chapter, starting with the best characterized protein known to aggregate in mental illness, disrupted in schizophrenia 1 (DISC1).

3.3 Disrupted in schizophrenia 1—the most studied protein to aggregate in mental illness

3.3.1 The *disrupted in schizophrenia 1* gene

The *DISC1* gene was discovered as disrupted by a balanced t(1;11)(q42.1;q14.3) chromosomal translocation in a large Scottish family with a high incidence of major mental illness (Millar et al., 2000). Follow-up of the family over a 20-year period revealed an increased incidence of major mental illnesses among relatives with the translocation, including SCZ, BPD, and MDD. In contrast, no incidence of these disorders was found among the healthy members of the family (St Clair et al., 1990; Blackwood et al., 2001).

Major efforts were undertaken to map and clone the t(1;11)(q42.1;q14.3) translocation breakpoints (Chubb et al., 2008), following the hypothesis that the rearrangement between the chromosomes directly disrupts gene function, thus resulting in psychosis. At a protein level, this would result in a loss of all amino acid (AA) residues C-terminal to AA 597 of the full-length protein (854 AA) (Millar et al., 2000). Breakpoint regions were identified on both chromosomes 1 and 11 (Muir et al., 1995); however, a long-range restriction map of chromosome 11q14.3 spanning one breakpoint did not yield any genes associated with mental illness (Blackwood et al., 2001; Millar et al., 1998). By contrast, the corresponding breakpoint on chromosome 1 was shown to be gene abundant. Cloning of this region resulted in the identification of two novel genes, provisionally named *DISC1* and *DISC2*, due to both genes being split by the translocation. While sequences of both genes overlap, *DISC2* seems to act as a noncoding structural RNA gene with a potential role in regulating *DISC1* expression. As such, alteration of the activity of these two genes has been proposed as a causal link to mental disorders in translocation carriers, either by production of a truncated DISC1 protein, or abnormal expression regulation (Millar et al., 2001).

Following its identification in this family, the *DISC1* gene became one of the most prominent early "candidate genes" for major mental illness, with multiple common and rare variants associated with such illnesses in a variety of specific populations (Chubb et al., 2008; Bradshaw and Porteous, 2012; Thomson et al., 2013), although no variant in it shows genome-wide association globally (Sullivan, 2013; Porteous et al., 2014).

3.3.2 The disrupted in schizophrenia 1 protein

DISC1 is a cytosolic, widely expressed multifunctional scaffold protein. As a scaffold, it lacks enzymatic activity and instead exerts its function by simultaneous interaction with multiple macromolecules (Yerabham et al., 2013; Shao et al., 2017; Teng et al., 2018). A large number of DISC1 interaction partners have been identified through yeast two-hybrid and coimmunoprecipitation studies (Millar et al., 2003; Morris et al., 2003; Millar et al., 2005; Brandon et al., 2005; Camargo et al., 2007). Due to its large interaction network, DISC1 has many reported roles, including, but not limited to, maturation, migration, and proliferation of neuronal cells, neurotransmission, and cytoskeleton organization. These roles suggest that DISC1 is generally important for normal brain development, while the production of aberrant protein, either due to genetic mutations, biological processes, or interactions with other proteins, can contribute to the etiology of one or more mental illnesses (Teng et al., 2018; Brandon et al., 2009). However, the mechanisms by which DISC1 exerts its functions remain largely unknown, in part due to a lack of enzymatic activity, as well as the complexity of its vast interaction network and lack of structural information (Bradshaw and Porteous, 2012; Soares et al., 2011).

While the *DISC1* gene was first linked to major mental illnesses more than 20 years ago, the crystal structure of the full-length DISC1 protein remains unknown at time of writing. For years, a lack of knowledge about the structure of DISC1 has impeded further research into its roles and functions within the pathology of mental disorders, partly due to a lack of homology with known proteins and unusually rapid sequence evolution (Sanchez-Pulido and Ponting, 2011; Yerabham et al., 2017). Moreover, the high susceptibility of recombinant DISC1 to aggregation and its low solubility further impeded its biophysical characterization (Bradshaw and Korth, 2019; Yerabham et al., 2017; Leliveld et al., 2009). As such, attempts at defining its structural organization were limited to in silico bioinformatics predictions, while the experimental approach was dependent upon shorter constructs and domain delineation of known sequence (Soares et al., 2011).

Any further characterization of the DISC1 structure was lacking until the utilization of a high-throughput Expression of Soluble Proteins by Random Incremental Truncation (ESPRIT) technique (Yerabham et al., 2017; Yumerefendi et al., 2010; Mas and Hart, 2017). Using this approach, four novel structural regions of the DISC1 protein were proposed (Fig. 3.2), named "D," "I," "S," and "C," respectively (Yerabham et al., 2017).

The D region represents the only soluble construct detected within the N-terminal half of DISC1, predominantly existing as a dimeric species with a high content of α-helices (Yerabham et al., 2017). Also shown to be dimeric, the I region can be expressed in Escherichia coli as a stable, soluble protein in a low-order oligomeric state. However, a purified construct encoding this region exhibited visible aggregation propensity at concentrations higher than 30 μM, making it the least stable of the four regions (Yerabham et al., 2017). Unlike other regions, the S region can be expressed in multiple different oligomeric states. Curiously, when expressed separately, the recombinant I and

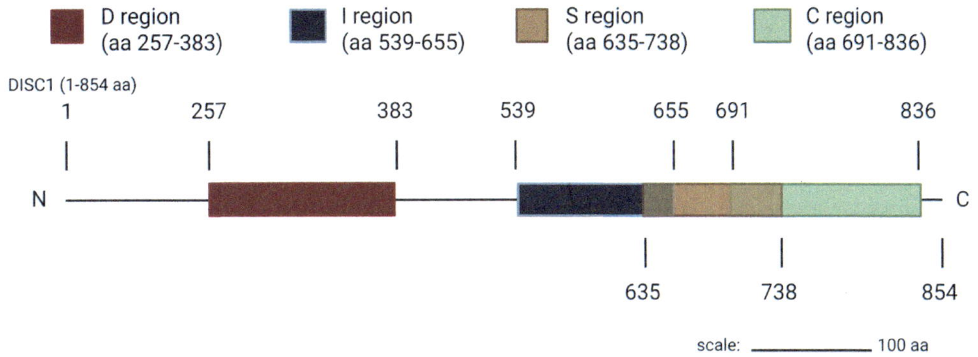

Figure 3.2 Domain structure of disrupted in schizophrenia 1 (DISC1). The major domains of DISC1, as determined using the ESPRIT system, along with their amino acid boundaries. Note that the I, S, and C regions, as currently defined, partially overlap each other. *Created in BioRender. Bradshaw, N. (2025) https://BioRender.com/gia3ybn. Based on data from Yerabham et al. (2017).*

S regions exist as stable and soluble proteins with differing oligomeric states. However, the constructs spanning both regions exhibited low solubility and were shown to be highly unstable in regard to their oligomerisation (Yerabham et al., 2017). The C region exists as a stable, elongated monomer, and it accounts for the random coil content of the protein (Soares et al., 2011; Yerabham et al., 2017).

Elucidation of the DISC1 domain architecture enabled creating region-specific constructs mimicking notable mutations within the gene, such as the American frameshift mutation (Sachs et al., 2005). At AA 807, this mutation lies within the C region of the DISC1 protein and, as such, could affect the structural integrity of the region. To mimic the truncation of the region as the result of the frameshift mutation, a corresponding construct was created by Yerabham and colleagues. Interestingly, while this construct was expressed at 50% of the level of the wild-type protein control, its solubility decreased by as much as 90%. It was further shown that the insoluble mutant undergoes aberrant multimerization, ultimately leading to accumulation of protein aggregates that disrupt the structure of the region (Yerabham et al., 2017). In vitro studies confirm the α-helical and coiled-coil content of this region (Cukkemane et al., 2021); however, it is heavily dependent on the presence of interaction partners and shows much higher disordered content in their absence (Cukkemane et al., 2025b). The C region of the protein also represents the only region with a partially solved three-dimensional structure to date (Ye et al., 2017; Wang et al., 2021).

3.3.3 Disrupted in schizophrenia 1 aggregation in major mental illnesses

As the protein product of a major gene implicated in the pathophysiology of multiple major mental illnesses, with a prominent role in behavioral control, DISC1 presented an excellent target for initial research of aberrant proteostasis into major mental illnesses (Korth, 2012). Investigation of *postmortem* brains of patients with SCZ, BPD, and MDD revealed that approximately 20% of patients had DISC1 aggregates within the cingulate cortex, while none were present in healthy controls (Leliveld et al., 2008), nor in patients suffering from neurodegenerative diseases (Ottis et al., 2011). Insoluble DISC1, a product of aggregates large enough to exceed solubility, also led to dysfunctional molecular interaction with NDEL1, leading to a loss-of-function phenotype similar to described familial cases carrying a DISC1 mutation (Leliveld et al., 2008). Inhibition of the autophagosomal pathway enhanced formation of such aggresomes, indicating that disruptions to the pathway can have negative effects on neuronal function and intracellular transport of organelle cargo (Atkin et al., 2012). The latter is of significant importance, considering the role of DISC1 in the regulation of synaptic vesicle transport (Flores et al., 2011).

3.3.4 Transgenic animal studies of disrupted in schizophrenia 1 aggregation

A transgenic rat model, overexpressing the full-length human *DISC1* transgene and common variants S704C and L607P (the tgDISC1 rat), displayed not only DISC1 aggregates within neurons, but also behavioral phenotypes consistent with disruption of dopamine neurotransmission. These include high sensitivity to amphetamine, hyper-exploratory behavior and motor deficits (Trossbach et al., 2016), as well as deficits in cognitive flexibility (Wang et al., 2022) and social performance (Wang et al., 2022; Seidisarouei et al., 2022). Many of these deficits have also been corroborated in the recent *Drosophila* tgDISC1 models (Samardžija et al., 2024). Furthermore, elevation in cytosolic dopamine levels increased DISC1 multimerization and insolubility, as well as complex formation with the dopamine transporter (Trossbach et al., 2016). Morphological aberrations reported in the same tgDISC1 model include a decrease in dopaminergic neurons in substantia nigra pars compacta (Hamburg et al., 2016) and dopamine content in the dorsal striatum (Trossbach et al., 2016). Both are a part of the nigrostriatal pathway responsible for locomotor behavior and reward functions.

Recently, the tgDISC1 rat model was employed to investigate social reward learning (Dören et al., 2025), suggesting a specific deficit that affects the key subcortical regions involved in dopamine signaling. In another study, DISC1 overexpression in a transgenic rat model led to protein misassembly that affected multiple processes, including glutamatergic and dopaminergic signaling (Sialana et al., 2018). Taken together, these results establish a causal connection between DISC1 aggregation and altered dopamine homeostasis, consequently supporting its role in the pathology of major mental illnesses.

Further research conducted on the tgDISC1 rat demonstrated that DISC1 aggregation leads to alterations in the CA1 region of the hippocampus proper.

Deficits include impairments in the encoding of location-independent information, as well as impaired neuronal response during sleep and/or rest and network synchrony, which have previously been reported for SCZ patients (Kaefer et al., 2019). Additionally, an investigation into the effects of DISC1 aggregation on social behavior, stress resilience, and spatial memory in the tgDISC1 rat suggests disturbances in cognitive flexibility and social behaviors (Wang et al., 2022).

3.3.5 Neuronal culture and cell studies of disrupted in schizophrenia 1 aggregation

Abnormalities in the glutamatergic pathway, including hypofunction of glutamate-gated N-methyl-D-aspartate (NMDA) receptors and their modulator D-serine (Hashimoto et al., 2003), are distinctly associated with psychosis in SCZ patients (Ma et al., 2013). Formation of D-serine is catalyzed by serine racemase (SR), which is stabilized by its binding to DISC1 (Ma et al., 2013). Curiously, SR has also been shown to associate with DISC1 aggresomes in cortical neurons. This interaction is enhanced in the presence of D-serine, which augments the formation of large DISC1 aggregates, facilitated by NMDA receptors (Jacobi et al., 2019). Such interaction provides an insight into the convergence of pathways between aberrant proteostasis and glutamatergic dysregulation, implicated in the etiology of SCZ. Separately, DISC1 has been shown to physically interact with dopamine in cells, and seemingly this interaction can induce DISC1 aggregation.

3.3.6 Disrupted in schizophrenia 1 condensation

While proteins perform a range of diverse functions in their native forms, they can also exist in other states. This includes a liquid condensed state, achieved through liquid–liquid phase separation. In theory, most proteins would be expected to achieve a liquid condensed state within the cell at high concentrations, via a process called the condensation pathway (Vendruscolo and Fuxreiter, 2022). During this process, proteins transition from their native state into a dense liquid state called "the droplet state," which can subsequently lead to the creation of irreversible toxic amyloids. Moreover, structural studies imply that the droplet state may also contain β-sheet structures, characteristic of amyloid aggregates (Vendruscolo and Fuxreiter, 2022). Although no research on DISC1 condensation has been performed to date, DISC1 coaggregation partner TDP-43 (Endo et al., 2018) is known to form condensates (Vendruscolo and Fuxreiter, 2022).

Phase separation has also been explored in proteins containing multiple repeats of small ubiquitin-related modifier (SUMO) domains (Banani et al., 2016), which can undergo a SUMOylation process. SUMOylation is a posttranslational modification known to regulate subcellular distribution and protein stability, and DISC1 contains several known SUMOylation sites (Tankou et al., 2016). As such, studying DISC1 within the framework of protein condensation may prove beneficial in future efforts to elucidate novel aggregation mechanisms.

3.3.7 Aggregation-critical region of disrupted in schizophrenia 1

There are several studies looking into the aggregation-critical region of DISC1 published to date. A study performed in HeLa cells found that a truncated DISC1 construct spanning AA 357–854 exhibits a "punctate" phenotype akin to smaller-sized aggregates. In terms of ESPRIT-determined regions, this would approximate a protein missing the N-terminal disordered region and a better part of the D region. Interestingly, this phenotype was lost with the truncation of what is now known as the I region (Brandon et al., 2005), thus considering it responsible for such expression. However, construct boundaries in this study were defined prior to the uncovering of the DISC1 domain structure, meaning that these constructs represented partial, thus unstable, domains.

The most recent study looking into the aggregation-critical DISC1 region(s) in neuroblastoma cells suggests that a short section of the protein (AA 384–415) found within the "linker" region between regions D and I is sufficient for aggregation. The deletion (Δ384–415) of this section resulted in the complete abolishment of DISC1 aggregation, thus defining it as aggregation-critical (Zaharija and Bradshaw, 2024). It was not found within any of the most stable DISC1 fragments yielded by the ESPRIT screening, but would be predicted to have α-helical content with coiled-coil potential (for AA 384–393), based on bioinformatics prediction (Soares et al., 2011). These findings are in accordance with the results obtained from the brains of patients with sporadic cases of chronic psychiatric conditions, in which an insoluble DISC1 species of approximately 72 kDa was detected using an antibody specific to part of the C region (Leliveld et al., 2008). The exact nature of this DISC1 species is unknown, but based on its size and the presence of the C-terminus, it would be expected to have lost the first N-terminal ~200 AA, possibly through the posttranslational modification process (Leliveld et al., 2008). In theory, such a species would contain all known structured regions, including the aggregation-critical region. While the artificial system of overexpressing purified recombinant DISC1 may be substantially different from the endogenous DISC1 expression in the brains of affected patients, it serves as a good initial model for the study of aggregation mechanisms.

Another region of DISC1 was also found to be capable of aggregation in this study, separate from the effects of the linker region (Zaharija and Bradshaw, 2024). This region encompasses the three C-terminal regions of DISC1, I, S, and C (AA 539–854). However, no such propensity was seen when the I and S (AA 539–738) or the S and C regions (AA 635–836) were expressed. These results broadly align with the study by Leliveld and colleagues, which demonstrated that a C-terminal region of DISC1, approximately corresponding to the S and C regions (AA 640–854), was capable of aggregation in cells (Leliveld et al., 2009). Purification of insoluble material from the brains of psychiatric patients also resulted in detection of C-terminal DISC1 fragments (Leliveld et al., 2008); however, while distinct, the mechanism by which this region of the protein aggregates is yet to be determined. While it cannot be formally ruled out that the extreme C-terminal tail (AA 836–854) of DISC1 plays a role in its aggregation, there is an indication that the I region may be complicit as well. The construct spanning AA 440–597, which includes the N-terminal part of the I region ahead of the breakpoint, was shown to produce a punctate expression pattern (Brandon et al., 2005). In contrast, further study of truncated DISC1 constructs with a "punctate" phenotype observed the lack of said phenotype in the construct bisecting the I region at the translocation breakpoint (AA 598–854). This construct includes the extreme C-terminal AAs but exhibits no sign of aggregate formation (Brandon et al., 2005). Other studies showed that this particular fragment of DISC1 is capable of self-association (Leliveld et al., 2009, 2008) and can exist as dimeric, octameric, and multimeric species under physiologically relevant conditions (Leliveld et al., 2009). Both the dimerization (AA 765–854) and oligomerisation (AA 668–747) domains can be found predominantly within the C region of DISC1 (Leliveld et al., 2009). The latter is a part of the well-conserved coiled-coil domain, which contains another self-association site at AA 403–504 (Kamiya et al., 2005), corresponding to a part of the linker found between regions D and I. While oligomerisation may represent a natural self-interaction, accumulation of higher-order species can also lead to aberrant multimerisation and formation of insoluble aggregates. These results indicate the existence of at least two distinct mechanisms of DISC1 aggregation—one driven by the central linker region and the other arising from the C-terminus of DISC1.

Biophysical characterization performed in an in vitro bacterial system indicates that the C region alone is capable of forming fibrillar aggregates (Cukkemane et al., 2021; Cukkemane et al., 2025a). This self-association and aggregation of the C region were driven by a β-core (AA 716–761) region, a tightly packed region of β-fibrils, which can act as a nucleus for DISC1 aggregation (Cukkemane et al., 2021; Cukkemane et al., 2025a), in a way similar to what has been observed in the interaction of DISC1 with HTT (Tanaka et al., 2017). This process is affected by genetic variants implicated in mental illness, as well as by simulation of a phosphorylation event, known to be critical for DISC1 function (Sachs et al., 2005; Cukkemane et al., 2025a; Ishizuka et al., 2011). Intriguingly, this aggregation process can be attenuated, at least in vitro, through the addition of peptides based on interaction partners of the DISC1 C domain (Cukkemane et al., 2025b).

While the two different aggregating regions of DISC1 are not necessarily conflicting, given the difference between the expression systems, the linker region between regions D and I shows the highest aggregation capacity of any individual section of DISC1 examined in neuroblastoma cells, diminishing any effect of aggregation of the C region alone (Zaharija and Bradshaw, 2024). Expression and purification of the linker region from *E. coli* is comparatively harder to achieve under the same conditions than is the case for the C region, which is sufficiently stable and easy to express. It is therefore reasonable to conclude that the linker region has an inherently larger aggregation propensity when studied in isolation (Zaharija and Bradshaw, 2024). The research done in mammalian cell systems to date demonstrates that the aggregation of DISC1 is dependent upon a short linker region found between D and I. Moreover, removing this region directly abolishes the aggregation propensity of DISC1 in its entirety (Fig. 3.3). Considering previous knowledge that DISC1 aggregation could arise through the recombinant C region of the protein (Cukkemane et al., 2021), this implicates the existence of more than one distinct molecular mechanism of aggregation. Therefore novel approaches, such as the study of protein condensation, need to be considered in future research.

Taken together, the insights gained into possible pathological mechanisms of DISC1 aggregation may serve as a stepping stone towards the discovery of alternative therapy options for at least a subset of patients suffering from chronic and debilitating psychiatric disorders, such as SCZ, BPD, and depression, as well as toward creating better model systems for future studies of DISC1 aggregation, separate from the effects of

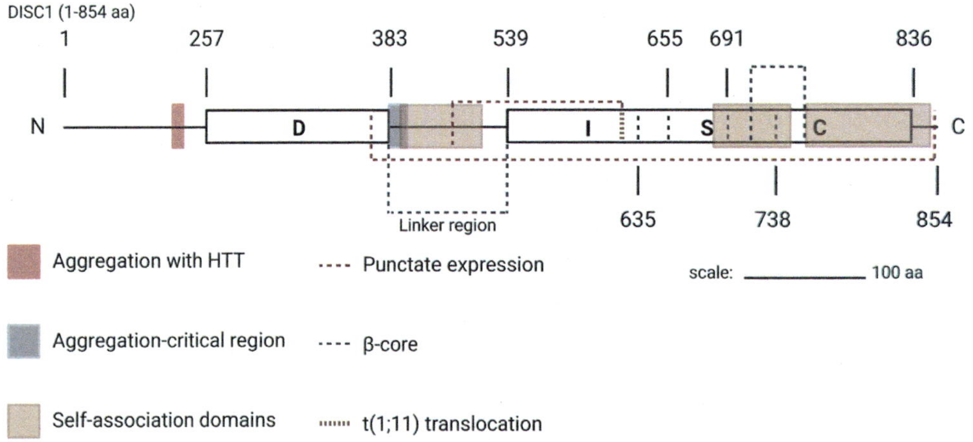

Figure 3.3 Aggregation-critical regions of the disrupted in schizophrenia 1 (DISC1) protein. Relative positions of regions critical for DISC1 aggregation, as identified in previous studies, overlaid on its domain structure. *Created in BioRender. Bradshaw, N. (2025) https://BioRender.com/gia3ybn. Based on data from Brandon et al. (2005), Leliveld et al. (2009), Tanaka et al. (2017), Yerabham et al. (2017), Cukkemane et al. (2021), Zaharija & Bradshaw (2024).*

overexpression. While much research still needs to be done, DISC1 represents by far the most characterized protein that aggregates in major mental illness to date.

3.4 Other proteins whose aggregation was identified by insolubility in patient brain samples

3.4.1 Dysbindin-1

Like DISC1, dysbindin-1 is the protein product of an early candidate gene for mental illness, in this case *dystrobrevin binding protein 1* (DTNBP1), which was variously associated with SCZ in various populations (Straub et al., 2002; Van Den Bogaert et al., 2003; Funke et al., 2004; Li et al., 2005). Also like DISC1, dysbindin was implicated as aggregating in major mental illness based on it being found in the purified insoluble protein fractions of a subset of patients with SCZ, BPD, and MDD, but not in matched control samples (Ottis et al., 2011). Functionally, dysbindin-1 exists in the brain as part of several different protein complexes, most notably the biogenesis of lysosome-related organelles complex-1, or BLOC-1 (Ghiani and Dell'Angelica, 2011; Mullin et al., 2011). BLOC-1 is a highly conserved protein complex best known for its role in transporting melanin and related proteins from early endosomes to mature ones; however, it also seems to play analogous roles in many cell types, including at the synapse (De Pace et al., 2025).

The *DTNBP1* gene is subjected to alternate splicing, leading to the expression of three major dysbindin-1 isoforms: 1A, 1B, and 1C. Of these, dysbindin-1A represents the full-length protein, while dysbindin-1B and 1C lack regions from the C-terminus or N-terminus, respectively (Talbot et al., 2009) (Fig. 3.4). Due to the similar sizes of these isoforms, all predicted in the range of 36–40 kDa, and the known phosphorylation of dysbindin-1 (Oyama et al., 2009; Fei et al., 2022) that will alter its apparent molecular weight, it is not clear which isoform(s) are seen to be insoluble in brain samples from patients (Ottis et al., 2011). Of the three isoforms, only dysbindin-1B was seen to spontaneously form aggregates when expressed in cell lines (Zhu et al., 2015; Xu et al., 2015). This difference between the isoforms appears to arise from the presence of a PEST (proline, glutamic acid, serine, and threonine-rich) domain at the C-terminus of dysbindin-1A and 1C, which is lacking from dysbindin-1B, as deletion of this region from dysbindin-1A increases its aggregation propensity (Xu et al., 2015). It was also shown that dysbindin-1A can be actively degraded by the proteasome, while dysbindin-1B cannot (Xu et al., 2015).

Aggregating dysbindin-1B appears to exist in aggresomes, as indicated by the presence of vimentin, γ-tubulin, and the 20S proteasome subunit in its aggregates (Xu et al., 2015). Notably, and unlike most other proteins investigated in the context of major mental illness, dysbindin-1B aggregates were seen to be neurotoxic (Zhu et al., 2015). They were also seen to be able to propagate between cells through exosomes,

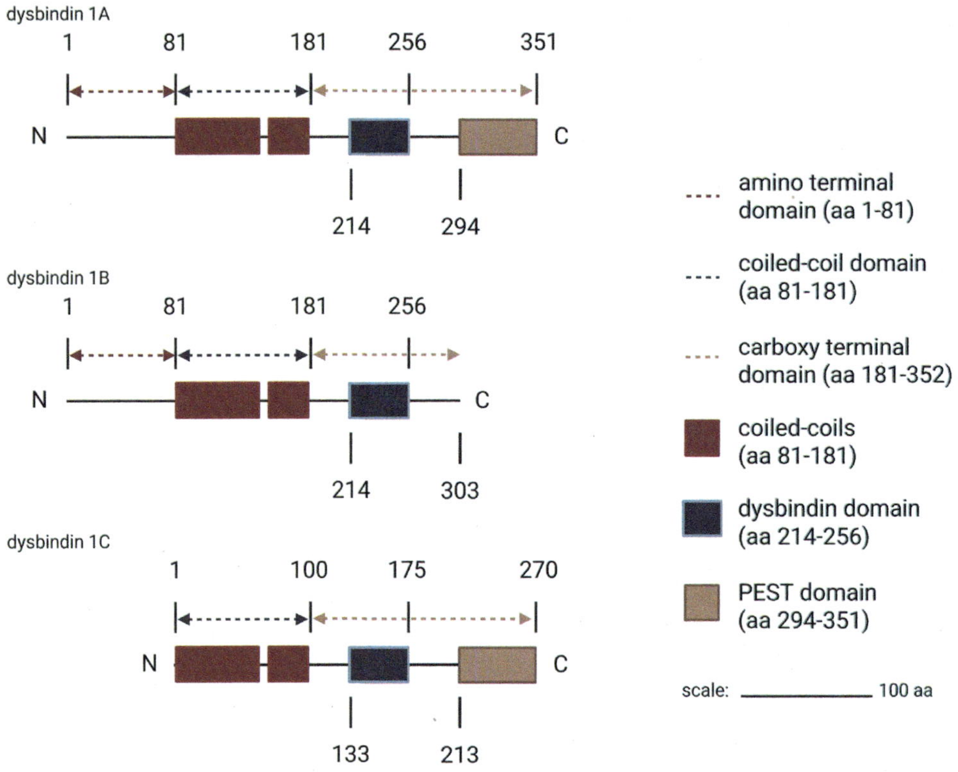

Figure 3.4 Domain structure of the dysbindin-1 isoforms. Alignment and comparative domain structures of dysbindin-1A, 1B, and 1C. Amino acid locations of individual regions are indicated. *Created in BioRender. Bradshaw, N. (2025) https://BioRender.com/gia3ybn.*

which was demonstrated using both in vitro primary neurons and in vivo in a mouse brain (Zhu et al., 2015). In a mouse model, overexpressing dysbindin-1B increased levels of apoptosis in the hippocampus, especially in the dentate gyrus (Yang and Xu, 2017). This was accompanied by behavioral phenotypes suggestive of anxiety as well as spatial memory learning and memory defects, highlighting the potential for dysbindin-1B aggregates to have a causative role in mental illness (Yang and Xu, 2017; Yang et al., 2016b). At a molecular level, the overexpressed dysbindin-1 suppressed expression of several other members of the BLOC-1 complex, leading to impaired synaptic vesicle trafficking (Yang et al., 2016b).

3.4.2 Collapsin response element mediator proteins 1 and 2

While DISC1 and dysbindin-1 are both the products of genes already implicated in major mental illness, collapsin response element mediator protein 1 (CRMP1) was the first example of a protein to be identified as aggregating in the brain through a hypothesis-generating approach, without prior knowledge of involvement in mental illness. Specifically, 15 samples of the posterior cingulate cortex (BA23) from patients with SCZ had their insoluble protein fraction purified, before these fractions were pooled and used to inoculate a mouse (Bader et al., 2012). This animal then generated antibodies against the insoluble proteins of the SCZ patients, which could then be extracted and isolated. Monoclonal antibodies raised from this mouse were then investigated to determine whether any of them showed specific binding to the SCZ insoluble fraction pool, and then compared to a directly equivalent pool derived from individuals without mental illness. One such antibody was identified and found to interact with CRMP1 (Bader et al., 2012). CRMP1 is known to be important for neurodevelopment and neuronal repair (Charrier et al., 2006; Lukomska et al., 2023). It, along with other members of the CRMP family, are mostly associated with semaphorin 3A signaling (Uchida et al., 2005; Makihara et al., 2016; Yamane et al., 2017), although it also works in other pathways such as reelin signaling.

CRMP1 was confirmed, using an alternative antibody, to indeed be insoluble in the brain homogenates of subgroups of patients with SCZ, BPD, and MDD, overlapping significantly with individuals that also showed insoluble DISC1 in the brain (Bader et al., 2012). This potential link with DISC1 was then reinforced by two further lines of evidence. Firstly, haplotypes of the *CMRP1* gene were associated with specific endophenotypes of SCZ, but only on the background of risk variants of the *DISC1* gene (Bader et al., 2012). Secondly, while a longer splice variant of CRMP1 (Fig. 3.5) was seen to be capable of aggregating alone, the shorter version was only seen to do so preferentially when it could interact with aggregating molecules of DISC1 or other aggregating proteins (Bader et al., 2012; Samardžija et al., 2023).

CRMP2, which is functionally similar to, but distinct from CRMP1 (Makihara et al., 2016), is also implicated as aggregating, based on data from induced pluripotent stem cells of a SCZ patient (Toyoshima et al., 2019). In these cells, in which a mutation in the GLO1 enzyme leads to carbonyl stress, the CRMP2 protein undergoes hyper-carbonyl posttranslational modification, forming cross-links within and between CRMP2 molecules, leading to aggregate-like accumulation of larger multimeric proteins (Toyoshima et al., 2019). Whether this process parallels the reported aggregation of CRMP1 or not remains to be confirmed.

3.4.3 Trio and F-actin binding protein isoform 1

Trio and F-actin binding protein isoform 1 (TRIOBP-1), also known as Tara or TAP68, was originally implicated as aggregating in SCZ using the same epitope-based proteomics approach as CRMP1 (Bader et al., 2012; Bradshaw et al., 2014). There was no prior genetic evidence to directly associate the *TRIOBP* gene with major mental illness, although one polymorphism in it has subsequently been associated with pain in

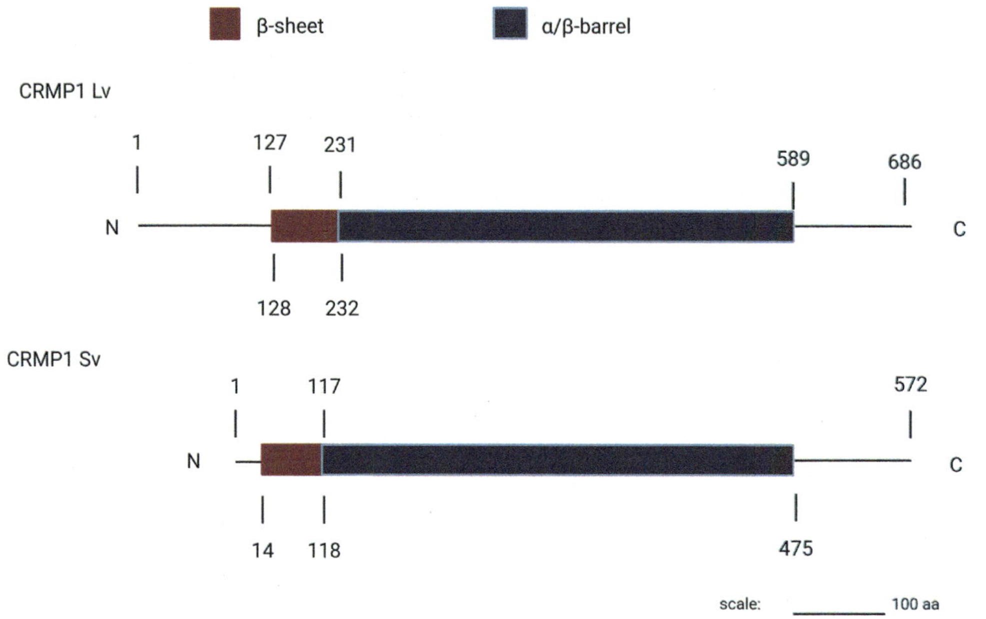

Figure 3.5 Domain structure of collapsin response element mediator protein 1 (CRMP1). Schematics of the domain structures of CRMP1 long variant (Lv) and short variant (Sv), aligned with each other. *Created in BioRender. Bradshaw, N. (2025) https://BioRender.com/gia3ybn.*

MDD patients at genome-wide significance (Zhang et al., 2022). The presence of insoluble TRIOBP-1 in brain samples of SCZ patients was subsequently confirmed in an independent sample, being seen in approximately 40% of anterior cortex samples from patients, but less than 5% of control individuals (Zaharija et al., 2022). Similar levels of insoluble TRIOBP-1 were seen in the same study in patients with MDD (Zaharija et al., 2022). Insoluble TRIOBP-1 was also seen in at least one individual who died by suicide, for whom no mental illness was recorded (Samardžija et al., 2023).

TRIOBP-1 is encoded by one of several splice variants of the *TRIOBP* gene, many of which do not overlap with each other (Shahin et al., 2006; Riazuddin et al., 2006; Zaharija et al., 2020). Transcripts from the 5' end of the gene, such as *TRIOBP-4,* encode protein isoforms critical for hearing through regulation of stereocilia in the inner ear (Kitajiri et al., 2010; Katsuno et al., 2019; Krey et al., 2022; Babahosseini et al., 2022); however, it is only TRIOBP-1 from the 3' end of the gene, as well as some uncharacterized shorter species, that have so far been seen to aggregate in mental illness (Zaharija et al., 2022). At least one longer *TRIOBP* splice variant, *TRIOBP-5,* which overlaps with *TRIOBP-1,* encodes a protein capable of aggregating in cell culture (Bradshaw et al., 2014) and in blood (Samardžija et al., 2021), but its relevance to mental illness is not yet determined.

TRIOBP-1 is expressed broadly across the human body and cell types (Seipel et al., 2001) and canonically functions as an interaction partner of F-actin and as an inhibitor of its depolymerization (Seipel et al., 2001; Bradshaw et al., 2017). It is implicated in a wide array of cellular functions, including mitotic progression (Zhu et al., 2012; Lan et al., 2014), cadherin-based adherent junction formation (Yano et al., 2011), filopodia formation in wound healing (Hong et al., 2016), neurite outgrowth (Woo et al., 2019), and β-catenin modulation (Wang et al., 2024). Many of these are seemingly tied to its role in actin modulation, while for others, any link remains unstudied.

Analysis of TRIOBP-1 aggregation has mostly been performed in cultured cell lines, with some confirmations in primary neurons (Bradshaw et al., 2014), and has focused on revealing parts of the protein involved in its aggregation. TRIOBP-1 can exist as either a 593 or 652-AA protein, with the N-terminal 59 AAs being optionally expressed depending on which of two start codons and Kozak sequences are used (Bradshaw et al., 2017) (Fig. 3.6). This is immediately followed by a predicted PH domain (Zaharija et al., 2022; Seipel et al., 2001), the function of which is unstudied. The C-terminal half of the protein consists of coiled-coil structures that appear to form into two distinct domains, with the larger central coiled-coil domain being responsible for multimerization and actin interactions (Bradshaw et al., 2017; Seipel et al., 2001; Lan et al., 2014). Aggregation of TRIOBP-1 can seemingly occur through two distinct mechanisms, dependent on two different sections of the protein. Deletion of both the N-terminal 59 optionally translated AAs and a short 8-AA motif near the center of the

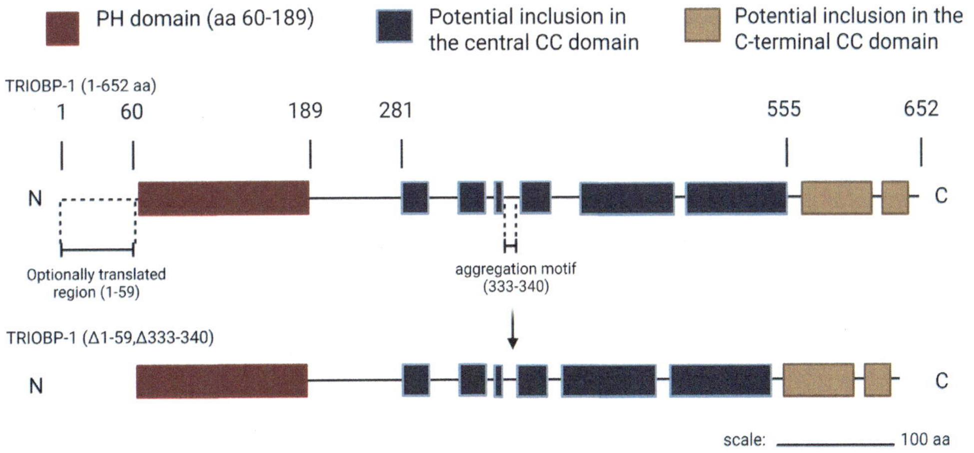

Figure 3.6 The domain structure of TRIOBP-1 and deletions that block its aggregation. The domain structure of TRIOBP-1. On the bottom row is shown a truncated version of this protein, lacking the N-terminal optionally translated domain and several amino acids from the center of the protein, which has been shown to have a significantly reduced propensity to aggregate. *Created in BioRender. Bradshaw, N. (2025) https://BioRender.com/gia3ybn. Based on data from Lan et al. (2014), Bradshaw et al. (2017), Zaharija et al. (2022).*

protein (in the central coiled-coil domain) abolishes the aggregation propensity of TRIOBP-1 in cell culture, while including either one of these regions still allows the protein to spontaneously aggregate (Bradshaw et al., 2017; Zaharija et al., 2022). Interestingly, while a mutant version of TRIOBP-1 lacking these regions does not aggregate by itself in cells, it does do so when coexpressed with DISC1 (Samardžija et al., 2023).

The consequences of TRIOBP-1 aggregation are largely unknown, but may include disruption of neurite outgrowth (Bradshaw et al., 2014). Curiously, aggregate-like structures have also been reported in stereocilia of the ear (Kazmierczak et al., 2017) and heart tissue (Jones et al., 2018), while insoluble protein was found in blood (Samardžija et al., 2021), suggesting that TRIOBP-1 aggregation may not be limited to the central nervous system.

3.5 Proteins whose aggregation in mental illness is implied through rare mutations

3.5.1 Neuronal Per-Amt-Sim-domain protein 3

Neuronal PAS-domain protein 3 (NPAS3) is a transcription factor expressed in the brain that belongs to the wider family of transcription factors based around a basic helix-loop-helix region and Per-Amt-Sim (PAS) domain (Nurnberger et al., 2014). Studies in knockout mice have shown NPAS3 to be critical for the development of the hippocampus, and the dentate gyrus in particular (Brunskill et al., 2005; Sha et al., 2012). Similar developmental locations and trajectories within the brain are also seen in humans (Gould and Kamnasaran, 2011; Wong et al., 2013). Functionally, NPAS3 acts through regulating the transcription of a range of different genes, including many that encode critical neurodevelopmental proteins, such as VGF and reelin (Sha et al., 2012; Yang et al., 2016a; Michaelson et al., 2017). NPAS3 was first directly implicated in major mental illness through the presence of a chromosomal translocation, splitting the *NPAS3* gene, in a mother and daughter with schizophreniform psychosis (Kamnasaran et al., 2003; Pickard et al., 2005). Genetic association studies, including some genome-wide association studies, then later implicated SNPs of the *NPAS3* gene in risk of SCZ and/or BPD (Nurnberger et al., 2014; Pickard et al., 2009; Macintyre et al., 2010; Huang et al., 2010; Weber et al., 2011).

The suggestion that NPAS3 may aggregate in major mental illness originates from a study of a small family with SCZ, which segregated with a mutation in the *NPAS3* gene that caused a V304I AA substitution (Yu et al., 2014). As a subtle difference in AAs, it was not clear how this mutation could affect or disrupt the function of the NPAS3 protein. Further research demonstrated, however, that the mutation led to increased insolubility of the NPAS3 protein in a variety of in vitro and cell culture systems (Nucifora et al., 2016). Furthermore, the mutation lies near the region, or possibly regions, of the NPAS3 protein that seem to drive its aggregation (Samardžija et al., 2021; Nucifora et al., 2016; Fig. 3.7).

A more recent study shows that insolubility of NPAS3 in the brain is present much more generally than just in this original family, suggesting that the V304I mutation is not the only way to induce NPAS3 aggregation (Samardžija et al., 2021). This is backed up by cell culture-based studies, which suggested the effect of oxidative stress on NPAS3 aggregation to be greater than that of the V304I mutation in isolation (Samardžija et al., 2021). To our knowledge, no group has yet investigated the prevalence of NPAS3 aggregation in a sufficiently large group of *postmortem* brain samples from patients and controls to verify if NPAS3 aggregation is generally associated with major mental illness. Increased quantities of NPAS3 protein are seen in the blood of SCZ patients, however, which implies at least a link between disease and altered NPAS3 proteostasis (Samardžija et al., 2021; Pavešić Radonja et al., 2023).

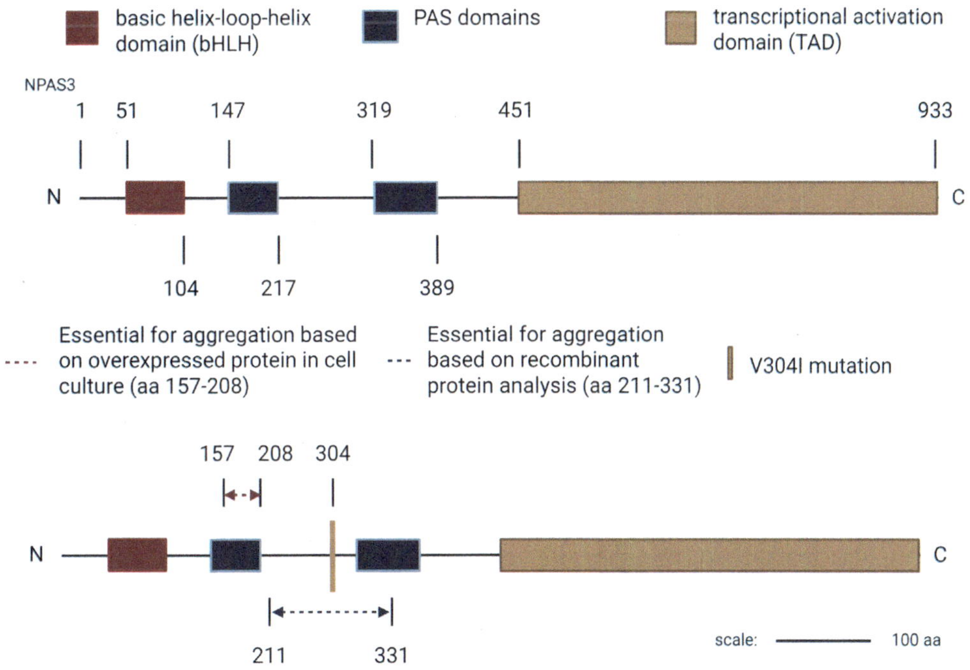

Figure 3.7 Domain structure of neuronal PAS-domain protein 3 (NPAS3). Schematics of the domain structure of NPAS3. The upper one shows the exact locations of the domains in terms of amino acid position. The lower shows the relative locations of two regions predicted, independently, to be essential for its aggregation, along with a V304I point mutation, found in a family with major mental illness, that is reported to increase its aggregation propensity. *Created in BioRender. Bradshaw, N. (2025) https://BioRender.com/gia3ybn. Based on data from Nucifora et al. (2016), Samardžija et al (2021).*

3.5.2 Fatty acid binding proteins (FABP3 and FABP7)

The fatty acid binding proteins (FABPs) are a family of small proteins involved in the regulation and transport of long-chain polyunsaturated fatty acids (Smathers and Petersen, 2011). Of these, three FABPs show significant expression in the brain, including FABP3 and FABP7, which are also known as heart-type and brain-type FABP, respectively (Liu et al., 2010). FABP7 is associated with astrocytes, where it acts as an important regulator of cortical development (Kipp et al., 2011; Ebrahimi et al., 2016), while FABP3 is expressed in neurons and is known to promote autophagy and apoptosis (Liu et al., 2021; Jang et al., 2024; Zhong et al., 2024).

As part of a more general investigation of *FABP* genes in autism and SCZ (Maekawa et al., 2010), sequencing revealed several seemingly *de novo* mutations in patients with SCZ, including two frameshift mutations: one in *FABP3* (E132fs) and one in *FABP7* (N80fs) (Shimamoto et al., 2014) (Fig. 3.8). Of these, the FABP3 frameshift occurs at the extreme C-terminus of the protein and adds an additional 68 nonspecific amino acids, while the FABP7 mutation would remove one of the two β-pleated sheet structures of FABP7 (Balendiran et al., 2000). Follow-up analysis showed that, while the wild-type FABP3 and FABP7 proteins were expressed both in the nucleus and cytoplasm, both of the frameshift mutants were predominantly outside the nucleus and formed aggregate-like structures (Shimamoto et al., 2014). Both of these mutants are expressed at a lower level than their wild-type equivalents, seemingly due to proteasomal degradation (Shimamoto et al., 2014). Like NPAS3, it is therefore plausible that the mutations cause aggregation in the brain of at least some SCZ patients.

Evidence already exists associating variants in the *FABP7* locus with both SCZ and BPD, while similar experiments on FABP3 have so far yielded negative results (Watanabe et al., 2007; Iwayama et al., 2010). In contrast, altered levels of *FABP3* have been detected in the brains of SCZ patients (BA8, frontal eye field), while *FABP7* did not vary significantly (Hamazaki et al., 2016).

3.6 Recurring themes between the proteins

3.6.1 Coaggregation and parallel aggregation of proteins in mental illness

While traditionally most neurodegenerative diseases are studied with an implicit hypothesis that each patient will show aggregation of one specific protein (e.g., TDP-43 or FUS), it is increasingly acknowledged that patients can and often do display more than one aggregating protein (Lashuel, 2021; Konstantoulea et al., 2022). As a result, the potential existence of multiple proteins aggregating in the same patient with mental illness is something that has been considered since some of the first studies on the topic (Ottis et al., 2011; Bader et al., 2012). This can occur through a process of "co-

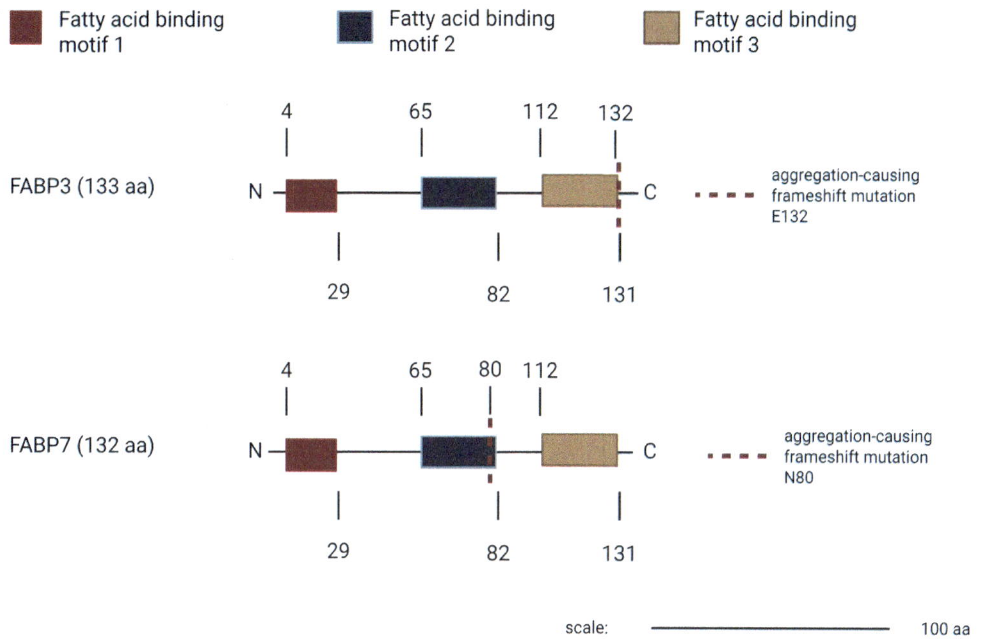

Figure 3.8 Domain structure of fatty acid binding protein 3 (FABP3) and FABP7. Schematic of the structure of FABP3 and FABP7, including their fatty acid binding domains and frameshift mutations that were detected in patients with schizophrenia and cause aggregation. *Created in BioRender. Bradshaw, N. (2025) https://BioRender.com/gia3ybn. Based on data in Smathers and Petersen (2011), Shimamoto et al. (2014).*

aggregation," in which one protein directly induces aggregation of another, or parallel aggregation, where each aggregates individually (Samardžija et al., 2023) (Fig. 3.1).

The presence of multiple insoluble, and therefore likely aggregating, proteins in the brain of the same patient appears to be relatively common, with insoluble DISC1 in particular often coexisting with at least one other insoluble protein (Ottis et al., 2011; Bader et al., 2012; Samardžija et al., 2023). While only relatively small numbers of brains have been investigated so far, the numbers seem to suggest that mechanisms of parallel aggregation are more common than instances of coaggregation, as the presence of insoluble DISC1 (for example) does not obviously increase the chance of other aggregating proteins being present, although this cannot be formally discounted without larger studies (Samardžija et al., 2023). Whether or not proteins can coaggregate or parallel aggregate with each other has also been investigated in mammalian cell lines. For example, in the case of both dysbindin-1 and CRMP1, more aggregation-prone versions of the proteins can coaggregate with other, less aggregation-prone isoforms (Zhu et al., 2015; Xu et al., 2015; Bader et al., 2012; Samardžija et al., 2023). Coaggregation between different proteins has also been reported based on cell culture experiments, with one of the proteins involved consistently being DISC1 (Ottis et al., 2011; Bader et al., 2012; Samardžija et al., 2023; Fig. 3.9). In all of these instances, it appears that DISC1 can act as a "seed" to cause its partners to coaggregate. It has not been investigated whether these proteins can also induce coaggregation of DISC1.

Caution must be taken when assuming that physical interaction of protein molecules in a cell culture system reflects the activity of the equivalent molecules in the brain, especially when the cell systems make use of overexpression to induce aggregation. Nevertheless, when proper negative controls are used to ensure that such overexpressed proteins do not just coaggregate with everything, these approaches can show the potential for such interactions, particularly when they correspond to the known coexistence of the same proteins aggregating in the brain.

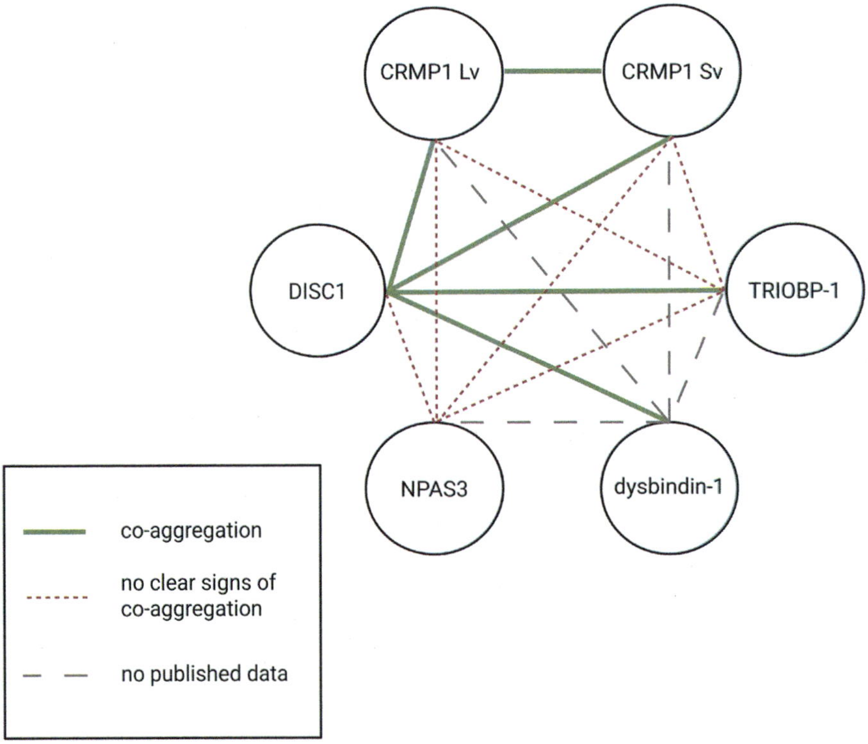

Figure 3.9 Known coaggregation between proteins involved in major mental illness. Schematic showing which proteins, of those discussed in this paper, are known to aggregate, and which do not seem to, based on analysis in cell culture. To our knowledge, no similar studies have looked at fatty acid binding protein 3 (FABP3) or FABP7. *Created in BioRender. Bradshaw, N. (2025) https://BioRender.com/gia3ybn. Based on data from Ottis et al. (2011), Bader et al. (2012), Samardžija et al. (2023).*

3.6.2 Partial overlap with other neurological disorders

One argument that potentially stood against protein aggregation playing a role in major mental illnesses is that none of the proteins known to aggregate in neurodegenerative conditions were known to also do so in major mental illness, meaning that any proteins that aggregate in mental illness must be ones not previously studied in the context of proteinopathy (Bradshaw and Korth, 2019). In recent years, several lines of evidence have emerged to show that the reverse may be true, with several of these "novel" proteins that aggregate in mental illness seemingly playing roles in neurodegenerative disease as well.

DISC1 aggregates were detected in murine models and brains of patients suffering from Huntington's disease. DISC1 and mutant huntingtin (HTT) form pathological complexes, resulting in dysregulation of the interaction between soluble DISC1 and PDE4. Loss of interaction leads to an aberrant increase in PDE4 activity, which in turn affects cAMP levels and cellular signaling. A decrease in cAMP levels can be detected in the cerebrospinal fluid of Huntington's disease patients, suggesting the role of PDE4 in disease development. Moreover, the loss of the physiological control DISC1 exerts over PDE4 due to the formation of the DISC1-HTT complex may also underlie anhedonia seen in a murine model of Huntington's disease, further confirming the role of abnormal proteostasis in disease development (Tanaka et al., 2017).

DISC1 coaggregates with another protein of importance in neurodegenerative disorders, TDP-43 (TAR DNA-binding protein 43 kDa). In frontotemporal lobar degeneration (FTLD), TDP-43 aggregation leads to progressive neuronal atrophy, causing severe mental and behavioral deficits. Coaggregation of TDP-43 and DISC1 results in impaired local translation in dendrites, as well as behavioral and mental deficits in both murine models and *postmortem* brains of FTLD patients. Behavioral deficits were improved upon exogenous DISC1 expression in the frontal cortex, implicating that the loss of soluble DISC1 is directly responsible for aberrant behavior relevant to psychiatric conditions (Endo et al., 2018).

CRMP1 was also seen to be an interaction partner of HTT and, very interestingly, was seen both to coaggregate with HTT, and also to limit the subsequent growth and toxicity of HTT aggregation, a process that could be seen in multiple experimental systems, including cells, mice, and flies (Stroedicke et al., 2015). Initial experiments have not shown any evidence so far that CRMP1 similarly counteracts aggregation of DISC1 (Samardžija et al., 2023). Additionally, at least four heterozygous *CRMP1* coding mutations have been reported in individuals with low IQ, developmental delay, and/or autism spectrum disorders (Ravindran et al., 2022; Liu et al., 2024). Two of these, P457L and T313M (according to the AA numbering of the CRMP1 short variant) led to the formation of larger multimeric complexes than wild-type CRMP1 (Ravindran et al., 2022). It is therefore plausible that one or more of these mutations may cause CRMP1 to aggregate in neurodevelopmental disorders. Additionally, while not known to be directly involved in aggregation, phosphorylation of CRMP1 has been shown to be a component of neuronal pathology in amyotrophic lateral sclerosis (Asano et al., 2022; Kawamoto et al., 2022).

FABP3 is well documented to play a role in α-synuclein pathology and has been seen to colocalise with α-synuclein aggregates, implying coaggregation, in the brains of patients with Parkinson's disease (Oizumi et al., 2021). Cell culture models show that formation of α-synuclein oligomers and aggregates is enhanced when FABP3 is expressed (Shioda et al., 2014), with α-synuclein pathology consistently reduced in neurons of mice that lack the *FABP3* gene (Shioda et al., 2014; Kawahata et al., 2019; Yabuki et al., 2020; Matsuo et al., 2021). Interactions between FABP7 and α-synuclein have only been studied to a limited extent, although they seem to coaggregate in cell culture models (Cheng et al., 2022). Effects of FABP proteins on other aggregates are less established, although levels of FABP3 are consistently seen to be upregulated in the cerebrospinal fluid of Alzheimer's disease patients and show correlation with other disease-related phenotypes (Desikan et al., 2013; Chiasserini et al., 2017; Brosseron et al., 2021; Dulewicz et al., 2021; Dhiman et al., 2022). Intriguingly, a recent study in *Drosophila* showed that the fabp protein, corresponding to both mammalian FABP3 and FABP7, can counteract Aβ pathology, seemingly through promoting autophagy (Jang et al., 2024).

3.6.3 Aggregation in tissues other than the brain

Another important emerging theme in the study of these proteins is the potential presence of protein aggregates, or biological correlates of them, outside of the brain. This is important for the translational value of this research field as, while protein aggregates in the brain are very interesting from a pathogenesis perspective, it is not practical to sample a patient's brain for the purposes of diagnosis or personalized medicine. If, however, these aggregates also formed in more accessible tissues, or if other biomarkers could be determined that predicted their presence, then these could have immediate clinical value.

Experiments to identify such correlating biomarkers were undertaken using the tgDISC1 transgenic rat, based on the hypothesis that it can model aspects of human mental illness associated with DISC1 aggregation (Trossbach et al., 2016, 2019). Microarray analysis of peripheral blood mononuclear cells from these animals, compared to those of wild-type littermates, identified various transcripts that were underexpressed in the transgenic animals, including several associated with the immune system (Trossbach et al., 2019). Indeed, the two most downregulated transcripts from the mice, *RG1* and *CCL4*, were similarly downregulated in two cohorts of SCZ patients. In both cohorts, the specificity of detection was good, but dysregulation of the transcripts was only associated with a subgroup of patients (Trossbach et al., 2019), which is logical given that DISC1 aggregation also only occurs in distinct subgroups (Leliveld et al., 2008). No specific symptoms were strongly associated with the presence of this putative biomarker, although there was a suggestive, nonsignificant association of *RG1/CCL4* downregulation with the prescription of clozapine, which is normally only given to patients who are resistant to at least two other antipsychotics (Trossbach et al., 2019).

Given that disrupted proteostasis will often arise as a result of genetic mutations or environmental factors that affect, to some degree, the entire organism, it is plausible that aggregating proteins in the brain may cooccur with some form of disturbed proteostasis in other tissues. Attempts have therefore been made to identify protein aggregates themselves in blood serum from patients with SCZ and MDD. Aggregates of both TRIOBP and dysbindin-1 were detected, although the low number of positive samples made it impossible to determine whether these were diagnostically relevant (Samardžija et al., 2021). NPAS3 was also seen in the blood of many individuals and showed some specificity for SCZ (Samardžija et al., 2021). Curiously, the presence of any of these three proteins was also associated with higher scores on the PANSS scale for SCZ symptoms (Kay et al., 1987), and there was a suggestive trend towards not being prescribed clozapine (Pavešić Radonja et al., 2023). Another study, not looking specifically at aggregation, saw levels of FABP3 in the plasma of patients with SCZ to be predictive of the effectiveness of the atypical antipsychotic olanzapine. Specifically, both negative and general PANSS scores dropped more dramatically in patients with higher baseline FABP3 (Tomasik et al., 2016). Finally, one study investigated CRMP1 in lymphoblastoid cell lines derived from patients with SCZ or control individuals, showing raised levels of CRMP1 in patient cell lines, although again this did not look specifically at aggregation of the protein (Ottis et al., 2011).

As an alternative to blood, one study has reported the presence of aggregates of DISC1 in cerebrospinal fluid, using surface-based fluorescence intensity distribution analysis, a highly sensitive technique that can distinguish between monomers of a protein and larger oligomeric states, which are assumed to be aggregates (Birkmann et al., 2007; Herrmann et al., 2017; Pils et al., 2023). Low femtomolar levels of DISC1 aggregates were detected in individuals diagnosed with SCZ or other related disorders. Subsequent analysis showed elevated levels of DISC1 aggregates in CSF, specifically in a subset of patients suffering from SCZ or schizoaffective disorder (Pils et al., 2023). There was also an association of DISC1 aggregate levels with a SNP previously implicated in SCZ and some of its endophenotypes (Pils et al., 2023; Hennah et al., 2009; Tomppo et al., 2009).

Finally, and most recently, the Nucifora group made use of advantages in taking nasal biopsies and establishing olfactory neuronal cells from them (Tajinda et al., 2010; Mihaljevic et al., 2023) to study protein aggregation in neurons of living SCZ patients

(Nucifora et al., 2025). As this group had previously seen in *postmortem* brain tissue (Nucifora et al., 2019), they found a subset of patients to display high levels of protein insolubility and ubiquitination in these cells, strongly suggestive of a general increase in protein aggregation when compared to both controls and patients outside of this subset (Nucifora et al., 2025). Furthermore, this subset of patients with disturbed proteostasis showed significantly lower ideational fluency and executive function and showed greater attention impairment, suggesting an effect of protein aggregation on aspects of neurocognitive function (Nucifora et al., 2025).

While data on protein aggregation in nonbrain tissue remains relatively new, these initial studies suggest that there is potential to translate the study of proteostasis and protein aggregation into a diagnostically useful procedure. Significantly more information is required, however, notably about whether proteostasis imbalance in peripheral tissues directly corresponds to aggregation in the brain, and about how variation in the proteostasis of individual proteins ties to specific clinical phenotypes and diagnoses.

3.6.4 Expression in the brain

Based on data from the Human Protein Atlas (Uhlén et al., 2015; Sjöstedt et al., 2020), all of the putative aggregating proteins show expression across the human brain, with relatively few instances of predominant expression in specific regions. The major exception is FABP7, which is prominent in the cerebellum, amygdala, and cerebral cortex, as well as dysbindin-1, which is prominent in the basal ganglia (www.proteinatlas.org). Data from the Human Brain Transcriptome Project (Kang et al., 2011), however, shows that the developmental trajectories of these genes in the brain vary, with *DTNBP1* (encoding dysbindin-1) and *FABP3* being expressed mostly postnatally, *CRMP1*, *TRIOBP*, *NPAS3*, and *FABP7* peaking during embryonic development, and *DISC1* being more consistent across the lifetime of an individual (www.hbatlas.org). It is therefore difficult to identify a common role or pathways for all of these proteins based on their expression pattern. A caveat to this data, though, is that it is only based on one splice variant of each gene, with *TRIOBP* and *DISC1* in particular known to have complicated patterns of alternate splicing leading to many protein isoforms (Shahin et al., 2006; Riazuddin et al., 2006; Nakata et al., 2009).

3.7 Conclusions and future directions

While still a relatively young field of study, and certainly not one that has mainstream acceptance as yet, the study of disturbed proteostasis and protein aggregation in major mental illness has shown significant progress in recent years, with DISC1 being by far the most studied aggregating protein in these conditions. Much work still remains to be done, however.

This includes learning more about both the causes and consequences of aggregation of these proteins. Based on protein aggregation in neurodegenerative diseases, it is likely that proteins in major mental illness aggregate on the basis of a combination of inherited genetic factors, de novo mutations, and environmental risk factors. Examples of mutations causing aggregation are reported for NPAS3, FABP3, and FABP7 (Nucifora et al., 2016; Yu et al., 2014; Shimamoto et al., 2014), but a thorough investigation has not yet been made, to the best of our knowledge. Identifying such mutations would be invaluable for developing future models of aggregation-related mental illness. Similarly, it remains an open question whether aggregation of proteins (either general or specific) is a consequence of major mental illness or a potential causative factor. Hints towards this come from the fact that cellular systems involved in proteostasis are impaired in some patients with mental illness (Rubio et al., 2013; Scott and Meador-Woodruff, 2020), arguing that protein aggregation may be an effect; however, initial animal models showing that protein overexpression leads to both aggregation in the brain and neurological and/or behavioral phenotypes argue more in favor of causality (Yang and Xu, 2017; Trossbach et al., 2016; Samardžija et al., 2024).

Another open question regards the nature of these aggregates. To date, all of the proteins have been characterized as aggregating based on their insolubility, in many cases within patient brain tissue, and more specifically by an increase in insolubility compared to control individuals and systems (Nucifora et al., 2016; Bader et al., 2012; Leliveld et al., 2008; Ottis et al., 2011; Bradshaw et al., 2014; Shimamoto et al., 2014). The definition of these insoluble species as aggregates is then supported by the presence of punctate or irregular species, resembling classical protein aggregates, in cell culture systems and/or rodent model systems, which correlate with the presence of insoluble protein (Bader et al., 2012; Ottis et al., 2011; Bradshaw et al., 2014; Trossbach et al., 2016; Yang and Xu, 2017; Yang et al., 2016b; Samardžija et al., 2021; Shimamoto et al., 2014). The definitive demonstration of visible aggregates in human tissue still remains to be performed, however. The form of the aggregates is also unclear. Protein aggregates in neurodegenerative disease can take the form of misfolded, β-strand-rich amyloid, amorphous aggregates, and/or protein condensates (Juković et al., 2024). To our knowledge, experiments to investigate the nature of such aggregates in mental illness have only been attempted for DISC1 using recombinant protein. These experiments have shown evidence of amyloid formation and unfolded aggregates in at least some circumstances (Tanaka et al., 2017; Cukkemane et al., 2025a; Cukkemane et al., 2021).

The biggest outstanding challenge, however, will be large-scale replication studies to determine the presence of aggregates of both specific proteins and proteins more generally in the wider patient population. As well as serving to replicate and reinforce previous work, such studies will allow the stratification of patients based on their protein aggregation status. Studies performed to date have shown that less than 50% of SCZ patients display either general protein aggregation or aggregation of specific proteins

(Bader et al., 2012; Leliveld et al., 2008; Ottis et al., 2011; Zaharija et al., 2022; Nucifora et al., 2025, 2019). It is therefore plausible that disturbed proteostasis and protein aggregation characterize one or more specific biological subgroups within SCZ and other major mental illnesses. From a long-term perspective, the ultimate goal of research in this field must be to generate new therapies (either treatment or preventative) based on the protein aggregation process. In the medium term, however, identification and characterization of these aggregation-related patient subgroups is an achievable goal with immediate translational value in terms of streamlining the diagnosis and treatment process, and working towards a personalized medicine approach to psychiatry, at least for these patient subgroups.

Acknowledgments

This work was supported by the Croatian Science Foundation (HRZZ, project numbers IP-2022-10-2745, DOK-2018-09-5395 and DOK-NPOO-2023-10-6409), the Alexander von Humboldt Foundation (1142747-HRV-IP), and the University of Rijeka (Uniri-iskusni-23–183).

References

Albert, P.R., 2015. Why is depression more prevalent in women? Journal of Psychiatry and Neuroscience 40 (4), 219–221. Available at https://doi.org/10.1503/jpn.150205.

Andrews, J.L., Goodfellow, F.J., Matosin, N., Snelling, M.K., Newell, K.A., Huang, X.F., et al., 2017. Alterations of ubiquitin related proteins in the pathology and development of schizophrenia: evidence from human and animal studies. Journal of Psychiatric Research 90, 31–39. Available at https://doi.org/10.1016/j.jpsychires.2017.01.009.

Asano, T., Nakamura, H., Kawamoto, Y., Tada, M., Kimura, Y., Takano, H., et al., 2022. Inhibition of Crmp1 phosphorylation at Ser522 ameliorates motor function and neuronal pathology in amyotrophic lateral sclerosis model mice. Eneuro 9 (3). Available at https://doi.org/10.1523/eneuro.0133-22.2022.

Atkin, T.A., Brandon, N.J., Kittler, J.T., 2012. Disrupted in Schizophrenia 1 forms pathological aggresomes that disrupt its function in intracellular transport. Human Molecular Genetics 21 (9), 2017–2028. Available at https://doi.org/10.1093/hmg/dds018.

APA, 2013. Diagnostic and Statistical Manual of Mental Disorders, fifth ed. American Psychiatric Association. Available at https://doi.org/10.1176/appi.books.9780890425596.

Aydınoğlu, Ü., Yazla, E., Çetin, İ., Kayadibi, H., 2024. Bipolar disorder and ubiquitin proteasome system dysfunction: peripheral blood levels of molecules playing a role in ubiquitination and their relationship to sleep quality. Discovery Medicine 36 (190), 2244. Available at https://doi.org/10.24976/discov.med.202436190.206.

Babahosseini, H., Belyantseva, I.A., Yousaf, R., Tona, R., Hadi, S., Inagaki, S., et al., 2022. Unbalanced bidirectional radial stiffness gradients within the organ of Corti promoted by TRIOBP. Proceedings of the National Academy of Sciences 119 (26). Available at https://doi.org/10.1073/pnas.2115190119.

Bader, V., Tomppo, L., Trossbach, S.V., Bradshaw, N.J., Prikulis, I., Leliveld, S.R., et al., 2012. Proteomic, genomic and translational approaches identify CRMP1 for a role in schizophrenia and its underlying traits. Human Molecular Genetics 21 (20), 4406–4418. Available at https://doi.org/10.1093/hmg/dds273.

Balendiran, G.K., Schnütgen, F., Scapin, G., Börchers, T., Xhong, N., Lim, K., et al., 2000. Crystal structure and thermodynamic analysis of human brain fatty acid-binding protein. Journal of Biological Chemistry 275 (35), 27045–27054. Available at https://doi.org/10.1016/s0021-9258(19)61478-x.

Banani, S.F., Rice, A.M., Peeples, W.B., Lin, Y., Jain, S., Parker, R., et al., 2016. Compositional control of phase-separated cellular bodies. Cell 166 (3), 651–663. Available at https://doi.org/10.1016/j.cell.2016.06.010.

Birkmann, E., Henke, F., Weinmann, N., Dumpitak, C., Groschup, M., Funke, A., et al., 2007. Counting of single prion particles bound to a capture-antibody surface (surface-FIDA). Veterinary Microbiology 123 (4), 294–304. Available at https://doi.org/10.1016/j.vetmic.2007.04.001.

Blackwood, D.H.R., Fordyce, A., Walker, M.T., St. Clair, D.M., Porteous, D.J., Muir, W.J., 2001. Schizophrenia and affective disorders—cosegregation with a translocation at chromosome 1q42 that directly disrupts brain-expressed genes: clinical and P300 findings in a family. The American Journal of Human Genetics 69 (2), 428–433. Available at https://doi.org/10.1086/321969.

Bo, S., Abu-Akel, A., Kongerslev, M., Haahr, U.H., Simonsen, E., 2011. Risk factors for violence among patients with schizophrenia. Clinical Psychology Review 31 (5), 711–726. Available at https://doi.org/10.1016/j.cpr.2011.03.002.

Bousman, C.A., Chana, G., Glatt, S.J., Chandler, S.D., May, T., Lohr, J., et al., 2010. Positive symptoms of psychosis correlate with expression of ubiquitin proteasome genes in peripheral blood. American Journal of Medical Genetics, Part B: Neuropsychiatric Genetics 153 (7), 1336–1341. Available at https://doi.org/10.1002/ajmg.b.31106.

Bousman, C.A., Chana, G., Glatt, S.J., Chandler, S.D., Lucero, G.R., Tatro, E., et al., 2010. Preliminary evidence of ubiquitin proteasome system dysregulation in schizophrenia and bipolar disorder: convergent pathway analysis findings from two independent samples. American Journal of Medical Genetics, Part B: Neuropsychiatric Genetics 153 (2), 494–502. Available at https://doi.org/10.1002/ajmg.b.31006.

Bradshaw, N.J., Porteous, D.J., 2012. DISC1-binding proteins in neural development, signalling and schizophrenia. Neuropharmacology 62 (3), 1230–1241. Available at https://doi.org/10.1016/j.neuropharm.2010.12.027.

Bradshaw, N.J., Korth, C., 2019. Protein misassembly and aggregation as potential convergence points for non-genetic causes of chronic mental illness. Molecular Psychiatry 24 (7), 936–951. Available at https://doi.org/10.1038/s41380-018-0133-2.

Bradshaw, N.J., Bader, V., Prikulis, I., Lueking, A., Müllner, S., Korth, C., et al., 2014. Aggregation of the protein TRIOBP-1 and its potential relevance to schizophrenia. PLoS One 9 (10), e111196. Available at https://doi.org/10.1371/journal.pone.0111196.

Bradshaw, N.J., Yerabham, A.S.K., Marreiros, R., Zhang, T., Nagel-Steger, L., Korth, C., 2017. An unpredicted aggregation-critical region of the actinpolymerizing protein TRIOBP-1/Tara, determined by elucidation of its domain structure. Journal of Biological Chemistry 292 (23), 9583–9598. Available at https://doi.org/10.1074/jbc.M116.767939.

Brandon, N.J., Schurov, I., Camargo, L.M., Handford, E.J., Duran-Jimeniz, B., Hunt, P., et al., 2005. Subcellular targeting of DISC1 is dependent on a domain independent from the Nudel binding site. Molecular and Cellular Neuroscience 28 (4), 613–624. Available at https://doi.org/10.1016/j.mcn.2004.11.003.

Brandon, N.J., Millar, J.K., Korth, C., Sive, H., Singh, K.K., Sawa, A., 2009. Understanding the role of DISC1 in psychiatric disease and during normal development. Journal of Neuroscience 29 (41), 12768–12775. Available at https://doi.org/10.1523/JNEUROSCI.3355-09.2009.

Brosseron, F., Kleemann, K., Kolbe, C.C., Santarelli, F., Castro-Gomez, S., Tacik, P., et al., 2021. Interrelations of Alzheimer´s disease candidate biomarkers neurogranin, fatty acid-binding protein 3 and ferritin to neurodegeneration and neuroinflammation. John Journal of Neurochemistry 157 (6), 2210–2224. Available at https://doi.org/10.1111/jnc.15175.

Brunskill, E.W., Ehrman, L.A., Williams, M.T., Klanke, J., Hammer, D., Schaefer, T.L., et al., 2005. Abnormal neurodevelopment, neurosignaling and behaviour in Npas3-deficient mice. European Journal of Neuroscience 22 (6), 1265–1276. Available at https://doi.org/10.1111/j.1460-9568.2005.04291.x.

Camargo, L.M., Collura, V., Rain, J.C., Mizuguchi, K., Hermjakob, H., Kerrien, S., et al., 2007. Disrupted in Schizophrenia 1 interactome: Evidence for the close connectivity of risk genes and a potential synaptic basis for schizophrenia. Molecular Psychiatry 12 (1), 74–86. Available at https://doi.org/10.1038/sj.mp.4001880.

Charrier, E., Mosinger, B., Meissirel, C., Aguera, M., Rogemond, V., Reibel, S., et al., 2006. Transient alterations in granule cell proliferation, apoptosis and migration in postnatal developing cerebellum of CRMP1-/- mice. Genes to Cells: Devoted to Molecular & Cellular Mechanisms 11 (12), 1337–1352. Available at 10.1111/j.1365-2443.2006.01024.x.

Cheng, A., Wang, Y.F., Shinoda, Y., Kawahata, I., Yamamoto, T., Jia, W.B., et al., 2022. Fatty acid-binding protein 7 triggers α-synuclein oligomerization in glial cells and oligodendrocytes associated with oxidative stress. Acta Pharmacologica Sinica 43 (3), 552–562. Available at 10.1038/s41401-021-00675-8.

Chiasserini, D., Biscetti, L., Eusebi, P., Salvadori, N., Frattini, G., Simoni, S., et al., 2017. Differential role of CSF fatty acid binding protein 3, α-synuclein, and Alzheimer's disease core biomarkers in Lewy body disorders and Alzheimer's dementia. Alzheimer's Research & Therapy 9 (1). Available at https://doi.org/10.1186/s13195-017-0276-4.

Cho, W.K., Shin, W.S., An, I., Bang, M., Cho, D.Y., Lee, S.H., 2019. Biological aspects of aggression and violence in Schizophrenia. Clinical Psychopharmacology and Neuroscience 17 (4), 475–486. Available at https://doi.org/10.9758/cpn.2019.17.4.475.

Chubb, J.E., Bradshaw, N.J., Soares, D.C., Porteous, D.J., Millar, J.K., 2008. The DISC locus in psychiatric illness. Molecular Psychiatry 13 (1), 36–64. Available at https://doi.org/10.1038/sj.mp.4002106.

Craddock, N., Owen, M.J., 2010. The Kraepelinian dichotomy - going, going. but still not gone. The British journal of psychiatry: the journal of mental science 196 (2), 92–95. Available at https://doi.org/10.1192/bjp.bp.109.073429.

Cukkemane, A., Becker, N., Zielinski, M., Frieg, B., Lakomek, N.A., Heise, H., et al., 2021. Conformational heterogeneity coupled with β-fibril formation of a scaffold protein involved in chronic mental illnesses. Translational Psychiatry 11 (1). Available at https://doi.org/10.1038/s41398-021-01765-1.

Cukkemane, A., Dingley, A.J., Mohrlüder, J., Santiago-Schübel, B., Weiergräber, O.H., Willbold, D., 2025b. A peptide mimetic therapeutic strategy targeting dysfunction of the scaffold protein DISC-1 in psychiatric disorders. European Journal of Pharmaceutical Sciences 211, 107148. Available at https://doi.org/10.1016/j.ejps.2025.107148.

Cukkemane, A., Becker, N., Kupreichyk, T., Heise, H., Willbold, D., Weiergräber, O.H., 2025a. Tracing the aggregation pathway of the scaffold protein DISC1: structural implications for chronic mental illnesses. Journal of Structural Biology: X 11, 100128. Available at https://doi.org/10.1016/j.yjsbx.2025.100128.

De Pace, R., Ghosh, S., Williamson, C.D., Bonifacino, J.S., 2025. BLOC-1 and BORC: complex regulators of endolysosomal dynamics. Cell Chemical Biology 32 (9), 1106–1124. Available at https://doi.org/10.1016/j.chembiol.2025.08.001.

Desikan, R.S., Thompson, W.K., Holland, D., Hess, C.P., Brewer, J.B., Zetterberg, H., et al., 2013. Heart fatty acid binding protein and Aβ-associated Alzheimer's neurodegeneration. Molecular Neurodegeneration 8, 39. Available at https://doi.org/10.1186/1750-1326-8-39.

Dhiman, K., Villemagne, V.L., Fowler, C., Bourgeat, P., Li, Q.X., Collins, S., et al., 2022. Cerebrospinal fluid levels of fatty acid–binding protein 3 are associated with likelihood of amyloidopathy in cognitively healthy individuals. Alzheimer's and Dementia: Diagnosis, Assessment and Disease Monitoring 14 (1). Available at https://doi.org/10.1002/dad2.12377.

Di Manno, L., Macdonald, J.A., Knight, T., 2015. Family dissolution and offspring depression and depressive symptoms: a systematic review of moderation effects. Journal of Affective Disorders 188, 68–79. Available at https://doi.org/10.1016/j.jad.2015.08.017.

Dören, J., Kupriyanova, Y., Schäble, S., Troßbach, S., McGuire, B., Vernon, A.C., Roden, M., Korth, C., Kalenscher, T., 2025. Social Reward Learning Deficits and Concordant Brain Alterations in Rats Overexpressing Disrupted-in-Schizophrenia 1 (DISC1). The Journal of Neuroscience 45, e1067252025. Available at https://doi.org/10.1523/jneurosci.1067-25.2025.

Dulewicz, M., Kulczyńska-Przybik, A., Słowik, A., Borawska, R., Mroczko, B., 2021. Fatty acid binding protein 3 (FABP3) and apolipoprotein E4 (ApoE4) as lipid metabolism-related biomarkers of Alzheimer's disease. Journal of Clinical Medicine 10 (14), 3009. Available at https://doi.org/10.3390/jcm10143009.

Ebrahimi, M., Yamamoto, Y., Sharifi, K., Kida, H., Kagawa, Y., Yasumoto, Y., et al., 2016. Astrocyte-expressed FABP 7 regulates dendritic morphology and excitatory synaptic function of cortical neurons. Glia 64 (1), 48–62. Available at https://doi.org/10.1002/glia.22902.

Endo, R., Takashima, N., Nekooki-Machida, Y., Komi, Y., Hui, K.K.W., Takao, M., et al., 2018. TAR DNA-binding protein 43 and disrupted in schizophrenia 1 coaggregation disrupts dendritic local translation and mental function in frontotemporal lobar degeneration. Biological Psychiatry 84 (7), 509–521. Available at https://doi.org/10.1016/j.biopsych.2018.03.008.

Fei, E., Chen, P., Zhang, Q., Zhong, Y., Zhou, T., 2022. Protein kinase B/Akt1 phosphorylates dysbindin-1A at serine 10 to regulate neuronal development. Neuroscience 490, 66–78. Available at https://doi.org/10.1016/j.neuroscience.2022.01.025.

Flores, R., Hirota, Y., Armstrong, B., Sawa, A., Tomoda, T., 2011. DISC1 regulates synaptic vesicle transport via a lithium-sensitive pathway. Neuroscience Research 71 (1), 71–77. Available at https://doi.org/10.1016/j.neures.2011.05.014.

Funke, B., Finn, C.T., Plocik, A.M., Lake, S., DeRosse, P., Kane, J.M., et al., 2004. Association of the DTNBP1 locus with schizophrenia in a U.S. population. American Journal of Human Genetics 75 (5), 891–898. Available at https://doi.org/10.1086/425279.

García-Fuster, M.J., Díez-Alarcia, R., Ferrer-Alcón, M., La Harpe, R., Meana, J.J., García-Sevilla, J.A., 2014). FADD adaptor and PEA-15/ERK1/2 partners in major depression and schizophrenia postmortem brains: basal contents and effects of psychotropic treatments. Neuroscience 277, 541–551. Available at https://doi.org/10.1016/j.neuroscience.2014.07.027.

Geddes, J.R., Miklowitz, D.J., 2013. Treatment of bipolar disorder. Lancet (London, England) 381 (9878), 1672–1682. Available at https://doi.org/10.1016/S0140-6736(13)60857-0.

Ghiani, C.A., Dell'Angelica, E.C., 2011. Dysbindin-containing complexes and their proposed functions in brain: from zero to (too) many in a decade. ASN Neuro 3 (2), 109–124. Available at https://doi.org/10.1042/AN20110010.

Gould, P., Kamnasaran, D., 2011. Immunohistochemical analyses of NPAS3 expression in the developing human fetal brain. Anatomia, Histologia, Embryologia 40 (3), 196–203. Available at https://doi.org/10.1111/j.1439-0264.2010.01059.x.

Hamazaki, K., Maekawa, M., Toyota, T., Iwayama, Y., Dean, B., Hamazaki, T., et al., 2016. Fatty acid composition and fatty acid binding protein expression in the postmortem frontal cortex of patients with schizophrenia: a case–control study. Schizophrenia Research 171 (1-3), 225–232. Available at https://doi.org/10.1016/j.schres.2016.01.014.

Hamburg, H., Trossbach, S.V., Bader, V., Chwiesko, C., Kipar, A., Sauvage, M., et al., 2016. Simultaneous effects on parvalbumin-positive interneuron and dopaminergic system development in a transgenic rat model for sporadic schizophrenia. Scientific Reports 6. Available at https://doi.org/10.1038/srep34946.

Hashimoto, K., Fukushima, T., Shimizu, E., Komatsu, N., Watanabe, H., Shinoda, N., et al., 2003. Decreased serum levels of D-serine in patients with schizophrenia: evidence in support of the N-methyl-D-aspartate receptor hypofunction hypothesis of schizophrenia. Archives of General Psychiatry 60 (6), 572–576. Available at https://doi.org/10.1001/archpsyc.60.6.572.

Hennah, W., Thomson, P., McQuillin, A., Bass, N., Loukola, A., Anjorin, A., et al., 2009. DISC1 association, heterogeneity and interplay in schizophrenia and bipolar disorder. Molecular Psychiatry 14 (9), 865–873. Available at https://doi.org/10.1038/mp.2008.22.

Herrmann, Y., Kulawik, A., Kühbach, K., Hülsemann, M., Peters, L., Bujnicki, T., et al., 2017. sFIDA automation yields sub-femtomolar limit of detection for Aβ aggregates in body fluids. Clinical Biochemistry 50 (4-5), 244–247. Available at https://doi.org/10.1016/j.clinbiochem.2016.11.001.

Hertzberg, L., Maggio, N., Muler, I., Yitzhaky, A., Majer, M., Haroutunian, V., et al., 2021. Comprehensive gene expression analysis detects global reduction of proteasome subunits in schizophrenia. Schizophrenia Bulletin 47 (3), 785–795. Available at https://doi.org/10.1093/schbul/sbaa160.

Hong, J.H., Kwak, Y., Woo, Y., Park, C., Lee, S.A., Lee, H., et al., 2016. Regulation of the actin cytoskeleton by the Ndel1-Tara complex is critical for cell migration. Scientific Reports 6. Available at https://doi.org/10.1038/srep31827.

Hou, L., Heilbronner, U., Degenhardt, F., Adli, M., Akiyama, K., Akula, N., et al., 2016. Genetic variants associated with response to lithium treatment in bipolar disorder: a genome-wide association study. Lancet (London, England) 387 (10023), 1085–1093.

Huang, J., Perlis, R.H., Lee, P.H., Rush, A.J., Fava, M., Sachs, G.S., et al., 2010. Cross-disorder genomewide analysis of schizophrenia, bipolar disorder, and depression. American Journal of Psychiatry 167 (10), 1254–1263. Available at https://doi.org/10.1176/appi.ajp.2010.09091335.

Hui, K.K., Endo, R., Sawa, A., Tanaka, M., 2022. A perspective on the potential involvement of impaired proteostasis in neuropsychiatric disorders. Biological Psychiatry 91 (4), 335–345. Available at https://doi.org/10.1016/j.biopsych.2021.09.001.

Ishizuka, K., Kamiya, A., Oh, E.C., Kanki, H., Seshadri, S., Robinson, J.F., et al., 2011. DISC1-dependent switch from progenitor proliferation to migration in the developing cortex. Nature 473 (7345), 92–96. Available at https://doi.org/10.1038/nature09859.

Iwayama, Y., Hattori, E., Maekawa, M., Yamada, K., Toyota, T., Ohnishi, T., et al., 2010. Association analyses between brain-expressed fatty-acid binding protein (FABP) genes and schizophrenia and bipolar disorder. American Journal of Medical Genetics Part B: Neuropsychiatric Genetics 153B (2), 484–493. Available at https://doi.org/10.1002/ajmg.b.31004.

Jacobi, A.A., Halawani, S., Lynch, D.R., Lin, H., 2019. Neuronal serine racemase associates with disrupted-in-schizophrenia-1 and DISC1 agglomerates: implications for schizophrenia. Neuroscience Letters 692, 107–114. Available at https://doi.org/10.1016/j.neulet.2018.10.055.

Jang, S., Choi, B., Lim, C., Kim, M., Lee, J.E., Lee, H., et al., 2024. Neuronal fatty acid-binding protein enhances autophagy and suppresses amyloid-β pathology in a Drosophila model of Alzheimer's disease. PLoS Genetics 20 (11). Available at https://doi.org/10.1371/journal.pgen.1011475.

Jarskog, L.F., Gilmore, J.H., Selinger, E.S., Lieberman, J.A., 2000. Cortical Bcl-2 protein expression and apoptotic regulation in schizophrenia. Biological Psychiatry 48 (7), 641–650. Available at https://doi.org/10.1016/S0006-3223(00)00988-4.

Jarskog, L.F., Selinger, E.S., Lieberman, J.A., Gilmore, J.H., 2004. Apoptotic proteins in the temporal cortex in schizophrenia: high bax/Bcl-2 ratio without caspase-3 activation. American Journal of Psychiatry 161 (1), 109–115. Available at https://doi.org/10.1176/appi.ajp.161.1.109.

Jones, D.K., Johnson, A.C., Roti Roti, E.C., Liu, F., Uelmen, R., Ayers, R.A., et al., 2018. Localization and functional consequences of a direct interaction between TRIOBP-1 and hERG proteins in the heart. Journal of Cell Science 131 (6). Available at https://doi.org/10.1242/jcs.206730.

Juković, M., Ratkaj, I., Kalafatovic, D., Bradshaw, N.J., 2024. Amyloids, amorphous aggregates and assemblies of peptides – assessing aggregation. Biophysical Chemistry 308, 107202. Available at https://doi.org/10.1016/j.bpc.2024.107202.

Kaefer, K., Malagon-Vina, H., Dickerson, D.D., O'Neill, J., Trossbach, S.V., Korth, C., et al., 2019. Disrupted-in-schizophrenia 1 overexpression disrupts hippocampal coding and oscillatory synchronization. Hippocampus 29 (9), 802–816. Available at https://doi.org/10.1002/hipo.23076.

Kamiya, A., Kubo, K.-ichiro, Tomoda, T., Takaki, M., Youn, R., Ozeki, Y., et al., 2005. A schizophrenia-associated mutation of DISC1 perturbs cerebral cortex development. Nature Cell Biology 7 (12), 1167–1178. Available at https://doi.org/10.1038/ncb1328.

Kamnasaran, D., Muir, W.J., Ferguson-Smith, M.A., Cox, D.W., 2003. Disruption of the neuronal PAS3 gene in a family affected with schizophrenia. Journal of Medical Genetics 40 (5), 325–332. Available at https://doi.org/10.1136/jmg.40.5.325.

Kang, H.J., Kawasawa, Y.I., Cheng, F., Zhu, Y., Xu, X., Li, M., et al., 2011. Spatio-temporal transcriptome of the human brain. Nature 478 (7370), 483–489. Available at https://doi.org/10.1038/nature10523.

Katsuno, T., Belyantseva, I.A., Cartagena-Rivera, A.X., Ohta, K., Crump, S.M., Petralia, R.S., et al., 2019. TRIOBP-5 sculpts stereocilia rootlets and stiffens supporting cells enabling hearing. JCI Insight 4 (12). Available at https://doi.org/10.1172/jci.insight.128561.

Kawahata, I., Bousset, L., Melki, R., Fukunaga, K., 2019. Fatty acid-binding protein 3 is critical for α-synuclein uptake and MPP+-induced mitochondrial dysfunction in cultured dopaminergic neurons. International Journal of Molecular Sciences 20 (21), 5358. Available at https://doi.org/10.3390/ijms20215358.

Kawamoto, Y., Tada, M., Asano, T., Nakamura, H., Jitsuki-Takahashi, A., Makihara, H., et al., 2022. Phosphorylated CRMP1, axon guidance protein, is a component of spheroids and is involved in axonal pathology in amyotrophic lateral sclerosis. Frontiers in Neurology 13. Available at https://doi.org/10.3389/fneur.2022.994676.

Kay, S.R., Fiszbein, A., Opler, L.A., 1987. The positive and negative syndrome scale (PANSS) for schizophrenia. Schizophrenia Bulletin 13 (2), 261–276. Available at https://doi.org/10.1093/schbul/13.2.261.

Kazmierczak, M., Kazmierczak, P., Peng, A.W., Harris, S.L., Shah, P., Puel, J.L., et al., 2017. Pejvakin, a candidate stereociliary rootlet protein, regulates hair cell function in a cell-autonomous manner. Journal of Neuroscience 37 (13), 3447–3464. Available at https://doi.org/10.1523/JNEUROSCI.2711-16.2017.

Kipp, M., Clarner, T., Gingele, S., Pott, F., Amor, S., Van Der Valk, P., et al., 2011. Brain lipid binding protein (FABP7) as modulator of astrocyte function. Physiological Research 60 (1), S49.

Kitajiri, S.I., Sakamoto, T., Belyantseva, I.A., Goodyear, R.J., Stepanyan, R., Fujiwara, I., et al., 2010. Actin-bundling protein TRIOBP forms resilient rootlets of hair cell stereocilia essential for hearing. Cell 141 (5), 786–798. Available at https://doi.org/10.1016/j.cell.2010.03.049.

Konradi, C., Eaton, M., MacDonald, M.L., Walsh, J., Benes, F.M., Heckers, S., 2004. Molecular evidence for mitochondrial dysfunction in bipolar disorder. Archives of General Psychiatry 61 (3), 300–308. Available at https://doi.org/10.1001/archpsyc.61.3.300.

Konstantoulea, K., Louros, N., Rousseau, F., Schymkowitz, J., 2022. Heterotypic interactions in amyloid function and disease. The FEBS Journal 289 (8), 2025–2046. Available at https://doi.org/10.1111/febs.15719.

Koromina, M., Ravi, A., Panagiotaropoulou, G., Schilder, B.M., Humphrey, J., Braun, A., et al., 2025. Fine-mapping genomic loci refines bipolar disorder risk genes. Nature Neuroscience 28 (7), 1393–1403. Available at https://doi.org/10.1038/s41593-025-01998-z.

Korth, C., 2012. Aggregated proteins in schizophrenia and other chronic mental diseases: DISC1opathies. Prion 6 (2), 134–141. Available at https://doi.org/10.4161/pri.18989.

Krey, J.F., Liu, C., Belyantseva, I.A., Bateschell, M., Dumont, R.A., Goldsmith, J., et al., 2022. ANKRD24 organizes TRIOBP to reinforce stereocilia insertion points. Journal of Cell Biology 221 (4). Available at https://doi.org/10.1083/jcb.202109134.

Lan, J., Zhu, Y., Xu, L., Yu, H., Yu, J., Liu, X., et al., 2014. The 68-kDa telomeric repeat binding factor 1 (TRF1)-associated protein (TAP68) interacts with and recruits TRF1 to the spindle pole during mitosis. Journal of Biological Chemistry 289 (20), 14145–14156. Available at https://doi.org/10.1074/jbc.m113.526244.

Lashuel, H.A., 2021. Rethinking protein aggregation and drug discovery in neurodegenerative diseases: why we need to embrace complexity? Current Opinion in Chemical Biology 64, 67–75. Available at https://doi.org/10.1016/j.cbpa.2021.05.006.

Leliveld, S.R., Bader, V., Hendriks, P., Prikulis, I., Sajnani, G., Requena, J.R., et al., 2008. Insolubility of disrupted-in-schizophrenia 1 disrupts oligomer-dependent interactions with nuclear distribution element 1 and is associated with sporadic mental disease. Journal of Neuroscience 28 (15), 3839–3845. Available at https://doi.org/10.1523/JNEUROSCI.5389-07.2008.

Leliveld, S.R., Hendriks, P., Michel, M., Sajnani, G., Bader, V., Trossbach, S., et al., 2009. Oligomer assembly of the C-terminal DISC1 domain (640-854) is controlled by self-association motifs and disease-associated polymorphism S704C. Biochemistry 48 (32), 7746–7755. Available at https://doi.org/10.1021/bi900901e.

Li, T., Zhang, F., Liu, X., Sun, X., Sham, P.C., Crombie, C., et al., 2005. Identifying potential risk haplotypes for schizophrenia at the DTNBP1 locus in Han Chinese and Scottish populations. Molecular Psychiatry 10 (11), 1037–1044. Available at https://doi.org/10.1038/sj.mp.4001718.

Liu, J., Wang, Q., Chen, J., 2024. Case report: a de novo variant of CRMP1 in an individual with a neurodevelopmental disorder. Frontiers in Neuroscience 18. Available at https://doi.org/10.3389/fnins.2024.1490731.

Liu, R.-Z., Mita, R., Beaulieu, M., Gao, Z., Godbout, R., 2010. Fatty acid binding proteins in brain development and disease. The International Journal of Developmental Biology 54 (8-9), 1229–1239. Available at https://doi.org/10.1387/ijdb.092976rl.

Liu, Z.Z., Hong, C.G., Hu, W.B., Chen, M.L., Duan, R., Li, H.M., et al., 2021. Autophagy receptor OPTN (optineurin) regulates mesenchymal stem cell fate and bone-fat balance during aging by clearing FABP3. Autophagy 17 (10), 2766–2782. Available at https://doi.org/10.1080/15548627.2020.1839286.

Lukomska, A., Theune, W.C., Xing, J., Frost, M.P., Damania, A., Gupta, M., et al., 2023. Experimental gene expression of developmentally downregulated Crmp1, Crmp4, and Crmp5 promotes axon regeneration and retinal ganglion cell survival after optic nerve injury. Brain Research 1809, 148368. Available at 10.1016/j.brainres.2023.148368.

Ma, T.M., Abazyan, S., Abazyan, B., Nomura, J., Yang, C., Seshadri, S., et al., 2013. Pathogenic disruption of DISC1-serine racemase binding elicits schizophrenia-like behavior via D-serine depletion. Molecular Psychiatry 18 (5), 557–567. Available at https://doi.org/10.1038/mp.2012.97.

Macintyre, G., Alford, T., Xiong, L., Rouleau, G.A., Tibbo, P.G., Cox, D.W., 2010. Association of NPAS3 exonic variation with schizophrenia. Schizophrenia Research 120 (1-3), 143–149. Available at https://doi.org/10.1016/j.schres.2010.04.002.

Maekawa, M., Iwayama, Y., Arai, R., Nakamura, K., Ohnishi, T., Toyota, T., et al., 2010. Polymorphism screening of brain-expressed FABP7, 5 and 3 genes and association studies in autism and schizophrenia in Japanese subjects. Journal of Human Genetics 55 (2), 127–130. Available at https://doi.org/10.1038/jhg.2009.133.

Makihara, H., Nakai, S., Ohkubo, W., Yamashita, N., Nakamura, F., Kiyonari, H., et al., 2016. CRMP1 and CRMP2 have synergistic but distinct roles in dendritic development. Genes to Cells: Devoted to Molecular & Cellular Mechanisms 21 (9), 994–1005. Available at https://doi.org/10.1111/gtc.12399.

Mas, P.J., Hart, D.J., 2017. ESPRIT: a method for defining soluble expression constructs in poorly understood gene sequences. In: Burgess-Brown, N. (Ed.), Heterologous Gene Expression in E.coli. Methods in Molecular Biology, vol 1586. Humana Press, New York, NY, pp. 45–63. Available at https://doi.org/10.1007/978-1-4939-6887-9_4.

Matsuo, K., Yabuki, Y., Melki, R., Bousset, L., Owada, Y., Fukunaga, K., 2021. Crucial role of FABP3 in αSyn-induced reduction of septal GABAergic neurons and cognitive decline in mice. International Journal of Molecular Sciences 22 (1), 400. Available at https://doi.org/10.3390/ijms22010400.

McIntosh, A.M., Lewis, C.M., Adams, M.J, 2024. Genome-wide study of half a million individuals with major depression identifies 697 independent associations, infers causal neuronal subtypes and biological targets for novel pharmacotherapies. medRxiv [Preprint] 2024. Available at https://doi.org/10.1101/2024.04.29.24306535.

Merikangas, K.R., Akiskal, H.S., Angst, J., Greenberg, P.E., Hirschfeld, R.M.A., Petukhova, M., et al., 2007. Lifetime and 12-month prevalence of bipolar spectrum disorder in the National Comorbidity Survey replication. Archives of General Psychiatry 64 (5), 543–552.

Michaelson, J.J., Shin, M.K., Koh, J.Y., Brueggeman, L., Zhang, A., Katzman, A., et al., 2017. Neuronal PAS domain proteins 1 and 3 are master regulators of neuropsychiatric risk genes. Biological Psychiatry 82 (3), 213–223. Available at https://doi.org/10.1016/j.biopsych.2017.03.021.

Middleton, F.A., Mirnics, K., Pierri, J.N., Lewis, D.A., Levitt, P., 2002. Gene expression profiling reveals alterations of specific metabolic pathways in schizophrenia. Journal of Neuroscience 22 (7), 2718–2729. Available at https://doi.org/10.1523/jneurosci.22-07-02718.2002.

Mihaljevic, M., Lam, M., Ayala-Grosso, C., Davis-Batt, F., Schretlen, D.J., Ishizuka, K., et al., 2023. Olfactory neuronal cells as a promising tool to realize the "druggable genome" approach for drug discovery in neuropsychiatric disorders. Frontiers in Neuroscience 16. Available at https://doi.org/10.3389/fnins.2022.1081124.

Millar, J.K., Christie, S., Porteous, D.J., 2003. Yeast two-hybrid screens implicate DISC1 in brain development and function. Biochemical and Biophysical Research Communications 311 (4), 1019–1025. Available at https://doi.org/10.1016/j.bbrc.2003.10.101.

Millar, J.K., Brown, J., Maule, J.C., Shibasaki, Y., Christie, S., Lawson, D., et al., 1998. A long-range restriction map across 3 Mb of the chromosome 11 breakpoint region of a translocation linked to schizophrenia: localization of the breakpoint and the search for neighbouring genes. Psychiatric Genetics 8 (3), 175–181. Available at https://doi.org/10.1097/00041444-199800830-00007.

Millar, J.K., Wilson-Annan, J.C., Anderson, S., Christie, S., Taylor, M.S., Semple, C.A.M., et al., 2000. Disruption of two novel genes by a translocation co-segregating with schizophrenia. Human Molecular Genetics 9 (9), 1415–1423. Available at https://doi.org/10.1093/hmg/9.9.1415.

Millar, J.K., Christie, S., Anderson, S., Lawson, D., Loh, D.H.-W., Devon, R.S., et al., 2001. Genomic structure and localisation within a linkage hotspot of disrupted in schizophrenia 1, a gene disrupted by a translocation segregating with schizophrenia. Molecular Psychiatry 6 (2), 173–178. Available at https://doi.org/10.1038/sj.mp.4000784.

Millar, J.K., Pickard, B.S., Mackie, S., James, R., Christie, S., Buchanan, S.R., et al., 2005. Genetics: DISC1 and PDE4B are interacting genetic factors in schizoprenia that regulate cAMP signaling. Science (New York, N.Y.) 310 (5751), 1187–1191. Available at https://doi.org/10.1126/science.1112915.

Morris, J.A., Kandpal, G., Ma, L., Austin, C.P., 2003. DISC1 (Disrupted-In-Schizophrenia 1) is a centrosome-associated protein that interacts with MAP1A, MIPT3, ATF4/5 and NUDEL: regulation and loss of interaction with mutation. Human Molecular Genetics 12 (13), 1591–1608. Available at https://doi.org/10.1093/hmg/ddg162.

Muir, W.J., Gosden, C.M., Brookes, A.J., Fantes, J., Evans, K.L., Maguire, S.M., et al., 1995. Direct microdissection and microcloning of a translocation breakpoint region, t(1;11) (q42.2;q21), associated with schizophrenia. Cytogenetics and Cell Genetics 70 (1-2), 35–40. Available at https://doi.org/10.1159/000133986.

Mullin, A.P., Gokhale, A., Larimore, J., Faundez, V., 2011. Cell biology of the BLOC-1 complex subunit dysbindin, a schizophrenia susceptibility gene. Molecular Neurobiology 44 (1), 53–64. Available at https://doi.org/10.1007/s12035-011-8183-3.

Nakata, K., Lipska, B.K., Hyde, T.M., Ye, T., Newburn, E.N., Morita, Y., et al., 2009. DISC1 splice variants are upregulated in schizophrenia and associated with risk polymorphisms. Proceedings of the National Academy of Sciences of the United States of America 106 (37), 15873–15878. Available at https://doi.org/10.1073/pnas.0903413106.

Nierenberg, A.A., Agustini, B., Köhler-Forsberg, O., Cusin, C., Katz, D., Sylvia, L.G., et al., 2023. Diagnosis and treatment of bipolar disorder: a review. JAMA: the Journal of the American Medical Association 330 (14), 1370–1380. Available at https://doi.org/10.1001/jama.2023.18588.

Nucifora, L.G., Wu, Y.W.C., Lee, B.J., Sha, L., Margolis, R.L., Ross, C.A., et al., 2016. A mutation in NPAS3 that segregates with schizophrenia in a small family leads to protein aggregation. Complex Psychiatry 2 (3), 133–144. Available at https://doi.org/10.1159/000447358.

Nucifora, L.G., MacDonald, M.L., Lee, B.J., Peters, M.E., Norris, A.L., Orsburn, B.C., et al., 2019. Increased protein insolubility in brains from a subset of patients with schizophrenia. American Journal of Psychiatry 176 (9), 730–743. Available at https://doi.org/10.1176/appi.ajp.2019.18070864.

Nucifora, L.G., Ishizuka, K., El Demerdash, N., Lee, B.J., Imai, M.T., Ayala-Grosso, C., et al., 2025. Protein aggregation identified in olfactory neuronal cells is associated with cognitive impairments in a subset of living schizophrenia patients. Molecular Psychiatry 30 (8), 3673–3685. Available at https://doi.org/10.1038/s41380-025-02956-8.

Nurnberger, J.I., Koller, D.L., Jung, J., Edenberg, H.J., Foroud, T., Guella, I., et al., 2014. Identification of pathways for bipolar disorder: a meta-analysis. JAMA Psychiatry 71 (6), 657–664. Available at https://doi.org/10.1001/jamapsychiatry.2014.176.

Ochneva, A., Zorkina, Y., Abramova, O., Pavlova, O., Ushakova, V., Morozova, A., et al., 2022. Protein misfolding and aggregation in the brain: common pathogenetic pathways in neurodegenerative and mental disorders. International Journal of Molecular Sciences 23 (22), 14498. Available at https://doi.org/10.3390/ijms232214498.

Oizumi, H., Yamasaki, K., Suzuki, H., Hasegawa, T., Sugimura, Y., Baba, T., et al., 2021. Fatty acid-binding protein 3 expression in the brain and skin in human synucleinopathies. Frontiers in Aging Neuroscience 13. Available at https://doi.org/10.3389/fnagi.2021.648982.

Ottis, P., Bader, V., Trossbach, S.V., Kretzschmar, H., Michel, M., Leliveld, S.R., et al., 2011. Convergence of two independent mental disease genes on the protein level: Recruitment of dysbindin to cell-invasive disrupted-in-schizophrenia 1 aggresomes. Biological Psychiatry 70 (7), 604–610. Available at https://doi.org/10.1016/j.biopsych.2011.03.027.

Oyama, S., Yamakawa, H., Sasagawa, N., Hosoi, Y., Futai, E., Ishiura, S., et al., 2009. Dysbindin-1, a schizophrenia-related protein, functionally interacts with the DNA- dependent protein kinase complex in an isoform-dependent manner. PLoS One 4 (1), e4199. Available at https://doi.org/10.1371/journal.pone.0004199.

Pickard, B.S., Malloy, M.P., Porteous, D.J., Blackwood, D.H.R., Muir, W.J., 2005. Disruption of a brain transcription factor, NPAS3, is associated with schizophrenia amid learning disability. American Journal of Medical Genetics - Neuropsychiatric Genetics 136 (1), 26–32. Available at https://doi.org/10.1002/ajmg.b.30204.

Pickard, B.S., Christoforou, A., Thomson, P.A., Fawkes, A., Evans, K.L., Morris, S.W., et al., 2009. Interacting haplotypes at the NPAS3 locus alter risk of schizophrenia and bipolar disorder. Molecular Psychiatry 14 (9), 874–884. Available at https://doi.org/10.1038/mp.2008.24.

Pils, M., Rutsch, J., Eren, F., Engberg, G., Piehl, F., Cervenka, S., et al., 2023. Disrupted-in-schizophrenia 1 (DISC1) protein aggregates in cerebrospinal fluid are elevated in first-episode psychosis patients. medRxiv. Available at https://doi.org/10.1101/2023.04.17.23288687.

Porteous, D.J., Thomson, P.A., Millar, J.K., Evans, K.L., Hennah, W., Soares, D.C., et al., 2014. DISC1 as a genetic risk factor for schizophrenia and related major mental illness: response to Sullivan. Molecular Psychiatry 19 (2), 141–143. Available at https://doi.org/10.1038/mp.2013.160.

Pavešić Radonja, A., Blažević Zelić, S., Rubeša, G., Bradshaw, N.J., 2023. Symptom severity in schizophrenia patients with NPAS3, dysbindin-1 and/or TRIOBP protein pathology in their blood serum: a PANSS-based follow up study. Psychiatria Danubina 35 (2), 180–186. Available at https://doi.org/10.24869/psyd.2023.180.

Ravindran, E., Arashiki, N., Becker, L.-L., Takizawa, K., Lévy, J., Rambaud, T., et al., 2022. Monoallelic CRMP1 gene variants cause neurodevelopmental disorder. eLife 11. Available at https://doi.org/10.7554/elife.80793.

Riazuddin, S., Khan, S.N., Ahmed, Z.M., Ghosh, M., Caution, K., Nazli, S., et al., 2006. Mutations in TRIOBP, which encodes a putative cytoskeletal-organizing protein, are associated with nonsyndromic recessive deafness. American Journal of Human Genetics 78 (1), 137–143. Available at https://doi.org/10.1086/499164.

Robinson, N., Ploner, A., Leone, M., Lichtenstein, P., Kendler, K.S., Bergen, S.E., 2023. Impact of early-life factors on risk for schizophrenia and bipolar disorder. Schizophrenia Bulletin 49 (3), 768–777. Available at https://doi.org/10.1093/schbul/sbac205.

Rubio, M.D., Wood, K., Haroutunian, V., Meador-Woodruff, J.H., 2013. Dysfunction of the ubiquitin proteasome and ubiquitin-like systems in schizophrenia. Neuropsychopharmacology: Official Publication of the American College of Neuropsychopharmacology 38 (10), 1910–1920. Available at https://doi.org/10.1038/npp.2013.84.

Sachs, N.A., Sawa, A., Holmes, S.E., Ross, C.A., DeLisi, L.E., Margolis, R.L., 2005. A frameshift mutation in disrupted in schizophrenia 1 in an American family with schizophrenia and schizoaffective disorder. Molecular Psychiatry 10 (8), 758–764. Available at https://doi.org/10.1038/sj.mp.4001667.

Samardžija, B., Pavešić Radonja, A., Zaharija, B., Bergman, M., Renner, É., Palkovits, M., et al., 2021. Protein aggregation of NPAS3, implicated in mental illness, is not limited to the V304I mutation. Journal of Personalized Medicine 11 (11), 1070. Available at https://doi.org/10.3390/jpm11111070.

Samardžija, B., Juković, M., Zaharija, B., Renner, É., Palkovits, M., Bradshaw, N.J., 2023. Co-aggregation and parallel aggregation of specific proteins in major mental illness. Cells 12 (14), 1848. Available at https://doi.org/10.3390/cells12141848.

Samardžija, B., Petrović, M., Zaharija, B., Medija, M., Meštrović, A., Bradshaw, N.J., et al., 2024. Showing effects on social interaction networks. Current Issues in Molecular Biology 46 (8), 8526–8549. Available at https://doi.org/10.3390/cimb46080502.

Sanchez-Pulido, L., Ponting, C.P., 2011. Structure and evolutionary history of DISC1. Human Molecular Genetics 20 (R2), R175. Available at https://doi.org/10.1093/hmg/ddr374.

Sartorius, N., 2007. Physical illness in people with mental disorders. World Psychiatry: Official Journal of the World Psychiatric Association (WPA) 6 (1), 3–4.

Schlechter, P., Rohde, P., Seeley, J.R., Klein, D.N., Olino, T.M., 2024. Examining the influence of episode number and age of onset on individual depressive symptoms across episodes of major depression. Journal of Psychiatric Research 175, 405–410. Available at https://doi.org/10.1016/j.jpsychires.2024.05.039.

Scott, M.R., Meador-Woodruff, J.H., 2020. Intracellular compartment-specific proteasome dysfunction in postmortem cortex in schizophrenia subjects. Molecular Psychiatry 25 (4), 776–790. Available at https://doi.org/10.1038/s41380-019-0359-7.

Scott, M.R., Rubio, M.D., Haroutunian, V., Meador-Woodruff, J.H., 2016. Protein expression of proteasome subunits in elderly patients with schizophrenia. Neuropsychopharmacology: Official Publication of the American College of Neuropsychopharmacology 41 (3), 896–905. Available at https://doi.org/10.1038/npp.2015.219.

Seidisarouei, M., Schäble, S., van Wingerden, M., Trossbach, S.V., Korth, C., Kalenscher, T., 2022. Social anhedonia as a disrupted-in-schizophrenia 1-dependent phenotype. Scientific Reports 12, 10182.

Seipel, K., O'Brien, S.P., Iannotti, E., Medley, Q.G., Streuli, M., 2001. Tara, a novel F-actin binding protein, associates with the Trio guanine nucleotide exchange factor and regulates actin cytoskeletal organization. Journal of Cell Science 114 (2), 389–399. Available at https://doi.org/10.1242/jcs.114.2.389.

Sha, L., MacIntyre, L., MacHell, J.A., Kelly, M.P., Porteous, D.J., Brandon, N.J., et al., 2012. Transcriptional regulation of neurodevelopmental and metabolic pathways by NPAS3. Molecular Psychiatry 17 (3), 267–279. Available at https://doi.org/10.1038/mp.2011.73.

Shahin, H., Walsh, T., Sobe, T., Abu Sa'ed, J., Abu Rayan, A., Lynch, E.D., et al., 2006. Mutations in a novel isoform of TRIOBP that encodes a filamentous-actin binding protein are responsible for DFNB28 recessive nonsyndromic hearing loss. The American Journal of Human Genetics 78 (1), 144–152. Available at https://doi.org/10.1086/499495.

Shao, L., Lu, B., Wen, Z., Teng, S., Wang, L., Zhao, Y., et al., 2017. Disrupted-in-schizophrenia-1 (DISC1) protein disturbs neural function in multiple disease-risk pathways. Human Molecular Genetics 26 (14), 2634–2648. Available at https://doi.org/10.1093/hmg/ddx147.

Shimamoto, C., Ohnishi, T., Maekawa, M., Watanabe, A., Ohba, H., Arai, R., et al., 2014. Functional characterization of FABP3, 5 and 7 gene variants identified in schizophrenia and autism spectrum disorder and mouse behavioral studies. Human Molecular Genetics 23 (24), 6495–6511. Available at https://doi.org/10.1093/hmg/ddu369.

Shioda, N., Yabuki, Y., Kobayashi, Y., Onozato, M., Owada, Y., Fukunaga, K., 2014. FABP3 protein promotes α-synuclein oligomerization associated with 1-methyl-1,2,3,6-tetrahydropiridine-induced neurotoxicity. Journal of Biological Chemistry 289 (27), 18957–18965. Available at https://doi.org/10.1074/jbc.m113.527341.

Sialana, F.J., Wang, A.-L., Fazari, B., Kristofova, M., Smidak, R., Trossbach, S.V., et al., 2018. Quantitative proteomics of synaptosomal fractions in a rat overexpressing human DISC1 gene indicates profound synaptic dysregulation in the dorsal striatum. Frontiers in Molecular Neuroscience 11. Available at https://doi.org/10.3389/FNMOL.2018.00026.

Sjöstedt, E., Zhong, W., Fagerberg, L., Karlsson, M., Mitsios, N., Adori, C., et al., 2020. An atlas of the protein-coding genes in the human, pig, and mouse brain. Science (New York, N.Y.) 367 (6482). Available at https://doi.org/10.1126/science.aay5947.

Smathers, R.L., Petersen, D.R., 2011. The human fatty acid-binding protein family: evolutionary divergences and functions. Human Genomics 5 (3), 170–191. Available at https://doi.org/10.1186/1479-7364-5-3-170.

Soares, D.C., Carlyle, B.C., Bradshaw, N.J., Porteous, D.J., 2011. DISC1: structure, function, and therapeutic potential for major mental illness. ACS Chemical Neuroscience 2 (11), 609–632. Available at https://doi.org/10.1021/cn200062k.

Solmi, M., Seitidis, G., Mavridis, D., Correll, C.U., Dragioti, E., Guimond, S., et al., 2023. Incidence, prevalence, and global burden of schizophrenia - data, with critical appraisal, from the Global Burden of Disease (GBD) 2019. Molecular Psychiatry 28 (12), 5319–5327. Available at https://doi.org/10.1038/s41380-023-02138-4.

St Clair, D., Blackwood, D., Muir, W., Walker, M., St Clair, D., Muir, W., et al., 1990. Association within a family of a balanced autosomal translocation with major mental illness. The Lancet 336 (8706), 13–16. Available at https://doi.org/10.1016/0140-6736(90)91520-k.

Straub, R.E., Jiang, Y., MacLean, C.J., Ma, Y., Webb, B.T., Myakishev, M.V., et al., 2002. Genetic variation in the 6p22.3 Gene DTNBP1, the human ortholog of the mouse dysbindin gene, is associated with schizophrenia. American Journal of Human Genetics 71 (2), 337–348. Available at https://doi.org/10.1086/341750.

Stroedicke, M., Bounab, Y., Strempel, N., Klockmeier, K., Yigit, S., Friedrich, R.P., et al., 2015. Systematic interaction network filtering identifies CRMP1 as a novel suppressor of huntingtin misfolding and neurotoxicity. Genome Research 25 (5), 701–713. Available at https://doi.org/10.1101/gr.182444.114.

Sullivan, P.F., 2013. Questions about DISC1 as a genetic risk factor for schizophrenia. Molecular Psychiatry 18 (10), 1050–1052. Available at https://doi.org/10.1038/mp.2012.182.

Tajinda, K., Ishizuka, K., Colantuoni, C., Morita, M., Winicki, J., Le, C., et al., 2010. Neuronal biomarkers from patients with mental illnesses: a novel method through nasal biopsy combined with laser-captured microdissection. Molecular Psychiatry 15 (3), 231–232. Available at https://doi.org/10.1038/mp.2009.73.

Talbot, K., Ong, W.- Y., Blake, D.J., Tang, J., Louneva, N., Carlson, G.C., et al., 2009. Dysbindin-1 and its protein family. In: Lajtha, A., Javitt, D., Kantrowitz, J. (Eds.), Handbook of Neurochemistry and Molecular Neurobiology. Springer, Boston, MA, pp. 107–241. Available at https://doi.org/10.1007/978-0-387-30410-6_5.

Tanaka, M., Ishizuka, K., Nekooki-Machida, Y., Endo, R., Takashima, N., Sasaki, H., et al., 2017. Aggregation of scaffolding protein DISC1 dysregulates phosphodiesterase 4 in Huntington's disease. Journal of Clinical Investigation 127 (4), 1438–1450. Available at https://doi.org/10.1172/JCI85594.

Tankou, S., Ishii, K., Elliott, C., Yalla, K.C., Day, J.P., Furukori, K., et al., 2016. SUMOylation of DISC1: a potential role in neural progenitor proliferation in the developing cortex. Complex Psychiatry 2 (1), 20–27. Available at https://doi.org/10.1159/000444257.

Teng, S., Thomson, P.A., McCarthy, S., Kramer, M., Muller, S., Lihm, J., et al., 2018. Rare disruptive variants in the DISC1 interactome and regulome: association with cognitive ability and schizophrenia. Molecular Psychiatry 23 (5), 1270–1277. Available at https://doi.org/10.1038/mp.2017.115.

Thomson, P.A., Malavasi, E.L.V., Grünewald, E., Soares, D.C., Borkowska, M., Millar, J.K., 2013. DISC1 genetics, biology and psychiatric illness. Frontiers in Biology 8 (1), 1–31. Available at https://doi.org/10.1007/s11515-012-1254-7.

Tomasik, J., Schwarz, E., Lago, S.G., Rothermundt, M., Leweke, F.M., van Beveren, N.J.M., Guest, P.C., Rahmoune, H., Steiner, J., Bahn, S., 2016. Pretreatment levels of the fatty acid handling proteins H-FABP and CD36 predict response to olanzapine in recent-onset schizophrenia patients. Brain, Behavior, and Immunity 52, 178–186. Available at https://doi.org/10.1016/j.bbi.2015.10.019.

Tomppo, L., Hennah, W., Miettunen, J., Jarvelin, M.R., Veijola, J., Ripatti, S., et al., 2009. Association of variants in DISC1 with psychosis-related traits in a large population cohort. Archives of General Psychiatry 66 (2), 134–141. Available at https://doi.org/10.1001/archgenpsychiatry.2008.524.

Torrey, E.F., 2011. Stigma and violence: isn't it time to connect the dots? Schizophrenia Bulletin 37 (5), 892–896. Available at https://doi.org/10.1093/schbul/sbr057.

Toyoshima, M., Jiang, X., Ogawa, T., Ohnishi, T., Yoshihara, S., Balan, S., et al., 2019. Enhanced carbonyl stress induces irreversible multimerization of CRMP2 in schizophrenia pathogenesis. Life Science Alliance 2 (5), e201900478. Available at https://doi.org/10.26508/lsa.201900478.

Trossbach, S.V., Bader, V., Hecher, L., Pum, M.E., Masoud, S.T., Prikulis, I., et al., 2016. Misassembly of full-length Disrupted-in-Schizophrenia 1 protein is linked to altered dopamine homeostasis and behavioral deficits. Molecular Psychiatry 21 (11), 1561–1572. Available at https://doi.org/10.1038/mp.2015.194.

Trossbach, S.V., Hecher, L., Schafflick, D., Deenen, R., Popa, O., Lautwein, T., et al., 2019. Dysregulation of a specific immune-related network of genes biologically defines a subset of schizophrenia. Translational Psychiatry 9 (1). Available at https://doi.org/10.1038/s41398-019-0486-6.

Trubetskoy, V., Pardiñas, A.F., Qi, T., Panagiotaropoulou, G., Awasthi, S., Bigdeli, T.B., et al., 2022. Mapping genomic loci implicates genes and synaptic biology in schizophrenia. Nature 604 (7906), 502–508. Available at https://doi.org/10.1038/s41586-022-04434-5.

Uchida, Y., Ohshima, T., Sasaki, Y., Suzuki, H., Yanai, S., Yamashita, N., et al., 2005. Semaphorin3A signalling is mediated via sequential Cdk5 and GSK3beta phosphorylation of CRMP2: implication of common phosphorylating mechanism underlying axon guidance and Alzheimer's disease. Genes to Cells: Devoted to Molecular & Cellular Mechanisms 10 (2), 165–179.

Uhlén, M., Fagerberg, L., Hallström, B.M., Lindskog, C., Oksvold, P., Mardinoglu, A., et al., 2015. Tissue-based map of the human proteome. Science (New York, N.Y.) 347 (6220). Available at https://doi.org/10.1126/science.1260419.

Van Den Bogaert, A., Schumacher, J., Schulze, T.G., Otte, A.C., Ohlraun, S., Kovalenko, S., et al., 2003. The DTNBP1 (dysbindin) gene contributes to schizophrenia, depending on family history of the disease. The American Journal of Human Genetics 73 (6), 1438–1443. Available at https://doi.org/10.1086/379928.

Vawter, M.P., Barrett, T., Cheadle, C., Sokolov, B.P., Wood, W.H., Donovan, D.M., et al., 2001. Application of cDNA microarrays to examine gene expression differences in schizophrenia. Brain Research Bulletin 55 (5), 641–650. Available at https://doi.org/10.1016/S0361-9230(01)00522-6.

Vawter, M.P., Crook, J.M., Hyde, T.M., Kleinman, J.E., Weinberger, D.R., Becker, K.G., et al., 2002. Microarray analysis of gene expression in the prefrontal cortex in schizophrenia: a preliminary study. Schizophrenia Research 58 (1), 11–20. Available at https://doi.org/10.1016/S0920-9964(01)00377-2.

Vendruscolo, M., Fuxreiter, M., 2022. Protein condensation diseases: therapeutic opportunities. Nature Communications 13, 5550. Available at https://doi.org/10.1038/s41467-022-32940-7.

Wang, A.L., Chao, O.Y., Nikolaus, S., Lamounier-Zepter, V., Hollenberg, C.P., Lubec, G., et al., 2022. Disrupted-in-schizophrenia 1 protein misassembly impairs cognitive flexibility and social behaviors in a transgenic rat model. Neuroscience 493, 41–51. Available at https://doi.org/10.1016/j.neuroscience.2022.04.013.

Wang, L., Zhao, W., Xia, C., Ma, S., Li, Z., Wang, N., et al., 2024. TRIOBP modulates β-catenin signaling by regulation of miR-29b in idiopathic pulmonary fibrosis. Cellular and Molecular Life Sciences 81, 13. Available at https://doi.org/10.1007/s00018-023-05080-4.

Wang, X., Ye, F., Wen, Z., Guo, Z., Yu, C., Huang, W.K., et al., 2021. Structural interaction between DISC1 and ATF4 underlying transcriptional and synaptic dysregulation in an iPSC model of mental disorders. Molecular Psychiatry 26 (4), 1346–1360. Available at https://doi.org/10.1038/s41380-019-0485-2.

Ward, H.B., Nemeroff, C.B., Carpenter, L., Grzenda, A., McDonald, W.M., Rodriguez, C.I., et al., 2023. Substance use disorders in schizophrenia: prevalence, etiology, biomarkers, and treatment. Personalized Medicine in Psychiatry 39-40, 100106. Available at https://doi.org/10.1016/j.pmip.2023.100106.

Watanabe, A., Toyota, T., Owada, Y., Hayashi, T., Iwayama, Y., Matsumata, M., et al., 2007. Fabp7 maps to a quantitative trait locus for a schizophrenia endophenotype. PLoS Biology 5 (11), e297. Available at https://doi.org/10.1371/journal.pbio.0050297.

Weber, H., Kittel-Schneider, S., Gessner, A., Domschke, K., Neuner, M., Jacob, C.P., et al., 2011. Cross-disorder analysis of bipolar risk genes: further evidence of dGKH as a risk gene for bipolar disorder, but also unipolar depression and adult ADHD. Neuropsychopharmacology: Official Publication of the

American College of Neuropsychopharmacology 36 (10), 2076–2085. Available at https://doi.org/10.1038/npp.2011.98.

WHO, 2025. International Classification of Diseases for Mortality and Morbitity Statistics. World Health Organization.

Wong, J., Duncan, C.E., Beveridge, N.J., Webster, M.J., Cairns, M.J., Shannon Weickert, C., 2013. Expression of NPAS3 in the human cortex and evidence of its posttranscriptional regulation by miR-17 during development, with implications for schizophrenia. Schizophrenia Bulletin 39 (2), 396–406. Available at https://doi.org/10.1093/schbul/sbr177.

Woo, Y., Kim, S.J., Suh, B.K., Kwak, Y., Jung, H.J., Nhung, T.T.M., et al., 2019. Sequential phosphorylation of NDEL1 by the DYRK2-GSK3β complex is critical for neuronal morphogenesis. eLife 8. Available at https://doi.org/10.7554/eLife.50850.

Xu, Y., Sun, Y., Ye, H., Zhu, L., Liu, J., Wu, X., et al., 2015. Increased dysbindin-1B isoform expression in schizophrenia and its propensity in aggresome formation. Cell Discovery 1 (1). Available at https://doi.org/10.1038/celldisc.2015.32.

Yabuki, Y., Matsuo, K., Kawahata, I., Fukui, N., Mizobata, T., Kawata, Y., et al., 2020. Fatty acid binding protein 3 enhances the spreading and toxicity of α-synuclein in mouse brain. International Journal of Molecular Sciences 21 (6), 2230. Available at https://doi.org/10.3390/ijms21062230.

Yamane, M., Yamashita, N., Hida, T., Kamiya, Y., Nakamura, F., Kolattukudy, P., et al., 2017. 1.7 for retrograde propagation of semaphorin3A signaling. Journal of Cell Science 130 (8), 1393–1403. Available at https://doi.org/10.1242/jcs.199737.

Yang, D., Zhang, W., Padhiar, A., Yue, Y., Shi, Y., Zheng, T., et al., 2016a. NPAS3 regulates transcription and expression of VGF: implications for neurogenesis and psychiatric disorders. Frontiers in Molecular Neuroscience 9 (2016). Available at https://doi.org/10.3389/fnmol.2016.00109.

Yang, W., Xu, Q., 2017. Hippocampal apoptosis and cognitive function in dysbindin-1B +/+ mice. Acta Academiae Medicinae Sinicae 39 (3), 307–311. Available at https://doi.org/10.3881/j.issn.1000-503X.2017.03.002.

Yang, W., Zhu, C., Shen, Y., Xu, Q., 2016b. The pathogenic mechanism of dysbindin-1B toxic aggregation: BLOC-1 and intercellular vesicle trafficking. Neuroscience 333, 78–91. Available at https://doi.org/10.1016/j.neuroscience.2016.07.008.

Yano, T., Yamazaki, Y., Adachi, M., Okawa, K., Fort, P., Uji, M., et al., 2011. Tara up-regulates E-cadherin transcription by binding to the Trio RhoGEF and inhibiting Rac signaling. Journal of Cell Biology 193 (2), 319–332. Available at https://doi.org/10.1083/jcb.201009100.

Ye, F., Kang, E., Yu, C., Qian, X., Jacob, F., Yu, C., et al., 2017. DISC1 regulates neurogenesis via modulating kinetochore attachment of Ndel1/Nde1 during mitosis. Neuron 96 (5), 1041. Available at https://doi.org/10.1016/j.neuron.2017.10.010.

Yerabham, A.S.K., Weiergräber, O.H., Bradshaw, N.J., Korth, C., 2013. Revisiting disrupted-in-schizophrenia 1 as a scaffold protein. Biological Chemistry 394 (11), 1425–1437. Available at https://doi.org/10.1515/hsz-2013-0178.

Yerabham, A.S.K., Mas, P.J., Decker, C., Soares, D.C., Weiergräber, O.H., Nagel-Steger, L., et al., 2017. A structural organization for the disrupted in schizophrenia 1 protein, identified by high-throughput screening, reveals distinctly folded regions, which are bisected by mental illness-related mutations. Journal of Biological Chemistry 292 (16), 6468–6477. Available at https://doi.org/10.1074/jbc.M116.773903.

Yu, L., Arbez, N., Nucifora, L.G., Sell, G.L., DeLisi, L.E., Ross, C.A., et al., 2014. A mutation in NPAS3 segregates with mental illness in a small family. Molecular Psychiatry 19 (1), 7–8. Available at https://doi.org/10.1038/mp.2012.192.

Yumerefendi, H., Tarendeau, F., Mas, P.J., Hart, D.J., 2010. ESPRIT: An automated, library-based method for mapping and soluble expression of protein domains from challenging targets. Journal of Structural Biology 172 (1), 66–74. Available at https://doi.org/10.1016/j.jsb.2010.02.021.

Zaharija, B., Bradshaw, N.J., 2024. Aggregation of disrupted in schizophrenia 1 arises from a central region of the protein. Progress in Neuro-Psychopharmacology and Biological Psychiatry 130, 110923. Available at https://doi.org/10.1016/j.pnpbp.2023.110923.

Zaharija, B., Samardžija, B., Bradshaw, N.J., 2020. The TRIOBP isoforms and their distinct roles in actin stabilization, deafness, mental illness, and cancer. Molecules (Basel, Switzerland) 25 (21), 4967. Available at https://doi.org/10.3390/molecules25214967.

Zaharija, B., Odorčić, M., Hart, A., Samardžija, B., Marreiros, R., Prikulis, I., et al., 2022. TRIOBP-1 protein aggregation exists in both major depressive disorder and schizophrenia, and can occur through two distinct regions of the protein. International Journal of Molecular Sciences 23 (19), 11048. Available at https://doi.org/10.3390/ijms231911048.

Zhang, Z., Liu, L., Zhang, H., Li, C., Chen, Y., Zhang, J., et al., 2022. The genetic structure of pain in depression patients: a genome-wide association study and proteome-wide association study. Journal of Psychiatric Research 156, 547–556. Available at https://doi.org/10.1016/j.jpsychires.2022.10.059.

Zhong, F.F., Wei, B., Bao, G.X., Lou, Y.P., Wei, M.E., Wang, X.Y., et al., 2024. FABP3 induces mitochondrial autophagy to promote neuronal cell apoptosis in brain ischemia-reperfusion injury. Neurotoxicity Research 42 (4). Available at https://doi.org/10.1007/s12640-024-00712-4.

Zhu, C.-Y., Shen, Y., Xu, Q., 2015. Propagation of dysbindin-1B aggregates: exosome-mediated transmission of neurotoxic deposits. Neuroscience 291, 301–316. Available at https://doi.org/10.1016/j.neuroscience.2015.02.016.

Zhu, Y., Wang, C., Lan, J., Yu, J., Jin, C., Huang, H., 2012. Phosphorylation of Tara by Plk1 is essential for faithful chromosome segregation in mitosis. Experimental Cell Research 318 (18), 2344–2352. Available at https://doi.org/10.1016/j.yexcr.2012.07.001.

CHAPTER 4

Protein aggregation in progressive myoclonus epilepsies: focus on progressive myoclonic epilepsy type 1

Eva Žerovnik[1,2]
[1]Department of Biochemistry and Molecular and Structural Biology, Jožef Stefan Institute, Ljubljana, Slovenia,
[2]Jožef Stefan International Postgraduate School (IPS), Ljubljana, Slovenia

4.1 Introduction

Proteinopathies are diseases of protein misfolding and aggregation. Misfolded proteins form amyloid-type aggregates or various condensed bodies (Vendruscolo and Fuxreiter, 2022). Proteostasis is the balance between protein synthesis, folding, misfolding, aggregation (Hipp et al., 2014), and finally degradation by the ubiquitin-proteasome system or autophagy. Impairment of the mitochondrial system can also occur, providing less energy and raising oxidative stress (OS). With the gut-brain axis imbalance, the brain becomes affected with neuroinflammation, OS, cytokine release, and glial cell activation. This latter causes excitotoxicity and hyperexcitability of neurons. A subclass of proteinopathies is neurodegenerative disease (ND), such as Alzheimer's, Huntington's, and Parkinson's disease. Progressive myoclonus epilepsies (PMEs) resemble proteinopathies, often characterized by autophagy and mitochondrial systems impairments (Sanz and Serratosa, 2020). We assume that protein aggregates and various inclusions observed in these types of epileptic syndromes contribute to a "gain in toxic function" as also observed in other neurodegenerative conditions. As noted, these commonly observed biological processes impacted are: OS, mitochondrial function, and autophagy. As for the "loss of function" of the genes affected, either by their reduced expression or mutations promoting aggregation, even though they have different functions, they mediate common molecular and cellular processes across PMEs.

4.2 What can we learn from familial forms of progressive myoclonus epilepsies?

There are several types of PMEs. The best known PMEs are Lafora disease (LD) and Unverricht–Lundborg disease (ULD), the latter also called progressive myoclonic

Molecular Insights into Proteinopathies
DOI: https://doi.org/10.1016/B978-0-443-34176-2.00007-9
© 2027 Elsevier Inc. All rights are reserved, including those for text and data mining, AI training, and similar technologies.

epilepsy type 1 (EPM1). Other, less common forms of PMEs include mitochondrial encephalopathies, neuronal ceroid lipofuscinoses (NCLs), sialidoses, myoclonus epilepsy with ragged-red fibers (MERRF), neuronopathic Gaucher disease, Farber disease, and Dentatorubral-pallidoluysian atrophy. The causal genes are known for ULD, LD, NCLs, MERRF, Farber disease, and sialidoses (Zimmern and Minassian, 2024). They are all characterized clinically by myoclonic and tonic-clonic seizures. Progressive symptoms include muscle rigidity, problems with balance, and, in some PMEs, mental decline and even behavioral and psychiatric problems. Those affected can become wheelchair-bound.

LD (also known as EPM2) is caused by mutations in the *EPM2A* and *NHLRC1* genes, which encode laforin and malin, respectively (Chan et al., 2003; Aguado et al., 2010). Laforin is a dual specificity phosphatase, and malin is an E3 ubiquitin ligase. Both proteins are involved in glycogen processing; therefore due to their mutations and consequent loss of function, insoluble bodies of glycogen (Lafora bodies) accumulate in the brain and some other tissues such as liver and skin. Lohi et al. (2005) proposed that laforin might direct two negative feedback loops: polyglucosan-laforin-glycogen synthase kinase-3 beta-glycogen synthase (GSK3β-GS), inhibiting GS activity, and polyglucosan-laforin-malin-GS, removing GS through proteasomal degradation. In addition, autophagy was found to be impaired in LD (Knecht et al., 2012). OS and autophagy impairment follow increased protein aggregation, due to less efficient ubiquitin-proteasome mediated clearance (Romá-Mateo et al., 2015), (Gentry et al., 2013), see Fig. 4.1. The disease onset is in late childhood or adolescence. It is characterized by progressive neurologic deterioration and mental health symptoms, leading to death within 10 years. Recently, using modern gene management techniques, several options for treating LD in the future have been proposed (Nitschke et al., 2018) (see Fig. 4.1).

NCLs are a group of monogenic, autosomal recessive neurodegenerative disorders (Pan et al., 2021). The majority of the 14 NCL genes (Gardner and Mole, 2021; Mole and Cotman, 2015) discovered thus far encode proteins that are part of endosomal/lysosomal pathways. CLN1/PPT1, CLN2/TPP1, CLN5, CLN10/CTSD, CLN13/CTSF, and CLN11/GRN are lysosomal proteins, more specifically: lysosomal enzymes (CLN1, CLN2, CLN10, and CLN13), a soluble protein (CLN5), and a protein in the secretory pathway (CLN11), whereas CLN3, CLN7/MFSD8, CLN12/ATP13A2 are membrane proteins (Gardner and Mole, 2021; Mole and Cotman, 2015). Two NCL genes encode endoplasmic reticulum membrane proteins (CLN6, CLN8) (Gardner and Mole, 2021; Mole and Cotman, 2015). The phenotype of lipofuscinoses depends somewhat on the mutated gene and the site of mutation. Apart from myoclonus, they present with psychomotor delay, progressive loss of vision, and dementia. Patients die early. The name comes from lipofuscin, an indigestible mixture of fluorescent cross-linked proteins and lipids. The debris affects the neuronal cytoskeleton and cellular

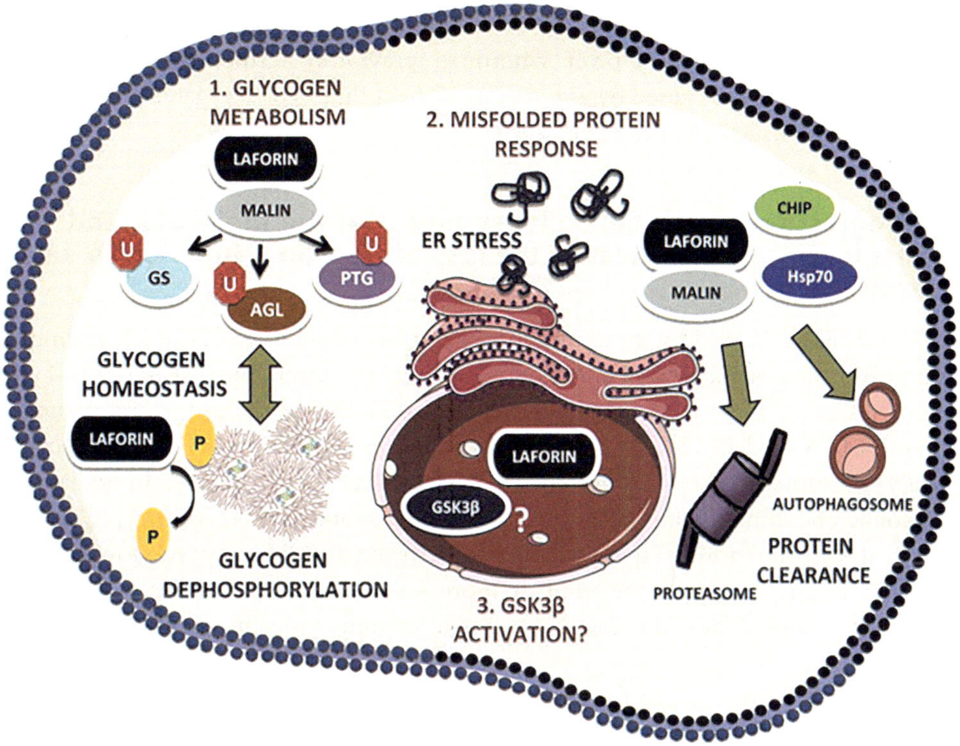

Figure 4.1 Laforin functions in cell physiology. Schematic view of the different functions of laforin: glycogen synthase (GS); glycogen debranching enzyme; heat shock protein 70 (HSP70); ubiquitin (U); phosphate (P). The figure was produced with the server medical art platform. *Taken from Gentry, M.S., Romá-Mateo, C., Sanz, P., 2013. Laforin, a protein with many faces: glucan phosphatase, adapter protein, et alii. FEBS Journal. 280 (2), 525–537. https://doi.org/10.1111/j.1742-4658.2012.08549.x. Copyright © 1999-2024 John Wiley & Sons, Incl. Open access under a Creative Commons Attribution 4.0 International Licence.*

trafficking. It triggers lysosomal membrane permeabilisation and subsequent atypical necroptosis.

MERRF is caused by mutations in the mitochondrial DNA (mtDNA), with the A8344G mutation in the tRNA(Lys) gene being the most common variant (Larsson et al., 1995; Brackmann et al., 2012). mtDNA is transmitted vertically from the mother to the offspring. The disease onset is in childhood and early adolescence. MERRF is a rare disease, a multisystem mitochondrial syndrome characterized by progressive myoclonus and generalized epileptic seizures. Cerebellar ataxia, muscle myopathy, cardiac arrhythmia, hearing loss, optic nerve atrophy, and dementia occur concomitantly.

Myoclonus epilepsy EPM7, albeit usually more severe than EPM1, results from mutations in the *KCNC1* gene (Muona et al., 2015) encoding the potassium ion channel subunit of the Kv3 subfamily of voltage-gated tetrameric channels. A further rare myoclonus epilepsy is Dravet syndrome, previously termed severe myoclonic epilepsy of infancy. It is caused by a mutation of the sodium voltage-gated channel alpha subunit (*SCN1A*) gene (Claes et al., 2001).

4.3 Progressive myoclonic epilepsy type 1 at the molecular and cellular level: consequences of the loss of cystatin B function to neural physiology

ULD or EPM1 is caused by mutations in the cystatin B gene (*CSTB* gene), coding for cystatin B (Lalioti et al., 1997; Pennacchio et al., 1996)—sometimes called stefin B (also assigned as cysB, stB), whose main function is to inhibit cysteine proteases such as cathepsins L, S, and B[1]. ULD, in most cases starts in childhood or early adolescence. Usually, an earlier onset of symptoms predicts a more severe course. In addition to myoclonus and tonic-clonic seizures, gait instability and muscle rigidity, symptoms include dysarthria (trouble speaking clearly), dysphagia (difficulty swallowing), and tremor. The phenotype can be mild or more severe, whereas patients may become wheelchair-bound. They do not have major cognitive decline, but depression is common.

At least 14 genetic mutations affecting the *CSTB* gene (Lalioti et al., 1997; Pennacchio et al., 1996) have been reported. The group of Lehesjoki has determined many other mutations apart from dodecamer repeat expansions, among them are missense, nonsense, frameshift, and deletion mutants (Joensuu et al., 2007). An update of the mutations and splice-variants in the *CSTB* gene has been gathered recently by Singh and Hämäläinen (2024). Patients heterozygous for a dodecamer repeat and a point mutation, either frameshift, stop, or missense with loss of function, present with differently severe pathologies (Gorski et al., 2023). Patients homozygous for two null alleles manifest the most severe neonatal-onset progressive encephalopathy (O'Brien et al., 2017).

Our group has shown that some of the mutations change stability, folding/unfolding, and propensity to aggregate of the stefin B protein. Mutant proteins may attain cytotoxic function similar to other amyloid-forming proteins—as we have proposed already in 2005 (Čeru et al., 2005). Our later studies have shown differences in inclusions upon expression of selected EPM1 mutants of stefin B in cell culture, which proved more or less toxic, depending on the unfolding intermediate and type of oligomers formed (Čeru et al., 2010a; Polajnar et al., 2014a). We have also studied primary astrocytes from stefin B knock-out (KO) mice, and have observed more general protein aggregation and an impaired autophagic flux (Polajnar et al., 2014b). As possible confirmation of gain of toxic function, patients bearing one allele of the usual dodecamer repeat expansion and one allele of R68X in the *CSTB* gene show more severe pathology and clinical symptoms (Koskenkorva et al., 2011). The truncating mutation leads to an unfolded protein, which aggregates extensively (Rabzelj et al., 2005); therefore it is reasonable that cells would degrade it rapidly. Patients who are heterozygous for the dodecamer repeat and Q71P or G50E missense mutation, which form aggresomes in cell culture (Polajnar et al., 2014a), may be expected to show more severe symptoms due to the gain of toxicity. However, most of the mutations also affect the protein's function. Thus G4R, Q71P, and G50E mutations showed loss of lysosomal localization (Joensuu et al., 2007; Alakurtti et al., 2005), which may be reflected in a normal course of the EPM1 disease, which is due to decreased inhibitory activity of this protein. In this context, "normal" relates to the most common mutation in EPM1 associated with the *CSTB* gene; dodecamer repeat extensions in the gene promoter, leading to much lower expression of the protein.

Even though human cystatin B (stefin B) (Jerala et al., 1988) is a cysteine protease inhibitor (Turk and Bode, 1991; Turk et al., 2008), it is likely that this protein may exert alternate functions (Žerovnik, 2017b, 2022). We have proposed that it may act as an amateur chaperone, helping other proteins to fold or prevent their unfolding (Žerovnik, 2017a; Taler-Verčič and Žerovnik, 2010). It readily forms domain-swapped dimers, similarly to stefin A (Jerala and Zerovnik, 1999; Staniforth et al., 2001), and tetramers composed of two domain-swapped dimers (Jenko Kokalj et al., 2007). We have shown that stefin B tetramers interact with Alzheimer's associated amyloid-beta peptide and reduce its amyloid fibril formation (Škerget et al., 2010). Importantly, the group of Melli has confirmed the oligomeric nature of this protein in cells (Cipollini et al., 2008) and that the oligomers depend on the redox state of the cell (Cipollini et al., 2008; Rispoli et al., 2013).

Furthermore, Italian groups have reported that cystatin B has a role in vesicular transport (Penna et al., 2019) as well as in neural proliferation and interneuron migration (Di Matteo et al., 2020), in synapse physiology (Penna et al., 2019; Gorski et al., 2020), and in neurogenesis (Di Matteo et al., 2020). That cystatin B has important roles in the brain (Singh and Hämäläinen, 2024), contributing to synaptic plasticity, has been confirmed more recently in studies of human cerebral organoids (Pizzella et al., 2024). Already in 2002, Di Giaimo et al. (2002) showed that cystatin B in the rat cerebellum interacts with proteins that play a role in cellular growth, proliferation, and differentiation. Using the yeast two-hybrid system, cystatin B was shown to interact with brain β-spectrin, a cytoskeletal protein, with neurofilament light chain (NF-L) which is important for cytoskeleton dynamics, and with receptor for activated C kinase 1 (RACK-1), which mediates the interaction of the activated C kinase embedded in the cell membrane with the cytoskeleton. The three proteins—RACK-1, β-spectrin, and NF-L colocalize in Purkinje cells and in Bergmann glia, where, in complexes with cystatin B, they contribute to cell development and differentiation (Singh and Hämäläinen, 2024).

In stefin B KO cells (Trstenjak Prebanda et al., 2019), cystatin B KO animal models and patients (Lehtinen et al., 2009), OS was found to be increased. Maher et al. (2014) showed that murine stefin B-deficient macrophages were significantly more sensitive to the lethal lipopolysaccharide (LPS)-induced sepsis due to increased caspase-11 expression and NLRP3 inflammasome activation (Maher et al., 2014). They also showed that stefin B deficiency resulted in the destablization of the mitochondrial membrane and elevated mitochondrial reactive oxygen species (ROS) generation (Pennacchio et al., 1998), leading to cell death. Recently, the same group has shown that stefin B inhibits NLRP3 inflammasome activation via 5' adenosine monophosphate-activated protein kinase (AMPK)/mammalian target of rapamycin kinase (mTOR) signaling (Trstenjak-Prebanda et al., 2023). In addition, nuclear localization and function(s) (Čeru et al., 2010b; Riccio et al., 2001) of stefin B are important.

4.4 Other gene mutations causing progressive myoclonic epilepsy type 1-like symptoms

Another mutation producing similar symptoms to EPM1 was identified affecting the Prickle1 protein (Bassuk et al., 2008; Berkovic et al., 2005). Apart from progressive myoclonus syndrome, Prickle paralogs have been associated with several neurological and NDs, including autism spectrum disorder and Alzheimer's disease, and with autoimmune disorders, such as rheumatoid arthritis. The Prickle family of proteins mediate planar cell polarity (PCP) signaling (Radaszkiewicz et al., 2024). Prickle binds to the transmembrane protein Vangl, and the Vangl-Prickle complexes accumulate at the plasma membrane, where they regulate the actin cytoskeleton (Tang and Gerlach, 2017; Dreyer et al., 2022). In addition, Prickle inhibits Disheveled (Dsh) (Radaszkiewicz et al., 2024) in the cytoplasm and the transmembrane protein Frizzled (Fz) (see Fig. 4.2).

Another mutation producing symptoms of progressive myoclonus epilepsy was found in the *SCARB2* gene (Dibbens et al., 2009). The function(s) of the *SCARB2* gene coding for lysosomal integral membrane protein—LIMP2, are only partially known. It is known that it functions in membrane transport at the endosomal/lysosomal compartment (Gonzalez et al., 2014). However, it has been shown that LIMP2, as well as at lysosomal membranes, resides at intercalated disks of the heart, where it interacts with N-cadherin. Knockdown of LIMP2 with RNA interference (RNAi) decreased the binding of N-cadherin to the phosphorylated form of β-catenin, and LIMP2 overexpression reversed this effect (Schroen et al., 2007).

4.5 Signaling pathways in common between progressive myoclonus epilepsies and neurodegenerative diseases

In EPM1, EPM2, and MERRF, increased OS and mitochondrial impairment are observed. For example, mitochondrial impairment with increased OS occurs in

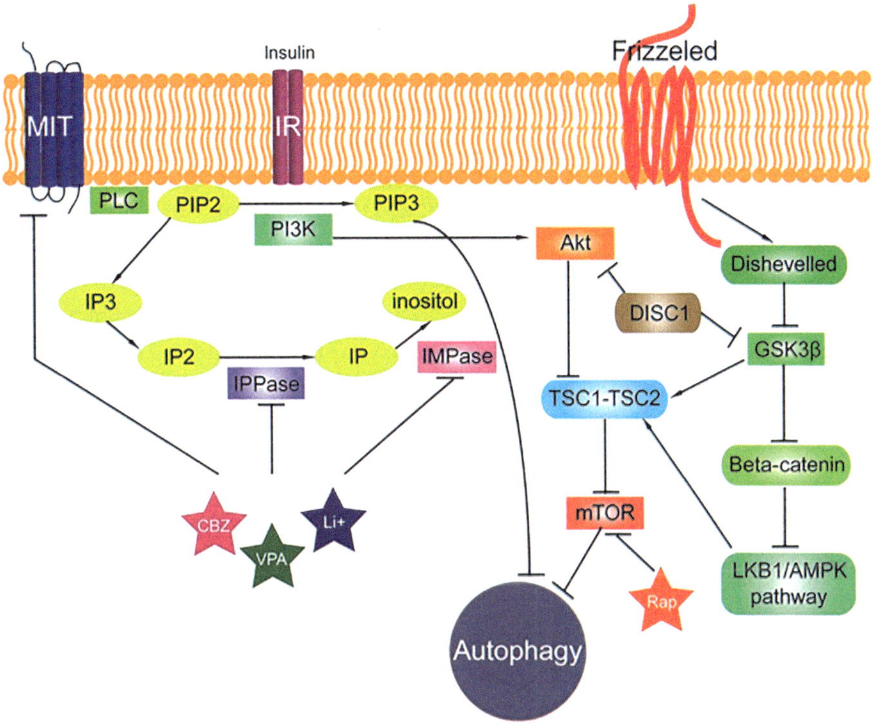

Figure 4.2 The signaling pathways connected to autophagy. The phosphatidylinositol signaling pathway is regulated by Class I phosphoinositide 3-kinases (PI3Ks), which are activated by kinase receptors like insulin receptors (IR) and are responsible for the production of phosphatidylinositol (3,4,5)-triphosphate (PIP3) from phosphatidylinositol 4,5-bisphosphate (PIP2). Phospholipase C (PLC) cleaves the PIP2 into diacyl glycerol and inositol 1,4,5-trisphosphate (IP3). Inositol polyphosphate 1-phosphatase (IPPase) catalyzes inositol bisphosphate (IP2) to inositol monophosphate (IP), which is further dephosphorylated by inositol monophosphatase (IMPase) to inositol. High-affinity inositol transport is additionally catalyzed by the active myo-inositol/H^+ transporter (MIT). MIT, IPPase, and IMPase are all inhibited by carbamazepine (CBZ), valproic acid (VPA), and lithium (Li^+). PI3Ks have been linked to an extraordinarily diverse group of cellular functions through regulation of the Akt/tuberous sclerosis proteins 1 and 2 (TSC1-TSC2)/mammalian target of rapamycin kinase (mTOR) pathway. Disrupted in Schizophrenia 1 (DISC1) in its wild-type form also negatively regulates both glycogen synthase kinase-3 beta (GSK3β) and Akt (also known as protein kinase B [PKB]). TSC1/2 form a complex that, like rapamycin (Rap) inhibits mTOR. Wnt signaling activation is mediated through binding of a Wnt-protein ligand to a Frizzled family receptor, which passes the biological signal to the protein Disheveled (Dsh). Dsh negatively regulatesGSK3β, which alternatively inhibits β-catenin, one of the central proteins of the Wnt signaling pathway. β-catenin, however, negatively regulates the liver kinase B1/5' adenosine monophosphate-activated protein kinase (LKB1/AMPK) pathway that indirectly regulates autophagy. *Taken from Polajnar, M., Žerovnik, E., 2014. Impaired autophagy: a link between neurodegenerative and neuropsychiatric diseases. Journal of Cellular and Molecular Medicine. 18 (9), 1705–1711. https://doi.org/10.1111/jcmm.12349. Copyright to authors and Elsevier. Open access: Creative Commons Attribution License.*

stefin B KO mice upon LPS challenge (Prebanda et al., 2019). Neuroinflammation is a feature of EPM1 and also some other PMEs such as NCLs and LD (Sanz and Serratosa, 2020). In EPM1, represented by model stefin B KO mice, neuroinflammation is a key hallmark (Okuneva et al., 2016). As well, innate immune responses and inflammasome activation were observed (Kopitar-Jerala, 2015a, 2015b). Activated glia with accompanying excitotoxicity are also common to the above conditions (Sanz and Serratosa, 2020).

As proposed by Sanz and Serratosa (2020) in EPM1, EPM2 and NCLs these common processes have been observed: reactive glia, OS, release of cytokines/chemokines, neuronal hyperexcitability and synaptic remodeling (synaptic plasticity), defined as "long-term potentiation and long-term depression, which are enduring changes in synaptic strength, induced by specific patterns of synaptic activity" (Bliss and Cooke, 2011).

Together, these pathological phenomena: mitochondrial impairment, increased OS, neuroinflammation, and activated (reactive) glia, likely end-points of other upstream events, are common to several PMEs but also to many other neurodegenerative conditions.

A common signaling pathway impaired in PMEs, NDs, Alzheimer's disease, as well as in neuropsychiatric diseases, is Wnt signaling, as suggested for PMEs by Polajnar and Žerovnik (2014, 2011). Wnt signaling is an autocrine paracrine signal transduction pathway. It plays a role in cytoskeleton and midbrain development, amongst others. Ablation of Wnt-1 (in mice) results in severe defects of the midbrain, the cerebellum, and the spinal cord, while ablation of Wnt-3a results in a total loss of the hippocampus. One can distinguish canonical and noncanonical Wnt signaling pathways. The canonical (classical) pathway is described in the legend to Fig. 4.2. In short, the Wnt-protein ligand binds to a Fz family receptor, which passes the signal to the protein Dsh. Dsh negatively regulates GSK3β, which inhibits β-catenin. At least two noncanonical Wnt pathways exist: the PCP pathway, in which Fz acts through monomeric GTPase and Jun N-terminal kinase to regulate the cytoskeleton (Montcouquiol et al., 2006), and the Wnt/Ca^{2+} signaling pathway, in which Fz activation leads to increased intracellular Ca^{2+} (Inestrosa et al., 2012) (see Fig. 4.2).

mTOR signaling is indeed affected in LD and in EPM1, where, in addition to the Prickle gene regulating PCP signaling (Radaszkiewicz et al., 2024) stefin B is also involved in PCP signaling (Žerovnik, 2022) and in autophagy (Polajnar et al., 2014b). A recent paper reports that stefin B overexpression downregulated mitochondrial ROS and lowered inflammasome activation. This occurred concomitantly with AMPK phosphorylation and downregulation of mTOR activity, leading to the induction of autophagy (Trstenjak-Prebanda et al., 2023).

4.6 What are the prospects for a cure or early treatment of various monogenic progressive myoclonus epilepsies?

Treatments for PMEs thus far have been supportive and palliative. With recent advances in enzyme replacement methods, short RNA-based therapies, stem cell and gene therapies, prospects are brighter (Zimmern and Minassian, 2024). More specifically, enzyme replacement therapy by providing active enzymes into the affected cells seems a likely way forward for ULD, NCLs, Farber disease, and sialidoses. To replace the impaired gene either by viral delivery or by mRNA embedded in liposomes might soon become possible for ULD (Gumusgoz et al., 2024) and Batten disease (Johnson et al., 2019), and if side effects are manageable, this would represent a cure. Gene replacement has been successfully applied in an LD mouse model (Zafra-Puerta et al., 2024).

As autophagy is impaired in LD, we suggested this might be ameliorated by autophagy-enhancing molecules, such as rapamycin analogs (Polajnar and Žerovnik, 2011). Recently, indeed, such a preclinical analysis for LD mice was conducted using trehalose (Sinha et al., 2021b). This treatment resulted in lower susceptibility to pentylenetetrazole-induced seizures but did not reduce the burden of Lafora bodies (Zimmern and Minassian, 2024).

In LD and ULD, proteostasis is also impaired (Polajnar and Žerovnik, 2011); therefore heat shock protein targeted therapies would seem beneficial. A proof-of-concept study was performed in an LD mouse model, where heat shock factor-1 was restored using dexamethasone treatment (Sinha et al., 2021a).

Neuroinflammation, which is a hallmark of most of PMEs, can be decreased by using known antiinflammatory drugs, but also by other modulators of inflammation (Mollá et al., 2021), as was also shown in a study of LD mice.

Another common impairment of PMEs is OS. Antioxidants, such as N-acetyl cysteine (NAC) (Sinha et al., 2021a), (Raghu et al., 2021), and different vitamins have been applied to improve EPM1, EPM2, and MERRF; however, often with no clear benefit (Edwards et al., 2002). The reason might be that oxidative damage has already occurred. An early treatment would likely bring more benefit; however, it depends on early diagnosis. Based on case studies, NAC proved beneficial for ULD (Selwa, 1999; Deepmala et al., 2015).

Some antioxidant molecules can also decrease the number of protein aggregates, which may increase along with the decrease of autophagic flux and increase in OS (this creates a vicious circle with negative feedback). We performed an in vitro study where the effect of polyphenols and some vitamins on the aggregation of stefin B was shown (Hasanbašić et al., 2018).

4.7 Conclusions

PMEs recently entered into an exciting era where disease cures may be expected in the next decade. By deepening our understanding of the underlying molecular and cellular mechanisms, we can pave the way for earlier interventions, more effective supportive therapies, and ultimately, curative approaches for familial PMEs. These may include stem cell therapies, enzyme replacement (through liposome-encapsulated proteins or mRNA-based delivery), gene expression suppression (via RNAi), or gene replacement to correct the underlying mutation. Similar developments are seen with some of the so-called orphan diseases (Kioutchoukova et al., 2023).

The chapter is a shortened and changed version of the paper Žerovnik E, 2024, Explor Neurosci. 2024;3:295–308 https://doi.org/10.37349/en.2024.00051.

Abbreviations

***CSTB* gene**	cystatin B gene
Dsh	disheveled
EPM1	progressive myoclonic epilepsy type 1
KO	knock-out
LD	Lafora disease
MERRF	myoclonus epilepsy with ragged-red fibers
mtDNA	mitochondrial DNA
NAC	N-acetyl cysteine
NCLs	neuronal ceroid lipofuscinoses
NDs	neurodegenerative diseases
OS	oxidative stress
PCP	planar cell polarity
PMEs	progressive myoclonus epilepsies
PTZ	pentylenetetrazole
ULD	Unverricht–Lundborg disease

Acknowledgement

My work was supported throughout by several projects (J7-4050, J1-8742) and the program P1-0140 by the Slovenian Research and Innovation agency (ARIS).

References

Aguado, C., Sarkar, S., Korolchuk, V.I., Criado, O., Vernia, S., Boya, P., et al., 2010. Laforin, the most common protein mutated in Lafora disease, regulates autophagy. Human Molecular Genetics 19 (14), 2867–2876. Available at https://doi.org/10.1093/hmg/ddq190.

Alakurtti, K., Weber, E., Rinne, R., Theil, G., Haan, G.J.de, Lindhout, D., et al., 2005. Loss of lysosomal association of cystain B proteins representing progressive myoclonus epilepsy, EPM1, mutations. European Journal of Human Genetics 13 (2), 208–215. Available at https://doi.org/10.1038/sj.ejhg.5201300.

Bassuk, A.G., Wallace, R.H., Buhr, A., Buller, A.R., Afawi, Z., Shimojo, M., et al., 2008. A homozygous mutation in human PRICKLE1 causes an autosomal-recessive progressive myoclonus epilepsy-ataxia syndrome. American Journal of Human Genetics 83 (5), 572–581. Available at https://doi.org/10.1016/j.ajhg.2008.10.003.

Berkovic, S.F., Mazarib, A., Walid, S., Neufeld, M.Y., Manelis, J., Nevo, Y., et al., 2005. A new clinical and molecular form of Unverricht-Lundborg disease localized by homozygosity mapping. Brain 128 (3), 652–658. Available at https://doi.org/10.1093/brain/awh377.

Bliss, T.V.P., Cooke, S.F., 2011. Long-term potentiation and long-term depression: a clinical perspective. Clinics (Sao Paulo, Brazil) 66 (1), 3–17. Available at https://doi.org/10.1590/S1807-59322011001300002.

Brackmann, F., Abicht, A., Ahting, U., Schröder, R., Trollmann, R., 2012. Classical MERRF phenotype associated with mitochondrial tRNALeu (m.3243A>G) mutation. European Journal of Pediatrics 171 (5), 859–862. Available at https://doi.org/10.1007/s00431-011-1662-8.

Čeru, S., Rabzelj, S., Kopitar-Jerala, N., Turk, V., Žerovnik, E., 2005. Protein aggregation as a possible cause for pathology in a subset of familial Unverricht-Lundborg disease. Medical Hypotheses 64 (5), 955–959. Available at https://doi.org/10.1016/j.mehy.2004.11.038.

Čeru, S., Layfield, R., Zavašnik-Bergant, T., Repnik, U., Kopitar-Jerala, N., Turk, V., et al., 2010a. Intracellular aggregation of human stefin B: confocal and electron microscopy study. Biology of the Cell 102 (6), 319–334. Available at https://doi.org/10.1042/BC20090163.

Čeru, S., Konjar, Š., Maher, K., Repnik, U., Križaj, I., Benčina, M., et al., 2010b. Stefin B interacts with histones and cathepsin L in the nucleus. Journal of Biological Chemistry 285 (13), 10078–10086. Available at https://doi.org/10.1074/jbc.m109.034793.

Chan, E.M., Bulman, D.E., Paterson, A.D., Turnbull, J., Andermann, E., Andermann, F., et al., 2003. Genetic mapping of a new Lafora progressive myoclonus epilepsy locus (EPM2b) on 6p22. Journal of Medical Genetics 40 (9), 671–675. Available at https://doi.org/10.1136/jmg.40.9.671.

Cipollini, E., Riccio, M., Di Giaimo, R., Dal Piaz, F., Pulice, G., Catania, S., et al., 2008. Cystatin B and its EPM1 mutants are polymeric and aggregate prone in vivo. Biochimica et Biophysica Acta (BBA) - Molecular Cell Research 1783 (2), 312–322. Available at https://doi.org/10.1016/j.bbamcr.2007.08.007.

Claes, L., Del-Favero, J., Ceulemans, B., Lagae, L., Van Broeckhoven, C., De Jonghe, P., 2001. De novo mutations in the sodium-channel gene SCN1A cause severe myoclonic epilepsy of infancy. The American Journal of Human Genetics 68 (6), 1327–1332. Available at https://doi.org/10.1086/320609.

Deepmala, Slattery, J., Kumar, N., Delhey, L., Berk, M., Dean, O., et al., 2015. Clinical trials of N-acetylcysteine in psychiatry and neurology: a systematic review. Neuroscience & Biobehavioral Reviews 55, 294–321. Available at https://doi.org/10.1016/j.neubiorev.2015.04.015.

Dibbens, L.M., Michelucci, R., Gambardella, A., Andermann, F., Rubboli, G., Bayly, M.A., et al., 2009. SCARB2 mutations in progressive myoclonus epilepsy (PME) without renal failure. Annals of Neurology 66 (4), 532–536. Available at https://doi.org/10.1002/ana.21765.

Dreyer, C.A., VanderVorst, K., Carraway, K.L., 2022. Vangl as a master scaffold for WNT/planar cell polarity signaling in development and disease. Frontiers in Cell and Developmental Biology 10. Available at https://doi.org/10.3389/fcell.2022.887100.

Edwards, M.J.J., Hargreaves, I.P., Heales, S.J.R., Jones, S.J., Ramachandran, V., Bhatia, K.P., et al., 2002. N -acetylcysteine and Unverricht–Lundborg disease Variable response and possible side effects. Neurology 59 (9), 1447–1449. Available at https://doi.org/10.1212/wnl.59.9.1447.

Gardner, E., Mole, S.E., 2021. The genetic basis of phenotypic heterogeneity in the neuronal ceroid lipofuscinoses. Frontiers in Neurology 12. Available at https://doi.org/10.3389/fneur.2021.754045.

Gentry, M.S., Romá-Mateo, C., Sanz, P., 2013. Laforin, a protein with many faces: glucan phosphatase, adapter protein, et alii. FEBS Journal 280 (2), 525–537. Available at https://doi.org/10.1111/j.1742-4658.2012.08549.x.

Di Giaimo, R., Riccio, M., Santi, S., Galeotti, C., Ambrosetti, D.C., Melli, M., 2002. New insights into the molecular basis of progressive myoclonus epilepsy: a multiprotein complex with cystatin B. Human Molecular Genetics 11 (23), 2941–2950.

Gonzalez, A., Valeiras, M., Sidransky, E., Tayebi, N., 2014. Lysosomal integral membrane protein-2: a new player in lysosome-related pathology. Molecular Genetics and Metabolism 111 (2), 84–91. Available at https://doi.org/10.1016/j.ymgme.2013.12.005.

Gorski, K., Spoljaric, A., Nyman, T.A., Kaila, K., Battersby, B.J., Lehesjoki, A.E., 2020. Quantitative changes in the mitochondrial proteome of cerebellar synaptosomes from preclinical cystatin B-deficient mice. Frontiers in Molecular Neuroscience 13. Available at https://doi.org/10.3389/fnmol.2020.570640.

Gorski, K., Jackson, C.B., Nyman, T.A., Rezov, V., Battersby, B.J., Lehesjoki, A.E., 2023. Progressive mitochondrial dysfunction in cerebellar synaptosomes of cystatin B-deficient mice. Frontiers in Molecular Neuroscience 16. Available at https://doi.org/10.3389/fnmol.2023.1175851.

Gumusgoz, E., Kasiri, S., Verma, M., Wu, J., Villarreal Acha, D., Marriam, U., et al., 2024. CSTB gene replacement improves neuroinflammation, neurodegeneration and ataxia in murine type 1 progressive myoclonus epilepsy. Gene Therapy 31 (5-6), 234–241. Available at https://doi.org/10.1038/s41434-023-00433-x.

Hasanbašić, S., Jahić, A., Berbić, S., Žnidarič, M.T., Žerovnik, E., Flora, S.J.S., 2018. Inhibition of protein aggregation by several antioxidants. Oxidative Medicine and Cellular Longevity 2018 (1). Available at https://doi.org/10.1155/2018/8613209.

Hipp, M.S., Park, S.H., Hartl, U.U., 2014. Proteostasis impairment in protein-misfolding and -aggregation diseases. Trends in Cell Biology 24 (9), 506–514. Available at https://doi.org/10.1016/j.tcb.2014.05.003.

Inestrosa, N.C., Montecinos-Oliva, C., Fuenzalida, M., 2012. Wnt signaling: role in Alzheimer disease and schizophrenia. Journal of Neuroimmune Pharmacology 7 (4), 788–807. Available at https://doi.org/10.1007/s11481-012-9417-5.

Jenko Kokalj, S., Gunčar, G., Štern, I., Morgan, G., Rabzelj, S., Kenig, M., et al., 2007. Essential role of proline isomerization in stefin b tetramer formation. Journal of Molecular Biology 366 (5), 1569–1579. Available at https://doi.org/10.1016/j.jmb.2006.12.025.

Jerala, R., Zerovnik, E., 1999. Accessing the global minimum conformation of stefin A dimer by annealing under partially denaturing conditions. Journal of Molecular Biology 291 (5), 1079–1089. Available at https://doi.org/10.1006/jmbi.1999.3045.

Jerala, R., Trstenjak, M., Lenarčič, B., Turk, V., 1988. Cloning a synthetic gene for human stefin B and its expression in *E. coli*. FEBS Letters 239 (1), 41–44. Available at https://doi.org/10.1016/0014-5793(88)80541-6.

Joensuu, T., Kuronen, M., Alakurtti, K., Tegelberg, S., Hakala, P., Aalto, A., et al., 2007. Cystatin B: mutation detection, alternative splicing and expression in progressive myclonus epilepsy of Unverricht-Lundborg type (EPM1) patients. European Journal of Human Genetics 15 (2), 185–193. Available at https://doi.org/10.1038/sj.ejhg.5201723.

Johnson, T.B., Cain, J.T., White, K.A., Ramirez-Montealegre, D., Pearce, D.A., Weimer, J.M., 2019. Therapeutic landscape for Batten disease: current treatments and future prospects. Nature Reviews Neurology 15 (3), 161–178. Available at https://doi.org/10.1038/s41582-019-0138-8.

Kioutchoukova, I.P., Foster, D.T., Thakkar, R.N., Foreman, M.A., Burgess, B.J., Toms, R.M., et al., 2023. Neurologic orphan diseases: emerging innovations and role for genetic treatments. World Journal of Experimental Medicine 13 (4), 59–79. Available at https://doi.org/10.5493/wjem.v13.i4.59.

Knecht, E., Criado-García, O., Aguado, C., Gayarre, J., Duran-Trio, L., Garcia-Cabrero, A.M., et al., 2012. Malin knockout mice support a primary role of autophagy in the pathogenesis of Lafora disease. Autophagy 8 (4), 701–703. Available at https://doi.org/10.4161/auto.19522.

Kopitar-Jerala, N., 2015a. Innate immune response in brain, NF-kappa B signaling and cystatins. Frontiers in Molecular Neuroscience 8. Available at https://doi.org/10.3389/fnmol.2015.00073.

Kopitar-Jerala, N., 2015b. The role of stefin B in neuro-inflammation. Frontiers in Cellular Neuroscience 9. Available at https://doi.org/10.3389/fncel.2015.00458.

Koskenkorva, P., Hyppönen, J., Aikiä, M., Mervaala, E., Kiviranta, T., Eriksson, K., et al., 2011. Severer phenotype in Unverricht-Lundborg disease (EPM1) patients compound heterozygous for the dodecamer repeat expansion and the c.202C>T mutation in the CSTB gene. Neuro-degenerative Diseases 8 (6), 515–522. Available at https://doi.org/10.1159/000323470.

Lalioti, M.D., Scott, H.S., Buresi, C., Rossier, C., Bottani, A., Morris, M.A., et al., 1997. Dodecamer repeat expansion in cystatin B gene in progressive myoclonus epilepsy. Nature 386 (6627), 847–851. Available at https://doi.org/10.1038/386847a0.

Larsson, N.-, Tulinius, M.H., Holme, E., Oldfors, A., 1995. Pathogenetic aspects of the A8344G mutation of mitochondrial DNA associated with MERRF syndrome and multiple symmetric lipomas. Muscle & Nerve 18 (14 S), S102. Available at https://doi.org/10.1002/mus.880181421.

Lehtinen, M.K., Tegelberg, S., Schipper, H., Su, H., Zukor, H., Manninen, O., et al., 2009. Cystatin B deficiency sensitizes neurons to oxidative stress in progressive myoclonus epilepsy, EPM1. Journal of Neuroscience 29 (18), 5910–5915. Available at https://doi.org/10.1523/JNEUROSCI.0682-09.2009.

Lohi, H., Ianzano, L., Zhao, X.C., Chan, E.M., Turnbull, J., Scherer, S.W., et al., 2005. Novel glycogen synthase kinase 3 and ubiquitination pathways in progressive myoclonus epilepsy. Human Molecular Genetics 14 (18), 2727–2736. Available at https://doi.org/10.1093/hmg/ddi306.

Maher, K., Jerič Kokelj, B., Butinar, M., Mikhaylov, G., Menček-Keber, M., Stoka, V., et al., 2014. A role for stefin B (cystatin B) in inflammation and endotoxemia. Journal of Biological Chemistry 289 (46), 31736–31750. Available at https://doi.org/10.1074/jbc.m114.609396.

Di Matteo, F., Pipicelli, F., Kyrousi, C., Tovecci, I., Penna, E., Crispino, M., et al., 2020. Cystatin B is essential for proliferation and interneuron migration in individuals with EPM 1 epilepsy. EMBO Molecular Medicine 12 (6). Available at https://doi.org/10.15252/emmm.201911419.

Mole, S.E., Cotman, S.L., 2015. Genetics of the neuronal ceroid lipofuscinoses (Batten disease). Biochimica et Biophysica Acta - Molecular Basis of Disease 1852 (10), 2237–2241. Available at https://doi.org/10.1016/j.bbadis.2015.05.011.

Mollá, B., Heredia, M., Sanz, P., 2021. Modulators of neuroinflammation have a beneficial effect in a Lafora disease mouse model. Molecular Neurobiology 58 (6), 2508–2522. Available at https://doi.org/10.1007/s12035-021-02285-1.

Montcouquiol, M., Crenshaw, E.B., Kelley, M.W., 2006. Noncanonical Wnt signaling and neural polarity. Annual Review of Neuroscience 29, 363–386. Available at https://doi.org/10.1146/annurev.neuro.29.051605.112933.

Muona, M., Berkovic, S.F., Dibbens, L.M., Oliver, K.L., Maljevic, S., Bayly, M.A., et al., 2015. A recurrent de novo mutation in KCNC1 causes progressive myoclonus epilepsy. Nature Genetics 47 (1), 39–46. Available at https://doi.org/10.1038/ng.3144.

Nitschke, F., Ahonen, S.J., Nitschke, S., Mitra, S., Minassian, B.A., 2018. Lafora disease — from pathogenesis to treatment strategies. Nature Reviews Neurology 14 (10), 606–617. Available at https://doi.org/10.1038/s41582-018-0057-0.

O'Brien, A., Marshall, C.R., Blaser, S., Ray, P.N., Yoon, G., 2017. Severe neurodegeneration, progressive cerebral volume loss and diffuse hypomyelination associated with a homozygous frameshift mutation in CSTB. European Journal of Human Genetics 25 (6), 775–778. Available at https://doi.org/10.1038/ejhg.2017.39.

Okuneva, O., Li, Z., Körber, I., Tegelberg, S., Joensuu, T., Tian, L., et al., 2016. Brain inflammation is accompanied by peripheral inflammation in Cstb −/− mice, a model for progressive myoclonus epilepsy. Journal of Neuroinflammation 13 (1). Available at https://doi.org/10.1186/s12974-016-0764-7.

Pan, C., Banerjee, K., Lehmann, G.L., Almeida, D., Hajjar, K.A., Benedicto, I., et al., 2021. Lipofuscin causes atypical necroptosis through lysosomal membrane permeabilization. Proceedings of the National Academy of Sciences of the United States of America 118 (47). Available at https://doi.org/10.1073/pnas.2100122118, https://doi.org/10.1073/pnas.2100122118.

Penna, E., Cerciello, A., Chambery, A., Russo, R., Cernilogar, F.M., Pedone, E.M., et al., 2019. Cystatin B involvement in synapse physiology of rodent brains and human cerebral organoids. Frontiers in Molecular Neuroscience 12. Available at https://doi.org/10.3389/fnmol.2019.00195.

Pennacchio, L.A., Lehesjoki, A.E., Stone, N.E., Willour, V.L., Virtaneva, K., Miao, J., et al., 1996. Mutations in the gene encoding cystatin B in progressive myoclonus epilepsy (EPM1). Science (New York, N.Y.) 271 (5256), 1731–1734. Available at https://doi.org/10.1126/science.271.5256.1731.

Pennacchio, L.A., Bouley, D.M., Higgins, K.M., Scott, M.P., Noebels, J.L., Myers, R.M., 1998. Progressive ataxia, myoclonic epilepsy and cerebellar apoptosis in cystatin B-deficient mice. Nature Genetics 20 (3), 251–258. Available at https://doi.org/10.1038/3059.

Pizzella, A., Penna, E., Abate, N., Frenna, E., Canafoglia, L., Ragona, F., et al., 2024. Pathological deficit of cystatin B impairs synaptic plasticity in EPM1 human cerebral organoids. Molecular Neurobiology 61 (7), 4318–4334. Available at https://doi.org/10.1007/s12035-023-03812-y.

Polajnar, M., Žerovnik, E., 2011. Impaired autophagy: a link between neurodegenerative diseases and progressive myoclonus epilepsies. Trends in Molecular Medicine 17 (6), 293–300. Available at https://doi.org/10.1016/j.molmed.2011.02.005.

Polajnar, M., Žerovnik, E., 2014. Impaired autophagy: a link between neurodegenerative and neuropsychiatric diseases. Journal of Cellular and Molecular Medicine 18 (9), 1705–1711. Available at https://doi.org/10.1111/jcmm.12349.

Polajnar, M., Zavašnik-Bergant, T., Kopitar-Jerala, N., Tušek-Žnidarič, M., Žerovnik, E., 2014a. Gain in toxic function of stefin B EPM1 mutants aggregates: correlation between cell death, aggregate number/size and oxidative stress. Biochimica et Biophysica Acta - Molecular Cell Research 1843 (9), 2089–2099. Available at https://doi.org/10.1016/j.bbamcr.2014.05.018.

Polajnar, M., Zavašnik-Bergant, T., Škerget, K., Vizovišek, M., Vidmar, R., Fonović, M., et al., 2014b. Human stefin B role in cell's response to misfolded proteins and autophagy. PLoS One 9 (7), e102500. Available at https://doi.org/10.1371/journal.pone.0102500.

Prebanda, M.T., Završnik, J., Turk, B., Jerala, N.K., 2019. Upregulation of mitochondrial redox sensitive proteins in LPS-treated stefin B-deficient macrophages. Cells 8 (12), 1476. Available at https://doi.org/10.3390/cells8121476.

Rabzelj, S., Turk, V., Žerovnik, E., 2005. In vitro study of stability and amyloid-fibril formation of two mutants of human stefin B (cystatin B) occurring in patients with EPM1. Protein Science 14 (10), 2713–2722. Available at https://doi.org/10.1110/ps.051609705.

Radaszkiewicz, K.A., Sulcova, M., Kohoutkova, E., Harnos, J., 2024. The role of prickle proteins in vertebrate development and pathology. Molecular and Cellular Biochemistry 479 (5), 1199–1221. Available at https://doi.org/10.1007/s11010-023-04787-z.

Raghu, G., Berk, M., Campochiaro, P.A., Jaeschke, H., Marenzi, G., Richeldi, L., et al., 2021. The multifaceted therapeutic role of N-acetylcysteine (NAC) in disorders characterized by oxidative stress. Current Neuropharmacology 19 (8), 1202–1224. Available at https://doi.org/10.2174/1570159x19666201230144109.

Riccio, M., Di Giaimo, R., Pianetti, S., Palmieri, P.P., Melli, M., Santi, S., 2001. Nuclear localization of cystatin B, the cathepsin inhibitor implicated in myoclonus epilepsy (EPM1). Experimental Cell Research 262 (2), 84–94. Available at https://doi.org/10.1006/excr.2000.5085.

Rispoli, A., Cipollini, E., Catania, S., Di Giaimo, R., Pulice, G., van Houte, S., et al., 2013. Insights in progressive myoclonus epilepsy: HSP70 promotes cystatin B polymerization. Biochimica et Biophysica Acta (BBA) - Proteins and Proteomics 1834 (12), 2591–2599. Available at https://doi.org/10.1016/j.bbapap.2013.09.011.

Romá-Mateo, C., Aguado, C., García-Giménez, J.L., Ibáñez-Cabellos, J.S., Seco-Cervera, M., Pallardó, F.V., et al., 2015. Increased Oxidative Stress and Impaired Antioxidant Response in Lafora Disease. Molecular Neurobiology 51 (3), 932–946. Available at https://doi.org/10.1007/s12035-014-8747-0.

Sanz, P., Serratosa, J.M., 2020. Neuroinflammation and progressive myoclonus epilepsies: From basic science to therapeutic opportunities. Expert Reviews in Molecular Medicine 22. Available at https://doi.org/10.1017/erm.2020.5.

Schroen, B., Leenders, J.J., Erk, A.V., Bertrand, A.T., Loon, M.V., Leeuwen, R.E.V., et al., 2007. Lysosomal integral membrane protein 2 is a novel component of the cardiac intercalated disc and vital for load-induced cardiac myocyte hypertrophy. Journal of Experimental Medicine 204 (5), 1227–1235. Available at https://doi.org/10.1084/jem.20070145.

Selwa, L.M., 1999. N-acetylcysteine therapy for Unverricht-Lundborg disease. Neurology 52 (2), 426–427. Available at https://doi.org/10.1212/wnl.52.2.426.

Singh, S., Hämäläinen, R.H., 2024. The roles of cystatin B in the brain and pathophysiological mechanisms of progressive myoclonic epilepsy type 1. Cells 13 (2), 170. Available at https://doi.org/10.3390/cells13020170.

Sinha, P., Verma, B., Ganesh, S., 2021a. Trehalose ameliorates seizure susceptibility in lafora disease mouse models by suppressing neuroinflammation and endoplasmic reticulum stress. Molecular Neurobiology 58 (3), 1088–1101. Available at https://doi.org/10.1007/s12035-020-02170-3.

Sinha, P., Verma, B., Ganesh, S., 2021b. Dexamethasone-induced activation of heat shock response ameliorates seizure susceptibility and neuroinflammation in mouse models of Lafora disease.

Experimental Neurology 340, 113656. Available at https://doi.org/10.1016/j.expneurol.2021.113656.

Škerget, K., Taler-Verčič, A., Bavdek, A., Hodnik, V., Čeru, S., Tušek-Žnidarič, M., et al., 2010. Interaction between oligomers of stefin B and amyloid-β in vitro and in cells. Journal of Biological Chemistry 285 (5), 3201–3210. Available at https://doi.org/10.1074/jbc.M109.024620.

Staniforth, R.A., Giannini, S., Higgins, L.D., Conroy, M.J., Hounslow, A.M., Jerala, R., et al., 2001. Three-dimensional domain swapping in the folded and molten-globule states of cystatins, an amyloid-forming structural superfamily. EMBO Journal 20 (17), 4774–4781. Available at https://doi.org/10.1093/emboj/20.17.4774.

Taler-Verčič, A., Žerovnik, E., 2010. Binding of amyloid peptides to domain-swapped dimers of other amyloid-forming proteins may prevent their neurotoxicity. Bioessays: News and Reviews in Molecular, Cellular and Developmental Biology 32 (12), 1020–1024. Available at https://doi.org/10.1002/bies.201000079.

Tang, D.D., Gerlach, B.D., 2017. The roles and regulation of the actin cytoskeleton, intermediate filaments and microtubules in smooth muscle cell migration. Respiratory Research 18 (1). Available at https://doi.org/10.1186/s12931-017-0544-7.

Trstenjak-Prebanda, M., Biasizzo, M., Dolinar, K., Pirkmajer, S., Turk, B., Brault, V., et al., 2023. Stefin B Inhibits NLRP3 Inflammasome Activation via AMPK/mTOR Signalling. Cells 12 (23), 2731. Available at https://doi.org/10.3390/cells12232731.

Turk, V., Bode, W., 1991. The cystatins: Protein inhibitors of cysteine proteinases. FEBS Letters 285 (2), 213–219. Available at https://doi.org/10.1016/0014-5793(91)80804-c.

Turk, V., Stoka, V., Turk, D., 2008. Cystatins: Biochemical and structural properties, and medical relevance. Frontiers in Bioscience 13 (14), 5406–5420. Available at https://doi.org/10.2741/3089.

Vendruscolo, M., Fuxreiter, M., 2022. Protein condensation diseases: therapeutic opportunities. Nature Communications 13 (1). Available at https://doi.org/10.1038/s41467-022-32940-7.

Zafra-Puerta, L., Iglesias-Cabeza, N., Burgos, D.F., Sciaccaluga, M., González-Fernández, J., Bellingacci, L., et al., 2024. Gene therapy for Lafora disease in the Epm2a−/− mouse model. Molecular Therapy 32 (7), 2130–2149. Available at https://doi.org/10.1016/j.ymthe.2024.05.032.

Zimmern, V., Minassian, B., 2024. Progressive myoclonus epilepsy: a scoping review of diagnostic, phenotypic and therapeutic advances. Genes 15 (2), 171. Available at https://doi.org/10.3390/genes15020171.

Žerovnik, E., 2017. Putative alternative functions of human stefin B (cystatin B): binding to amyloid-beta, membranes, and copper. Journal of Molecular Recognition 30 (1). Available at https://doi.org/10.1002/jmr.2562.

Žerovnik, E., 2017b. Co-chaperoning by amyloid-forming proteins: cystatins vs. crystallins. European Biophysics Journal 46 (8), 789–793. Available at https://doi.org/10.1007/s00249-017-1214-x.

Žerovnik, E., 2022. Human stefin B: from its structure, folding, and aggregation to its function in health and disease. Frontiers in Molecular Neuroscience 15. Available at https://doi.org/10.3389/fnmol.2022.1009976.

PART 2

Common signaling pathways and biomarkers of proteinopathies

CHAPTER 5

Neuroinflammation and neurodegeneration in schizophrenia: potential biomarkers and novel therapeutic strategies

Madeleine King
Division of Physiology Pharmacology and Neuroscience, School of Life Sciences, Queen's Medical Centre, University of Nottingham Medical School, Nottingham, United Kingdom

5.1 Introduction

Schizophrenia is a complex psychiatric illness that affects approximately 0.43% of the adult population (Institute for Health Metrics and Evaluation, 2021) and is ranked among the top 10 causes of global disability (Velligan and Rao, 2023). It typically presents in early adulthood and is marginally more common in males than in females. It is characterized by positive symptoms (including hallucinations and delusions), negative symptoms (like avolition and blunted affect), and impairments across multiple cognitive domains (such as attention, working memory, and executive function). Pharmacological treatment predominantly involves the use of first-generation (typical) antipsychotics to selectively block dopamine D_2 receptors, or second-generation (atypical) antipsychotics that are also antagonists, inverse agonizts, or partial agonizts of multiple additional subtypes, including 5-HT_{2A}, 5-HT_6, 5-HT_7, dopamine D_3, and/or histamine H_1. These approaches have limited effects on the negative or cognitive symptoms, and a significant proportion of patients also gain inadequate relief from positive symptoms or discontinue treatment due to side effects (Barnes et al., 2020). The first nondopaminergic treatment for schizophrenia was approved by the FDA in September 2024 (Cobenfy™, formerly Kar-XT) (Dolgin, 2024). It consists of xanomeline, an M_1 and M_4 receptor agonist that provides more effective relief from negative symptoms (Sharma et al., 2025), alongside tropsium, a nonbrain-penetrant muscarinic antagonist that limits peripherally mediated side effects. While this represents a welcome advance, xanomeline and other existing approaches only aim to normalize aberrant neurotransmission rather than addressing underlying disease neurobiology. It is therefore essential that the search for novel therapeutics continues. It is anticipated that machine learning will help identify patient subgroups who would benefit from more personalized treatment approaches (Coutts et al., 2025), and in time these may be informed by individual genetic, environmental, and lifestyle risk factors (Milic et al., 2025).

Molecular Insights into Proteinopathies
DOI: https://doi.org/10.1016/B978-0-443-34176-2.00006-7
© 2027 Elsevier Inc. All rights are reserved, including those for text and data mining, AI training, and similar technologies.

There are numerous reports of brain-regional transcriptomic changes in schizophrenia, but because the correlation between mRNA and protein expression is generally only around 40% (de Sousa Abreu et al., 2009; Vogel and Marcotte, 2012), the current chapter will focus predominantly on cellular, proteomic, and imaging data published in the last ~5 years. These demonstrate upregulation of proteins involved in inflammation, dysregulation of those involved in autophagy, and downregulation of those associated with synapses, mitochondria, and perineuronal nets (PNNs). In some cases, these findings vary across stages of the illness, so demographic details from cited studies are included throughout the chapter to provide additional important context. It should be noted that much of the work reviewed here comes from postmortem studies, where patient samples are almost always from those with some history of antipsychotic use. This is obviously not the case for so-called "control" groups. Nevertheless, subgroup analyses in positron emission tomography (PET) studies (where unmedicated first-episode patients are encountered) and complementary preclinical data suggest that medication confounds are unlikely to underlie the vast majority of "disease-related" alterations (Halff et al., 2021). The chapter finishes by considering advances in identifying protein biomarkers for schizophrenia in peripheral fluids and exploring the potential of novel therapeutic approaches. These include new antiinflammatory agents, autophagy inducers, TrkB receptor modulators, and matrix metalloproteinase (MMP) inhibitors.

5.2 Etiology of schizophrenia and risk for other illnesses

Schizophrenia has a multifactorial etiology involving a mix of genetic and environmental risk factors. In contrast to classical proteinopathies like Huntington's disease, there is no single causative gene mutation or any individual dysregulated protein. Instead, there is an array of relatively common single nucleotide polymorphisms (SNPs), each associated with a relatively small fold increase in risk, plus some rarer copy number variants (CNVs) with larger impacts. A 2022 genome-wide association study highlighted 342 independent SNPs spanning 287 loci, with 106 protein-coding genes proposed to underpin these associations (Trubetskoy et al., 2022). Some encode neurotransmitter receptors (*GABBR2*, *GRIN2A*, *GRM1*) and ion channels or transporters (*ATP2A2*, *CACNA1C*, *CLCN3*, *TRPC4*). However, the majority encode a diverse array of proteins linked to neuronal development (*AKT3*, CNTN4, *DCC*, *KLF6*, *NEGR1*, *PTPRD*, *SP4*, *ZNF804A*), chromatin modification (*KANSL1*, *SETD6*), synaptic organization or synaptic vesicle recycling (*DLGAP2*, *GPM6A*, *SNAP91*), protein folding or degradation (*CUL9*, *DNAJA3*, *PDIA3*, *PJA1*, *PSMA4*), mitochondrial function (*DARS2*, *IMMP2L*), autophagy and mitophagy (*AMBRA1*, *ATG13*, *BNIP3L*), immune function (*BCL11B*, *CISD2*, *CSMD1*), cytokine and interferon signaling (*IL1RAPL1, IRF3*), and redox homeostasis (*ASPHD1, ENOX1, TXNRD1*). CNVs also predominantly affect genes involved in neurodevelopment or neuronal and synaptic activity (Marshall et al., 2017). Examples include *MKRN1, MYH9,* and *NRXN1* deletions, as well as *AMN1, KNDC1,* and *PCMTD2* duplications (Prata et al., 2019; Zarrei et al., 2019; Maury et al., 2023). A very recent multiomics integration of SNP genotyping, RNA sequencing (RNA-Seq), ribosomal profiling (Ribo-Seq), and proteomic datasets from prefrontal cortical samples showed that, even though the number of significant quantifiable trait loci (QTL) decreased along the path of the central dogma, there were still 657 QTL at the protein level and 161 genes with protein-specific QTL (Liang et al., 2025a).

Just as there is no single genetic contributor, there is also no single environmental trigger. Instead, multiple potential influences span prenatal development through to adolescence. The earliest risks are maternal stress, infection, or malnutrition during the first trimester of gestation. For example, individuals whose mothers experienced bereavement or famine within that window have a twofold higher likelihood of developing schizophrenia, and those whose mothers contracted influenza have a sevenfold higher likelihood. These increases in disease risk have been attributed to epigenetic changes that reshape developmental trajectories (Álvarez-Mejía et al., 2025). Obstetric complications leading to hypoxia and neonatal infection with *Toxoplasma gondii* are other recognized early contributors (Dalman et al., 2001; Mortensen et al., 2007), and there is interest in the potential impact of microplastics or flame retardants (Doolittle and Duque-Wilckens, 2025). One of the most prominent risks throughout childhood and adolescence is social adversity, with physical or emotional neglect before the age of 18 producing a 3.3-fold increase in the likelihood of illness (Zhou et al., 2025). Adolescent cannabis use leads to a twofold increase (Pourebrahim et al., 2025). Multiple factors associated with urbanicity, such as air pollution (Attademo and Bernardini, 2020) and increased light at night (Deprato et al., 2025), are also implicated, although it is hard to disentangle their individual contributions from accompanying differences in social cohesion.

Whilst schizophrenia is widely regarded to stem from altered neurodevelopment, it is interesting that several risks for this illness are also linked to neurodegenerative proteinopathies, and patients with schizophrenia are more likely to develop these. For example, Mendelian randomization analyses (Baranova et al., 2024; Tao et al., 2025) associate schizophrenia with increased odds of all-cause dementia, Alzheimer's disease (AD), vascular dementia, and potentially also amyotrophic lateral sclerosis (ALS). Cohort studies in Taiwan, the USA, Australia, and Finland support this pattern for all-cause dementia, AD, or vascular dementia (Tapiainen et al., 2017; Lin et al., 2018; Almeida et al., 2019; Stroup et al., 2021; Liou et al., 2023), and a recent umbrella review found all-cause dementia to be 2.29 times more common in patients with schizophrenia (Zhao et al., 2022). Separate analyses of medical records suggest patients hospitalized due to schizophrenia have a higher rate of new ALS diagnoses across the following year than those without recent psychosis, and there are also higher instances of schizophrenia after an initial ALS diagnosis (Turner et al., 2016; Longinetti et al., 2017). Intriguingly, there are also reports of altered motor neuron function in schizophrenia. This might suggest some overlapping pathophysiology (Kisby et al., 2024), although neither illness seems to influence the progression or prognosis of the other. Some neuropathological studies suggest a higher prevalence of AD-like neuropathology in elderly patients with schizophrenia than those without (Prohovnik et al., 1993), but other findings do not support worsening of core AD hallmarks. For example, multiple studies from relatively large cohorts of older (mean $\pm$ SD age 62 ± 11 years, $n = 125$) or elderly schizophrenia patients (79 ± 11 years, $n = 100$) found no evidence for more extensive neurofibrillary tangles or amyloid plaques than in age-matched individuals with other psychiatric illnesses or controls without any diagnosis (Niizato et al., 1998; Purohit et al., 1998). A 2024 metaanalysis involving 318 patients from 14 studies supports this (Wilson et al., 2024). Importantly, this does not preclude other degenerative changes in schizophrenia, and indeed this illness shares gray (Ulugut et al., 2023) and white (Kochunov et al., 2021) matter deficits with neurodegenerative illnesses. Volumetric analyses show that these are progressive, being relatively subtle in first-episode patients and more widespread in chronic cases. Affected regions in the latter include the orbital, middle frontal, anterior cingulate, and middle temporal cortices, as well as the insula (Olabi et al., 2011; Torres et al., 2016). Schizophrenia and neurodegenerative illnesses also share molecular changes indicative of cellular stress and apoptosis (Durrenberger et al., 2015), as well as parallels in terms of genomic instability and aberrant activation of DNA damage response pathways (Kisby et al., 2024).

5.3 Neurobiological hypotheses

Initial hypotheses for schizophrenia focused on dysregulated dopaminergic neurotransmission (Meltzer and Stahl, 1976) and NMDA receptor hypofunction (Olney et al., 1999) before evolving to include neuroinflammation (Watanabe et al., 2010). The current master version links these (Howes and Onwordi, 2023). It proposes that multiple early life events trigger aberrant microglial-mediated synaptic pruning, which leads to cortical excitation–inhibition imbalance (underlying the negative and cognitive symptoms), plus disinhibition of striatal and midbrain dopaminergic neurons (causing psychotic symptoms, the stress of which further exacerbates aberrant synaptic pruning). Continued evolution of neurobiological hypotheses is to be expected, for example, to reflect involvement of the microbiota–gut–brain axis (Lee et al., 2025b), gut microbiota-derived metabolites (Karska et al., 2025), and gut microbiota-derived extracellular vesicles (EVs) (Wang et al., 2025).

5.4 Brain regional proteomic changes relevant to neuroinflammatory and neurodegenerative aspects of schizophrenia

5.4.1 Upregulation of proteins involved in immune function and inflammation

An extensive body of literature has focused on neuroimmune and neuroinflammatory changes in schizophrenia, from imaging studies of microglial activation at different stages to a wealth of postmortem measures. PET studies have focused on an 18 kDa translocator protein (TSPO) that is present on mitochondrial membranes and highly expressed in microglia. Initial scans using the first-generation TSPO ligand [^{11}C]-PK11195 revealed increased total binding to gray matter in young adults within their first 5 years of illness (24 ± 2 years) (van Berckel et al., 2008). There are similar reports in the hippocampus (Doorduin et al., 2009), and with the second-generation TSPO ligands [^{18}F]-PBR111 (Ottoy et al., 2018) and [^{11}C]-PBR28. These extend to patients with established schizophrenia (47 ± 9 years; Bloomfield et al., 2016) and even unmedicated individuals at ultra-high risk for psychosis (24 ± 5 years). However, there are also a range of negative findings in multiple brain regions and patient cohorts (Coughlin et al., 2016; Holmes et al., 2016; van der Doef et al., 2016; Hafizi et al., 2017). One 2019 metaanalysis (involving 190 cases and 200 controls across 12 studies; Marques et al., 2019) detected elevations in sensitive outcome measures that reflect nondisplaceable specific binding to TSPO (i.e., binding potential), but no change in combined indices of specific plus nonspecific binding (i.e., volume of distribution). There are two important considerations here. Firstly, several studies support the existence of different patient subgroups, including a "high inflammatory" subtype that represents ~30%–45% of cases and also appears to have more pronounced decreases in cortical gray matter volume (Fillman et al., 2013; Zhang et al., 2016; Purves-Tyson et al., 2021). The "low" and "high" inflammation subgroups were originally defined on the basis of postmortem transcriptomic data from part of the dorsolateral prefrontal cortex (Brodmann area 46; BA46), so it is unclear how they align with subjects in the TSPO-PET studies, and it is conceivable that some negative neuroimaging findings may involve cohorts where the "high" inflammation subgroup is underrepresented. Secondly, the TSPO marker has been criticized because its expression is not specific to microglial mitochondria, and there is no TSPO upregulation in activated microglial cultures or in postmortem samples from patients with schizophrenia (Sneeboer et al., 2020). It has therefore been argued that TSPO-PET signals actually reflect the general density of inflammatory cells rather than any changes in their activation state (Nutma et al., 2023). The absence of any PET ligands for other microglial activation markers prevents additional insights into the role of this process in disease onset or progression.

Postmortem findings are obviously unable to reveal if brain-regional inflammation precedes and contributes to the onset of schizophrenia (as current hypotheses propose; Howes and Onwordi, 2023), or whether it is secondary to other neurobiological disruptions.

Despite this, there is evidence for changes to established postmortem microglial markers (CD68, HLADR, or Iba-1 expression, morphological appearance) by the later stages of the illness. Accordingly, a 2017 metaanalysis (involving 181 cases and 159 controls from 11 studies) confirmed an increase in the density of activated microglia. This was most consistent in the temporal cortex (van Kesteren et al., 2017), a region implicated in auditory hallucinations and social/emotional processing deficits. There were also increases in levels of proinflammatory proteins and the transcripts encoding them in a pooled mRNA and protein analysis (330 cases and 323 controls across 14 studies). Many component studies showed increases in multiple inflammatory proteins across the dorsolateral prefrontal (BA9), rostrolateral prefrontal (BA10), and ventral anterior cingulate (BA24) cortices of various age ranges, and there have been similar reports subsequently. For example, levels of TNFα, IL-6, and lymphotoxin A were increased in BA9 of patients aged 43 ± 16 years (Pandey et al., 2018), and TNFα, IL-1β, IL-6, and IL-10 increases were described in pooled BA9/10/24 samples from those aged 77 ± 9 years (Kim et al., 2016). At similar life stages, those aged 46 ± 14 years showed increased expression of the lipopolysaccharide-activated Toll-like receptor 4 (TLR4) and its downstream adapter protein myeloid differentiation factor 88 (MyD88) in BA9 (García-Bueno et al., 2016). And while those 74 ± 9 years actually exhibited lower TLR4 and MyD88 expression, they did show other indices of increased inflammation, including higher levels of the prostaglandin-producing cyclooxygenase-2 (COX-2) enzyme and of a toxic product of lipid peroxidation, malondialdehyde (MacDowell et al., 2017). The latter points towards increased oxidative stress, and indeed some studies have shown lower levels of the antioxidant glutathione in schizophrenia, either at postmortem (e.g., BA10 of those aged 45 ± 14 years; Gawryluk et al., 2011) or using magnetic resonance spectroscopy (MRS), for example, anterior cingulate cortex of those aged 27 ± 7 years (Kumar et al., 2020). A metaanalysis of MRS data from the anterior cingulate or medial prefrontal cortices showed no change in schizophrenia patients as a whole (639 cases and 704 controls from 20 studies), but when clinical high-risk and first-episode studies were excluded, the remaining dataset (~370 cases and ~400 controls from 14 studies) did support a decrease (Murray et al., 2024). Further support for redox dysregulation in schizophrenia is provided by a proteomic analysis of BA24 gray matter. Of 42 proteins that were differentially expressed in schizophrenia cases (45 ± 14 years), four are involved in oxidative stress. Patient samples showed lower levels of dimethylarginine dimethylaminohydrolase 1 and hydroxyacylglutathione hydrolase (Glx II) as well as higher levels of superoxide dismutase and annexin A1 (Clark et al., 2006). Receptor for advanced glycation end-product (RAGE) and MMP9 have been proposed as possible substrates for the interaction between oxidative stress and neuroinflammation in schizophrenia (Dwir et al., 2020). Other studies of individual proteins at later stages of the illness (60 ± 10 and 77 ± 9 years) revealed increased expression of the inflammasome-forming NLR family pyrin domain-containing one protein in the orbitofrontal cortex BA11/12 (Španić Popovački et al., 2024), and of the caspase 1 enzyme (that cleaves inactive IL-1β and IL-18 precursors into mature, active forms) within the mitochondrial fraction of pooled BA9/10/24 (Kim et al., 2016). Interestingly, a larger peptidomic analysis of the anterior temporal lobe (BA38) in later illness (67 ± 14 years) showed reduced levels of a fragment termed PepH, which is derived from the histone H2B type 1-H protein. PepH increased survival and prevented lipopolysaccharide-induced endotoxicity in a cultured neuronal cell line, so it is plausible that PepH downregulation in schizophrenia may be one factor that confers increased vulnerability to TLR4-mediated inflammation and neurodegeneration (Café-Mendes et al., 2017). Because the combined immune and inflammatory changes in schizophrenia show multiple overlaps with those in AD (including microglial activation, elevated cytokine levels, and redox dysregulation), it is not surprising that some similar downstream effects (including dysregulation of mammalian target of rapamycin [mTOR] signaling and autophagy) also feature in both illnesses (Russo et al., 2025).

5.4.2 Dysregulation of proteins involved in autophagy

Autophagy plays key roles throughout neurodevelopment and healthy aging by promoting cytoskeletal changes during differentiation, pruning redundant or inappropriate synapses, and eliminating dysfunctional mitochondria that are major sources of reactive oxygen species. It is well-recognized that deficient autophagy contributes to neurodegenerative illnesses by permitting accumulation of neurotoxic protein aggregates or misfolded proteins and triggering increased oxidative stress. Conversely, excessive autophagy can also have neurodegenerative consequences, including excessive synaptic loss (Zapata-Muñoz et al., 2021). Interestingly, region-specific transcriptomic and proteomic changes in schizophrenia appear to point towards the possibility of both scenarios at different stages of the illness. But autophagic changes in schizophrenia are currently less well-characterized than inflammatory ones and potentially less severe than in neurodegenerative illnesses. As outlined below, there are gaps in knowledge arising from the quantification of different endpoints at different time points or in different patient cohorts, and there also appear to be some potentially conflicting changes within the same patient groups.

The superior temporal gyrus (BA22) is implicated in the positive symptoms of schizophrenia, and in older patients (72 ± 17 years), it has reduced levels of the mRNAs encoding two proautophagic proteins, autophagy-related 3 and 7 (ATG3 and ATG7) that are essential for autophagosome formation. Curiously, this is accompanied by a potentially opposing reduction in mRNA for Bcl-2-like protein 1 (Bcl-xL), which inhibits autophagy by binding and sequestering the autophagy initiator beclin-1. The same study found none of these changes in BA10, which is one of the regions implicated in negative symptoms (Barnes et al., 2011), and there were also no changes to ATG3, ATG7, or Bcl-xL transcript levels in BA46 of a middle-aged cohort (45 ± 9 years). However, middle-aged individuals did show widespread reductions in beclin-1 mRNA across the frontal cortex, hippocampus, and striatum (Lanz et al., 2019), and a similar hippocampal decrease occurred in a middle-older-age cohort (52 ± 17 years; Merenlender-Wagner et al., 2015), where only that region was examined. Despite this, none of the aforementioned regions appear to show differential expression of transcripts encoding other autophagy regulators like Bax or p62 (Barnes et al., 2011; Merenlender-Wagner et al., 2015).

There do not seem to have been follow-up assessments of brain-regional ATG3, ATG7, beclin-1, or Bcl-xL protein levels, but there are changes in the expression or activation of other autophagy-related proteins (Barnes et al., 2011). Middle-aged patients (43 ± 8 years; 21 ± 10 years since diagnosis) exhibit gender-specific effects in BA46, which point towards increased autophagy in females with schizophrenia. This group also has increased activity of the autophagy-promoting enzyme AMP-activated protein kinase (AMPK; shown by increased phosphorylation at Thr172), plus a corresponding increase in beclin-1 activity (shown by increased phosphorylation at Ser93; Bjornson et al., 2024). It remains to be seen whether these effects persist into later life or begin to affect males. Aside from one older mixed-sex cohort (68 ± 14 years; 29 ± 16 years since diagnosis) that showed increased expression of the autophagy-promoting RB1-inducuble coil protein 1 (RB1CC1) within the dentate gyrus (Lappas et al., 2025), other changes in older patients generally point towards deficient autophagy. For example, there is reduced RB1CC1 expression in the CA1 subfield, along with reduced expression of autophagy-related protein 16-like 1 (ATG16L1), which initiates autophagosome formation (Lappas et al., 2025). Another elderly cohort (68 ± 11 years; 42 ± 10 years since diagnosis) showed increased activation of the mTOR pathway, which is a key suppressor of autophagy. These changes include increased total levels of mTOR protein in BA10 and increased activation of a downstream effector, the ribosomal protein S6, in BA22 (shown by increased phosphorylation at Ser235/236; Izumi et al., 2022). mTOR activation is triggered by TLR4 activation (Lashgari et al., 2024), and under normal circumstances, the resulting suppression of autophagy plays an important role in reducing inflammation by facilitating the removal of damaged cell components. But as noted above, prolonged suppression can lead to abnormal patterns of protein aggregation, which are reported in schizophrenia.

Insoluble accumulations of the DISC1 scaffold protein have been detected in BA23 (42 ± 9 years; Leliveld et al., 2008), and in vitro studies show that these aggregates interfere with the normal role of DISC1 in intracellular transport of mitochondria (Atkin et al., 2012). In 80% of cases with insoluble accumulations, dysbindin family members coprecipitate with DISC1 (Ottis et al., 2011), which is noteworthy because dysbindin-1B has neurotoxic effects in vitro (Zhu et al., 2015), and its overexpression in mouse hippocampus promotes dendritic shrinkage and loss of dendritic spines (Yang et al., 2016). Similarly, insoluble cytoskeletal aggregates of trio-binding protein 1 have been detected in BA23 (Zaharija et al., 2022), and aggregates of neuronal PAS domain protein 3 have been found in the insular cortex (Samardžija et al., 2021).

Abnormal patterns of solubility are not restricted to these few examples. Postmortem samples from multiple brain banks, including the Harvard Brain Tissue Resource Center (BA10; 54 ± 4 years), University of Pittsburgh (BA22; 49 ± 13 years), and University of Texas Southwestern Medical Center (BA46; 51 ± 15 years), all include a subgroup of schizophrenia patients (approximately 50%) with increased levels of insoluble and ubiquitinated proteins (Nucifora et al., 2019). The high insolubility brains have a lower pH (6.4 ± 0.3 rather than 6.7 ± 0.3) and are more likely to be from females, but it is currently unclear how this insolubility group might map to the "high" and "low" inflammatory subtypes described above, or to other subgroups (Fillman et al., 2013; Zhang et al., 2016; Purves-Tyson et al., 2021). The insoluble fraction comprises nearly 1000 proteins, many of which are associated with cell assembly and organization, as well as several stages of the ubiquitination process. Single-cell RNA-Seq indicates that the insoluble proteins are enriched in neurons, specifically parvalbumin-positive GABAergic neurons and excitatory neurons throughout cortical layers 2–6 (Nucifora et al., 2019), which are key contributors to the excitatory–inhibitory imbalance in schizophrenia.

5.4.3 Downregulation of proteins associated with synapses, mitochondria, and perineuronal nets

Inflammation and dysregulation of autophagy both have the potential to reduce synaptic integrity (de Bartolomeis et al., 2022), and there is extensive evidence that both presynaptic and postsynaptic perturbations do occur in schizophrenia. This includes several metaanalyses within the last ~5 years. For example, postmortem findings support decreased expression of the synaptic vesicle protein synaptophysin in the hippocampus, prefrontal (BA9–10 and 46–47), and cingulate cortices (BA24 and 32–33; Osimo et al., 2019). Despite the same authors finding no overall changes in presynaptic SNARE complex proteins like SNAP-25, VAMP, or syntaxin (that are involved in vesicle fusion and exocytosis), separate support for presynaptic deficits comes from PET studies using ligands for synaptic vesicle protein 2A (SV2A). In chronic cases (42 ± 11, 41 ± 12, or 41 ± 11 years; scanned 17 ± 12 years after symptom onset), there are consistent widespread reductions in [^{11}C]UCB-J volume of distribution, which is decreased by 8%–13% in the frontal and/or anterior cingulate cortices (Cohen's d effect sizes 0.8–0.9) (Onwordi et al., 2020; Onwordi et al., 2021; Radhakrishnan et al., 2021). First-episode cases (27 ± 8, scanned 3 ± 2 years after onset) show no change in volume of distribution (Onwordi et al., 2024), but there are 18%–19% decreases in binding potential in the orbitofrontal cortex and right middle frontal gyrus (Cohen's d 1.5; Yoon et al., 2023). One recent study using the alternative tracer [^{18}F]SynVesT-1 even showed reduced binding potential in the anterior cingulate and prefrontal cortex at an earlier stage in first-episode psychosis (26 ± 5 years, equivalent to 0.9 ± 1 years after onset), plus the anterior cingulate of those at clinical high risk (21 ± 3 years), albeit with smaller effect sizes (Cohen's d 0.29–0.35) (Blasco et al., 2025). Increasing effect sizes and the emergence of additional decreases in [^{11}C]UCB-J volume of distribution with disease progression appear to point towards increasing presynaptic deficits across the course of the illness, rather than an initial failure of development. Unfortunately, SV2A PET studies are unable to reveal whether decreases in tracer binding reflect loss of entire neurons, cells with fewer synaptic contacts, reduced numbers of synaptic vesicles, and/or downregulation of the SV2A protein. It is also unclear whether the presynaptic deficits detected with this approach vary according to neurotransmitter phenotype. SV2A expression typically colocalizes more strongly with GABAergic than glutamatergic terminals (Mendoza-Torreblanca et al., 2019; Pazarlar et al., 2022), but there are region-specific variations. For example, the polymorphic layer of the dentate gyrus shows the highest colocalization of SV2A with glutamatergic markers and the lowest colocalization with GABAergic ones https://www.biorxiv.org/content/10.1101/2024.07.15.603608v1.full.pdf. There is even a restricted population of neurons that are entirely devoid of SV2A expression, including a glutamatergic subset that relies solely on the alternative family member SV2B, and a GABAergic population that relies on SV2C (Pazarlar et al., 2022). There are no PET ligands capable of capturing presynaptic changes affecting these cells or that enable temporal analysis of global postsynaptic changes. A 2020 metaanalysis of postmortem data on dendritic spine density, postsynaptic density number, and postsynaptic protein expression did find an overall decrease in schizophrenia (Berdenis van Berlekom et al., 2020). This was attributed to reductions in cortical regions, with the impact on dendritic spine density being more robust. Additional subanalyses revealed deficits in the frontal cortical layer 3, and these may not have been captured by previous studies that did not include layer-specific assessments. As with presynaptic changes, there is a distinct possibility that deficits might arise from changes at the cellular, morphological, or individual protein level, and vary according to neurotransmitter phenotype or patient subgroup.

There are reports of reduced M_1/M_4 receptor expression across the striatum, hippocampus, and frontal cortical regions (Saint-Georges et al., 2025). This is noteworthy here because these receptors are normally localized to both presynaptic and postsynaptic sites on dopaminergic, glutamatergic, and cholinergic neurons (Levey et al., 1995). A subgroup of schizophrenia patients (~25%) have pronounced muscarinic receptor deficits ~75% loss (Scarr et al., 2009); however, it is unknown whether this population is distinct from or overlapping with other subgroups. The implications for therapeutic responses to muscarinic agents are also unclear, since there was no patient stratification in the trials that detected positive effects of xanomeline–trospium (Cobenfy™ (Mohammed et al., 2025)) nor in those that failed to detect any beneficial effect of the M_4 positive allosteric modulator emraclidine (https://www.prnewswire.com/news-releases/abbvie-provides-update-on-phase-2-results-for-emraclidine-in-schizophrenia-302301190.html).

There are also pronounced GABAergic changes in schizophrenia (Simmonite et al., 2023), with a 2019 metaanalysis supporting frontal cortical reductions in the density of interneurons that express the calcium-binding protein parvalbumin (Kaar et al., 2019). One component study attributes this to a "low GABA" subgroup that represents ~50% of cases and has a ~30% decrease in transcripts encoding parvalbumin and the GABA synthesis enzyme GAD_{67} (Volk et al., 2016). Some authors have been able to detect deficits in GABAergic transcripts without accompanying decreases in neuronal numbers (Akbarian et al., 1995), whereas others (including work in older patients; 60 ± 11 years) also find reduced interneuron numbers across multiple layers of the cingulate and prefrontal cortices (Benes et al., 1991). This highlights the need for better characterization of patient subgroups to enable stratified analysis and reporting in future studies, whether these be imaging and postmortem assessments of disease-related changes or clinical evaluations of novel therapeutics. This is especially important for small sample sizes where unbalanced recruitment is inevitable, and a 2025 study by Lorincz et al. illustrates how inclusion of additional pertinent data can provide vital contextual information in such cases. The authors analysed BA9 tissue from a small number of elderly patients (71 ± 10 years; $n = 6$) that demonstrated a 30% reduction in the intensity of parvalbumin immunoreactivity without any change in the density of labeled cells. In these samples, there was no change in expression of two presynaptic active zone proteins, bassoon and munc13-1, nor any postsynaptic differences for excitatory junctions as a whole using measures like PSD-95 labeling. However, specific excitatory synapses on parvalbumin-expressing interneurons did contain lower NMDA receptor densities (Lorincz et al., 2025). This is consistent with the established hypothesis that reduced glutamatergic input to these cells leads to reduced GABAergic tone, followed by disinhibition of glutamatergic pyramidal cells and associated excitotoxicity (Deutsch et al., 2001).

Parvalbumin-positive interneurons are particularly vulnerable to excitotoxic damage for several reasons (Ruden et al., 2021). These fast-spiking cells require high levels of ATP to maintain ionic gradients, so they normally have approximately twice as many mitochondria as nonparvalbumin neurons (Glausier et al., 2017). As a consequence, they generate higher levels of reactive oxygen species and are more susceptible to redox dysregulation and loss of protective PNN, which are a specialized form of extracellular matrix that surrounds their soma and dendrites to support synaptic contacts in these regions (Cabungcal et al., 2013). Furthermore, parvalbumin-positive interneurons receive a strong excitatory drive and are rich in AMPA receptors that enable excessive Ca^{2+} and Zn^{2+} entry during intense stimulation, triggering mitochondrial damage and apoptotic cell death (Ruden et al., 2021). There is evidence that the pathological changes in schizophrenia impinge on multiple aspects of this vulnerability to generate a vicious cycle of increasing excitatory–inhibitory imbalance (Cuenod et al., 2022). Ultrastructural studies indicate that both inhibitory and excitatory neurons contain fewer mitochondria (for review, see Roberts, 2021). Separate metaanalyses show the downregulation of 18 genes encoding mitochondrial ribosomes (211 cases and 211 controls from 10 studies; Bartal et al., 2023), and multiple alterations to the mitochondrial electron transport chain (201 cases and 206 controls across 13 studies; Holper et al., 2019). The first enzyme complex in this chain (NADH dehydrogenase) contributes to the proton gradient that drives ATP synthesis. It consists of 45 subunits, and there is sufficient brain-regional expression data to enable metaanalysis on four of these. Levels of NDUFS1, NDUFV1, and NDUFV2 decreased in the striatum, NDUFV1 decreased in the frontal cortex (BA46/49), and NDUFV2 increased in the parietal cortex (BA19). The fourth enzyme complex, cytochrome c oxidase, catalyzes the rate-limiting step of the chain. It consists of 13 subunits, and although their expression appears unaltered in schizophrenia, the activity of complex IV is reduced in the frontal cortex and increased in the striatum. There are similar decreases in frontal cortical complex IV activity in both AD and Parkinson's disease (Holper et al., 2019). Examination of patient demographics in the schizophrenia studies showed that changes to complex I protein expression are detectable in middle-aged (43 ± 8, 45 ± 13, or 46 ± 12 years; Iwamoto et al., 2005; Ben-Shachar and Karry, 2008; Karry et al., 2004), whereas complex IV activity measures all involved tissue from elderly individuals (67 ± 12, 78 ± 7, 80 ± 12, or 82 ± 11 years; Maurer et al., 2001; Prince et al., 1999; Cavelier et al., 1995; Prince et al., 2000). Postmortem tissue samples from patients with schizophrenia also show widespread reductions in PNN density across BA9 and the entorhinal cortex (BA28), amygdala, thalamic reticular nucleus, and inferior colliculus (Lisboa et al., 2024). Again, these are detectable in both middle-aged 46 ± 9 or 46 ± 10 years (Alcaide et al., 2019; Enwright et al., 2016) and older cohorts 62 ± 15 or 62 ± 17 years (Steullet et al., 2018; Pantazopoulos et al., 2010). It is entirely plausible that immune and inflammatory changes could contribute to the onset of synaptic, mitochondrial, and PNN deficits in schizophrenia, and equally likely that the neuronal deficits perpetuate a cycle of worsening inflammation.

5.5 Peripheral fluid biomarkers relevant to neuroinflammatory and neurodegenerative aspects of schizophrenia

Psychiatric disorders like schizophrenia are currently diagnosed on the basis of clinical symptoms alone. This is because there are no validated molecular tests to help identify patients, predict treatment responsiveness, or provide objective measures of treatment efficacy. And although there is progress towards identifying functional neuroimaging biomarkers (Di Camillo et al., 2024; Gkintoni et al., 2026), their reliance on specialist facilities available at relatively few sites will limit incorporation into routine clinical practice. Molecular biomarkers in readily accessible peripheral samples (i.e., saliva, serum, plasma, peripheral blood mononuclear cells [PBMCs]) or cerebrospinal fluid (CSF) are therefore highly desirable. But despite approximately 5000 articles on the subject in the past 25 years, including multiomics studies showing hallmarks of accelerated aging in schizophrenia (Campeau et al., 2022; Qi et al., 2025), there is still no widely accepted individual marker or consensus panel that appears to have the necessary levels of diagnostic accuracy or specificity across all stages of the illness (Berdeville et al., 2025). The challenges in this field are highlighted by a 2015 study, albeit of transcript levels, where only 1% of the 1247 genes dysregulated in brain or blood samples were differentially expressed in the same direction across both datasets (Bergon et al., 2015). Furthermore, several proteins that do show the same direction of change in brain and peripheral samples are also differentially expressed in other illnesses. For example, lower levels of brain-derived neurotrophic factor (BDNF) throughout the course of schizophrenia (Huo et al., 2021; Nunes Noto et al., 2021) are now thought to reflect depressive symptoms in this illness (Manchia et al., 2022), as well as in asthma (Kawamoto et al., 2025) and chronic pain states (Yang et al., 2025). Likewise, patients with schizophrenia and dementia both show overlap in terms of lower serum MMP7 and elevated tissue inhibitor of metalloproteinases 2 (Tylec et al., 2021).

5.5.1 Peripheral protein biomarkers relevant to immune function and inflammation

Some of the most commonly reported peripheral alterations in schizophrenia relate to complement and coagulation systems (Byrne et al., 2024; Hong et al., 2025). A 2022 metaanalysis of mass spectrometry data identified 217 differentially expressed proteins in schizophrenia patients compared to those without a psychiatric diagnosis. These were enriched across lipid metabolism, complement or coagulation cascades, and immune response pathways. Changes to individual proteins include upregulation of ficolin-3 and downregulation of the apolipoproteins ApoA1, ApoA2, ApoC1, and ApoC3 (Rodrigues et al., 2022), but unfortunately, most of these overlap with depression and bipolar disorder (Song et al., 2015; Caspani et al., 2021). This message is reinforced in a recent analysis of inflammatory proteins, which showed similar plasma changes to IL-8, IL-12, and IL-13 in both schizophrenia and bipolar disorder (Naifar et al., 2025).

Encouragingly, this picture may improve with proteome profiling of plasma-derived EVs rather than whole plasma. For example, the complement proteins C4 and C4 binding protein A, as well as the anticoagulant protein S, were all upregulated in the EV-derived fraction from two sets of patients with chronic schizophrenia (39 ± 12 or 38 ± 13 years; 14 ± 8 or 13 ± 9 years since diagnosis, $n = 20$ or 26) yet downregulated or unaltered in depression and bipolar. Using a panel of four EV-derived proteins, an XGBoost-based machine-learning approach was able to identify chronic schizophrenia cases with an area under the curve (AUC) of 0.895, overall accuracy of 84%, sensitivity (for true positives) of 85%,and specificity (for true negatives) of 82%. It was also able to predict antipsychotic responses in a follow-up cohort (44 ± 16 years; 15 ± 13years since diagnosis, $n = 27$), and further examinations in larger and more diverse populations are now awaited. Disappointingly, the same panel provided no reliable separation between controls and first-episode unmedicated patients (30 ± 11 years; 0.7 ± 0.8 years since diagnosis, n = 35; Xue et al., 2024), who are the very population where diagnostic and prognostic biomarkers would be of greatest value. Another proteomic study of plasma-derived EVs from a small cohort of first-episode unmedicated patients (39 ± 11 years; 2 ± 2 years since diagnosis, n = 10) did identify 121 differentially expressed proteins (enriched in domains related to cytokines and complement components, as well as a category related to cholinergic synapses; Sun et al., 2025), but any diagnostic or prognostic potential has yet to be reported.

One alternative strategy to maximize the utility of whole plasma proteomic datasets involves adopting more predictive machine-learning algorithms. Gradient-boosting approaches (e.g., XGBoost) that have limited capacity for handling high-dimensional molecular interactions achieve limited discrimination (area under the precision–recall curve, AUPRC, ~0.74), whereas neural network-based approaches (e.g., CNN-BiLSTM) that capture nonlinear interactions are far superior (AUPRC ~0.96; Hong et al., 2025). A second refinement strategy involves integrating plasma proteomic datasets with those on posttranslational modifications and metabolomics. Two algorithms achieved good accuracies (>90%) using this multiomics approach, with the LightGBMXT gradient-boosting approach being superior across multiple metrics. Whilst molecular signatures differentiating schizophrenia cases from controls were related to immune responses and coagulation processes, it is important to note that posttranslational modifications were amongst the key discriminators. These include carbamylation at immunoglobulin constant region sites IGKC_K20 and IGHG1_K8, and oxidation of coagulation factor F10 at residue M8 (Hong et al., 2025). A separate area of promise involves refining the initial acquisition of proteomic datasets. Four-dimensional data-independent acquisition (4D-DIA) proteomics improves on traditional mass spectrometry by including ion-mobility separation instead of just the standard three dimensions (mass-to-charge ratio, retention time, and intensity). In a chronic schizophrenia population (44 ± 11 years, n = 40), this approach identified 79 differentially expressed proteins, enriched in chemokine signaling pathways and cytokine receptor interactions. Secondary screening and weighting of these markers revealed a top three comprising the cytoskeletal proteins profilin-1 and beta-actin-like protein 2, plus the glycolytic enzyme glyceraldehyde-3-phosphate dehydrogenase, all of which were upregulated in schizophrenia. Based on these three plasma proteins, machine-learning algorithms were able to achieve a diagnostic model with an AUC of 0.97, accuracy of 88%, sensitivity of 83%, and specificity of 92% (Shen et al., 2024). With the exception of sensitivity, these outperform metrics for the panel of four EV-derived proteins described above (Xue et al., 2024), but again, diagnostic capabilities still need to be evaluated in more diverse populations encompassing different geographical locations and earlier stages of the illness.

There has been some recent progress towards identifying potential inflammatory biomarkers in first-episode psychosis. Although elevated MMP9 levels have been detected at early stages and remain into chronic illness (Schoretsanitis et al., 2021; Liang et al., 2025b), similar changes occur in Fragile X syndrome (Laroui et al., 2022) and migraine (Imamura et al., 2008), so any use of this marker would need to be as part of a wider panel. A small proof-of-concept study examining peptides in PBMCs (from n = 6 patients) proposed a 22-residue fragment from galectin-related protein (LDDGHLN-NSLSSPVQADVYFPR) plus 10- and 11-residue peptides from fibrinogen alpha chain protein (QFTSSTSYNR and GSESGIFTNTK) as high priorities for future investigation (Santa et al., 2024). In a larger cohort, levels of the antioxidant enzyme peroxiredoxin 2 were higher in the first-episode group (22 ± 5 years, n = 49) than in controls or those at ultra-high risk of psychosis (Lee et al., 2024). When a focused panel of 92 immune markers was assessed in plasma and CSF from first-episode patients (28 ± 1 years, n = 77), one antiinflammatory marker, IL-10 receptor subunit alpha (IL-10RA), was downregulated in plasma and 12 were upregulated. These include axin-1 (a component of the beta-catenin destruction complex), IL-7, and the chemokines CXCL1, CXCL5, and CXCL6. The only change in CSF that survived corrections for multiple comparisons was a reduction in levels of CD5 (a surface marker for most T cells and a subset of B cells). These changes remained apparent at follow-up 1.5 years later (Eren et al., 2023), but now require confirmation at additional locations.

5.5.2 Peripheral protein biomarkers relevant to autophagy

Brain regional autophagic changes in schizophrenia are less well-characterized than inflammatory ones, and this pattern extends to peripheral markers of autophagy. In a multiomics characterization of multiple psychiatric disorders, it was possible to classify the 327 patients (26 ± 10 years) into three subgroups on the basis of differential plasma metabolites; S1 had consistent downregulation (of amino acids, organic acids, carbohydrates, and fatty acids), S2 had mixed upregulation (of organic acids) and downregulation (of steroids, bile acids, fatty acids, and amino acids) and S3 showed little metabolomic change. Of these groups, S1 showed more differentially expressed proteins (enriched in complement and coagulation cascades) and more differentially methylated genes, which included several hypomethylations enriched in the regulation of autophagy (Hao et al., 2023). The metabolite-based clusters did not map onto current clinical diagnoses, but the multiple levels of information provided by such studies are likely to aid emerging transdiagnostic approaches to classifying psychiatric illnesses (Farías Venegas et al., 2025).

There are altered peripheral levels of several autophagy-related proteins in schizophrenia-only studies, and as with the brain-regional findings, some directions of change appear contradictory. Chronic cases (39 ± 10 or 39 ± 13 years) show increased serum levels of two pro-autophagic mediators: microtubule-associated protein 1 light chain 3 beta

(LC3B, +40%; Zhang et al., 2025) and high mobility group box 1 protein (HMGB1, +30%; Yildiz et al., 2024). LC3B is central to the initiation and execution of autophagy and is considered one of the most reliable autophagic markers. HMGB1 plays a dual role, triggering autophagy via extracellular binding to RAGE receptors and promoting autophagosome formation via intracellular binding to beclin 1. Both of these changes appear to align with brain-regional evidence for increased autophagy in middle-aged schizophrenia patients (Bjornson et al., 2024). Yet they are accompanied by decreased serum levels of another proautophagic mediator, beclin 1 itself (-24%; Yildiz et al., 2024). Although these observations are open to interpretation, they are not unique to schizophrenia (Naguib and Rashed, 2018; Mallarpu et al., 2025), so any potential utility would need to be part of a wider biomarker panel. In first-episode cases (30 ± 8 years), it has been possible to detect DISC1 protein aggregates in CSF (Pils et al., 2023), which again aligns with BA23 observations in a similar age group (Leliveld et al., 2008) and currently appears to have less overlap with other illnesses than other autophagy-related markers.

5.5.3 Peripheral protein biomarkers relevant to synapses, mitochondria, and perineuronal nets

Some proteomic studies revealing inflammatory changes in peripheral fluids from schizophrenia patients have also provided support for differential expression of synaptic proteins in the same samples. For example, there was enrichment of those related to cholinergic synapses in plasma-derived EVs from a small cohort (n = 10) of first-episode unmedicated patients (39 ± 11 years; 2 ± 2 years since diagnosis; Sun et al., 2025). Similarly, in a multiomics characterization of a large cohort (n = 327) spanning multiple psychiatric disorders, the S1 subgroup (that featured metabolomic decreases and hypomethylation of genes relevant to autophagy) showed accompanying hypermethylation of genes enriched in biological processes related to neurotransmitter transport, the synaptic vesicle cycle, modulation of synaptic transmission, and regulation of trans-synaptic signaling. Hypermethylated genes were enriched in cellular compartments involving asymmetric synapses, glutamatergic synapses, dendritic spines, and neuronal spines. There was also hypomethylation of other genes involved in the synaptic vesicle cycle, as well as vesicle-mediated transport in synapses (Hao et al., 2023), and implications of this work for transdiagnostic psychiatry (Farías Venegas et al., 2025) have already been noted.

In terms of studies that have not yet been considered in this chapter, plasma EVs from patients with schizophrenia (26 ± 3 years; 3 ± 4 years since diagnosis) showed lower expression of one cytoskeletal protein, α-II-spectrin, that is essential for inhibitory synapse formation and plays critical roles in stabilizing synapses and regulating synaptic transmission. However, synaptophysin levels in the same samples were unaltered (Ranganathan et al., 2022). Serum samples from a reasonably similar population (31 ± 4 years; 13 ± 3 years since diagnosis) showed differential expression of four vesicle coat proteins, and whilst these changes appear capable of differentiating simple schizophrenia from paranoid schizophrenia, schizotypal disorder, and acute polymorphic psychotic disorder, the utility of this information is limited by the fact that protein identities were not provided (Dmitrieva et al., 2022). CSF samples from those with first-episode psychosis (28 ± 7 years; 1 ± 1 years since diagnosis) showed upregulation of 35 proteins, including neuropilin and tolloid-like protein 1, which is involved in regulating the abundance of NMDA receptor subunits in the postsynaptic density, and the metalloproteinase ADAMTS-4 that breaks down PNN proteins like aggrecan and brevican. The same CSF samples also showed downregulation of 37 proteins, including neuron-derived neurotrophic factor and cell adhesion molecule 3, which both help to regulate neuronal development and synapse formation, G protein-coupled receptor 158, which is involved in regulating synaptic transmission and neurotransmitter secretion, and glyoxalase 1, which detoxifies reactive glycolysis products (Haroon et al., 2024).

The metaanalysis that showed brain-regional downregulation of mitochondrial ribosome genes in schizophrenia also showed 16 downregulations in PBMCs (53 cases and 37 controls from 2 studies; Bartal et al., 2023). Of these, the two genes encoding mitochondrial ribosomal proteins L4 (MRPL4) and S7 (MRPS7) were downregulated in both brain and PBMCs. Likewise, the metaanalysis that showed brain-regional alterations to the mitochondrial electron transport chain also supported changes in blood elements (Holper et al., 2019). These include lower levels of complex I subunit NDUFV2 (mirroring findings in the striatum but opposing those in BA19) and higher levels of NDUFS1 (opposing those in the striatum). An additional recent study showed decreases in mitochondrial DNA copy number in peripheral blood leukocytes, and an accompanying metaanalysis showed that this effect was also detectable in whole blood or blood cells but not plasma (Wang et al., 2025). Lastly, the 35 proteins downregulated in CSF samples from those with first-episode psychosis (28 ± 7 years; 1 ± 1 years since diagnosis) include the mitochondrial enzyme hexokinase 1, which is crucial for the first step of glycolysis, and mitochondrial chaperonin 10, which assists with the folding and assembly of mitochondrial proteins (Haroon et al., 2024).

In summary, many of the peripheral protein changes in schizophrenia have very plausible links to the underlying neurobiology of this illness. In terms of peripheral fluid evidence for neuronal injury, there are no changes to plasma levels of neurofilament light chain protein (NfL) in those with treatment-resistant schizophrenia (40 ± 10 years; 18 ± 9 yearssince diagnosis), which is a more severe subtype affecting ~30% of patients and characterized by the persistence of symptoms after two antipsychotic trials (Eratne et al., 2022). In contrast, those with ultra-treatment-resistant schizophrenia (40 ± 10 years; 18 ± 9 years since diagnosis), which is an even more severe subtype defined by a lack of response to clozapine, do have elevated plasma levels of the neuroaxonal injury markers NfL and ubiquitin C-terminal hydrolase L1 (UCH-L1), as well as the astroglial injury marker glial fibrillary acidic protein (GFAP). Furthermore,

NfL, UCH-L1, and GFAP elevations had good accuracy (AUC 0.85–0.913) in distinguishing this neuroprogressive subgroup from controls (de Oliveira et al., 2025). It could be hypothesized that these patients exhibit more profound brain-regional neurodegeneration than treatment-responsive individuals, but confirmation will require neuroimaging or postmortem studies with appropriate levels of stratification.

5.6 Novel therapeutic approaches to counteract neuroinflammatory and neurodegenerative changes in schizophrenia

Pharmacological agents designed to normalize glutamatergic or GABAergic neurotransmission did not produce the expected clinical benefits in schizophrenia patients (for example, see Morozova et al., 2017; Surti et al., 2023), but there is still considerable interest in pursuing potential disease-modifying approaches. Proteomic insights have been invaluable to this process, and there are a number of emerging areas of promise (Lieberman et al., 2025; Muszyński et al., 2025), including those targeting inflammatory, autophagic, synaptic, and PNN pathology.

5.6.1 Potential treatments that influence immune function and inflammation

Given the wealth of data on immune and inflammatory changes in schizophrenia, it is no surprise that there have been many clinical trials of antiinflammatory agents, mostly as adjuncts to antipsychotics (for reviews, see Almulla and Maes, 2025; Huang et al., 2025). By 2020, their numbers were sufficient to enable a metaanalysis involving 70 randomized controlled trials with a combined total of over 4000 participants. This found an overall benefit of adjunctive agents on the Positive and Negative Syndrome Scale (PANSS), with no supremacy of those purposefully developed for immunomodulation (e.g., nonsteroidal antiinflammatory drugs like celecoxib and aspirin, the tetracycline antibiotic minocycline, and monoclonal antibodies like tocilizumab that inhibit cytokine signaling) over those subsequently found to have potential immunomodulatory properties (e.g., steroid hormones like estrogen, statins used to regulate cholesterol levels, and glitazones used to manage diabetes) (Jeppesen et al., 2020). Additional metaanalyses restricted to specific pharmacological agents have also found beneficial effects of adjunctive therapy. The COX-2 inhibitor celecoxib and antioxidant *N*-acetylcysteine both improved overall PANSS scores, as well as those on positive and negative subscales (Yolland et al., 2020; Farag et al., 2025); minocycline and statins both improved overall and negative subscores (Panizzutti et al., 2023; Peng et al., 2024), and sulforaphane, which is an antioxidant and antiinflammatory isothiocyanate found in brassicas, had modest effects on negative subscores (Kassar et al., 2025). There are not yet sufficient data for metaanalyses on other antiinflammatory approaches, but in individual trials, the phosphodiesterase inhibitor pentoxifylline (Shamabadi et al., 2025), sphingosine-1-phosphate mimetic fingolimod (Karbalaee et al., 2023), PPAR-α receptor activator palmitoylethanolamide (Salehi et al., 2022), and antioxidant resveratrol (Samaei et al., 2020) all improved PANSS negative subscores, whereas the anti-IL-1β monoclonal antibody canakinumab improved positive ones (Weickert et al., 2024). However, there have also been negative findings, for example, the BeneMin trial found no effect of minocycline on negative symptoms in recent-onset psychosis (Deakin et al., 2018), and *N*-acetylcysteine had no effect on any PANSS scores in a similar population (Conus et al., 2018). Retrospective stratification of patients in BeneMin or simvastatin trials into inflammatory or noninflammatory phenotypes (using plasma C-reactive protein levels or monocyte gene expression signatures, respectively) had minimal effects on the outcome (Krynicki et al., 2021; Aichholzer et al., 2022), so additional work is required to identify which patients and stages of illness are most likely to respond to which particular antiinflammatory agent.

Interestingly, the atypical antipsychotic clozapine is recommended for the management of treatment-resistant schizophrenia, and an emerging line of thought now links its efficacy in these cases to immunomodulatory effects and regulation of microglial inflammatory responses (Amerio et al., 2024). Along similar lines, a recent preclinical study showed that the typical antipsychotic chlorpromazine is able to reduce activity and expression of the Kv1.3 voltage-gated potassium channels that are upregulated in activated microglia, reduce lipopolysaccharide-induced IL-1β and IL-6 accumulation in mouse prefrontal cortex, and also prevent accompanying changes in anxiety- and depressive-like behavior (Lee et al., 2025a). However, it is not known whether clinically meaningful effects on microglial activation occur with current treatment regimens. Potential novel mechanisms for targeting immune and inflammatory changes in schizophrenia include nanoantibody-based complement C4b inhibitors (De la O Becerra et al., 2022) and antibodies or polysialylated ligands to target abnormal sialylation or sialic acid-binding immunoglobulin-like lectins receptor signaling that might underlie inappropriate microglial responses to host synapses (Wißfeld et al., 2024). These approaches are still at a relatively early stage of development, so their ability to actually modulate proposed early contributors to schizophrenia pathology (Howes and Onwordi, 2023) has yet to be examined in relevant preclinical models.

5.6.2 Potential treatments that influence autophagy

Existing antipsychotics are capable of inducing autophagy to varying extents and via different mechanisms. For example, chlorpromazine increases autophagy by inhibiting mTOR signaling, whereas other typical and atypical agents (e.g., pimozide and clozapine) activate the alpha-1 catalytic subunit of AMPK1 to promote autophagy through an mTOR-independent route involving unc-51-like autophagy activating kinase 1 (Ryskalin et al., 2018). But, as with its effects on microglial Kv1.3 channels, it is not known whether clinical chlorpromazine regimens actually achieve meaningful effects on autophagy. mTOR inhibitors like rapamycin (that are already licensed to prevent organ transplant rejection and for use in oncology) have shown encouraging effects on cognition in preclinical models for schizophrenia (Meffre et al., 2012). It is therefore plausible that they may be of benefit in the later stages of illness, where there is evidence for increased mTOR activation and deficient autophagy (Izumi et al., 2022; Lappas et al., 2025). Other approaches that have the potential to normalize deficient autophagy include beclin 1 activators such as spermidine or beclin 1 mimetics (Sainani et al., 2022), but care would be needed to avoid any use of autophagy inducers in the earlier stages of schizophrenia or in particular subgroups where autophagy might actually be already increased (Bjornson et al., 2024).

5.6.3 Potential treatments that influence synapses, mitochondria, and perineuronal nets

Compounds that increase BDNF levels or activate TrkB receptors would be expected to enhance synaptogenesis. Preliminary behavioral assessments of selective TrkB modulators have been completed, and accompanying synaptic data are now required. For example, acute administration of a TrkB agonistic antibody (Ab4B19) reversed sensorimotor gating deficits in a multiple-hit mouse model involving genetic BDNF deficiencies plus postnatal stressors (Chen et al., 2022), but chronic effects are yet to be reported. More recently, repeated administration of a combined $GABA_A$ receptor and TSPO agonist (etifoxine) achieved TrkB receptor-dependent reversal of maternal immune activation-induced cognitive, affective, and parvalbumin-positive interneuron deficits in mice (Mao et al., 2025). Clinical trials of this compound have already been conducted in anxiety (Vicente et al., 2020) and are planned in depression (Brunner et al., 2024), so hopefully those in schizophrenia will follow. Serotonergic and glutamatergic psychedelic agents also influence TrkB and mTOR signaling (Aleksandrova and Phillips, 2021), and initial contraindications in those with psychotic disorders may be overcome by the development of newer analogs with reduced hallucinogenic potential (Tuck et al., 2025), stimulating recent interest in whether microdosing regimes might restore synaptic density in schizophrenia (Sapienza et al., 2025).

Targeting of perisynaptic metalloproteinases represents another angle to overcome synaptic deficits in schizophrenia. MMP9 regulates NMDA-dependent LTP at excitatory synapses (Wiera and Mozrzymas, 2021), and *N*-acetylcysteine administration during critical developmental windows rescues parvalbumin interneuron deficits in glutathione-deficient mice via an MMP9/RAGE-dependent mechanism (Dwir et al., 2021). In the same model, repeated treatment with the MMP9 inhibitor SB-3CT at a similar developmental stage normalized microglial activation. It also prevented reductions in absolute numbers of parvalbumin interneurons and numbers of those interneurons enwrapped by PNNs (Dwir et al., 2020). These preclinical findings are perhaps some of the most encouraging in terms of potential disease-modifying approaches in schizophrenia, and the results of future clinical studies are now eagerly awaited. Furthermore, machine-learning approaches are beginning to identify patient subgroups and predict who would benefit from particular treatment approaches, and it is anticipated that this will improve patient outcomes in the short- to mid-term.

References

Aichholzer, M., Gangadin, S.S., Sommer, I.E.C., Wijkhuis, A., de Witte, L.D., Kahn, R.S., et al., 2022. Inflammatory monocyte gene signature predicts beneficial within group effect of simvastatin in patients with schizophrenia spectrum disorders in a secondary analysis of a randomized controlled trial. Brain, Behavior, & Immunity - Health 26, 100551. Available at https://doi.org/10.1016/j.bbih.2022.100551.

Akbarian, S., Kim, J.J., Potkin, S.G., Hagman, J.O., Tafazzoli, A., Bunney, W.E., 1995. Gene expression for glutamic acid decarboxylase is reduced without loss of neurons in prefrontal cortex of schizophrenics. Archives of General Psychiatry 52 (4), 258–266. Available at https://doi.org/10.1001/archpsyc.1995.03950160008002.

Álvarez-Mejía, D., Rodas, J.A., Leon-Rojas, J.E., 2025. From womb to mind: prenatal epigenetic influences on mental health disorders. International Journal of Molecular Sciences 26 (13), 6096. Available at https://doi.org/10.3390/ijms26136096.

Alcaide, J., Guirado, R., Crespo, C., Blasco-Ibáñez, J.M., Varea, E., Sanjuan, J., et al., 2019. Alterations of perineuronal nets in the dorsolateral prefrontal cortex of neuropsychiatric patients. International Journal of Bipolar Disorders 7 (1). Available at https://doi.org/10.1186/s40345-019-0161-0.

Aleksandrova, L.R., Phillips, A.G., 2021. Neuroplasticity as a convergent mechanism of ketamine and classical psychedelics. Trends in Pharmacological Sciences 42 (11), 929–942. Available at https://doi.org/10.1016/j.tips.2021.08.003.

Almeida, O.P., Ford, A.H., Hankey, G.J., Yeap, B.B., Golledge, J., Flicker, L., 2019. Risk of dementia associated with psychotic disorders in later life: the health in men study (HIMS). Psychological Medicine 49 (2), 232–242. Available at https://doi.org/10.1017/S003329171800065X.

Almulla, A.F., Maes, M., 2025. Peripheral immune-inflammatory pathways in major depressive disorder, bipolar disorder, and schizophrenia: exploring their potential as treatment targets. CNS Drugs 39 (8), 739–762. Available at https://doi.org/10.1007/s40263-025-01195-3.

Amerio, A., Magnani, L., Arduino, G., Fesce, F., de Filippis, R., Parise, A., et al., 2024. Immunomodulatory effects of clozapine: more than just a side effect in schizophrenia. Current Neuropharmacology 22 (7), 1233–1247. Available at https://doi.org/10.2174/1570159X22666231128101725.

Atkin, T.A., Brandon, N.J., Kittler, J.T., 2012. Disrupted in schizophrenia 1 forms pathological aggresomes that disrupt its function in intracellular transport. Human Molecular Genetics 21 (9), 2017–2028. Available at https://doi.org/10.1093/hmg/dds018.

Attademo, L., Bernardini, F., 2020. Air pollution as risk factor for mental disorders: in search for a possible link with Alzheimer's disease and schizophrenia. Journal of Alzheimer's Disease JAD 76 (3), 825–830. Available at https://doi.org/10.3233/JAD-200289.

Baranova, A., Zhao, Q., Cao, H., Chandhoke, V., Zhang, F., 2024. Causal influences of neuropsychiatric disorders on Alzheimer's disease. Translational Psychiatry 14 (1). Available at https://doi.org/10.1038/s41398-024-02822-1.

Barnes, M.R., Huxley-Jones, J., Maycox, P.R., Lennon, M., Thornber, A., Kelly, F., et al., 2011. Transcription and pathway analysis of the superior temporal cortex and anterior prefrontal cortex in schizophrenia. Journal of Neuroscience Research 89 (8), 1218–1227. Available at https://doi.org/10.1002/jnr.22647.

Barnes, T.R., Drake, R., Paton, C., Cooper, S.J., Deakin, B., Ferrier, I.N., et al., 2020. Evidence-based guidelines for the pharmacological treatment of schizophrenia: updated recommendations from the British association for psychopharmacology. Journal of Psychopharmacology (Oxford, England) 34 (1), 3–78. Available at https://doi.org/10.1177/0269881119889296.

Bartal, G., Yitzhaky, A., Segev, A., Hertzberg, L., 2023. Multiple genes encoding mitochondrial ribosomes are downregulated in brain and blood samples of individuals with schizophrenia. The World Journal of Biological Psychiatry the Official Journal of the World Federation of Societies of Biological Psychiatry 24 (9), 829–837. Available at https://doi.org/10.1080/15622975.2023.2211653.

Berdeville, C.H.d S.F., Silva-Amaral, D., Dalgalarrondo, P., Banzato, C.E.M., Martins-de-Souza, D., 2025. A scoping review of protein biomarkers for schizophrenia: state of progress, underlying biology, and methodological considerations. Neuroscience and Biobehavioral Reviews 168. Available at https://doi.org/10.1016/j.neubiorev.2024.105949.

Benes, F.M., McSparren, J., Bird, E.D., SanGiovanni, J.P., Vincent, S.L., 1991. Deficits in small interneurons in prefrontal and cingulate cortices of schizophrenic and schizoaffective patients. Archives of General Psychiatry 48 (11), 996–1001. Available at https://doi.org/10.1001/archpsyc.1991.01810350036005.

Ben-Shachar, D., Karry, R., 2008. Neuroanatomical pattern of mitochondrial complex I pathology varies between schizophrenia, bipolar disorder and major depression. PLoS One 3 (11). Available at https://doi.org/10.1371/journal.pone.0003676.

Berdenis van Berlekom, A., Muflihah, C.H., Snijders, G.J.L.J., MacGillavry, H.D., Middeldorp, J., Hol, E.M., et al., 2020. Synapse pathology in schizophrenia: a meta-analysis of postsynaptic elements in postmortem brain studies. Schizophrenia Bulletin 46 (2), 374–386. Available at https://doi.org/10.1093/schbul/sbz060.

Bergon, A., Belzeaux, R., Comte, M., Pelletier, F., Hervé, M., Gardiner, E.J., et al., 2015. CX3CR1 is dysregulated in blood and brain from schizophrenia patients. Schizophrenia Research 168 (1-2), 434–443. Available at https://doi.org/10.1016/j.schres.2015.08.010.

Bjornson, K.J., Vanderplow, A.M., Bhasker, A.I., Cahill, M.E., 2024. Increased regional activity of a pro-autophagy pathway in schizophrenia as a contributor to sex differences in the disease pathology. Cell Reports. Medicine. 5 (7). Available at https://doi.org/10.1016/j.xcrm.2024.101652.

Blasco, M.B., Nisha Aji, K., Ramos-Jiménez, C., Leppert, I.R., Tardif, C.L., Cohen, J., et al., 2025. Synaptic density in early stages of psychosis and clinical high risk. JAMA Psychiatry 82 (2), 171–180. Available at https://doi.org/10.1001/jamapsychiatry.2024.3608.

Bloomfield, P.S., Selvaraj, S., Veronese, M., Rizzo, G., Bertoldo, A., Owen, D.R., et al., 2016. Microglial activity in people at ultra high risk of psychosis and in schizophrenia: an [(11)C]PBR28 PET brain imaging study. The American Journal of Psychiatry 173 (1), 44–52. Available at https://doi.org/10.1176/appi.ajp.2015.14101358.

Brunner, L.-M., Riebel, M., Wein, S., Koller, M., Zeman, F., Huppertz, G., et al., 2024. The translocator protein 18kDa ligand etifoxine in the treatment of depressive disorders-a double-blind, randomized, placebo-controlled proof-of-concept study. Trials 25 (1). Available at https://doi.org/10.1186/s13063-024-08120-x.

Byrne, J.F., Healy, C., Föcking, M., Heurich, M., Susai, S.R., Mongan, D., et al., 2024. Plasma complement and coagulation proteins as prognostic factors of negative symptoms: an analysis of the NAPLS 2 and 3 studies. Brain, Behavior, and Immunity 119, 188–196. Available at https://doi.org/10.1016/j.bbi.2024.03.049.

Cabungcal, J.-H., Steullet, P., Morishita, H., Kraftsik, R., Cuenod, M., Hensch, T.K., et al., 2013. Perineuronal nets protect fast-spiking interneurons against oxidative stress. Proceedings of the National Academy of Sciences of the United States of America 110 (22), 9130–9135. Available at https://doi.org/10.1073/pnas.1300454110.

Café-Mendes, C.C., Ferro, E.S., Torrão, A.S., Crunfli, F., Rioli, V., Schmitt, A., et al., 2017. Peptidomic analysis of the anterior temporal lobe and corpus callosum from schizophrenia patients. Journal of Proteomics 151, 97–105. Available at https://doi.org/10.1016/j.jprot.2016.05.025.

Campeau, A., Mills, R.H., Stevens, T., Rossitto, L.-A., Meehan, M., Dorrestein, P., et al., 2022. Multi-omics of human plasma reveals molecular features of dysregulated inflammation and accelerated aging in schizophrenia. Molecular Psychiatry 27 (2), 1217–1225. Available at https://doi.org/10.1038/s41380-021-01339-z.

Caspani, G., Turecki, G., Lam, R.W., Milev, R.V., Frey, B.N., MacQueen, G.M., et al., 2021. Metabolomic signatures associated with depression and predictors of antidepressant response in humans: a CAN-BIND-1 report. Communications Biology 4 (1). Available at https://doi.org/10.1038/s42003-021-02421-6.

Cavelier, L., Jazin, E.E., Eriksson, I., Prince, J., Båve, U., Oreland, L., et al., 1995. Decreased cytochrome-c oxidase activity and lack of age-related accumulation of mitochondrial DNA deletions in the brains of schizophrenics. Genomics 29 (1), 217–224. Available at https://doi.org/10.1006/geno.1995.1234.

Chen, Y., Li, S., Zhang, T., Yang, F., Lu, B., 2022. Corticosterone antagonist or TrkB agonist attenuates schizophrenia-like behavior in a mouse model combining Bdnf-e6 deficiency and developmental stress. iScience 25 (7). Available at https://doi.org/10.1016/j.isci.2022.104609.

Clark, D., Dedova, I., Cordwell, S., Matsumoto, I., 2006. A proteome analysis of the anterior cingulate cortex gray matter in schizophrenia. Molecular Psychiatry 11 (5). Available at https://doi.org/10.1038/sj.mp.4001806.

Conus, P., Seidman, L.J., Fournier, M., Xin, L., Cleusix, M., Baumann, P.S., et al., 2018. N-acetylcysteine in a double-blind randomized placebo-controlled trial: toward biomarker-guided treatment in early psychosis. Schizophrenia Bulletin 44 (2), 317–327. Available at https://doi.org/10.1093/schbul/sbx093.

Coughlin, J.M., Wang, Y., Ambinder, E.B., Ward, R.E., Minn, I., Vranesic, M., et al., 2016. In vivo markers of inflammatory response in recent-onset schizophrenia: a combined study using [(11)C] DPA-713 PET and analysis of CSF and plasma. Translational Psychiatry 6 (4). Available at https://doi.org/10.1038/tp.2016.40.

Coutts, F., Mena, S., Ucur, E., Fleischhacker, W.W., Kahn, R., Lieberman, J., et al., 2025. Ensuring generalizability and clinical utility in mental health care applications: robust artificial intelligence-based treatment predictions in diverse psychosis populations. Psychiatry and Clinical Neurosciences. Available at https://doi.org/10.1111/pcn.13914.

Cuenod, M., Steullet, P., Cabungcal, J.-H., Dwir, D., Khadimallah, I., Klauser, P., et al., 2022. Caught in vicious circles: a perspective on dynamic feed-forward loops driving oxidative stress in schizophrenia. Molecular Psychiatry 27 (4), 1886–1897. Available at https://doi.org/10.1038/s41380-021-01374-w.

Dalman, C., Thomas, H.V., David, A.S., Gentz, J., Lewis, G., Allebeck, P., 2001. Signs of asphyxia at birth and risk of schizophrenia. Population-Based Case-control Study. The British Journal of Psychiatry the Journal of Mental Science 179, 403–408. Available at https://doi.org/10.1192/bjp.179.5.403.

de Bartolomeis, A., Barone, A., Vellucci, L., Mazza, B., Austin, M.C., Iasevoli, F., et al., 2022. Linking inflammation, aberrant glutamate-dopamine interaction, and post-synaptic changes: translational relevance for schizophrenia and antipsychotic treatment: a systematic review. Molecular Neurobiology 59 (10), 6460–6501. Available at https://doi.org/10.1007/s12035-022-02976-3.

De la O Becerra, K.I., Oosterheert, W., van den Bos, R.M., Xenaki, K.T., Lorent, J.H., Ruyken, M., et al., 2022. Multifaceted activities of seven nanobodies against complement C4b. Journal of immunology (Baltimore, Md. 1950) 208 (9), 2207–2219. Available at https://doi.org/10.4049/jimmunol.2100647.

de Oliveira, C.A., Pinto, J.P., de Pinho, C.S.N., Soares, M.V.R., Silva, M.F.S., Lucena, D.F., et al., 2025. Blood-based biomarkers of neuroaxonal and astroglial injury identify ultra-treatment-resistant schizophrenia. European Neuropsychopharmacology the Journal of the European College of Neuropsychopharmacology 99, 16–18.

de Sousa Abreu, R., Penalva, L.O., Marcotte, E.M., Vogel, C., 2009. Global signatures of protein and mRNA expression levels. Molecular Biosystems 5 (12), 1512–1526. Available at https://doi.org/10.1039/b908315d.

Deakin, B., Suckling, J., Barnes, T.R.E., Byrne, K., Chaudhry, I.B., Dazzan, P., et al., 2018. The benefit of minocycline on negative symptoms of schizophrenia in patients with recent-onset psychosis (BeneMin): a randomised, double-blind, placebo-controlled trial. The lancet. Psychiatry. 5 (11), 885–894. Available at https://doi.org/10.1016/S2215-0366(18)30345-6.

Deprato, A., Haldar, P., Navarro, J.F., Harding, B.N., Lacy, P., Maidstone, R., et al., 2025. Associations between light at night and mental health: a systematic review and meta-analysis. The Science of the Total Environment 974. Available at https://doi.org/10.1016/j.scitotenv.2025.179188.

Di Camillo, F., Grimaldi, D.A., Cattarinussi, G., Di Giorgio, A., Locatelli, C., Khuntia, A., et al., 2024. Magnetic resonance imaging-based machine learning classification of schizophrenia spectrum disorders: a meta-analysis. Psychiatry and Clinical Neurosciences 78 (12), 732–743. Available at https://doi.org/10.1111/pcn.13736.

Deutsch, S.I., Rosse, R.B., Schwartz, B.L., Mastropaolo, J., 2001. A revised excitotoxic hypothesis of schizophrenia: therapeutic implications. Clinical Neuropharmacology 24 (1), 43–49. Available at https://doi.org/10.1097/00002826-200101000-00008.

Dmitrieva, E., Smirnova, L., Seregin, A., Zgoda, V., Semke, A., Ivanova, S., 2022. Proteomic profile of serum from patients with schizophrenia spectrum disorders. PeerJ 10. Available at https://doi.org/10.7717/peerj.13907.

Dolgin, E., 2024. Revolutionary drug for schizophrenia wins US approval. Nature 634 (8033), 276–277. Available at https://doi.org/10.1038/d41586-024-03123-9.

Doolittle, T., Duque-Wilckens, N., 2025. Beyond infections: exploring immune-mediated pathways linking cannabis and emerging environmental contaminants to neurodevelopmental outcomes. Advances in Experimental Medicine and Biology 1477, 281–309. Available at https://doi.org/10.1007/978-3-031-89525-8_11.

Doorduin, J., de Vries, E.F.J., Willemsen, A.T.M., de Groot, J.C., Dierckx, R.A., Klein, H.C., 2009. Neuroinflammation in schizophrenia-related psychosis: a PET study. Journal of Nuclear Medicine Official Publication, Society of Nuclear Medicine 50 (11), 1801–1807. Available at https://doi.org/10.2967/jnumed.109.066647.

Durrenberger, P.F., Fernando, F.S., Kashefi, S.N., Bonnert, T.P., Seilhean, D., Nait-Oumesmar, B., et al., 2015. Common mechanisms in neurodegeneration and neuroinflammation: a BrainNet Europe gene expression microarray study. Journal of Neural Transmission (Vienna, Austria 1996) 122 (7), 1055–1068. Available at https://doi.org/10.1007/s00702-014-1293-0.

Dwir, D., Cabungcal, J.-H., Xin, L., Giangreco, B., Parietti, E., Cleusix, M., et al., 2021. Timely N-acetyl-cysteine and environmental enrichment rescue oxidative stress-induced parvalbumin interneuron impairments via MMP9/RAGE pathway: a translational approach for early intervention in psychosis. Schizophrenia Bulletin 47 (6), 1782–1794. Available at https://doi.org/10.1093/schbul/sbab066.

Dwir, D., Giangreco, B., Xin, L., Tenenbaum, L., Cabungcal, J.-H., Steullet, P., et al., 2020. MMP9/RAGE pathway overactivation mediates redox dysregulation and neuroinflammation, leading to inhibitory/excitatory imbalance: a reverse translation study in schizophrenia patients. Molecular Psychiatry 25 (11), 2889–2904. Available at https://doi.org/10.1038/s41380-019-0393-5.

Enwright, J.F., Sanapala, S., Foglio, A., Berry, R., Fish, K.N., Lewis, D.A., 2016. Reduced labeling of parvalbumin neurons and perineuronal nets in the dorsolateral prefrontal cortex of subjects with schizophrenia. Neuropsychopharmacology Official Publication of the American College of Neuropsychopharmacology 41 (9), 2206–2214. Available at https://doi.org/10.1038/npp.2016.24.

Eratne, D., Janelidze, S., Malpas, C.B., Loi, S., Walterfang, M., Merritt, A., et al., 2022. Plasma neurofilament light chain protein is not increased in treatment-resistant schizophrenia and first-degree relatives. The Australian and New Zealand Journal of Psychiatry 56 (10), 1295–1305. Available at https://doi.org/10.1177/00048674211058684.

Eren, F., Schwieler, L., Orhan, F., Malmqvist, A., Piehl, F., Cervenka, S., et al., 2023. Immunological protein profiling of first-episode psychosis patients identifies CSF and blood biomarkers correlating with disease severity. Brain, Behavior, and Immunity 111, 376–385. Available at https://doi.org/10.1016/j.bbi.2023.04.020.

Farag, N.A., Selim, A., Hassan, O., Makhlouf, H.A., Soliman, M., Taman, L., et al., 2025. Efficacy and safety of celecoxib in schizophrenia: a systematic review and meta-analysis of randomized controlled trials. Inflammopharmacology 33 (9), 5049–5065. Available at https://doi.org/10.1007/s10787-025-01908-6.

Farías Venegas, C., Varela Espinoza, L.F., Ramírez Matta, C., Barra Cañas, B., Araneda Maldonado, M., Sánchez Oñate, A., 2025. New developments in psychiatric classification: a transdiagnostic approach. Cureus 17 (5). Available at https://doi.org/10.7759/cureus.84580.

Fillman, S.G., Cloonan, N., Catts, V.S., Miller, L.C., Wong, J., McCrossin, T., et al., 2013. Increased inflammatory markers identified in the dorsolateral prefrontal cortex of individuals with schizophrenia. Molecular Psychiatry 18 (2), 206–214. Available at https://doi.org/10.1038/mp.2012.110.

García-Bueno, B., Gassó, P., MacDowell, K.S., Callado, L.F., Mas, S., Bernardo, M., et al., 2016. Evidence of activation of the Toll-like receptor-4 proinflammatory pathway in patients with schizophrenia. Journal of Psychiatry & Neuroscience JPN 41 (3). Available at https://doi.org/10.1503/jpn.150195.

Gawryluk, J.W., Wang, J.-F., Andreazza, A.C., Shao, L., Young, L.T., 2011. Decreased levels of glutathione, the major brain antioxidant, in post-mortem prefrontal cortex from patients with psychiatric disorders. The International Journal of Neuropsychopharmacology 14 (1), 123–130. Available at https://doi.org/10.1017/S1461145710000805.

Gkintoni, E., Telonis, G., Halkiopoulos, C., Boutsinas, B., 2026. Integrating neuroimaging and machine learning to predict mental disorder outcomes: a systematic review. Advances in Experimental Medicine and Biology 1487, 429–468. Available at https://doi.org/10.1007/978-3-032-03398-7_41.

Glausier, J.R., Roberts, R.C., Lewis, D.A., 2017. Ultrastructural analysis of parvalbumin synapses in human dorsolateral prefrontal cortex. The Journal of Comparative Neurology 525 (9), 2075–2089. Available at https://doi.org/10.1002/cne.24171.

Hafizi, S., Tseng, H.-H., Rao, N., Selvanathan, T., Kenk, M., Bazinet, R.P., et al., 2017. Imaging microglial activation in untreated first-episode psychosis: a PET study with [18F]FEPPA. The American Journal of Psychiatry 174 (2), 118–124. Available at https://doi.org/10.1176/appi.ajp.2016.16020171.

Halff, E.F., Cotel, M.-C., Natesan, S., McQuade, R., Ottley, C.J., Srivastava, D.P., et al., 2021. Effects of chronic exposure to haloperidol, olanzapine or lithium on SV2A and NLGN synaptic puncta in the rat frontal cortex. Behavioural Brain Research 405. Available at https://doi.org/10.1016/j.bbr.2021.113203.

Hao, M., Qin, Y., Li, Y., Tang, Y., Ma, Z., Tan, J., et al., 2023. Metabolome subtyping reveals multi-omics characteristics and biological heterogeneity in major psychiatric disorders. Psychiatry Research 330. Available at https://doi.org/10.1016/j.psychres.2023.115605.

Haroon, H., Ho, A.M.-C., Gupta, V.K., Dasari, S., Sellgren, C.M., Cervenka, S., et al., 2024. Cerebrospinal fluid proteomic signatures are associated with symptom severity of first-episode psychosis. Journal of Psychiatric Research 171, 306–315. Available at https://doi.org/10.1016/j.jpsychires.2024.02.002.

Holmes, S.E., Hinz, R., Drake, R.J., Gregory, C.J., Conen, S., Matthews, J.C., et al., 2016. In vivo imaging of brain microglial activity in antipsychotic-free and medicated schizophrenia: a [11C](R)-PK11195 positron emission tomography study. Molecular Psychiatry 21 (12), 1672–1679. Available at https://doi.org/10.1038/mp.2016.180.

Holper, L., Ben-Shachar, D., Mann, J.J., 2019. Multivariate meta-analyses of mitochondrial complex I and IV in major depressive disorder, bipolar disorder, schizophrenia, Alzheimer disease, and Parkinson disease. Neuropsychopharmacology Official Publication of the American College of Neuropsychopharmacology 44 (5), 837–849. Available at https://doi.org/10.1038/s41386-018-0090-0.

Hong, F., Chen, Q., Luo, X., Xie, S., Wei, Y., Li, X., et al., 2025. A multi-omics integration framework with automated machine learning identifies peripheral immune-coagulation biomarkers for schizophrenia risk stratification. International Journal of Molecular Sciences 26 (15), 7640. Available at https://doi.org/10.3390/ijms26157640.

Howes, O.D., Onwordi, E.C., 2023. The synaptic hypothesis of schizophrenia version III: a master mechanism. Molecular Psychiatry 28 (5), 1843–1856. Available at https://doi.org/10.1038/s41380-023-02043-w.

Huang, H., Luo, Z., Min, J., Luo, W., Zhou, X., Wang, C., 2025. Targeting neuroinflammation in schizophrenia: a comprehensive review of mechanisms and pharmacological interventions. International Immunopharmacology 159. Available at https://doi.org/10.1016/j.intimp.2025.114910.

Huo, L., Zheng, Z., Lu, X., Wu, F., Ning, Y., Zhang, X.Y., 2021. Decreased peripheral BDNF levels and cognitive impairment in late-life schizophrenia. Frontiers in psychiatry 12. Available at https://doi.org/10.3389/fpsyt.2021.641278.

Imamura, K., Takeshima, T., Fusayasu, E., Nakashima, K., 2008. Increased plasma matrix metalloproteinase-9 levels in migraineurs. Headache 48 (1), 135–139. Available at https://doi.org/10.1111/j.1526-4610.2007.00958.x.

Institute for Health Metrics and Evaluation (IHME). Global Burden of Disease 2021: Findings from the GBD 2021 Study. Seattle, WA: IHME, 2024.

Iwamoto, K., Bundo, M., Kato, T., 2005. Altered expression of mitochondria-related genes in postmortem brains of patients with bipolar disorder or schizophrenia, as revealed by large-scale DNA microarray analysis. Human Molecular Genetics 14 (2), 241–253. Available at https://doi.org/10.1093/hmg/ddi022.

Izumi, R., Hino, M., Nagaoka, A., Shishido, R., Kakita, A., Hoshino, M., et al., 2022. Dysregulation of DPYSL2 expression by mTOR signaling in schizophrenia: Multi-level study of postmortem brain. Neuroscience Research 175, 73–81. Available at https://doi.org/10.1016/j.neures.2021.09.004.

Jeppesen, R., Christensen, R.H.B., Pedersen, E.M.J., Nordentoft, M., Hjorthøj, C., Köhler-Forsberg, O., et al., 2020. Efficacy and safety of anti-inflammatory agents in treatment of psychotic disorders - a comprehensive systematic review and meta-analysis. Brain, Behavior, and Immunity 90, 364–380. Available at https://doi.org/10.1016/j.bbi.2020.08.028.

Kaar, S.J., Angelescu, I., Marques, T.R., Howes, O.D., 2019. Pre-frontal parvalbumin interneurons in schizophrenia: a meta-analysis of post-mortem studies. Journal of Neural Transmission (Vienna, Austria 1996) 126 (12), 1637–1651. Available at https://doi.org/10.1007/s00702-019-02080-2.

Karbalaee, M., Jameie, M., Amanollahi, M., TaghaviZanjani, F., Parsaei, M., Basti, F.A., et al., 2023. Efficacy and safety of adjunctive therapy with fingolimod in patients with schizophrenia: a randomized, double-blind, placebo-controlled clinical trial. Schizophrenia Research 254, 92–98. Available at https://doi.org/10.1016/j.schres.2023.02.020.

Karry, R., Klein, E., Ben Shachar, D., 2004. Mitochondrial complex I subunits expression is altered in schizophrenia: a postmortem study. Biological Psychiatry 55 (7), 676–684. Available at https://doi.org/10.1016/j.biopsych.2003.12.012.

Karska, J., Skonieczna-Żydecka, K., Jakubiak, N., Czarnecka, W., Misiak, B., 2025. Short-chain fatty acids in mood and schizophrenia spectrum disorders: evidence in the field and translational perspectives. Progress in Neuro-psychopharmacology & Biological Psychiatry 140. Available at https://doi.org/10.1016/j.pnpbp.2025.111430.

Kassar, O., Mansour, M.E.M., Farag, N.A., Selim, A., Kewiaa, Y., Yousef, O., et al., 2025. Efficacy and safety of sulforaphane in schizophrenia: a systematic review and meta-analysis of randomized controlled trials. BMC Psychiatry 25 (1). Available at https://doi.org/10.1186/s12888-025-07515-7.

Kawamoto, K., Iwamoto, H., Hirano, T., Amano, Y., Fukatsu-Chikumoto, A., Yamane, M., et al., 2025. Brain-derived neurotrophic factor levels are associated with depressive symptoms in patients with asthma. The Journal of Allergy and Clinical Immunology: In Practice. Available at https://doi.org/10.1016/j.jaip.2025.11.001.

Kim, H.K., Andreazza, A.C., Elmi, N., Chen, W., Young, L.T., 2016. Nod-like receptor pyrin containing 3 (NLRP3) in the post-mortem frontal cortex from patients with bipolar disorder: a potential mediator between mitochondria and immune-activation. Journal of Psychiatric Research 72, 43–50.

Kisby, G.E., Wilson, D.M., Spencer, P.S., 2024. Introducing the role of genotoxicity in neurodegenerative diseases and neuropsychiatric disorders. International Journal of Molecular Sciences 25 (13). Available at https://doi.org/10.3390/ijms25137221, http://www.mdpi.com/journal/ijms.

Kochunov, P., Zavaliangos-Petropulu, A., Jahanshad, N., Thompson, P.M., Ryan, M.C., Chiappelli, J., et al., 2021. A white matter connection of schizophrenia and Alzheimer's Disease. Schizophrenia Bulletin 47 (1), 197–206. Available at https://doi.org/10.1093/schbul/sbaa078.

Krynicki, C.R., Dazzan, P., Pariante, C.M., Barnes, N.M., Vincent, R.C., Roberts, A., et al., 2021. Deconstructing depression and negative symptoms of schizophrenia; differential and longitudinal immune correlates, and response to minocycline treatment. Brain, Behavior, and Immunity 91, 498–504.

Kumar, J., Liddle, E.B., Fernandes, C.C., Palaniyappan, L., Hall, E.L., Robson, S.E., et al., 2020. Glutathione and glutamate in schizophrenia: a 7T MRS study. Molecular Psychiatry 25 (4), 873–882. Available at https://doi.org/10.1038/s41380-018-0104-7.

Lanz, T.A., Reinhart, V., Sheehan, M.J., Rizzo, S.J.S., Bove, S.E., James, L.C., et al., 2019. Postmortem transcriptional profiling reveals widespread increase in inflammation in schizophrenia: a comparison of prefrontal cortex, striatum, and hippocampus among matched tetrads of controls with subjects diagnosed with schizophrenia, bipolar or major depressive disorder. Translational psychiatry 9 (1). Available at https://doi.org/10.1038/s41398-019-0492-8.

Lappas, A.S., Ioannou, M., Christodoulou, N.G., 2025. Histopathological evidence of cellular alterations in the dentate gyrus is associated with aberrant RB1CC1-ATG16L1 expression in the hippocampus among older adults with chronic schizophrenia: a pilot post-mortem study. Schizophrenia Research 275, 14–24.

Laroui, A., Galarneau, L., Abolghasemi, A., Benachenhou, S., Plantefève, R., Bouchouirab, F.Z., et al., 2022. Clinical significance of matrix metalloproteinase-9 in fragile X syndrome. Scientific reports 12 (1). Available at https://doi.org/10.1038/s41598-022-19476-y.

Lashgari, N.-A., Roudsari, N.M., Shamsnia, H.S., Shayan, M., Momtaz, S., Abdolghaffari, A.H., 2024. TLR/mTOR inflammatory signaling pathway: novel insight for the treatment of schizophrenia. Canadian Journal of Physiology and Pharmacology 102 (3), 150–160. Available at https://doi.org/10.1139/cjpp-2023-0107.

Lee, H.-Y., Lee, Y., Chung, C., Park, S.-I., Shin, H.J., Joe, E.-H., et al., 2025a. The antipsychotic chlorpromazine reduces neuroinflammation by inhibiting microglial voltage-gated potassium channels. Glia 73 (1), 210–227. Available at https://doi.org/10.1002/glia.24629.

Lee, S.-H., Han, C., Shin, C., 2025b. IUPHAR review: microbiota-gut-brain axis and its role in neuropsychiatric disorders. Pharmacological Research 216.

Lee, H., Kim, M., Kim, S.H., Lee, J., Lee, T.Y., Rhee, S.J., et al., 2024. Proteomic profiling in the progression of psychosis: analysis of clinical high-risk, first episode psychosis, and healthy controls. Journal of Psychiatric Research 169, 264–271.

Leliveld, S.R., Bader, V., Hendriks, P., Prikulis, I., Sajnani, G., Requena, J.R., et al., 2008. Insolubility of disrupted-in-schizophrenia 1 disrupts oligomer-dependent interactions with nuclear distribution element 1 and is associated with sporadic mental disease. The Journal of Neuroscience the Official Journal of the Society for Neuroscience 28 (15), 3839–3845. Available at https://doi.org/10.1523/JNEUROSCI.5389-07.2008.

Levey, A.I., Edmunds, S.M., Koliatsos, V., Wiley, R.G., Heilman, C.J., 1995. Expression of m1-m4 muscarinic acetylcholine receptor proteins in rat hippocampus and regulation by cholinergic innervation. The Journal of neuroscience the official journal of the Society for Neuroscience 15 (5 Pt 2), 4077–4092.

Liang, Q., Jiang, Y., Shieh, A.W., Zhou, D., Chen, R., Wang, F., et al., 2025a. Impact of common variants on brain gene expression from RNA to protein to schizophrenia risk. Nature Communications 16 (1). Available at https://doi.org/10.1038/s41467-025-65818-5.

Liang, Y., Pan, H., Yang, Z., Yu, C., Jiang, T., Li, Y., et al., 2025b. Alterations in the plasma concentrations of BDNF, proBDNF, and MMP-9 in patients with schizophrenia. European Journal of Medical Research 30 (1). Available at https://doi.org/10.1186/s40001-025-03153-7.

Lieberman, J.A., Mendelsohn, A., Goldberg, T.E., Emsley, R., 2025. Preventing disease progression in schizophrenia: what are we waiting for. Journal of Psychiatric Research 181, 716–727.

Lin, C.-E., Chung, C.-H., Chen, L.-F., Chi, M.-J., 2018. Increased risk of dementia in patients with Schizophrenia: a population-based cohort study in Taiwan. European psychiatry the journal of the Association of European Psychiatrists 53, 7–16.

Liou, Y.-J., Tsai, S.-J., Bai, Y.-M., Chen, T.-J., Chen, M.-H., 2023. Dementia risk in middle-aged patients with schizophrenia, bipolar disorder, and major depressive disorder: a cohort study of 84,824 subjects. European Archives of Psychiatry and Clinical Neuroscience 273 (1), 219–227. Available at https://doi.org/10.1007/s00406-022-01389-6.

Lisboa, J.R.F., Costa, O., Pakes, G.H., Colodete, D.A.E., Gomes, F.V., 2024. Perineuronal net density in schizophrenia: a systematic review of postmortem brain studies. Schizophrenia Research 271, 100–109.

Longinetti, E., Mariosa, D., Larsson, H., Ye, W., Ingre, C., Almqvist, C., et al., 2017. Neurodegenerative and psychiatric diseases among families with amyotrophic lateral sclerosis. Neurology 89 (6), 578–585. Available at https://doi.org/10.1212/WNL.0000000000004179.

Lorincz, A., Ashaber, M., Nusser, Z., 2025. Altered molecular composition of a specific subset of prefrontal cortical excitatory synapses in schizophrenia. The Journal of Neuroscience 45 (38). Available at https://doi.org/10.1523/jneurosci.0645-25.2025, e0645252025.

MacDowell, K.S., Pinacho, R., Leza, J.C., Costa, J., Ramos, B., García-Bueno, B., 2017. Differential regulation of the TLR4 signalling pathway in post-mortem prefrontal cortex and cerebellum in chronic schizophrenia: relationship with SP transcription factors. Progress in Neuro-Psychopharmacology & Biological Psychiatry 79 (Pt B), 481–492.

Mallarpu, C.S., Chelluri, S.I., Katragadda, T.K., Singarapu, M., Chelluri, L.K., Madiraju, C., 2025. Programmed cell death markers in COVID-19 survivors with and without sepsis. Frontiers in Immunology 16. Available at https://doi.org/10.3389/fimmu.2025.1535938.

Manchia, M., Isayeva, U., Collu, R., Primavera, D., Deriu, L., Caboni, E., et al., 2022. Converging evidence points to BDNF as biomarker of depressive symptoms in schizophrenia-spectrum disorders. Brain Sciences 12 (12) Available from. Available at https://doi.org/10.3390/brainsci12121666, http://www.mdpi.com/journal/brainsci.

Mao, M.-J., Yu, H.-L., Liu, Q.-Z., Wen, Y.-Z., Jiang, M., Yuan, H.-M., et al., 2025. TSPO activation ameliorates maternal immune activation induced PV interneuron deficits via BDNF/TrkB signaling. Psychopharmacology 242 (6), 1303–1319. Available at https://doi.org/10.1007/s00213-024-06728-0.

Marques, T.R., Ashok, A.H., Pillinger, T., Veronese, M., Turkheimer, F.E., Dazzan, P., et al., 2019. Neuroinflammation in schizophrenia: meta-analysis of in vivo microglial imaging studies. Psychological Medicine 49 (13), 2186–2196. Available at https://doi.org/10.1017/S0033291718003057.

Marshall, C.R., Howrigan, D.P., Merico, D., Thiruvahindrapuram, B., Wu, W., Greer, D.S., et al., 2017. Contribution of copy number variants to schizophrenia from a genome-wide study of 41,321 subjects. Nature Genetics 49 (1), 27–35. Available at https://doi.org/10.1038/ng.3725.

Maurer, I., Zierz, S., Möller, H., 2001. Evidence for a mitochondrial oxidative phosphorylation defect in brains from patients with schizophrenia. Schizophrenia Research 48 (1), 125–136.

Maury, E.A., Sherman, M.A., Genovese, G., Gilgenast, T.G., Kamath, T., Burris, S.J., et al., 2023. Disruptions. Cell Genomics 3 (8). Available at https://doi.org/10.1016/j.xgen.2023.100356.

Meffre, J., Chaumont-Dubel, S., Mannoury la Cour, C., Loiseau, F., Watson, D.J.G., Dekeyne, A., et al., 2012. 5-HT(6) receptor recruitment of mTOR as a mechanism for perturbed cognition in schizophrenia. EMBO molecular medicine 4 (10), 1043–1056. Available at https://doi.org/10.1002/emmm.201201410.

Meltzer, H.Y., Stahl, S.M., 1976. The dopamine hypothesis of schizophrenia: a review. Schizophrenia Bulletin 2 (1), 19–76.

Mendoza-Torreblanca, J.G., García-Cruz, M.E., Sánchez-Cruz, I., Gomez-Gonzalez, B., Juárez-Méndez, S., Gómez-Lira, G., 2019. Analysis of differential expression of synaptic vesicle protein 2A in the adult rat brain. Neuroscience 419, 108–120.

Merenlender-Wagner, A., Malishkevich, A., Shemer, Z., Udawela, M., Gibbons, A., Scarr, E., et al., 2015. Autophagy has a key role in the pathophysiology of schizophrenia. Molecular Psychiatry 20 (1), 126–132. Available at https://doi.org/10.1038/mp.2013.174.

Milic, J., Vucurovic, M., Jovic, D., Stankovic, V., Grego, E., Jankovic, S., et al., 2025. Exploring the potential of precision medicine in neuropsychiatry: a commentary on new insights for tailored treatments based on genetic, environmental, and lifestyle factors. Genes 16 (4), 371. Available at https://doi.org/10.3390/genes16040371.

Mohammed, H.E., Gomaa, M.A., Khalifa, Y.M., Shawky, A.A., 2025. Does KarXT (xanomeline-trospium) represent a novel approach to schizophrenia management? A GRADE-assessed systematic review and meta-analysis of randomized controlled clinical trials. BMC Psychiatry 25 (1). Available at https://doi.org/10.1186/s12888-025-06696-5.

Morozova, M., Burminskiy, D., Rupchev, G., Lepilkina, T., Potanin, S., Beniashvili, A., et al., 2017. 5-HT6 receptor antagonist as an adjunct treatment targeting residual symptoms in patients with schizophrenia: unexpected sex-related effects (Double-Blind Placebo-Controlled Trial). Journal of Clinical Psychopharmacology 37 (2), 169–175. Available at https://doi.org/10.1097/JCP.0000000000000673.

Mortensen, P.B., Nørgaard-Pedersen, B., Waltoft, B.L., Sørensen, T.L., Hougaard, D., Torrey, E.F., et al., 2007. Toxoplasma gondii as a risk factor for early-onset schizophrenia: analysis of filter paper blood samples obtained at birth. Biological Psychiatry 61 (5), 688–693.

Murray, A.J., Humpston, C.S., Wilson, M., Rogers, J.C., Zia Ul Haq Katshu, M., Liddle, P.F., et al., 2024. Measurement of brain glutathione with magnetic Resonance spectroscopy in schizophrenia-spectrum disorders - a systematic review and meta-analysis. Brain, Behavior, and Immunity 115, 3–12.

Muszyński, J., Bienert, A., Elsorady, R.W., Rybakowski, F., 2025. New pharmacological approaches in the treatment of schizophrenia. Pharmacological reports PR 77 (3), 561–575. Available at https://doi.org/10.1007/s43440-025-00722-9.

Naguib, M., Rashed, L.A., 2018. Serum level of the autophagy biomarker Beclin-1 in patients with diabetic kidney disease. Diabetes Research and Clinical Practice 143, 56–61.

Naifar, M., Ducatez, F., Guidara, W., Maalej, M., Lesueur, C., Pilon, C., et al., 2025. Plasma metabolic and inflammatory protein signatures in psychiatric disorders. International Journal of Molecular Sciences 26 (13), 6260. Available at https://doi.org/10.3390/ijms26136260.

Niizato, K., Arai, T., Kuroki, N., Kase, K., Iritani, S., Ikeda, K., 1998. Autopsy study of Alzheimer's disease brain pathology in schizophrenia. Schizophrenia Research 31 (2-3), 177–184.

Nucifora, L.G., MacDonald, M.L., Lee, B.J., Peters, M.E., Norris, A.L., Orsburn, B.C., et al., 2019. Increased protein insolubility in brains from a subset of patients with schizophrenia. The American Journal of Psychiatry 176 (9), 730–743. Available at https://doi.org/10.1176/appi.ajp.2019.18070864.

Nunes Noto, M., Maes, M., Vargas Nunes, S.O., Ota, V.K., Cavalcante, D., Oliveira, G., et al., 2021. BDNF in antipsychotic naive first episode psychosis: effects of risperidone and the immune-inflammatory response system. Journal of Psychiatric Research 141, 206–213.

Nutma, E., Fancy, N., Weinert, M., Tsartsalis, S., Marzin, M.C., Muirhead, R.C.J., et al., 2023. Translocator protein is a marker of activated microglia in rodent models but not human neurodegenerative diseases. Nature communications 14 (1). Available at https://doi.org/10.1038/s41467-023-40937-z.

Olabi, B., Ellison-Wright, I., McIntosh, A.M., Wood, S.J., Bullmore, E., Lawrie, S.M., 2011. Are there progressive brain changes in schizophrenia? A meta-analysis of structural magnetic resonance imaging studies. Biological Psychiatry 70 (1), 88–96. Available at https://doi.org/10.1016/j.biopsych.2011.01.032.

Olney, J.W., Newcomer, J.W., Farber, N.B., 1999. NMDA receptor hypofunction model of schizophrenia. Journal of Psychiatric Research 33 (6), 523–533.

Onwordi, E.C., Halff, E.F., Whitehurst, T., Mansur, A., Cotel, M.-C., Wells, L., et al., 2020. Synaptic density marker SV2A is reduced in schizophrenia patients and unaffected by antipsychotics in rats. Nature communications 11 (1). Available at https://doi.org/10.1038/s41467-019-14122-0.

Onwordi, E.C., Whitehurst, T., Mansur, A., Statton, B., Berry, A., Quinlan, M., et al., 2021. The relationship between synaptic density marker SV2A, glutamate and N-acetyl aspartate levels in healthy volunteers and schizophrenia: a multimodal PET and magnetic resonance spectroscopy brain imaging study. Translational psychiatry 11 (1). Available at https://doi.org/10.1038/s41398-021-01515-3.

Onwordi, E.C., Whitehurst, T., Shatalina, E., Mansur, A., Arumuham, A., Osugo, M., et al., 2024. Synaptic terminal density early in the course of schizophrenia: an in vivo UCB-J positron emission tomographic imaging study of SV2A. Biological Psychiatry 95 (7), 639–646.

Osimo, E.F., Beck, K., Reis Marques, T., Howes, O.D., 2019. Synaptic loss in schizophrenia: a meta-analysis and systematic review of synaptic protein and mRNA measures. Molecular Psychiatry 24 (4), 549–561. Available at https://doi.org/10.1038/s41380-018-0041-5.

Ottis, P., Bader, V., Trossbach, S.V., Kretzschmar, H., Michel, M., Leliveld, S.R., et al., 2011. Convergence of two independent mental disease genes on the protein level: recruitment of dysbindin to cell-invasive disrupted-in-schizophrenia 1 aggresomes. Biological Psychiatry 70 (7), 604–610. Available at https://doi.org/10.1016/j.biopsych.2011.03.027.

Ottoy, J., De Picker, L., Verhaeghe, J., Deleye, S., Wyffels, L., Kosten, L., et al., 2018. F-PBR111 PET imaging in healthy controls and schizophrenia: test-retest reproducibility and quantification of neuroinflammation. Journal of nuclear medicine official publication, Society of Nuclear Medicine 59 (8), 1267–1274. Available at https://doi.org/10.2967/jnumed.117.203315.

Pandey, G.N., Rizavi, H.S., Zhang, H., Ren, X., 2018. Abnormal gene and protein expression of inflammatory cytokines in the postmortem brain of schizophrenia patients. Schizophrenia Research 192, 247–254.

Panizzutti, B., Skvarc, D., Lin, S., Croce, S., Meehan, A., Bortolasci, C.C., et al., 2023. Minocycline as treatment for psychiatric and neurological conditions: a systematic review and meta-analysis. International Journal of Molecular Sciences 24 (6). Available at https://doi.org/10.3390/ijms24065250, http://www.mdpi.com/journal/ijms.

Pantazopoulos, H., Woo, T.-U.W., Lim, M.P., Lange, N., Berretta, S., 2010. Extracellular matrix-glial abnormalities in the amygdala and entorhinal cortex of subjects diagnosed with schizophrenia. Archives of General Psychiatry 67 (2), 155–166. Available at https://doi.org/10.1001/archgenpsychiatry.2009.196.

Pazarlar, B.A., Aripaka, S.S., Petukhov, V., Pinborg, L., Khodosevich, K., Mikkelsen, J.D., 2022. Expression profile of synaptic vesicle glycoprotein 2A, B, and C paralogues in temporal neocortex tissue from patients with temporal lobe epilepsy (TLE). Molecular brain 15 (1). Available at https://doi.org/10.1186/s13041-022-00931-w.

Peng, T.-R., Lin, H.-H., Lee, M.-C., Chen, S.-M., 2024. Statins as an adjuvant therapy for patients with schizophrenia: an up-to-date systematic review and meta-analysis. General Hospital Psychiatry 89, 75–83.

Pils, M., Rutsch, J., Eren, F., Engberg, G., Piehl, F., Cervenka, S., et al., 2023. Disrupted-in-schizophrenia 1 protein aggregates in cerebrospinal fluid are elevated in patients with first-episode psychosis. Psychiatry and Clinical Neurosciences 77 (12), 665–671. Available at https://doi.org/10.1111/pcn.13594.

Pourebrahim, S., Ahmad, T., Rottmann, E., Schulze, J., Scheller, B., 2025. Does cannabis use contribute to schizophrenia? A causation analysis based on epidemiological evidence. Biomolecules 15 (3). Available at https://doi.org/10.3390/biom15030368, http://www.mdpi.com/journal/biomolecules.

Prata, D.P., Costa-Neves, B., Cosme, G., Vassos, E., 2019. Unravelling the genetic basis of schizophrenia and bipolar disorder with GWAS: a systematic review. Journal of Psychiatric Research 114, 178–207.

Prince, J.A., Blennow, K., Gottfries, C.G., Karlsson, I., Oreland, L., 1999. Mitochondrial function is differentially altered in the basal ganglia of chronic schizophrenics. Neuropsychopharmacology Official Publication of the American College of Neuropsychopharmacology 21 (3), 372–379.

Prince, J.A., Harro, J., Blennow, K., Gottfries, C.G., Oreland, L., 2000. Putamen mitochondrial energy metabolism is highly correlated to emotional and intellectual impairment in schizophrenics. Neuropsychopharmacology Official Publication of the American College of Neuropsychopharmacology 22 (3), 284–292.

Prohovnik, I., Dwork, A.J., Kaufman, M.A., Willson, N., 1993. Alzheimer-type neuropathology in elderly schizophrenia patients. Schizophrenia Bulletin 19 (4), 805–816.

Purohit, D.P., Perl, D.P., Haroutunian, V., Powchik, P., Davidson, M., Davis, K.L., 1998. Alzheimer disease and related neurodegenerative diseases in elderly patients with schizophrenia: a postmortem neuropathologic study of 100 cases. Archives of General Psychiatry 55 (3), 205–211.

Purves-Tyson, T.D., Weber-Stadlbauer, U., Richetto, J., Rothmond, D.A., Labouesse, M.A., Polesel, M., et al., 2021. Increased levels of midbrain immune-related transcripts in schizophrenia and in murine offspring after maternal immune activation. Molecular Psychiatry 26 (3), 849–863. Available at https://doi.org/10.1038/s41380-019-0434-0.

Qi, D.D., Liu, P., Wang, Y.M., Tai, X.G., Ma, S.F., Wang, Y., 2025. A multi-omic dissection of molecular hallmarks of accelerated aging in schizophrenia. Scientific Reports 15 (1). Available at https://doi.org/10.1038/s41598-025-98181-y.

Radhakrishnan, R., Skosnik, P.D., Ranganathan, M., Naganawa, M., Toyonaga, T., Finnema, S., et al., 2021. In vivo evidence of lower synaptic vesicle density in schizophrenia. Molecular Psychiatry 26 (12), 7690–7698. Available at https://doi.org/10.1038/s41380-021-01184-0.

Ranganathan, M., Rahman, M., Ganesh, S., D'Souza, D.C., Skosnik, P.D., Radhakrishnan, R., et al., 2022. Analysis of circulating exosomes reveals a peripheral signature of astrocytic pathology in schizophrenia. The World Journal of Biological Psychiatry the Official Journal of the World Federation of Societies of Biological Psychiatry 23 (1), 33–45. Available at https://doi.org/10.1080/15622975.2021.1907720.

Roberts, R.C., 2021. Mitochondrial dysfunction in schizophrenia: with a focus on postmortem studies. Mitochondrion 56, 91–101.

Rodrigues, J.E., Martinho, A., Santa, C., Madeira, N., Coroa, M., Santos, V., et al., 2022. Systematic review and meta-analysis of mass spectrometry proteomics applied to human peripheral fluids to assess potential biomarkers of schizophrenia. International Journal of Molecular Sciences 23 (9), 4917. Available at https://doi.org/10.3390/ijms23094917.

Ruden, J.B., Dugan, L.L., Konradi, C., 2021. Parvalbumin interneuron vulnerability and brain disorders. Neuropsychopharmacology Official Publication of the American College of Neuropsychopharmacology 46 (2), 279–287. Available at https://doi.org/10.1038/s41386-020-0778-9.

Russo, A.P., Pastorello, Y., Dénes, L., Brînzaniuc, K., Krupinski, J., Slevin, M., 2025. The inflammatory nexus: unraveling shared pathways and promising treatments in Alzheimer's disease and schizophrenia. International Journal of Molecular Sciences 26 (13), 6237. Available at https://doi.org/10.3390/ijms26136237.

Ryskalin, L., Limanaqi, F., Frati, A., Busceti, C.L., Fornai, F., 2018. mTOR-related brain dysfunctions in neuropsychiatric disorders. International Journal of Molecular Sciences 19 (8), 2226. Available at https://doi.org/10.3390/ijms19082226.

Sainani, S.R., Pansare, P.A., Rode, K., Bhalchim, V., Doke, R., Desai, S., 2022. Emendation of autophagic dysfuction in neurological disorders: a potential therapeutic target. The International Journal of Neuroscience 132 (5), 466–482. Available at https://doi.org/10.1080/00207454.2020.1822356.

Saint-Georges, Z., MacDonald, J., Al-Khalili, R., Hamati, R., Solmi, M., Keshavan, M.S., et al., 2025. Cholinergic system in schizophrenia: a systematic review and meta-analysis. Molecular Psychiatry 30 (7), 3301–3315. Available at https://doi.org/10.1038/s41380-025-03023-y.

Salehi, A., Namaei, P., TaghaviZanjani, F., Bagheri, S., Moradi, K., Khodaei Ardakani, M.-R., et al., 2022. Adjuvant palmitoylethanolamide therapy with risperidone improves negative symptoms in patients with schizophrenia: a randomized, double-blinded, placebo-controlled trial. Psychiatry Research 316.

Samaei, A., Moradi, K., Bagheri, S., Ashraf-Ganjouei, A., Alikhani, R., Mousavi, S.B., et al., 2020. Resveratrol adjunct therapy for negative symptoms in patients with stable schizophrenia: a double-blind, randomized placebo-controlled trial. The international journal of neuropsychopharmacology 23 (12), 775–782. Available at https://doi.org/10.1093/ijnp/pyaa006.

Samardžija, B., Pavešić Radonja, A., Zaharija, B., Bergman, M., Renner, É., Palkovits, M., et al., 2021. Protein aggregation of NPAS3, implicated in mental illness, is not limited to the V304I mutation. Journal of Personalized Medicine 11 (11), 1070. Available at https://doi.org/10.3390/jpm11111070.

Santa, C., Rodrigues, J.E., Martinho, A., Mendes, V.M., Madeira, N., Coroa, M., et al., 2024. Proteomic analysis of peripheral blood mononuclear cells in first episode psychosis - Protein and peptide-centered approaches to elucidate potential diagnostic biomarkers. Journal of Proteomics 309.

Sapienza, J., Spangaro, M., Comai, S., Sabé, M., La Torre, J., Buonarroti, M., et al., 2025. Microdosing psychedelics to restore synaptic density in schizophrenia. International Journal of Molecular Sciences 26 (18), 8949. Available at https://doi.org/10.3390/ijms26188949.

Scarr, E., Cowie, T.F., Kanellakis, S., Sundram, S., Pantelis, C., Dean, B., 2009. Decreased cortical muscarinic receptors define a subgroup of subjects with schizophrenia. Molecular Psychiatry 14 (11), 1017–1023. Available at https://doi.org/10.1038/mp.2008.28.

Schoretsanitis, G., de Filippis, R., Ntogka, M., Leucht, S., Correll, C.U., Kane, J.M., 2021. Matrix metalloproteinase 9 blood alterations in patients with schizophrenia spectrum disorders: a systematic review and meta-analysis. Schizophrenia Bulletin 47 (4), 986–996. Available at https://doi.org/10.1093/schbul/sbab001.

Shamabadi, A., Rafiei-Tabatabaei, E.S., Kazemzadeh, K., Farahmand, K., Fallahpour, B., Khodaei Ardakani, M.R., et al., 2025. Pentoxifylline adjunct to risperidone for negative symptoms of stable schizophrenia: a randomized, double-blind, placebo-controlled trial. International Journal of Neuropsychopharmacology 28 (1). Available at https://doi.org/10.1093/ijnp/pyae051.

Sharma, N., Khatib, M.N., Roopashree, R., Kaur, M., Srivastava, M., Barwal, A., et al., 2025. Efficacy and safety of the muscarinic receptor agonist KarXT (Xanomeline-Trospium) in schizophrenia: a systematic review, meta-analysis and bayesian sensitivity analysis. International Journal of Neuropsychopharmacology. Available at https://doi.org/10.1093/ijnp/pyaf045.

Shen, H.-P., Dong, X., Li, Z.-B., Wu, J.-Z., Zheng, C.-M., Hu, X.-J., et al., 2024. Protein profiles and novel molecular biomarkers of schizophrenia based on 4D-DIA proteomics. Journal of Proteome Research 23 (7), 2376–2385. Available at https://doi.org/10.1021/acs.jproteome.4c00040.

Simmonite, M., Steeby, C.J., Taylor, S.F., 2023. Medial frontal cortex GABA concentrations in psychosis spectrum and mood disorders: a meta-analysis of proton magnetic resonance spectroscopy studies. Biological Psychiatry 93 (2), 125–136.

Sneeboer, M.A.M., van der Doef, T., Litjens, M., Psy, N.B.B., Melief, J., Hol, E.M., et al., 2020. Microglial activation in schizophrenia: is translocator 18 kDa protein (TSPO) the right marker? Schizophrenia Research 215, 167–172.

Song, Y.R., Wu, B., Yang, Y.T., Chen, J., Zhang, L.J., Zhang, Z.W., et al., 2015. Specific alterations in plasma proteins during depressed, manic, and euthymic states of bipolar disorder. Brazilian Journal of Medical and Biological Research=Revista brasileira de pesquisas medicas e biologica 48 (11), 973–982.

Španić Popovački, E., Vogrinc, D., Fuller, H.R., Langer Horvat, L., Mayer, D., Kopić, J., et al., 2024. Increased NLRP1 mRNA and protein expression suggests inflammasome activation in the dorsolateral prefrontal and medial orbitofrontal cortex in schizophrenia. Biomolecules 14 (3). Available at https://doi.org/10.3390/biom14030302.

Steullet, P., Cabungcal, J.-H., Bukhari, S.A., Ardelt, M.I., Pantazopoulos, H., Hamati, F., et al., 2018. The thalamic reticular nucleus in schizophrenia and bipolar disorder: role of parvalbumin-expressing neuron networks and oxidative stress. Molecular Psychiatry 23 (10), 2057–2065. Available at https://doi.org/10.1038/mp.2017.230.

Stroup, T.S., Olfson, M., Huang, C., Wall, M.M., Goldberg, T., Devanand, D.P., et al., 2021. Age-specific prevalence and incidence of dementia diagnoses among older us adults with schizophrenia. JAMA Psychiatry 78 (6), 632–641. Available at https://doi.org/10.1001/jamapsychiatry.2021.0042.

Sun, X., Qiu, Y., Ji, L., Li, M., Wang, J., Su, Q., et al., 2025. Proteomic characterization of plasma exosomes in drug-naïve schizophrenia: insights into psychosis and cognitive dysfunction. Psychiatry Research 351.

Surti, T.S., Ranganathan, M., Johannesen, J.K., Gueorguieva, R., Deaso, E., Kenney, J.G., et al., 2023. Randomized controlled trial of the glycine transporter 1 inhibitor PF-03463275 to enhance cognitive training and neuroplasticity in schizophrenia. Schizophrenia Research 256, 36–43.

Tao, F., Deng, S., Zhuo, B., Liu, J., Liang, X., Shi, J., et al., 2025. Causal relationship between schizophrenia and five types of dementia: a bidirectional two-sample Mendelian randomization study. PLoS One 20 (5). Available at https://doi.org/10.1371/journal.pone.0322752.

Tapiainen, V., Hartikainen, S., Taipale, H., Tiihonen, J., Tolppanen, A.-M., 2017. Hospital-treated mental and behavioral disorders and risk of Alzheimer's disease: a nationwide nested case-control study. European Psychiatry the Journal of the Association of European Psychiatrists 43, 92–98.

Torres, U.S., Duran, F.L.S., Schaufelberger, M.S., Crippa, J.A.S., Louzã, M.R., Sallet, P.C., et al., 2016. Patterns of regional gray matter loss at different stages of schizophrenia: a multisite, cross-sectional VBM study in first-episode and chronic illness. NeuroImage. Clinical 12, 1–15. Available at https://doi.org/10.1016/j.nicl.2016.06.002.

Trubetskoy, V., Pardiñas, A.F., Qi, T., Panagiotaropoulou, G., Awasthi, S., Bigdeli, T.B., et al., 2022. Mapping genomic loci implicates genes and synaptic biology in schizophrenia. Nature 604 (7906), 502–508. Available at https://doi.org/10.1038/s41586-022-04434-5.

Tuck, J.R., Dunlap, L.E., Khatib, Y.A., Hatzipantelis, C.J., Weiser Novak, S., Rahn, R.M., et al., 2025. Molecular design of a therapeutic LSD analogue with reduced hallucinogenic potential. Proceedings of the National Academy of Sciences of the United States of America 122 (16). Available at https://doi.org/10.1073/pnas.2416106122.

Turner, M.R., Goldacre, R., Talbot, K., Goldacre, M.J., 2016. Psychiatric disorders prior to amyotrophic lateral sclerosis. Annals of Neurology 80 (6), 935–938. Available at https://doi.org/10.1002/ana.24801.

Tylec, A., Skałecki, M., Kocot, J., Kurzepa, J., 2021. Activity of selected metalloproteinases in neurodegenerative diseases of the central nervous system as exemplified by dementia and schizophrenia. Psychiatria Polska 55 (6), 1221–1233.

Ulugut, H., Trieu, C., Groot, C., van 't Hooft, J.J., Tijms, B.M., Scheltens, P., et al., 2023. Overlap of neuroanatomical involvement in frontotemporal dementia and primary psychiatric disorders: a meta-analysis. Biological Psychiatry 93 (9), 820–828.

van Berckel, B.N., Bossong, M.G., Boellaard, R., Kloet, R., Schuitemaker, A., Caspers, E., et al., 2008. Microglia activation in recent-onset schizophrenia: a quantitative (R)-[11C]PK11195 positron emission tomography study. Biological Psychiatry 64 (9), 820–822. Available at https://doi.org/10.1016/j.biopsych.2008.04.025.

van der Doef, T.F., de Witte, L.D., Sutterland, A.L., Jobse, E., Yaqub, M., Boellaard, R., et al., 2016. In vivo (R)-[(11)C]PK11195 PET imaging of 18kDa translocator protein in recent onset psychosis. NPJ Schizophrenia 2. Available at https://doi.org/10.1038/npjschz.2016.31.

van Kesteren, C.F.M.G., Gremmels, H., de Witte, L.D., Hol, E.M., Van Gool, A.R., Falkai, P.G., et al., 2017. Immune involvement in the pathogenesis of schizophrenia: a meta-analysis on postmortem brain studies. Translational Psychiatry 7 (3). Available at https://doi.org/10.1038/tp.2017.4.

Velligan, D.I., Rao, S., 2023. The epidemiology and global burden of schizophrenia. Journal of Clinical Psychiatry 84 (1). Available at https://doi.org/10.4088/JCP.MS21078COM5.

Vicente, B., Saldivia, S., Hormazabal, N., Bustos, C., Rubí, P., 2020. Etifoxine is non-inferior than clonazepam for reduction of anxiety symptoms in the treatment of anxiety disorders: a randomized, double blind, non-inferiority trial. Psychopharmacology 237 (11), 3357–3367. Available at https://doi.org/10.1007/s00213-020-05617-6.

Vogel, C., Marcotte, E.M., 2012. Insights into the regulation of protein abundance from proteomic and transcriptomic analyses. Nature Reviews Genetics 13 (4), 227–232. Available at https://doi.org/10.1038/nrg3185.

Volk, D.W., Sampson, A.R., Zhang, Y., Edelson, J.R., Lewis, D.A., 2016. Cortical GABA markers identify a molecular subtype of psychotic and bipolar disorders. Psychological Medicine 46 (12), 2501–2512. Available at https://doi.org/10.1017/S0033291716001446.

Wang, G., Liu, S., Fan, X., Li, J., Xue, Q., Liu, K., et al., 2025. Mitochondrial dysfunction and cognitive impairment in schizophrenia: the role of inflammation. Schizophrenia Bulletin 51 (4), 1092–1103. Available at https://doi.org/10.1093/schbul/sbae196.

Wang, L., Liu, R., Wang, Y., 2025. The roles of extracellular vesicles in mental disorders: information carriers, biomarkers, therapeutic agents. Frontiers in Pharmacology 16. Available at https://doi.org/10.3389/fphar.2025.1591469.

Watanabe, Y., Someya, T., Nawa, H., 2010. Cytokine hypothesis of schizophrenia pathogenesis: evidence from human studies and animal models. Psychiatry and Clinical Neurosciences 64 (3), 217–230. Available at https://doi.org/10.1111/j.1440-1819.2010.02094.x.

Weickert, T.W., Jacomb, I., Lenroot, R., Lappin, J., Weinberg, D., Brooks, W.S., et al., 2024. Adjunctive canakinumab reduces peripheral inflammation markers and improves positive symptoms in people with schizophrenia and inflammation: a randomized control trial. Brain, Behavior, and Immunity 115, 191–200.

Wiera, G., Mozrzymas, J.W., 2021. Extracellular metalloproteinases in the plasticity of excitatory and inhibitory synapses. Cells 10 (8). Available at https://doi.org/10.3390/cells10082055.

Wilson, J.C., Liu, K.Y., Jones, K., Mahmood, J., Arya, U., Howard, R., 2024. Biomarkers of neurodegeneration in schizophrenia: systematic review and meta-analysis. BMJ Mental Health 27 (1), e301017. Available at https://doi.org/10.1136/bmjment-2024-301017.

Wißfeld, J., Abou Assale, T., Cuevas-Rios, G., Liao, H., Neumann, H., 2024. Therapeutic potential to target sialylation and SIGLECs in neurodegenerative and psychiatric diseases. Frontiers in Neurology 15. Available at https://doi.org/10.3389/fneur.2024.1330874.

Xue, T., Liu, W., Wang, L., Shi, Y., Hu, Y., Yang, J., et al., 2024. Extracellular vesicle biomarkers for complement dysfunction in schizophrenia. Brain a Journal of Neurology 147 (3), 1075–1086. Available at https://doi.org/10.1093/brain/awad341.

Yang, F.-C., Zhou, Y., Zhang, S.-Y., Ma, R., 2025. Depression and anxiety in patients with endometriosis-associated chronic pain: neuroimmune mechanisms mediated by inflammatory factors. World Journal of Psychiatry 15 (10). Available at https://doi.org/10.5498/wjp.v15.i10.107147.

Yang, W., Zhu, C., Shen, Y., Xu, Q., 2016. The pathogenic mechanism of dysbindin-1B toxic aggregation: BLOC-1 and intercellular vesicle trafficking. Neuroscience 333, 78–91.

Yildiz, S., Kazğan Kiliçaslan, A., Sirlier Emir, B., Uğur, K., Kiliç, F., 2024. Serum HMGB1 and beclin 1 levels in patients with a diagnosis of schizophrenia. Turk Psikiyatri Dergisi=Turkish Journal of Psychiatry 35 (1), 1–7. Available at https://doi.org/10.5080/u27030.

Yolland, C.O., Hanratty, D., Neill, E., Rossell, S.L., Berk, M., Dean, O.M., et al., 2020. -acetylcysteine in the treatment of schizophrenia. The Australian and New Zealand Journal of Psychiatry 54 (5), 453–466. Available at https://doi.org/10.1177/0004867419893439.

Yoon, J.H., Zhang, Z., Mormino, E., Davidzon, G., Minzenberg, M.J., Ballon, J., et al., 2023. Reductions in synaptic marker SV2A in early-course Schizophrenia. Journal of Psychiatric Research 161, 213–217.

Zaharija, B., Odorčić, M., Hart, A., Samardžija, B., Marreiros, R., Prikulis, I., et al., 2022. TRIOBP-1 protein aggregation exists in both major depressive disorder and schizophrenia, and can occur through two distinct regions of the protein. International Journal of Molecular Sciences 23 (19), 11048. Available at https://doi.org/10.3390/ijms231911048.

Zapata-Muñoz, J., Villarejo-Zori, B., Largo-Barrientos, P., Boya, P., 2021. Towards a better understanding of the neuro-developmental role of autophagy in sickness and in health. Cell Stress 5 (7), 99–118. Available at https://doi.org/10.15698/cst2021.07.253.

Zarrei, M., Burton, C.L., Engchuan, W., Young, E.J., Higginbotham, E.J., MacDonald, J.R., et al., 2019. A large data resource of genomic copy number variation across neurodevelopmental disorders. NPJ Genomic Medicine 4. Available at https://doi.org/10.1038/s41525-019-0098-3.

Zhang, J., Wang, H., Xue, X., Wu, X., Li, W., Lv, Z., et al., 2025. Human endogenous retrovirus W family envelope protein (ERVWE1) regulates macroautophagy activation and micromitophagy inhibition via NOXA1 in schizophrenia. Virologica Sinica 40 (3), 401–418.

Zhang, Y., Catts, V.S., Sheedy, D., McCrossin, T., Kril, J.J., Shannon Weickert, C., 2016. Cortical grey matter volume reduction in people with schizophrenia is associated with neuro-inflammation. Translational Psychiatry 6 (12). Available at https://doi.org/10.1038/tp.2016.238.

Zhao, Q., Xiang, H., Cai, Y., Meng, S.S., Zhang, Y., Qiu, P., 2022. Systematic evaluation of the associations between mental disorders and dementia: an umbrella review of systematic reviews and meta-analyses. Journal of Affective Disorders 307, 301–309.

Zhou, L., Sommer, I.E.C., Yang, P., Sikirin, L., van Os, J., Bentall, R.P., et al., 2025. What do four decades of research tell us about the association between childhood adversity and psychosis: an updated and extended multi-level meta-analysis. The American Journal of Psychiatry 182 (4), 360–372. Available at https://doi.org/10.1176/appi.ajp.20240456.

Zhu, C.-Y., Shen, Y., Xu, Q., 2015. Propagation of dysbindin-1B aggregates: exosome-mediated transmission of neurotoxic deposits. Neuroscience 291, 301–316.

CHAPTER 6

Disease-specific glycosylation in the neurodegenerative brain

Meredith Edwards[1], Natasha Vassileff[1], Joern R Steinert[2,*], and Jereme G Spiers[1,*]
[1]John Curtin School of Medical Research, College of Science and Medicine, The Australian National University, Canberra, ACT, Australia,
[2]School of Life Sciences, Faculty of Medicine and Health Sciences, University of Nottingham, Nottingham, United Kingdom

6.1 Introduction

The remarkable complexity of living organisms cannot be explained by gene count alone, but rather by intricate layers of molecular regulation (Craig Venter et al., 2001). In humans, a modest number of genes is compensated for by combinatorial regulation generated by co- and post-translational modifications (PTM) of functional molecules such as proteins, thereby creating biological diversity at a scale somewhat difficult to grasp. Glycosylation, one of several ubiquitous and dynamic modification pathways, exponentially expands proteome diversity (Moremen et al., 2012; Reily et al., 2019). This occurs by way of conjugation of a vast assortment of assembled monosaccharide structures of varying sizes and configurations, known as glycans or "sugar trees", to target proteins mainly within the endoplasmic reticulum (ER) and Golgi apparatus. The mature glycoproteins are trafficked to the cell membrane to form a complex coat, known as the glycocalyx. They adorn the surface of almost every cell of every living organism and have been likened by scientists to a dense forest or a sea kelp bed (Varki, 2017). The entire collection of all possible glycoconjugates forms what has been termed the glycome, and not surprisingly, what is known may be analogous perhaps to our understanding of space or even the oceans, as understanding of the glycome continues to be challenged and reshaped by new discoveries (Aminov and Aminova, 2023; Flynn et al., 2021; Reily et al., 2019; Varki, 2017). Indeed, recent findings have extended the field of glycobiology beyond proteins to include glycosylation of RNA (Flynn et al., 2021; Li et al., 2023a,b; Chai et al., 2023; Watson et al., 2019).

Glycoconjugates of all types are involved in system-wide functions within the cellular microenvironment and metabolic milieu including, but absolutely not limited to, receptor-mediated cell signaling, growth and differentiation, pathogen-host recognition, accurate protein folding, stability and activity, tumor-immune interactions, ion channel function, synaptogenesis, synaptic vesicle formation and function, neurotransmission, and neuroplasticity

[*]Joint Senior Authors.

Molecular Insights into Proteinopathies
DOI: https://doi.org/10.1016/B978-0-443-34176-2.00015-8
© 2027 Elsevier Inc. All rights are reserved, including those for text and data mining, AI training, and similar technologies.

(Craig Venter et al., 2001; Bradberry et al., 2023; Kleene and Schachner, 2004; Kruger and Isom, 2016; Li and Ding, 2019; Muller et al., 1996; Reily et al., 2019; Schmidt and Catterall, 1987; Smith and Bertozzi, 2021; Vitorakis and Piperi, 2024; Zhang et al., 2020). A growing body of evidence has revealed that significant alterations to glycosylation occur in aging and neurodegenerative diseases (NDDs), and that changes to glycan structures and modification levels may serve as useful disease biomarkers (Fig. 6.1).

Thus the advance of glycoproteomics, a large-scale study of glycoproteins, provides invaluable insights into the involvement of glycosylation in the pathogenesis of NDDs and has been instrumental in identifying novel potential treatment targets. In this review, primary protein glycosylation pathways will be discussed in the context of understanding the key roles of protein glycosylation and how it may be altered in NDDs such as Alzheimer's disease (AD), Parkinson's disease (PD), and prion diseases.

6.2 Glycobiology and glycosylation

6.2.1 Glycan sources and metabolism

In humans, glycan structures are initially assembled from approximately ten monosaccharide building blocks (Fig. 6.2, parent molecule glucose [*Glc*], galactose [*Gal*],

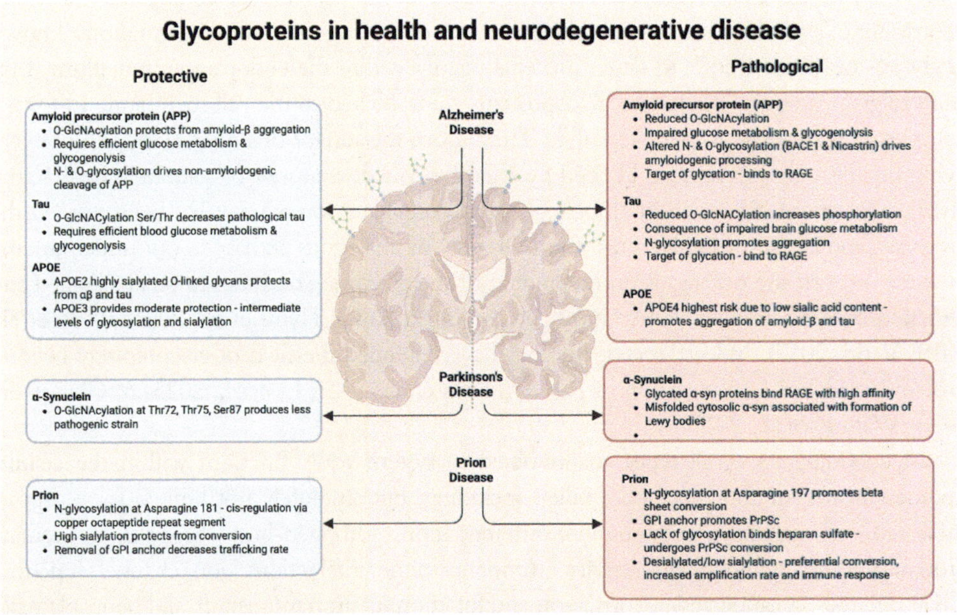

Figure 6.1 Summary of key neuroprotective protein glycosylation (*blue*) *versus* pathological alterations shown to promote neurodegenerative diseases (*red*). *Created in BioRender. Edwards, M. (2026) https://BioRender.com/k2fh84l.*

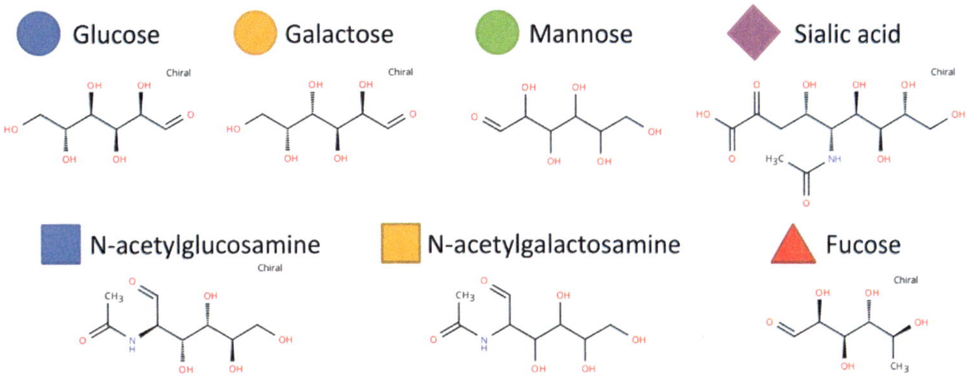

Figure 6.2 Major monosaccharide building blocks using glycan symbol nomenclature. *Glc*: glucose; *Gal*: galactose; *Man*: mannose; *Sia*: sialic acid; *GlcNAc*: N-acetylglucosamine; *GalNAc*: N-acetylgalactosamine; *Fuc*: fucose.

mannose [*Man*], N-acetylglucosamine [*GlcNAc*], glucuronic acid [*GlcUA*], N-acetylgalactosamine [*GalNAc*], sialic acid [*Sia*]) predominantly found as N-acetylneuraminic acid (*Neu5Ac*) and CMP-sialic acid in human cells, fucose (*Fuc*), xylose (*Xyl*) and iduronic acid (*IdoA*) (Reily et al., 2019; Scheper et al., 2023; Smith and Bertozzi, 2021). Sugars required for glycan biosynthesis are derived predominantly from dietary sources of carbohydrate, including fruits, vegetables, and grains. These are rapidly acted upon by several carbohydrate- degrading enzymes, including salivary and pancreatic α-amylases for polysaccharide and oligosaccharide digestion, and α-glucosidases located at the intestinal brush border for disaccharide digestion, to form absorbable monosaccharides for transport across the apical gut epithelium. This process is mediated by dedicated glucose transporters-2 and -5 (GLUT-2, -5) and sodium-glucose like transporter-1 (Barber et al., 2022). GLUT2 then mediates intracellular monosaccharide efflux across the basolateral membrane for entry into systemic circulation and delivery to target tissues. In addition to the exogenous supply of carbohydrate substrate, several endogenous intracellular sources of sugars exist whereby monosaccharides such as *Glu*, *Gal*, *Man*, *GlcNAc* and *Sia* can be reused from glycoconjugates enzymatically degraded by glycosidases and mannosidases in the ER, Golgi apparatus and lysosomes, with evidence suggesting that the contribution to the monosaccharide pool from these salvage pathways is quite considerable (Fujita et al., 2008). Essential glycan intermediates such as UDP-*GlcNAc* are produced *de novo* by the hexosamine biosynthetic pathway (HBP), which metabolizes in humans around 2%–5% of intracellular glucose for glycan synthesis.

Importantly, glycogen metabolism in the brain was recently shown to be the major reservoir of free glucosamine for the production of UDP-*GlcNAc*, a critical substrate required for protein *N*-glycosylation (Sun et al., 2021). Impairment of glycogen metabolism, as seen in aging, dementia, and demonstrated in mouse models of glycogen storage disease, results in a shortage of UDP-*GlcNAc*, lower levels of protein *N*-glycosylation, ER stress, and neuroinflammation in key brain regions, including the hippocampus (Sun et al., 2021).

6.2.2 Primary protein glycosylation pathways

Protein glycosylation, a complex PTM mostly occurring in the ER and Golgi apparatus, is mediated by a large family of glycosyltransferases, glycosidases, and mannosidase enzymes responsible for the covalent attachment and trimming of sugar moieties on specific amino acids of nascent proteins, thus forming a diverse array of linear, branched, and hybrid glycans (Narimatsu et al., 2019; Spiro, 2002). This unparalleled diversity is accomplished because multiple parameters of glycosylation can be adjusted depending on the cell, including glycan binding sites (single or multiple), glycosidic linkages (*N*- or *O*-), glycan composition and order, glycan structure (branched or linear), and length (short- or long-chain polysaccharide) (Reily et al., 2019; Spiro, 2002; Varki, 2017). Much like an advanced language, these modifications confer a much higher order of information at both the molecular and cellular levels, beyond that of genome-to-protein encoding and function (Moremen et al., 2012). It is estimated that around 2% of the human genome encodes glycosyltransferase and other related enzymes, and that approximately half of all translated proteins within a cell will undergo glycosylation of some kind for a specific function, dictated by the glycan structure and location of glycosylation (Pradeep et al., 2023). However, unlike proteins, which are products of gene transcription, glycans are dynamically assembled, trimmed, and transferred onto proteins in a nontemplate fashion (built through enzymatic processes) depending on enzyme availability, amino acid sequence, availability, and destination (Fig. 6.3) (Fisher et al., 2019b; Smith and Bertozzi, 2021). It is also thought that glycosylation of secreted and membrane-associated proteins reduces the risk of proteolysis and unwanted intercellular reactions in the extracellular milieu, and provides decoy sites for pathogens (Smith and Bertozzi, 2021).

Whilst there are numerous glycosylation types, two primary protein glycosylation pathways, *N*- and *O*-glycosylation are known to play a central role in regulating critical functions within the central nervous system and conversely, are involved in the pathogenesis of NDDs (Hargett and Renfrow, 2019; Kang et al., 2024; Krieg et al., 1998; Roller et al., 2020; Zhong et al., 2023).

6.2.3 N-glycosylation

6.2.3.1 Cytosolic N-glycan precursors

N-glycosylation is dependent on the accurate synthesis of a mature glycolipid precursor, $Glc_3Man_9GlcNAc_2$-PP-dolichol. Assembly of an immature precursor originates at the cytoplasmic face of the ER membrane, with the consecutive addition of two *GlcNAc*

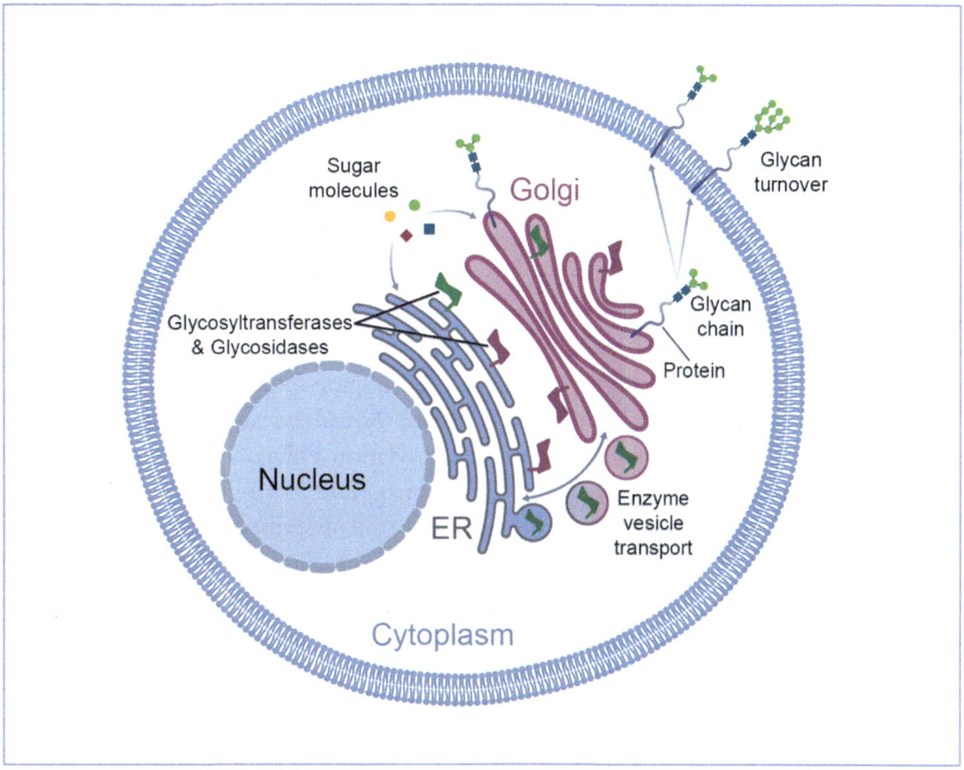

Figure 6.3 Graphic overview for major protein N- and O-glycosylation biosynthetic pathways. Nascent proteins are co- and post-translationally modified with monosaccharides to form initial glycan structures in the endoplasmic reticulum and transported to the Golgi apparatus for final processing. The mature glycoprotein is then secreted to its final destination either in the plasma membrane or is secreted.

and five *Man* residues to lipid anchor dolichyl-phosphate using UDP-*GlcNAc* and GDP-*Man* as glycosyl donors, catalyzed by glycosyltransferase complexes to form branched glycan $Man_5GlcNAc_2$-PP-dolichol (Rush, 2015; Sanyal et al., 2008). At the core of each *N*-glycan is a ubiquitous and conserved oligosaccharide, $GlcNAc_2Man_3$. Each enzyme that catalyzes *N*-glycan precursor synthesis is encoded by asparagine-linked glycosylation (ALG) genes (Wang et al., 2015).

6.2.3.2 N-glycosylation in the endoplasmic reticulum

$Man_5GlcNAc_2$-PP-dolichol is then rapidly translocated from the cytoplasmic face into the luminal space of the ER by a membrane-bound, ATP-independent, bidirectional flippase protein (Sanyal et al., 2008). Elongation of $Man_5GlcNAc_2$-PP-dolichol in the ER lumen is often referred to as the lipid-linked oligosaccharide assembly pathway. The lipid-linked oligosaccharide pathway is facilitated by mannosyl- and glucosyltransferases that require intermediates Dol-P, GDP-*Man*, and UDP-*Glc* as substrate donors for the addition of four mannose and three glucose sugars to $Man_5GlcNAc_2$-PP-dolichol, eventually completing synthesis of the mature *N*-glycan precursor, $Glc_3Man_9GlcNAc_2$-PP-dolichol (Aebi, 2013). The terminal step in this ordered process is the "capping" of the $Glc_3Man_9GlcNAc_2$-PP-dolichol α-antenna with an α-1,2-linked glucose, which acts as the recognition site for oligosaccharyltransferase (OST)/Sec61 complex to bind and transfer the glycan structure to the growing polypeptide chain once it reaches approximately 12–14 amino acids (Aebi, 2013; Esmail and Manolson, 2021). The OST complex attaches $Glc_3Man_9GlcNAc_2$-PP-dolichol via a β-glycosylamine link to a protein at the side chain of acceptor amino acid asparagine sequence motif, either Asn-X-threonine (Thr) or Asn-X-serine (Ser), where X can be any amino acid except proline, which has been shown to reduce glycosylation efficiency and interfere with peptide chain loop conformation (Fig. 6.4) (Spiro, 2002). *N*-glycosylation is often characterized as a PTM but may more accurately be described as a co-translational modification as nascent polypeptides can be *N*-glycosylated as they enter the ER lumen from the membrane-bound ribosomes during translation (Ruiz-Canada et al., 2009). OST isoforms mediate two distinct but complementary *N*-glycosylation outcomes, with OST catalytic subunit STT3A facilitating co-translational *N*-glycosylation and

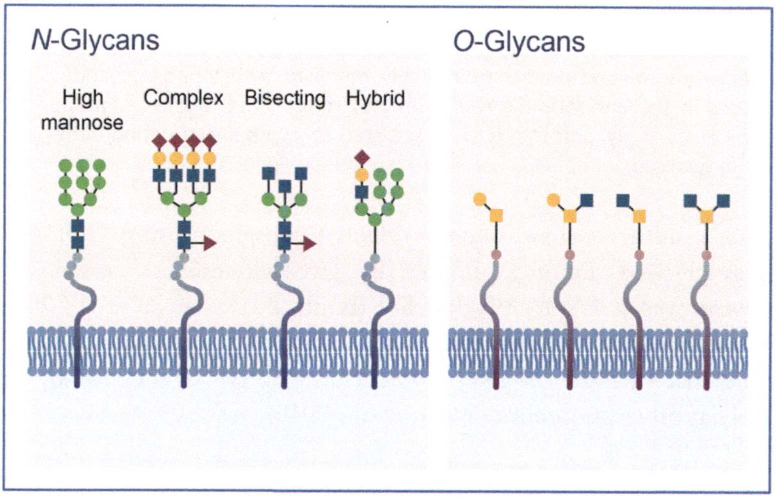

Figure 6.4 Example of *N*- and *O*-glycans. *N*-glycoproteins are decorated with a variety of glycan "sugar trees" that consist of a conserved core of $GlcNAc_2Man_3$, and can be extended with a range of sugars, including a terminal *Fuc* and a *Sia* cap. The *N*-glycans are linked to the Asn-X-Ser/Thr consensus sequence of the polypeptide. The *O*-glycoproteins commonly have a *GalNAc* initiating sugar linked to either Ser/Thr of the folded protein. *O*-glycans are extended with a variety of monosaccharides.

STT3B subunit mediating PTMs. Together, the OST isoforms scan unfolded polypeptides for acceptor sites, thereby maximizing the efficiency of *N*-glycosylation (Ruiz-Canada et al., 2009). Immediately upon attachment to the polypeptide, the terminal *Glc* "cap" and the subsequent *Glc* are removed by glucosidase I and II, respectively.

6.2.3.3 Endoplasmic reticulum glycoprotein folding and quality control mechanisms

The newly synthesized *N*-glycoproteins then need to be folded accurately to achieve their mature tertiary and quaternary structures. In humans, this process is mediated by several ER chaperone molecules, including BiP, a member of the heat shock protein family, Grp94, and the lectins calnexin and calreticulin stabilized by oxidoreductase ERp57, which bind to the monoglucosylated glycoprotein to promote protein folding. Glycoproteins are released if accurate, and the remaining *Glc* residue is removed by glucuronidase II. Newly synthesized glycoproteins undergo further trimming by mannosidases ERManI/ERManII prior to release from the ER and transport to the Golgi apparatus, as represented in (Frigerio and Lord, 2000; Reily et al., 2019; Wang et al., 2015). To ensure only correctly folded glycoproteins leave the ER, quality control mechanisms are in place that proofread to identify misfolded proteins and hold them in the ER for reprocessing or degradation. Misfolded glycoproteins that require reprocessing undergo a transient monoreglucosylation, facilitated by calcium- dependent UDP-*Glc*: glycoprotein glucosyltransferase, which acts as a retention signal for calnexin/calreticulin complex to bind in order to reattempt folding, a process that occurs in the order of seconds to minutes (Ellgaard and Helenius, 2003). The quality control cycle, that is, the glucosidase II-calnexin/calreticulin-glucosyltransferase cycle, repeats until a native conformation is achieved and only properly folded proteins are released from the ER (Ellgaard and Helenius, 2003; Frigerio and Lord, 2000).

However, the number of times the cycle can occur is limited, and this is proposed to be regulated in a temporal and structural manner by ERManI, which determines the entry of the persistently misfolded, incomplete, or mutated *N*-glycoproteins into the ER-associated degradation pathway via translocation by the Sec61 complex to the proteasome for degradation (Ellgaard and Helenius, 2003; Frigerio and Lord, 2000; Lederkremer and Glickman, 2005; Roth, 2002; Wilson et al., 2000). Despite quality control measures, it may be appreciated that, as in most biological processes, errors inevitably occur that have deleterious consequences. More recent evidence from studies in *C. elegans* suggests that half of all co- and post-translationally misfolded proteins may universally evade chaperone correction and degradation, and remain in a soluble state over timescales that extend beyond months, leading to accumulation of protein aggregates and cellular stress over time (Nissley et al., 2022; Santra et al., 2019). This remains to be observed in human studies but may have considerable implications for the pathogenesis of NDDs, where accumulation of intracellular protein aggregates is a pathological hallmark.

Nevertheless, comparable to other biological quality control systems including DNA and RNA repair pathways, the quality control cycle in the ER affords glycoproteins several chances at accurate maturation, thereby ensuring functional fidelity of glycoproteins entering the secretory pathways. In humans and other higher eukaryotes, properly folded glycoproteins that are cleared for transit to the Golgi have the terminal residue of mannose removed to form $Man_8GlcNAc_2$, which is transported to the *cis*-Golgi by coat protein (COP) II transport vesicle (Reily et al., 2019; Scheper et al., 2023).

6.2.3.4 Golgi glycan processing and maturation

The Golgi apparatus is the central processing center of the secretory pathway. Its intriguing structure comprises multilayered and laterally stacked flat cisternae, that have distinct but tethered subcompartments—*cis*-, *medial*- and *trans*-Golgi network—in which a specialist family of glycosylation enzymes is housed to ensure sequential, accurate and diverse glycan chain processing (Aebi, 2013; Zhang and Wang, 2016). The majority of the $Man_8GlcNAc_2$*N*-glycoproteins that arrive at the *cis*-Golgi are initially trimmed by α-mannosidases to generate a key intermediate for *medial*-Golgi processing, $Man_5GlcNAc_2$.

In the *medial*-Golgi, glucosyltransferase GlcNAcT-I transfers one *GlcNAc* residue to substrate $Man_5GlcNAc_2$, which is the initiating event for further synthesis of complex and hybrid *N*-glycans (Fig. 6.4) that follows. In the *trans*-Golgi, *N*-glycan diversity increases exponentially with the addition of each monosaccharide by dedicated enzymes, either *GlcNAc, GalNAc, GlcUA, Fuc, Xyl,* or *Gal,* and a final capping of the multiantennae with *Fuc* (fucosylation) or, most abundantly, *Sia* or *Neur5Ac* (sialylation). Thus sialylated glycan terminals make up the majority of the glycocalyx canopy. Various factors such as cell type and location, developmental stage, protein type, enzyme protein levels, and monosaccharide availability ultimately determine the exact type and heterogeneity of final glycan structures produced by the Golgi machinery, although physiological changes to cells that occur as a result of aging, inflammation, and disease alter the glycome in that less, more or different glycans are produced (Fisher et al., 2019b; Smith and Bertozzi, 2021).

6.2.4 *O*-glycosylation

Whilst not as common as *N*-glycosylation, protein *O*-glycosylation is essential for the formation of mucins, extracellular matrix components, immune structures including immunoglobulins (Ig) and antibodies, and the regulation of protein processing

(Akasaka-Manya and Manya, 2020; Zhao and Lang, 2023). In contrast to *N*-glycosylation, protein *O*-glycosylation occurs post-translationally in the *cis*-Golgi, with an initial monosaccharide attached to an oxygen atom of the hydroxyl group of either Ser or Thr residues on the folded protein (Fig. 6.3). Additionally, this can cooccur on a protein that has already been *N*-glycosylated in the ER.

Mucin-type *O*-glycosylation, the most common form, begins with the addition of a single *GalNAc* from UDP-*GalNAc* by polypeptide GalNAc-transferases, and this initiating *GalNAc* moiety is rapidly elongated throughout the *cis*-, *medial*-, and *trans*-Golgi by C3GnT or T-synthase with chaperone COSMC to create one of six core mucin-glycan structures. Core glycans can be further extended with *GalNAc*, *Gal*, *GlcNAc*, *Fuc*, *Man*, and *Sia* to form either a branched or linear mature *O*-*GalNAc* complex glycan containing up to 20–40 glycans, found in abundance on extracellular and secreted glycoproteins (Reily et al., 2019; Varki, 2017; Pradeep et al., 2023). Not expressed in healthy cells, a single truncated Ser/Thr-*GalNAc* glycan forms Tn antigen, robustly expressed in tumor cells, including glioblastoma, via the Tn antigen-macrophage galactose-type lectin axis, and drives an immune suppressive tumor microenvironment (Cornelissen et al., 2020; Dusoswa et al., 2020). The addition of an acetylated *Sia* residue to Tn antigen to form sialyl-Tn helps masks the underlying sugar, thereby preventing recognition by the immune system or antisialyl-Tn antibodies (Dube and Bertozzi, 2005; Dusoswa et al., 2020). There are several key *O*-glycoprotein subtypes, in which the initiating monosaccharide residue can either be *Man* (*O*-*Man*), *Fuc* (*O*-*Fuc*), or *Glc* (*O*-*Glc*), and are elongated with a variety of other sugars to form *O*-glycans known to play important roles in the central nervous system, and when dysregulated to be associated with neurological and neurodegenerative disorders (Haukedal and Freude, 2021; Pradeep et al., 2023).

6.2.5 *O*-GlcNAcylation

Another important type of *O*-glycosylation PTM is *O*-GlcNAcylation, which plays a pivotal role in nutrient sensing, cellular stress responses, gene expression, signaling transduction, cell cycling, protein homeostasis, and neuronal and synaptic function. Notably, *O*-GlcNAcylation is most prominent in the brain, highlighting its critical role in neuronal function in health and neurodegenerative disease (Hart et al., 2011; Park et al., 2021; Lee et al., 2021). It is also coupled to glucose metabolism, which can be impaired in aging and neurodegenerative processes (Li et al., 2023a; Underwood et al., 2024; Scherer et al., 2021). *O*-GlcNAcylation is unique in that it occurs outside the canonical ER-Golgi pathway, taking place in the cytosol, nucleus, and various other locations including mitochondria, the cytoskeleton, and the ER (Morales and Pratt, 2024). *O*-GlcNAcylation is also a rapid and reversible cycle, during which *O*-GlcNAc transferase (OGT) concurrently removes a phosphate and uses UDP-*GlcNAc* as a donor substrate to link *O-GlcNAc* to serine/threonine residues of target nucleocytoplasmic proteins via an O-β-glycosidic bond, and O-GlcNAcase (OGA) removes it via hydrolysis. Given the bidirectional relationship between O-GlcNAcylation and phosphorylation, as evidenced by OGT activity, it is not surprising that this ubiquitous PTM plays such a critical role in many human diseases, including NDDs (Hart et al., 2011; Yang and Qian, 2017).

6.2.6 Nonenzymatic glycation

Nonenzymatic glycosylation, known as glycation, involves the spontaneous covalent attachment of a reducing sugar moiety, usually *Glc* or fructose, to proteins on the arginine and lysine residues, and on adenine or guanine of nucleic acids (see Section 6.2.7) (Perrone et al., 2020). Initial steps of protein glycation, also known as the Maillard reaction, form Schiff-base adducts, subsequently converted to Amadori products, which are then irreversibly oxidized via reactive intermediates and oxyradicals to form toxic compounds, known as advanced glycation end-products (AGEs) (Choi et al., 2014). AGEs are a group of heterogeneous molecules known to interact with the receptor for advanced glycation end-products (RAGE), which are expressed on many cells in the periphery, including macrophages, epithelial and endothelial cells, and on neurons, astrocytes, and microglia in the brain (Derk et al., 2018). The AGE-RAGE axis is robustly implicated in brain aging and neurodegeneration as activation by AGE ligands are key drivers of blood-brain barrier disruption, oxidative damage, ER stress, and neuroinflammatory signaling pathways seen in AD and PD (Table 6.1) (Bhattacharya et al., 2023; Choi et al., 2014; Derk et al., 2018; Long et al., 2022; Lue et al., 2001; Vitorakis and Piperi, 2024).

Table 6.1 Summarizes major glycan receptors and ligands in the central nervous system, including cell types and alterations in aging and neurodegenerative diseases (BBB: blood brain barrier; SIGLEC: sialic acid-binding immunoglobulin-like lectin; NDDs: neurodegenerative diseases; NCAM: neural cell adhesion molecule; AGEs: advanced glycation end products; HMBG1: high-mobility box group 1; RAGE: receptor for advanced glycation end products; Aβ: amyloid beta).

Cellular AGEs are produced via the activity of triose-phosphate isomerase (TPI) a metabolic enzyme in the glycolytic pathway that catalyzes the conversion and adjusts the equilibrium between glyceraldehyde 3-phosphate (GAP) and dihydroxyacetone phosphate (DHAP). GAP and DHAP are products of the cleavage of fructose 1,6-bisphosphate catalyzed by aldolase. While DHAP has a role in fatty acid metabolism, GAP continues in the glycolytic pathway ultimately producing pyruvate. Although the isomerisation of DHAP to GAP is not required to produce pyruvate, it increases the efficiency of glycolysis (Roland et al., 2016). DHAP transforms spontaneously into the cytotoxic metabolic by-product methylglyoxal (Tajes et al., 2013). The enhanced levels

Table 6.1 Major glycan receptors, ligands, and glycosylation alterations in aging and neurodegenerative disease.

Receptors	Cells and brain regions	Glycan ligands	Alterations and mechanisms in NDDs	References
Selectins - P-selectin E-selectin L-selectin	Endothelial; activated astrocytes Endothelial; platelets Leukocytes	Sialic acid – sialyl-Lewis^{x} Fucose Fucosylated sialyl-Lewis^{x}	Aging: • ↑ fucosylation • Leukocyte adhesion and migration ↑ neuroinflammation NDD: • Alter BBB function • ↑ Immune cell infiltration	Li et al. (2018), McEver (2015), Yeini et al. (2021)
SIGLECs	Schwann cells Microglia Neurons/ myelinated Leukocytes GlycoRNAs	Sialic acid only	Aging: • Sialylation of neurons ↓ with age NDD: • ↑ CD33 (SIGLEC-3) risk factor for AD • CD33$^{\Delta E2}$ ↓ risk of AD • Sialylation of neurons also may ↑ in NDDs • SIGLEC-11 activation by polySia NCAM is neuroprotective	Bhattacherjee et al. (2021), Chai et al. (2023), Gonzalez-Gil et al. (2022), Griciuc et al. (2013), Liao et al. (2021)
RAGEs	Neurons Astrocytes Microglia Endothelial cells Hippocampus Cortex Cerebellum *Substantia nigra* Striatum	AGEs HMGB1 S100 Group Alpha-synuclein Amyloid-beta Tau	NDD: • RAGE expression ↑ in AD • AGE accumulation drives amyloidosis • RAGE involved in Aβ transport across BBB and intra-neuronally • AGE/tau colocalization • Neuroinflammation induced by α-synuclein binding in PD • Aβ -RAGE microglial activation	Lue et al., (2001), Choi et al. (2014), Deane et al. (2003), Long et al. (2022), Takuma et al. (2009), Bhattacharya et al. (2023)

of neuroinflammation in AD (Heneka et al., 2025), causing a nitro-oxidative environment, favor 3-nitrotyrosination of TPI (Guix et al., 2009) and as a result, inhibition of this enzyme leads to non-enzymatic methylglyoxal-mediated enhanced protein glycation and formation of AGE (Ahmed et al., 2003; Guix et al., 2009). AGE formation has also a causative role in the vascular complications of diabetes mellitus and several other neurodegenerative diseases, including AD, PD, and Huntington diseases (Li et al., 2012; Yamagishi, 2019). TPI deficiency is a rare human multisystem disorder with severe neuromuscular symptoms that arises exclusively from mutations within the *TPI1* gene (Myers et al., 2024). Mutant and dysfunctional TPI has been shown to cause neurological and vascular phenotypes (Stone et al., 2023; Oláh et al., 2002), illustrating the broad actions of AGEs.

However, exogenous AGEs are also acquired from sources including high-sugar consumption, soft drinks, fried foods, charred meats, heat-processed foods, cooking methods that use high and/or dry heat, browning of foods by the Maillard reaction, alcohol, and smoking (Aragno and Mastrocola, 2017). Interestingly, there is evidence to suggest the contribution of exogenous AGEs outweighs those that are endogenously produced during normal aging, which has implications for the impact of dietary and lifestyle behaviors on AGE production and disease (Table 6.2) (Rungratanawanich et al., 2021).

Table 6.2 Summarizes potential glycosylation and glycation-related modifiable factors for healthy aging and NDD prevention and/or adjunctive therapy, including dietary, lifestyle, and supplemental interventions as described in the evidence (PDGF-R1β: platelet derived growth factor receptor-1β; Cx-43: connexin-43; ECM: extracellular matrix; JAK/STAT: Janus kinase/signal transducers and activators of transcription; NF-kB: nuclear factor kappa-light-chain-enhancer; SOD1: superoxide dismutase; GPx: glutathione peroxidase; IgG: immunoglobulin G; LPS: lipopolysaccharide; SR: safe range).

6.2.7 Key glycosylation subtypes in the brain

6.2.7.1 Sialylation

Sialic acids are well-characterized and important components of all glycan types in the brain, serving many critical functions including regulating neural cell adhesion, migration, axon myelination, microglial homeostasis, synaptic transmission and development, and is involved in regulating the half-life of glycoproteins when removed (Muller et al., 1996; Rawal and Zhao, 2021). Sialylation is also important in brain aging, as decreased levels have been observed (Rawal and Zhao, 2021). Of particular interest, polysialylated *N*- and *O*-glycoproteins are found on neural cell adhesion molecule (NCAM) protein, as PSA-NCAM, which has been shown to contribute to the induction of neural development, synaptic plasticity, and regeneration in the

Table 6.2 Potential glycosylation and glycation-related modifiable factors for healthy aging and neurodegenerative disease (NDD) prevention.

Modification category	Modification type	Source	Dose/frequency	Mechanisms and major findings	References
Dietary	**Reduce exogenous sources of AGEs cooking methods**	Smoking, alcohol, heat processed foods, soft drinks, charred meats Swap BBQ, frying and grilling for steaming and slow cooking methods to reduce AGEs	Significantly reduce or eliminate to reduce burden of exogenous AGEs and glycation	Aging: • AGEs ↑ nonfunctional proteins NDD: • Disrupt BBB via ↓ integrin $\alpha 1$, PDGF-R1β and Cx-43 • Remodel ECM • ↑ neuroinflammation via JAK/STAT, Ras/Raf, PI3K/AKT/NF-kB • ↓ brain catalase, SOD1 and GPx	Bhattacharya et al. (2023), Rungratanawanich et al. (2021), Sparvero et al. (2009), Videira and Castro-Caldas (2018)
	Caloric restriction	Low calorie diet (LCD) followed by bariatric surgery in obese patients	Medical intervention	Aging: • ↓ IgG *N*-glycans with bisecting *GlcNAc*—markers of reduced inflammation	Greto et al. (2021)
Lifestyle	**Exercise**	High-intensity sprint training Moderate long-term exercise	6-week protocol 3 years	Aging: • ↓ IgG *N*-glycans suggestive of antiinflammatory shift • Glycan Age lower in moderately active individuals	Tijardović et al. (2019), Šimunić-Briški et al. (2024)

(Continued)

Table 6.2 (Continued)

Modification category	Modification type	Source	Dose/frequency	Mechanisms and major findings	References
	Blood glucose control	Keep HbA1c within optimal range	Range—4%–5.6%	Aging: • ↓ activation AGE/RAGE	Underwood et al. (2024)
Supplemental	**Curcumin**	Oral supplementation		Aging: • Inhibits AGE formation • ↓ AGE induced inflammation NDD: • Curcumin derivative GT863 inhibits Aβ production by suppressing *N*-glycosylation of Nicastrin (in vitro)	Alizadeh and Kheirouri (2019), Urano et al. (2020)
	Galacto-oligosaccharide (GOS)	Supplemental oral powder Dietary sources – • Cashews, pistachios, wheatbran, oats, wholegrain bread, legumes, soybeans	Clinically effective dose: 5.5 g oral powder	Aging: • ↓ Metabolic syndrome • ↑ Secretory IgA in GIT • NDD: • Improve populations of commensal gut microbes, ↓ LPS	Vulevic et al. (2013), Zhan et al. (2016)
	Melatonin	Supplemental form	Low dose 2–5 mg, medium term supplementation	Aging:	Sumsuzzman et al. (2021), Tseng et al. (2022),

		Food sources: oily fish, sour cherries, mushrooms Circadian rhythm improvements through lifestyle modifications.		• >50 years of age melatonin production declines • Neuroprotective; antioxidant • Improve insulin sensitivity NDD: • Lower levels in CSF of AD patients • ↓ Aβ production by antagonizing Aβ enzymes; ↑ clearance via glymphatic system (RCT) • Prevent tau phosphorylation • Significant improvement in cognitive function in mild AD (systematic review/metaanalysis)	Xu et al. (2020), Zhou et al. (2003), Zhang et al. (2025)
	Chicoric acid	Supplemental form: *Echinacea purpurea* Phenolic acid from chicory and purple coneflower plant, *Echinacea purpurea*	• 250 mg equiv. chicoric acid 1.5 mg • Mouse studies - 40 mg/kg	Aging: • Glucose metabolism regulation via GLUT1 NDD: • Inhibit tau phosphorylation (tp) via AMPK • ↑ tau glycosylation	Farag et al. (2021), Lee and Scagel (2013), Lv et al. (2021), Wang et al. (2022), Yang et al. (2022)

(*Continued*)

Table 6.2 (Continued)

Modification category	Modification type	Source	Dose/frequency	Mechanisms and major findings	References
		Foods sources with diverse concentrations: dandelion leaves, basil, lemon balm, endive, radicchio, lettuce.		• Delivery in ferritin nanocapsule improved effectiveness of tau inhibition • ↓ amyloidosis via inhibition of Nf-kB • ↑ BDNF; DA; 5-HT (mice in PD model) • ↓ TNF-α; IL1β; CASP3; BCL-2; Nf-kB (in vivo)	

hippocampus (Li and Ding, 2019; Muller et al., 1996). Polysialylation is a unique form of glycan modification catalyzed by polysialyltransferases ST8SIA2 and ST8SIA4, which can link 50–200 sialic acid residues in an α2,8-linkage, forming long, highly charged chains which afford PSA-NCAM its specialist capabilities in the brain including adult neurogenesis (Eckhardt et al., 2000; Kleene and Schachner, 2004; Muller et al., 1996; Quartu et al., 2008). Changes in PSA-NCAM expression have been linked with many neurological disorders and neurodegeneration, particularly in the hippocampus and neocortex, with some postulating that this may be a compensatory mechanism for loss of neurons (Jin et al., 2004; Rawal and Zhao, 2021).

6.2.7.2 Fucosylation

Fucose *(Fuc)* is the second most abundant nonreducing terminal saccharide expressed on *N*- and *O*-glycoproteins, although unlike sialic acid, *Fuc* is often concurrently expressed on subterminal and core sites (Li et al., 2018; Tosh et al., 2019). Fucosylation of glycoproteins is catalised by a family of fucosyltransferases (FUTs) which utilize GDP-fucose as a *Fuc* donor, sourced either *de novo* or via salvage pathways (Becker and Lowe, 2003). In a comparable manner to sialic acid, *Fuc* also appears to play a functionally significant role in the central nervous system, whereby fucosylated *N*-glycans participate in long-term potentiation, synapse formation and function, learning and memory, and cell-cell communication (Bradberry et al., 2023). Recent evidence also indicates that fucosylated microglia are involved in neurodegenerative processes, including inflammatory responses to Aβ oligomers (Jin et al., 2023).

6.2.8 Glycan receptors

Cell surface *N*- and *O*-glycan epitopes function as ligands for carbohydrate-specific receptors, which makes them attractive as potential targets for therapies aimed at treating underlying disease mechanisms in neurodegeneration. Cell surface receptors are also commonly glycosylated, with the extracellular domain containing glycan structures that either activate or block receptor activity and are involved in various cellular processes including immune responses, cell adhesion and pro- and antiinflammatory responses (Table 6.1) (Gao et al., 2021; Reily et al., 2019; Smith and Bertozzi, 2021; Wißfeld et al., 2024).

6.2.8.1 Selectins

Selectins are a glycoconjugate-specific membrane-bound receptor family of calcium-dependent (C-type) lectins comprised of three types, named according to their stimulus/expression patterns: P-selectin, expressed on endothelium and platelets when exposed to histamine, thrombin, and complement; E-selectin, expressed on endothelial cells on exposure to inflammatory stimuli including tumor necrosis factor-alpha (TNF-α), interleukin 1-beta (IL-1β), and bacterial lipopolysaccharide (LPS); and L-selectin, expressed on leukocytes and playing a major role in inflammatory responses. L-Selectin is also a ligand itself for E-selectin (McEver, 2015; Smith and Bertozzi, 2021; Yeini et al., 2021). Selectins require ligands that are sialylated or fucosylated in specific positions, and have the capacity to recognize and interact with both glycans and scaffold proteins, which increases binding affinity (Li et al., 2018; Smith and Bertozzi, 2021). Although selectins are known to play a role in neuroinflammation, their role in neurodegenerative disease is currently less clear.

6.2.8.2 Sialic acid binding-immunoglobulin lectin-type

Sialic acid binding-immunoglobulin lectin-type (SIGLEC) receptors are highly expressed on immune cells including microglia, and as the name suggests, recognize and bind to sialic acid terminals of glycoproteins. SIGLEC classification is based on signal transduction interactions, with the majority interacting with tyrosine phosphatase SHP-1 and SHP-2 via a cytoplasmic domain immunoreceptor tyrosine-based inhibitory motif (ITIM). Microglial expressed SIGLEC receptors including SIGLEC-3 and SIGLEC-11 regulate ITIM inhibitory actions and homeostatic endocytosis in the CNS by detecting intact neurons, meaning that the terminal sialic acid cap is still attached and microglia do not need to mount a response (Siew et al., 2023). In contrast, a small subset of SIGLECs including SIGLEC-14 contain immunoreceptor tyrosine-based activating motifs (ITAM) that are targeted by microglia when encountering damaged neurons that display a desialylated plasma membrane glycoprotein or glycolipid (Wißfeld et al., 2024). There are 14 known human SIGLECs, with several polymorphisms in SIGLEC-3, -11 and -14 being implicated in NDDs. SIGLEC-3, also known as CD33, has been widely studied, with specific isoforms and higher expression associated with an increased risk of AD (Gonzalez-Gil et al., 2022; Griciuc et al., 2013). Proposed mechanisms underlying the increased risk include enhanced ITAM signaling, reduced clearance of toxic Aβ42 proteins, and chronic neuroinflammation (Eskandari-Sedighi et al., 2023). Interestingly, the $CD33^{\Delta E2}$ isoform, which lacks the sialic acid binding domain, is associated with a reduced risk of AD (Wißfeld et al., 2024). SIGLEC-11 has demonstrated neuroprotective and antiinflammatory effects via one of its ligands, polySia, in vitro and in vivo, whereas genome-wide association studies have elucidated a SNP associated with a higher risk for NDD (Bellenguez et al., 2022; Liao et al., 2021; Wang and Neumann, 2010). SIGLEC-14 is less well characterized, it is thought to have an inverse relationship with ITIM SIGLEC-5, and that a decrease in SIGLEC-14 gives rise to increased expression of SIGLEC-5, which may dampen activation of microglia, thus impacting AD risk (Shaw et al., 2021). Recent evidence has also shown that glycoRNAs, which utilizes the canonical *N*-glycan biosynthesis pathway to produce sialic acid-enriched glycans, are able to interact with SIGLEC receptor family, and that some SIGLEC signaling is dependent on glycoRNA mediation (Flynn et al., 2021; Chai et al., 2023).

6.2.8.3 Receptor for advanced glycation end-products

RAGE are pattern recognition receptors that belong to the Ig superfamily of membrane-bound receptors and are highly associated with normal aging, and NDDs, including AD and PD (Choi et al., 2014; Deane et al., 2003; Derk et al., 2018; Long et al., 2022; Lue et al., 2001; Takuma et al., 2009; Videira and Castro-Caldas, 2018). The full-length RAGE V domain contains two *N*-glycosylation sites, which facilitate the majority of extracellular ligand binding (Sparvero et al., 2009). There are also several RAGE subtypes: (1) soluble RAGE (sRAGE), a circulating form of RAGE which lacks the transmembrane domain and can bind circulating AGEs, acting as a decoy receptor and facilitating their clearance via renal excretion and hepatic uptake; (2) endogenous secretory RAGE (esRAGE) and soluble RAGE (sRAGE), both C domain truncated forms; and (3) N-truncated RAGE which lacks the V domain and can be protective against neurodegeneration by attenuating inflammatory responses (Vitorakis and Piperi, 2024). RAGE are multiligand receptors, which also bind high-mobility group box 1 protein and several S100 proteins, including S100B, S100A8, and S100A9, which are increased in inflammation. AGE/RAGE axis activation induces an extensive list of molecular pathways in normal aging, which are balanced by CNS antioxidant defences, whereas under pathological conditions they contribute to oxidative stress, chronic neuroinflammation, damage to the BBB, synaptic dysfunction, microglial activation, and ultimately cognitive impairment and neurodegeneration (Derk et al., 2018; Long et al., 2022).

6.2.9 Glycosylation and neurodegeneration

Ultimately, as glycomic diversity is matched by its functions, it may well be appreciated that perturbations of glycosylation pathways, the biosynthesis machinery that facilitates them, and the receptors known to be activated by them, will profoundly impact cell physiology. Dysregulated glycosylation of proteins and lipids plays a central role in aging and neurodegenerative diseases like AD, PD, and prion disease, although it is only beginning to be understood. While a formidable task, understanding the role of glycosylation and glycoconjugate/receptor interactions in these diseases will hopefully provide new diagnostic and therapeutic targets for urgently needed disease-modifying treatments.

6.2.10 Glycosylation in Alzheimer's disease

AD is a progressive neurodegenerative disease typically affecting older adults and is characterized by cognitive decline, memory and behavioral impairment, loss of independence, and death. AD is also renowned for its hallmark neuropathological features of extracellular Aβ plaques and cytoplasmic hyperphosphorylated tau fibrils.

However, understanding of the complex pathophysiology of the disease remains incomplete. Consequently, results from emerging treatments that target AD processes have been controversial and associated with significant side effects, and current longstanding treatments are still only able to offer modest symptomatic relief (Cetin et al., 2022; Van Dyck et al., 2023). Hence, to date, there is still no effective means of prevention, treatment, or cure (Joshi et al., 2014; Kang et al., 2024; Zhao and Lang, 2023). Advances in glycomics and glycoproteomics have made it possible to characterize alterations in glycosylation associated with AD, not only expanding understanding of disease pathophysiology and challenging the Aβ "amyloid cascade" hypothesis of AD, but also invigorating new research for novel biomarkers for early disease detection, therapeutic targets, and treatments (Gaunitz et al., 2021a,b).

Increasing evidence suggests that *N*- and *O*-glycosylation play a complex and fundamental role in AD, highlighted by findings that show key proteins, including amyloid precursor protein (APP), tau, beta-site APP-cleaving enzyme (BACE1), γ-secretase (Nicastrin), and apolipoprotein E (APOE) are subject to aberrant glycosylation in key brain regions relevant to AD—the hippocampus, temporal, and frontal cortex (Akasaka-Manya and Manya, 2020; Conroy et al., 2021; Haukedal and Freude, 2021; Zhang et al., 2020). In normal brains, multiantennary sialofucosylated *N*-glycan species have been found to be the most abundant across all brain regions, along with oligomannose and bisected *GlcNAc* structures (Gaunitz et al., 2021a,b). Significantly altered bisecting *N*-glycans detected in the cerebrospinal fluid (CSF) of AD patients using protein mass spectrometry highlighted the role of glycosylation in AD, and suggested that monitoring glycosylation changes in patients may be a suitable diagnostic strategy for early detection (Palmigiano et al., 2016). Mapping of the *N*-glycome and *N*-glycoproteome in postmortem brains of elderly AD and hippocampal sclerosis subjects revealed distinct regional and *N*-glycan structural variations between subjects compared to controls (Tena et al., 2022). Notably, in AD brain samples, a decrease in sialofucosylated *N*-glycan species in the frontal cortex was found, in line with other findings, and was reported to indicate loss of cell-surface interactions in neurons.

Neuroinflammation is also shown to modify the *N*-glycome via dysregulation of sialylation and fucosylation, as demonstrated in a murine model (Rebelo et al., 2021). This shows a ~7.5% reduction in the relative abundance of sialylated structures within the inflamed striatum and a drop of core fucosylation levels by about 8.5% in the LPS-injected hemisphere. Fucosylation of Aβ oligomers by FUT8 was upregulated in human induced pluripotent stem cell-derived microglia (hiMG), which induced microglial activation of inflammatory cytokines IL-1β and TNF-α (Jin et al., 2023). Inflammatory activation of hiMG was reversed with inhibition of FUT8 activity, suggesting that dysregulated core fucosylation in AD contributes to neuroinflammation. This finding was validated by in vivo experiments using 5XFAD mice and analysis of human AD brain samples, which were found to contain increased FUT8 in the mid-temporal cortex compared to age-matched controls. Interestingly, the p53 signaling pathway was found to be involved as a regulator of FUT8 expression in Aβ-activated hiMG. Further research is needed to elucidate specific receptor mechanisms involved.

In a number of animal and human studies on aging, AD and other NDDs, *O*-GlcNAcylation has been positively associated with maintenance of cognitive function, protein quality and fidelity of folding, and inhibition of Aβ, tau and α-synuclein (α-Syn) protein aggregation both directly and indirectly via small heat shock proteins (Balana et al., 2021; Cole and Hart, 2001; Park et al., 2021; Ryan et al., 2019; Taylor et al., 2014; Wheatley et al., 2019). On the other hand, *O*-GlcNAcylation of cyclic AMP response element binding protein at Ser140 was shown to dampen long-term memory, though this was proposed to be part of a regulatory mechanism (Rexach et al., 2012). Broadly, it has been suggested that disturbed *N*-glycosylation may largely drive neurodegenerative processes, whereas *O*-glycosylation, in particular *O*-GlcNAcylation, may confer neuroprotection and regulation of higher-order neuronal functions, though further research will no doubt clarify this idea and resolve any caveats (Haukedal and Freude, 2021; Kang et al., 2024; Yuzwa et al., 2012).

6.2.10.1 Amyloid precursor protein and amyloid-beta

APP is a key molecule in the pathogenesis of AD. APP functions as a synaptic cell adhesion molecule that, when sequentially cleaved by BACE1 and γ-secretase (Nicastrin) in the amyloidogenic pathway, results in the release of toxic Aβ oligomers into the intracellular space and eventual accumulation of extracellular amyloid plaques (Hampel et al., 2021; Schilling et al., 2017). Glycosylation is a critical factor that determines whether APP goes down the route of amyloidogenic or nonamyloidogenic cleavage. Early evidence also showed that inhibition of *N*-glycosylation in WT hippocampal neurons with tunicamycin led to APP mis-sorting, highlighting the importance of glycosylation in APP axonal localization (Lin et al., 2022). A number of studies have demonstrated that APP is subject to both *N*- and *O*-glycosylation and that disruption to either has implications for AD pathophysiology (Haukedal and Freude, 2021; Tena et al., 2022; Zhang et al., 2020). Two *N*-glycosites proposed for APP are located at Asn467 and Asn496, and early evidence showed that targeted mutation of these residues disrupted its processing and decreased secretion, thus suggesting that APP *N*-glycosylation was important for its intracellular sorting (Yazaki et al., 1996). More recent evidence, in vitro and in human AD brains, validated these observations, showing that alteration of either of the *N*-glycosites results in upregulated amyloidogenic proteolytic processing and increased Aβ1-42 intracellular levels compared to controls (Lin et al., 2022; Boix et al., 2020; Schedin-Weiss et al., 2014).

Alterations to glycan structure may also contribute to AD processes. For example, BACE1 has four proposed *N*-glycosylation sites, Asn153, Asn172, Asn223, and Asn354, which are comprised of bisecting *GlcNAc* synthesized by the GnT-III enzyme. In AD, upregulation of bisecting *GlcNAc* modifications on BACE1 by GnT-III has been observed, suggesting that such disturbances modulate APP β-cleavage and drive Aβ production via higher BACE1 activity. Further, pathological levels of bisecting *GlcNAc* are suggested to block BACE1 from lysosomal trafficking in the brain. A knockout AD mouse model demonstrated that loss of the GnT-III enzyme enhanced BACE1 lysosomal degradation and resulted in significant downregulation of Aβ and improved cognitive impairment (Kizuka et al., 2015). BACE1-targeted therapies have been used previously in AD and glioblastoma with adverse effects; therefore, it was proposed that GnT-III may be a lower-risk disease-modifying target (Kizuka et al., 2015; McDade et al., 2021).

There is growing evidence for APP *O*-glycosylation and *O*-GlcNAcylation sites, in addition to alterations in site occupancies and glycan structures, and variations in levels of expression (Akasaka-Manya and Manya, 2020; Haukedal and Freude, 2021; Kang et al., 2024; Park et al., 2021; Reily et al., 2019; Zhu et al., 2014; Lee et al., 2021). Almost three decades ago, APP was found to be *O-GlcNAc*-modified in murine cerebellar neurons (Griffith et al., 1995). Since then, *O*-glycosylated sites with *GalNAc, GlcNAc, Gal,* and *Sia* at Thr291, Thr292, and Thr576 on sAPP695 isoform were found using Chinese hamster ovary cells (Perdivara et al., 2009) (Fig. 6.5). Further, Thr576 was reported as an important *O*-GlcNAcylation site on APP695 in HeLa cells, with implications for trafficking of APP, and nonamyloidogenic processing (Chun et al., 2017). The authors found that increasing *O*-GlcNAcylation attenuated Aβ secretion, which indicated that Thr576 may be a putative therapeutic target for AD. Kim et al. (2013) also found that increasing *O*-GlcNAcylation of γ-secretase (Nicastrin) at Ser708 using OGA inhibitor NButGT in 5XFAD mice, which show AD-like cerebral

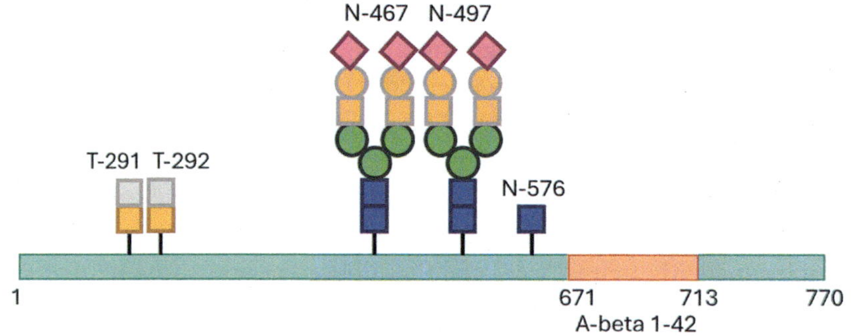

Figure 6.5 Schematic representation of the *O*-glycosylation sites on amyloid precursor protein at threonine-291, threonine-292, and asparagine-576, in addition to the *N*-glycosylation sites on asparagine-467 and asparagine-497.

amyloidosis, led to reduced enzymatic activity and lower levels of Aβ1-40 and Aβ1-42, accompanied by reductions in astrocyte and microglia activation and memory impairments (Kim et al., 2013). However, Aβ1-40 and Aβ1-42 levels were not measured in the in vitro arm of the study, which makes it difficult to reconcile with the in vivo results. Overall, the evidence suggests that *O-GlcNAc* modification promotes nonamyloidogenic processing of APP, thereby reducing toxic Aβ. In addition, *O*-glycosylation of APP with *GalNAc* by (polypeptide) ppGalNAc-T6 enzyme has been shown to decrease Aβ production by altering APP processing and trafficking (Akasaka-Manya et al., 2017; Akasaka-Manya and Manya, 2020). Taken together, preservation of *O*-glycosylation of APP, BACE1, and γ-secretase appears to exert protection against AD processes. Furthermore, findings such as these suggest that targeting glycosylation modifications may be feasible for the treatment of AD.

6.2.10.2 Tau

Zhang and coworkers produced a large-scale data set of the human *N*-glycoproteome comparing AD and healthy brains from dorsolateral prefrontal cortex tissue samples and were able to identify a large number of *N*-glycosites with differentially hyper- or hypo-*N*-glycosylated proteins including tau, APP, BACE1, and Nicastrin that were unique to AD samples (Zhang et al., 2020). Despite very early indications that tau is heavily *O-GlcNAc* modified, it is more recently reported not to be typically *N*-glycosylated in the normal brain, and if found, indicates abnormal glycosylation processes in AD (Griffith et al., 1995). The presence of *N*-glycosylation sites in AD brain samples of mice and human subjects suggested that abnormal glycosylation biosynthesis was allowing OST enzymes access to tau in the ER, driving hyper-*N*-glycosylation. This led the researchers to observe upregulated mRNA expression and activity of key OST subunits STT3A and STT3B, and altered expression of dolichol-oligosaccharide synthesis ALG family of enzymes, thereby disrupting donor substrate availability. Multifactorial mechanisms proposed to underlie these *N*-glycosylation impairments included tau and Aß accumulation and aggregation contributing to ER stress, thus affecting protein folding and disruption of the glycosylation machinery. Further analysis showed the specific site found to be aberrantly modified in AD was N410 on truncated tau441, corresponding to N727 on full-length tau758 isoform (Zhang et al., 2020). In several other studies, similar observations of tau hyper-*N*-glycosylation have been found in AD brains, with sites Asn359-Ile-Thr, Asn167-Ala-Thr, and Asn410-Val-Ser identified and proposed to be involved in tau aggregation (Losev et al., 2019, 2021; Sato et al., 2001; Tena et al., 2022) (Fig. 6.6).

Conversely, evidence has shown that a parallel decrease in tau O-GlcNAcylation occurs in AD brains and is associated with an increase in phosphorylation (Liu et al., 2004; Pinho et al., 2019). In AD brains, O-GlcNAcylation has been reported to be

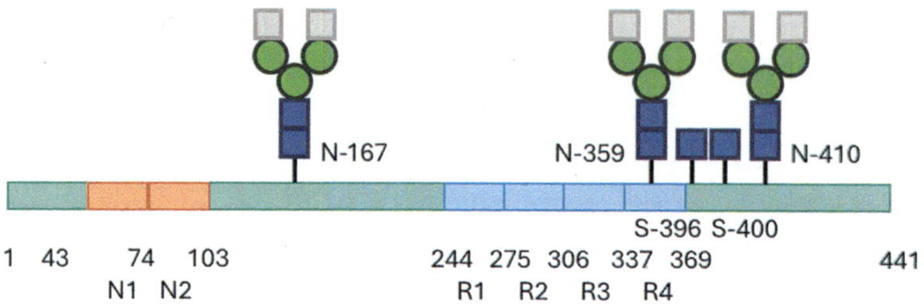

Figure 6.6 Schematic representation of the *N*-glycosylation sites on tau at asparagine-167, asparagine-359, and asparagine-410, and *O*-glycosylation sites at serine-396 and serine-400. "N" refers to the N-terminal inserts, and "R" refers to the C-terminal repeat domains.

reduced by approximately 50% compared to controls (Kim et al., 2013). Yuzwa et al. (2012) highlighted the importance of tau *O*-GlcNAcylation by using a transgenic mouse model to demonstrate that increasing levels of *O*-GlcNAc modifications by inhibition of OGA prevented the aggregation of tau and dampened tau-induced neurodegeneration. Diminished *O*-GlcNAcylation is also associated with impaired glucose metabolism, which can occur during aging and is an early event in the prodrome of AD in brain regions vulnerable to neurodegeneration (Scherer et al., 2021; Lee et al., 2006, 2021, 2023a; Liu et al., 2009; Zhu et al., 2014). One tau *O-GlcNAc* modification at Ser400 was found at low stoichiometry in one study, corroborating previous findings in rat brain by mass spectrometry and western blot using an *O-GlcNAc* antibody (Morris et al., 2015; Wang et al., 2010; Yuzwa et al., 2011). Importantly, Morris and colleagues point out that it is wise to keep in mind that tau is subject to complex PTM in vivo, including ubiquitination, acetylation, and methylation, and that only one *O-GlcNAc* site has been identified on tau to date. Though the glycosite was clearly identifiable, the authors also refuted the inverse relationship between *O*-GlcNAcylation and hyperphosphorylation of tau via site competition, and suggested that the protective effect of *O-GlcNAc* modification of tau may be driven by an unknown mechanism (Morris et al., 2015). Subsequently, using 3XTg-AD mice, which express human APP, PS1, and tau mutations, Gatta et al. (2016) found hippocampal hyperphosphorylated tau on Ser396 and Thr205, and frontal cortex Thr205, and significantly reduced hippocampal *O*-GlcNAcylated Ser396 tau, compared to WT mice. This was despite no observed reduction in glutamine fructose-6-phosphate amidotransferase, which regulates glucose flux into the HBP, and OGT/OGA enzyme expression. This potentially indicated the imbalance between hyperphosphorylated and hypo-*O*-GlcNAcylated tau was independent of the HBP and enzymatic activity of OGT and OGA (Gatta et al., 2016). It was concluded that defects in glucose uptake in the brain may be responsible for the reductions in *O*-GlcNAcylation, as regulatory enzymes were not perturbed.

As the field of glycobiology continues to expand, the association between hyperglycemia and glycosylation alterations in neurodegenerative disease is beginning to be elucidated. Since the Rotterdam study in the 1990s, a growing body of evidence has demonstrated that type 2 diabetes (T2D) is a major risk factor for dementia and AD, increasing risk of disease by over 50%, and why AD is often referred to as type 3 diabetes (Ott et al., 1996, 1999; Chatterjee and Mudher, 2018). With respect to *O*-GlcNAcylation, postulated underlying mechanisms include impaired neuronal glucose utilization driven by chronic hyperglycemia and insulin resistance, which downregulates GLUT1 and GLUT3 transporters, thus starving the HBP for substrate and limiting production of *O*-GlcNAcylation donor UDP-*GlcNAc*. Ultimately, relevant to AD pathology, this results in decreased ability for *O*-GlcNAcylation of tau, APP, and Nicastrin, and increased levels of hyperphosphorylated tau and amyloid plaques (Gatta et al., 2016; Haukedal and Freude, 2021; Yuzwa et al., 2012).

This would then suggest that tight blood glucose control over the lifespan supports ongoing neuroprotective processes such as *O*-GlcNAcylation and contributes to reducing the risk of neurodegenerative disease during aging. Another motivator for keeping blood glucose within a healthy range is the relationship between toxic AGEs, their receptor RAGE, and neurodegeneration (Vitorakis and Piperi, 2024; Otazu et al., 2021). A well-established biomarker regularly used in the general practitioner's office to determine long-term blood glucose control is hemoglobin A1c (HbA1c), which gives a percentage of hemoglobin that has been glycated over three months. A lower HbA1c is associated with a lower risk of AD and dementia in older adults (Underwood et al., 2024). HbA1c is a precursor for Hb-AGEs, which then form AGEs, first found in diabetes patients in the 1960s (Rahbar et al., 1969). As previously discussed, Aβ, tau, and α-syn in PD are targets of the AGE-RAGE axis and play a role in the development of senile plaques, neurofibrillary tangles, and Lewy bodies.

6.2.10.3 Apolipoprotein E

The human APOE gene is comprised of three major isoforms as a result of genetic polymorphisms: APOE2, associated with reduced risk of AD; APOE3, the most common genotype; and APOE4, which is associated with increased risk of AD, Aβ and tau pathology, Lewy body dementia, impaired brain glucose metabolism and increased severity of neurodegenerative disease pathology (Yamazaki et al., 2019). Seminal work by Wernette-Hammond and colleagues found that nascent APOE was *O*-glycosylated at Thr194, and it has since been found to be highly sialylated (Flowers et al., 2020; Wernette-Hammond et al., 1989). As APOE and Aβ are known to functionally interact, it has been proposed that the sialic acid moiety is integral in this interaction in the context of AD (Haukedal and Freude, 2021; Xu et al., 2006; Yamazaki et al., 2019; Kanekiyo et al., 2014). Notably, brain APOE has been shown to be more highly sialylated than in plasma, although it was found that *O*-glycosylation of plasma APOE was a useful biomarker of brain amyloidosis in a study using CSF and plasma samples from older individuals (Lawler et al., 2023). Flowers and coworkers conducted a glycoproteomic analysis of APOE3 and found additional glycosylation sites on Thr8 and in the N-terminal hinge region (aa192–206), and a putative Ser197 *O*-glycosylation site with *Neu5Acα2-3Galβ1-3GalNAcα1* being the most prevalent glycan (Flowers et al., 2020). Further sites on the C-terminal lipid-binding domain (aa 283–299) were observed at Thr289, Ser290, and Ser296, and CSF APOE was found to be more highly glycosylated and sialylated than plasma APOE (Fig. 6.7). Further, the C-terminal modification was proposed to be more important to the APOE4 isoform for lipid-binding than to the APOE3 isoform (Flowers et al., 2020). More recently, APOE4 carrier status was significantly negatively associated with reduced hinge region glycosylation, proposed to be linked to lower plasma APOE levels and significantly increased Aβ accumulation (Lawler et al., 2023). Hence, the glycosylation status of APOE may be a promising blood biomarker to detect early changes in AD pathogenesis. However, mechanisms underlying glycosylation in the APOE4 isoform are yet to be explored, and given the contribution of APOE4 and glycosylation status to AD, further studies are warranted to elucidate these details.

6.2.10.4 Alterations to the endoplasmic reticulum and Golgi in Alzheimer's disease

As previously stated, the fidelity of glycosylation processing is critical, as the consequences of inaccurately glycosylated proteins have been linked with a number of human diseases, including those associated with neurodegeneration (Conroy et al., 2021; Fisher et al., 2019b; Reily et al., 2019; Zhang et al., 2020). Golgi stacking factors, Golgi reassembly stacking protein of 55kD (GRASP55), and GRASP65 mediate Golgi stack formation, maintaining protein trafficking accuracy and efficiency. Disruption of GRASP proteins has been shown to result in Golgi fragmentation, which increases coat protein (COP) access to Golgi membranes, allowing vesicular budding, fusion, and premature exocytic transport (Zhang and Wang, 2016). Impaired glycosylation processes and accelerated protein trafficking are two consequences that hold significant implications for NDDs. Golgi fragmentation in neurons is a feature of AD, proposed to

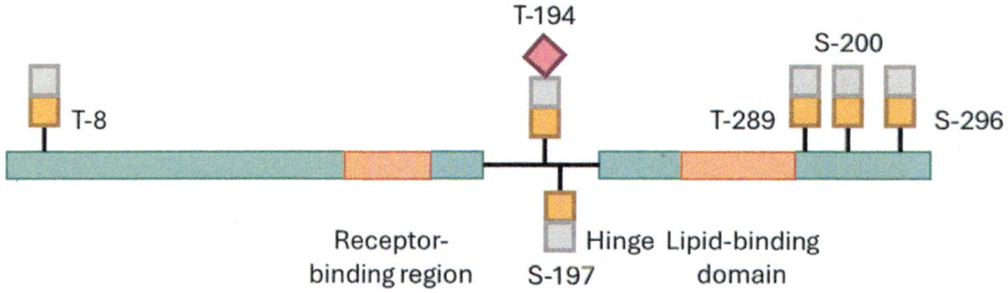

Figure 6.7 Schematic representation of the *O*-glycosylation sites on apolipoprotein E at threonine-8, threonine-194, serine-197, threonine-289, serine-200, and serine-296.

be a result of Aβ peptide accumulation causing GRASP65 phosphorylation by activated cyclin-dependent kinase-5, which is linked to increased Aβ production and disease progression (Joshi et al., 2014; Joshi and Wang, 2015; Zhang and Wang, 2016).

6.2.11 Glycosylation in Parkinson's disease

PD is an age-related progressive and complex neurodegenerative disease characterized by loss of dopaminergic (DA) neurons in the *substantia nigra* pars compacta (SNpc). Pathological mechanisms underpinning PD pathophysiology include misfolding and aggregation of α-syn, subsequent development of Lewy bodies and neurites in the CNS, and hallmarks of neuroinflammation, oxidative stress, impaired immunity, mitochondrial dysfunction, and impaired protein clearance (Jankovic and Tan, 2020). Cardinal clinical signs of PD at diagnosis include bradykinesia, rigidity, postural instability, and resting tremor, clear indications that the disease is already well established. Comparable to AD, the prodromal phase progresses over decades, may be asymptomatic to begin with, and manifests with early nonmotor symptoms, including gastrointestinal (GIT) changes, including constipation, and olfactory disturbances (Braak et al., 2003). While the etiology of PD is thought to be a result of interactions between genetic, epigenetic, and environmental risk factors, it is not yet fully understood (Shaheen et al., 2024). Recent advances in neuroimaging include the use of F-AV-133 vesicular monoamine transporter type 2 PET scans, which may aid early diagnosis and monitoring of disease progression, but they have yet to be used in clinical practice (Beauchamp et al., 2023). PD drug-based treatments include the mainstay Levodopa, which attenuates symptoms and is often coprescribed with other medications, and other methods such as deep brain stimulation to relieve severe tremors. However, no therapies to date influence disease progression nor offer a cure (Jankovic and Tan, 2020). Nonetheless, drug development and biomarker identification research continue to make progress and include the expanding field of glycomedicine and the potential of glycan-targeted therapies (Fisher et al., 2019a; McFarthing et al., 2024; Reily et al., 2019; Wang, 2023).

Further evidence to support the hypothesis that PD pathogenesis is associated with early upper GIT changes, such as mucosal damage, has recently been reported (Chang et al., 2024). The findings indicated that upper GIT disturbances were associated with increased risk of developing PD, monitoring patients with upper GIT erosion on gastroscopy was suggested to assist in early detection and intervention. There is a relatively large body of evidence devoted to investigating the involvement of distinct changes in the GIT, microbiome, and enteric nervous system (ENS) in PD pathogenesis, including the presence of Lewy bodies and misfolded α-syn in the ENS, and in the alimentary tract as far back as the late 1980s (Anis et al., 2022; Wakabayashi et al., 1988). The gut-brain hypothesis proposes that PD pathology begins in the GIT and is propagated to the ENS via the vagus nerve to the brain, spreading in a prion-like manner, thus leading to neurodegeneration and motor symptoms characteristic of PD. Early GIT disturbances, including constipation, are proposed to be clinical clues hinting at the peripheral pathophysiology occurring within, with some studies suggesting the presence of misfolded pathological α-syn and Lewy bodies in the GIT/ENS are involved (Anis et al., 2022; Bu et al., 2019; Pellegrini et al., 2022).

Recent studies have also revealed significant glycobiology alterations in PD, and identification of novel disease biomarkers, which may help improve the current approaches for diagnostic and PD therapeutic strategies utilizing glycans, glycoconjugates, and their receptors. There appears to be an emerging intersection between fields of the gut-brain axis hypothesis of PD, the canonical CNS pathology of PD, and glycobiology, which may begin to fill some of the gaps in understanding PD, in addition to identifying novel biomarkers for early detection and glycobiology-derived therapies.

Alterations in gene expression of glycosyltransferases, sialyltransferases, and sialidases were observed in the *substantia nigra* of PD patients using real-time PCR (Schneider and Singh, 2022). Specifically, glycosyltransferase genes *B3GALT2* and *B4GALT1* were found to be significantly upregulated, whereas polysialyltransferase genes *ST8SIA2* and *ST8SIA4* were found to be significantly downregulated in PD brains compared to age-matched controls. It can be appreciated how dysregulation of glycosylation-related genes contributes to neurodegenerative processes seen in PD, as B3GALT2 and B4GALT1 are involved in microglial activation and neuroinflammation via IL-6 and TNF-α. Downregulation of polysialyltransferase genes *ST8SIA2* and *ST8SIA4* also drives PD pathophysiology by reducing polySia/SIGLEC-11 antiinflammatory activity and microglial homeostasis, in addition to impaired neurotrophic factor signaling via brain-derived neurotrophic factor, and neurotransmission of dopamine and norepinephrine (Schneider and Singh, 2022; Shahraz et al., 2015).

Therapeutics utilizing antibodies and polysialylated ligands that target SIGLECs have been explored in experimental models of NDDs, including AD and PD (Bhattacherjee et al., 2021; Liao et al., 2021). Low-molecular-weight polysialic acid (polySia avDP20) was used in vitro and in vivo in a humanized SIGLEC-11 transgenic mouse model of PD to investigate its antiinflammatory effects (Liao et al., 2021). The LPS-induced activation of inflammatory cytokine Il1b, TLR4 genes *Cd14* and *Myd44*, and phagocytosis-related gene *Fcer1g* were attenuated by polySia avDP20 in SIGLEC-11 mice. Furthermore, reduced transcription levels of complement pathway genes C4 and *Itgam*, and oxidative burst *Cybb* were observed in the SIGLEC-11 mice, but to a lesser degree in WT. For both WT and SIGLEC-11 mice, LPS-induced loss of DA neurons in the SNpc was robustly diminished by polySia avDP20 treatment, suggesting its use as a plausible immunomodulatory and antiinflammatory therapy for NDDs like PD, and highlighted the importance of the SIGLEC-11 and polySia association. Further, the neuroinflammatory *milieu* alters glycosylation of neuroprotective proteins, reduces sialic acid levels in most instances in PD, resulting in downregulated expression of SIGLEC receptors and activation of complement 3 (Videira and Castro-Caldas, 2018). Triggering receptor expressed on myeloid cells 2 (TREM2), expressed on microglia, is a multiligand receptor for ligands including APOE and phosphatidylserine on desialylated neuronal membrane proteins, and exerts neuroprotective effects. TREM2 is shown to be glycosylated with sialylated and/or fucosylated glycans, which determine correct conformation and trafficking to the membrane. Hence, aberrant glycosylation leads to decreased TREM2 activity and stability in both PD and AD (Videira and Castro-Caldas, 2018; Park et al., 2016).

N-glycosylation alterations have been identified using glycoproteomics and proposed as potential serum biomarkers in PD patients (Xu et al., 2022). In this study, overall *N*-glycan increases in *Sia*, core *Fuc*, and bisecting *GlcNAc* modifications were observed in PD patients compared to healthy age and gender matched controls, with glycosylation changes relevant to PD found in ceruloplasmin, haptoglobin, kininogen-1, complement factor H, and clusterin, thus placing them as putative disease markers. The authors proposed that dysregulated expression of key glycosyltransferase enzymes FUT8, ST3GAL2, ST6GAL2, and Gnt-III may be due to neuroinflammatory processes or underlying epigenetic changes associated with PD, though further validation of the findings in larger patient cohorts would be required to assess any clinical utility of these selected proteins for disease diagnosis (Xu et al., 2022). In contrast, low *N*-sialylation and increased fucosylation on tri-antennary glycans decorated with 2 and 3 terminal sialic acids were found in the serum of male PD patients using CE-MS/MS glycomics, suggesting that some glycosylation changes may be more pronounced in males (Váradi et al., 2019).

Another plasma biomarker study demonstrated that immunoglobulin G (IgG) *N*-glycans were significantly altered in PD patients compared to healthy controls (Russell et al., 2017). IgG has two *N*-glycans at the conserved Asn297 residue in the Fc region, though these glycans can exhibit diverse microheterogeneity depending on the expression and activity of glycosylation enzymes and the cellular milieu; thus it may exert either pro-inflammatory or antiinflammatory activities. IgG is known to interact with α-syn aggregates in Lewy bodies and DA neurons, and has been detected in the CSF and blood of PD patients previously (Sui et al., 2014). In this study, PD patients exhibited reduced IgG sialylation, high-mannose, and agalactosylated glycans compared to controls. Importantly, decreased sialylation of IgG *N*-glycans was associated with a reduced capacity to dampen antibody-dependent cell cytotoxicity mediated through select Fc-gamma receptor IIIa binding and driven by α-syn, which agreed with earlier findings by Scallon and colleagues (Russell et al., 2017; Scallon et al., 2007). Together, these findings highlight the complexity of biosynthetic pathways and regulatory mechanisms governing glycosylation, even at a single site. Nonetheless, IgG *N*-glycome alterations could hold potential as a novel biomarker for PD.

A recent study sought to characterize the nigro-striatal *N*-glycome in human PD and healthy brain samples (Rebelo et al., 2024). Interestingly, polySia structures were significantly increased in the striatum and reduced in the *substantia nigra* in later-stage PD brains, compared to controls. More remarkably, in contrast to IgG sialylation and other findings, sialylation was found to be increased in both brain regions (Rebelo et al., 2024). Note that similar trends are observed in brain and other cancers, where hypersialylation confers a protective cloak for cells to evade immune detection, and increases binding to SIGLECs, thus promoting immunosuppressive signaling (Dobie and Skropeta, 2021). Moreover, they pose the question, are certain protein constituents like Lewy bodies also undergoing upregulated sialylation to evade degradation and elimination by the immune system via a comparable mechanism? Besides these intriguing findings and hypotheses, increases in *N*-glycosylated bisecting, hybrid, sialylated, and core-outer arm fucosylation were increased in the striatum, and oligomannosylation was decreased, which was proposed to drive neuroinflammatory and neurodegenerative processes. In contrast, although with the same outcomes, bisecting, hybrid, and polysialylation decreased, and core-outer arm fucosylation, sialylation, and oligomannosylation were increased in the *substantia nigra* in PD brains, implicating differential regulation of *N*-glycosylation traits, enzymes, and unfolded protein response-related proteins in region-specific areas (Rebelo et al., 2024).

The *O*-glycome of the nigro-striatum was profiled in brain samples of PD, incidental Lewy body disease (ILBD), and healthy controls for the first time by Wilkinson et al. (2021). Interestingly, overall sialylation was significantly increased in the *substantia nigra* of PD samples, particularly males, relative to ILBD and controls, which has been similarly observed in *N*-glycome reports. In the striatum of ILBD, decreases in mannose-core *O*-glycans were found, whereas in PD samples, increases in sialylated and decreases in sulfated *O*-glycans were found. Importantly, the increase seen in sialylation in the striatum of PD samples was attributed to higher levels of the disialylated Core 3 structure *NeuNAcα2-3Galβ1-3[NeuNAcα2-6]GalNAc,* which has also been implicated in cancer and neurodegeneration (Brockhausen, 1999; Wilkinson et al., 2021). As the majority of the *O*-glycan alterations were seen in the striatum, the authors suggest that this brain region may be a preferable focus for biomarker identification and therapeutic strategies.

6.2.11.1 α-Synuclein

A key presynaptic protein implicated in aging, PD and other neurodegenerative diseases that include "*α*-synucleinopathies," α-syn, is thought to be involved in synaptic plasticity and synaptic vesicle turnover (Giasson and Lee, 2001). In PD, α-syn fibrils induce neuroinflammation via microglial activation and expression of TNF-α, IL-1β and IL-6, partly due to high affinity as ligands for RAGE, which mediates α-syn binding to microglia. Inhibition of RAGE may be a potential target for PD, as inhibition or knockout of RAGE was shown to robustly ameliorate cytokine responses of microglia in response to α-syn aggregates (Long et al., 2022). Glycated α-syn proteins bind RAGE, in which expression is upregulated in the CNS and at the blood-brain barrier in PD, thus inducing aggregation, sustaining a feedback loop of neuroinflammation, and driving reactive oxygen species production (Padmaraju et al., 2011). Misfolded cytosolic α-syn has long been associated with the formation of Lewy body inclusions, in addition to being a primary structural component in the form of amyloid fibrils (Goedert et al., 2013).

However, more recent research has revealed that Lewy bodies are also comprised of a variety of membranous compartments, which indicate defects in vesicular trafficking, in addition to nonfibrillar α-syn (Shahmoradian et al., 2019). Other neurotoxic variants of α-syn, often products of incomplete autophagic degradation, are also known to be highly mitotoxic in PD, driving mitochondrial damage and mitophagy, and have been found in primary neurons, mouse, and human PD brains (Grassi et al., 2018). Not surprisingly, perhaps, *O*-GlcNAcylation of α-syn has been demonstrated to produce a less pathogenic strain and was shown in a recent mouse study to reduce α-syn seeding and Lewy body pathology (Balana et al., 2024). Further, *O*-GlcNAcylation within DA neurons has been shown to safeguard against α-syn induced pathology, which highlights the critical nature of this PTM for DA neuronal function and survival in PD (Lee et al., 2020). The neuroprotective effects of the *O*-GlcNAc modification to α-syn (Levine et al., 2019) relevant to aggregation (including Thr72, Thr75, Ser87) are consistently demonstrated in multiple in vitro and animal studies, suggesting that further research and development of *O*-GlcNAcylation therapies for PD is warranted (Lewis et al., 2017; Marotta et al., 2015) (Fig. 6.8).

6.2.11.2 Glycosylation, Mitochondrial dysfunction, and Parkinson's disease

DA neurons are highly sensitive to mitochondrial stress, as highlighted by the involvement of mitochondrial dysfunction and oxidative stress in the pathogenesis of PD, a common pathway shared with AD (Alghusen et al., 2024; Pickrell and Youle, 2015; Videira and Castro-Caldas, 2018). In a recent animal study, aberrant dopamine

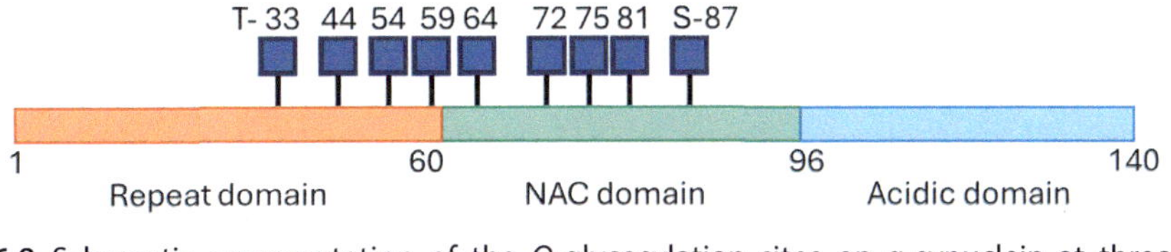

Figure 6.8 Schematic representation of the *O*-glycosylation sites on α-synuclein at threonine-33, threonine-44, threonine-54, threonine-64, threonine-72, threonine-75, and threonine-81 and serine-87.

transporter (DAT) glycosylation and Golgi membrane trafficking were augmented by glial cell-line derived neurotrophic factor, a key factor for DA neuron survival, which improved cognitive function in PD mice (Chengcheng et al., 2024). *N*-glycosylation is required for the accurate folding and stabilization of DAT, in addition to cell surface expression. Parkin, an E3 ubiquitin-protein ligase involved in the ubiquitin-proteasome pathway and ubiquitin-mediated autophagy (mitophagy), was shown in vitro to ubiquitinate misfolded or incorrectly glycosylated DAT for degradation, thereby ameliorating dopamine toxicity and improving dopamine utilization of functional DAT (Jiang et al., 2004). It has been proposed in many studies that defects in intracellular transport as a result of ER stress and/or mutations in major proteins including parkin, PTEN-induced kinase-1 (PINK1), DJ-1, GTPase Rab1, LRRK2, TREM2 and DAT involved in protein folding, mitochondrial function, degradation pathways, dopamine re-uptake, and inflammatory signaling pathways, underpin DA neuron death in PD (Abeliovich and Gitler, 2016). Further, Martinez-Menarguez et al. (2021) propose that the dense and highly arborised axonal morphology of nigrostriatal DA neurons to the striatum, required to facilitate significant secretory protein flux and efficient recycling, may contribute to the vulnerability and sensitivity of these neurons to mitochondrial stress and cell death (Martínez-Menárguez et al., 2021).

Maintenance of proper protein glycosylation and prevention of glycation-induced inflammation are shown to be critical for preservation of mitochondrial health and DA neurons. *O*-GlcNAcylation is an important PTM for mitochondrial morphology, function, stress response, and mitophagy, and is altered in AD and PD models (Alghusen et al., 2024; Burnham-Marusich and Berninsone, 2012; Kim et al., 2024). Mitophagy protein PINK1, a serine/threonine kinase, is modified by *O-GlcNAc* and phosphorylates Parkin, which is recruited for ubiquitination and degradation of damaged mitochondria via autophagy (Alghusen et al., 2024). Decreased levels of *O-GlcNAc* may impair the response to mitochondrial stress by PINK1, though further research is needed to clarify the mechanisms of this role. Parkin activity is also altered in PD, with decreased expression resulting in accumulation of damaged mitochondria, perpetuating a cycle of oxidative stress and neuroinflammation (Videira and Castro-Caldas, 2018).

Parkin also has an affinity for *O*-glycosylated isoforms of α-syn. This was shown to be in contrast to the non-glycosylated isoform of α-syn, which is inert to parkin (Giasson and Lee, 2001; Shimura et al., 2001). It is hypothesized that in the normal brain, parkin forms a complex with E2 enzyme UbcH7 and poly-ubiquinates a specialized *O*-glycosylated form of α-syn (αSp22) for degradation in the proteasome, which prevents accumulation of aggregated and ubiquitinated α-syn in Lewy bodies, and is therefore neuroprotective against PD. The importance of functional parkin is highlighted by the effect of loss-of-function parkin mutations in autosomal recessive juvenile parkinsonism, in which αSp22 can no longer be ubiquitinated and degraded, resulting in pathological increases in non-ubiquinated αSp22, which may contribute to neurodegeneration (Shimura et al., 2001). These results were consistent with a subsequent study by Chung et al. (2001). Correspondingly, the effectiveness of the parkin/αSp22 interaction appears to rely on the fidelity of O-GlcNAcylation of α-syn. Interestingly, whilst parkin is subject to a variety of PTM directly affecting structure and function, including phosphorylation, ubiquitination, acetylation, NEDDylation, S-nitrosylation, and SUMOylation, it is not reported to be directly glycosylated, only interacting with and regulating glycosylated proteins, including α-syn as described (Shimura et al., 2001; Chakraborty et al., 2017).

6.2.12 Glycosylation in prion disease

The physiological role of the cellular prion protein, PrPC, is essential for neuronal function (Panes et al., 2021); however, its misfolding and propagation across cells result in various pathologies. These so-called transmissible spongiform encephalopathies, or prion diseases, are fatal neurodegenerative disorders that include Creutzfeldt-Jakob disease (CJD) and related diseases in humans, scrapie in sheep, and bovine spongiform encephalopathy in cattle (Prusiner, 1989; Koshy et al., 2022). The misfolding of prions causes various pathological phenotypes, including enhanced neuroinflammation and elevated redox stress (Bourgognon et al., 2021; Bourgognon and Steinert, 2019; Hay et al., 2022, 2025; Spiers et al., 2019, 2022, 2023). Neuroinflammatory signaling is linked to the non-enzymatic glycation of PrPC, which is a critical PTM that influences prion strain diversity, infectivity, and disease progression (Bourgognon et al., 2021; Muronets et al., 2025). Interestingly, this neuroinflammation is in part due to an increase in nitric oxide signaling, which has been shown to enhance nitrergic stress in hippocampi by increasing expression of the RAGE, resulting in prion misfolding and compromised neuronal function (Bourgognon et al., 2021). Nitric oxide signaling has further been found to increase 3-nitrotyrosination of TPI, an enzyme involved in disease-associated glycation (Bourgognon et al., 2021; Spiers and Steinert, 2021). Metabolic analysis has revealed that this increased neuroinflammatory nitric oxide signaling is related to an upregulation of arginine-related metabolites, precursors to nitric oxide, in prion disease (Bourgognon et al., 2018). A feature that is observed in prion disease, in part as a result of increased conversion of PrPC to aggregation-prone and neurotoxic PrPSc, which impedes the ability of PrPC to enhance synaptic responses at the neuromuscular junction (Robinson et al., 2014). However, the most prominent PTM in prion disease are the glycans on the PrPC protein itself.

PrPC is a glycoprotein that is localized predominantly to detergent-resistant membrane microdomains via a glycosylphosphatidylinositol (GPI) anchor and contains two conserved *N*-linked glycosylation sites, and the resulting glycoforms -di, mono-, and unglycosylated which play distinct roles in prion biology (Ermonval et al., 2003). This includes the enrichment of monoglycosylated PrPSc in sporadic CJD cases, classified as type 1 and 2, and an abundance of diglycosylated PrPSc in variant CJD cases, classified as type 4 (Aguzzi and Heppner, 2000; Collinge et al., 1996; Wu et al., 2015). While glycosylation is not strictly required for prion propagation, it modulates the efficiency of transmission and the neuropathological profile of prion strains. For instance, transgenic mice expressing glycan-free PrPC can still sustain prion infection, but exhibit altered histopathology and increased deposition of disease-associated PrPSc (Tuzi et al., 2008). Moreover, the sialylation status of these glycans has been shown to influence neuroinflammation and strain-specific tropism, suggesting that glycan composition affects both immune evasion and replication dynamics (Makarava and Baskakov, 2023). Cell culture experiments have unambiguously established that not only the strain but also the cells in which the prion replication takes place can determine the PrPSc glycotype (Birkett et al., 2001; Paquet et al., 2007; Vorberg and Priola, 2002). Interestingly, oxidative stress increases disulfide bridge formation in PrP, resulting in elevated generation of the unglycosylated form (Capellari et al., 1999). Moreover, PrPC glycosylation may differ in individual cell types or brain regions (Beringue et al., 2003; Ermonval et al., 2009). The glycoform ratio associated with a defined strain also varies depending on the brain region or organ in which PrPSc is formed (Levavasseur et al., 2008; Somerville et al., 2005). Such observations have led to the speculation that the glycoform ratio of PrPC expressed by neuronal cells might be a determinant of the different brain-targeting selectivity manifested by prion strains (Katorcha et al., 2015).

Equally important is the intracellular trafficking of PrPC, which governs its conversion to PrPSc and subsequent spread. PrPC is synthesized in the ER, glycosylated in the Golgi apparatus, and anchored to the plasma membrane via a GPI anchor. Interestingly, the M232R mutation in the prion protein, associated with accelerated prion disease development, alters the protein's ER translocation by decreasing the GPI-attachment signal peptides' hydrophobicity, resulting in reduced *N*-linked and GPI glycosylation (Kobayashi et al., 2023). The protein resides predominantly in lipid rafts, where it interacts with various signaling molecules. From the membrane, PrPC undergoes endocytosis and is trafficked through early and late endosomes, a process that facilitates its encounter with PrPSc and conversion. Misrouting or altered trafficking of PrPC can enhance susceptibility to prion infection and influence strain-specific propagation (Alves et al., 2020). Furthermore, PrPSc can hijack exosomal pathways for intercellular transmission, contributing to the spread of infectivity across neural networks (Alves et al., 2020; Khadka et al., 2023; Vassileff et al., 2020). Together, glycosylation and trafficking define a complex regulatory landscape that shapes prion strain behavior, host susceptibility, and disease outcome. These processes are increasingly recognized as potential therapeutic targets and biomarkers for prion diseases. However, an obstacle in the detection of prions, evaluation of protease resistance, and discrimination of their strains, is the extent of their *N*-glycosylation, which conceals PrP epitopes from antibody recognition (Kang et al., 2020). Whilst interfering with detection, the glycosylation of PrP has been found to be neuroprotective in prion disease.

Extracellular plaques in prion diseases are composed of unglycosylated, ADAM metallopeptidase domain 10-cleaved PrP, which tightly binds heparan sulfate, a polysaccharide involved in accelerating PrPC to PrPSc conversion by promoting protein misfolding, oligomerisation, and fibrillation before stabilizing and providing proteolytic protection to the generated aggregates (Maïza et al., 2018). Interestingly, unglycosylated PrPC can induce spontaneous inward currents in cells, a feature of PrP-induced neurotoxicity (Schilling et al., 2023). Together, these results suggest that the glycans contribute structurally to autoregulation in the wild-type protein. Glycans on PrPC hinder plaque formation and significantly reduce spongiform degeneration in addition to reducing PrP binding to glycosaminoglycans, molecules that accelerate PrPC to PrPSc conversion (Sevillano et al., 2020). In vivo PrPC is *N*-glycosylated at asparagine 181 and 197 (180 and 196 in mouse), positions close to the location of many C-terminal mutations associated with prion diseases (Fig. 6.9) (Rudd et al., 2002; Schilling et al., 2023). Glycosylation at residue 181 (180 in mouse) is key for cis-regulation of the prion protein, a process where the unstructured *N*-terminus is structurally linked to the globular C-terminus, creating a tertiary structure that prevents conformational changes required for amyloid fibril formation (Mishra et al., 2019; Schilling et al., 2020). This neuroprotective action is achieved by the glycans, specifically on asparagine 180 localizing the copper-bound octapeptide repeat segment to histidine 176 and henceforth facilitating the cis-regulatory interaction (Schilling et al., 2023). Alternatively, glycosylation at residue 197 (196 in mouse) is involved in beta-sheet structure conversion within the amyloid fibrils (Mishra et al., 2019). Interestingly, the conformation of the input PrPSc seed primarily determines the cofactor and glycosylation preferences (Burke et al., 2020; Schilling et al., 2023). Whereas, the amplification efficiency of sporadic CJD PrPSc subtypes has been found to be inversely correlated with the degree of PrPC glycosylation and directly related to the PrP genotype (Camacho et al., 2019). In addition to the two C-terminal glycans, the GPI anchor of PrP contributes to prion disease pathology.

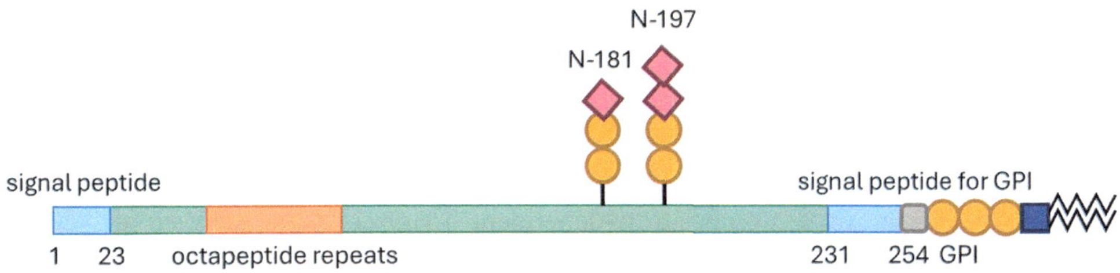

Figure 6.9 Schematic representation of the *N*-glycosylation sites on the prion protein at asparagine-181 and asparagine-197.

The GPI anchor is involved in providing PrPC a stable association with the lipid bilayer and is vital for the membrane anchorage required for the protein's *N*-glycosylation in the ER (Wu et al., 2015; Walmsley et al., 2001). However, the GPI anchor is also pivotal in driving aggregate conformation and prion pathology and promotes PrPSc replication, resulting in the generation of subfibrillar aggregates (Callender et al., 2020; Sevillano et al., 2020). Furthermore, removal of the GPI anchor decreases the trafficking rate of the protein by fourfold (Walmsley et al., 2001). Additionally, inoculation of GPI anchor-free PrPC with PrPSc leads to the development of fibrillar extracellular plaques, formation of which is facilitated through their interaction with heparan sulfate (Sevillano et al., 2020). Detection of these plaques is indicative of the GPI-anchorless proteins ability to transit through the interstitial fluid and blood more effectively to accumulate peripherally, as opposed to the traditional diffuse parenchymal brain deposits synonymous with the disease (Lv et al., 2021; Sevillano et al., 2020; Wu et al., 2015).

In addition to glycosylation, sialylation plays an important role in prion pathology. The PrP glycans are decorated with sialic acid residues, which are involved in immune regulation, prion conversion, and cross-species transmission (Lv et al., 2021). The terminal sialic acid residues are negatively charged at physiological pH, creating electrostatic repulsion that structurally constrains PrPSc replication (Katorcha et al., 2015; Katorcha et al., 2016). Notably, diglycosylated PrPC moieties, which have more sialic acid residues than monoglycosylated PrPC, are preferentially excluded from conversion, an action that is reversed when the substrate PrPC is partially desialylated (Katorcha et al., 2015). This partial desialylation results in an amplification rate increase from 10- to 10,000-fold in a strain-dependent manner (Katorcha et al., 2014; Katorcha et al., 2015). Furthermore, partial desialylation of substrate PrPC results in elimination of glycoform preference by the seeding PrPSc and significantly reduces the species barrier for prion infection. PrPSc sialylation is region-specific, with the thalamus and brainstem, some of the first regions to succumb to infection, exhibiting low sialylation (Mirabile et al., 2015; Makarava et al., 2020). Interestingly, the locus coeruleus, whose degradation is linked to the clinical manifestation of CJD, has been found to harbor prion protein deposition preceding the motor and autonomic centers of the brainstem; however, its sialylation status is yet to be investigated (Mirabile et al., 2015). The sialylation status is also required for PrPSc trafficking to the spleen and lymph nodes, with partially desialylated PrPSc predominantly migrating to the liver. Finally, partial desialylation of PrPSc increases the microglial inflammatory response evident through activation of the NF-κB pathway and increased IL-6, TNF-α, and inducible nitric oxide synthase expression in a desialylation-dependant manner (Srivastava et al., 2018). Similarly, an *N*-glycosylated PrPSc strain lacking sialylation was found to colocalize with microglia, which demonstrate upregulated phagocytic activity during prion disease (Makarava et al., 2021). This increased microglial activation and subsequent neuroinflammation are proposed to be triggered by the exposed galactose found on this PrPSc strain and the partially desialylated PrPSc proteins (Makarava and Baskakov, 2023). This body of work demonstrates the complicated role that N-glycosylation and sialylation play in the infectivity, immunity, and trafficking of PrP in prion disease.

Collectively, the evidence overwhelmingly demonstrates the fundamental role of glycosylation alterations in the pathophysiology of NDDs, with the findings providing several potential inroads for the development of novel treatments and disease biomarkers in the future.

6.2.13 Modifiable factors

Recent evidence suggests that age-related NDDs diseases such as AD and PD may be reduced or preventable through management of modifiable risk factors, including environmental exposure, lifestyle (exercise, sleep, smoking, and alcohol use), and diet, which together have been termed the "exposome" (Cavaliere and Gülöksüz, 2022). In addition, prevention and treatment of metabolic diseases, including T2D and cardiovascular disease, may help prevent NDDs, and mainstay treatment strategies often include nutritional and lifestyle modifications (Ley et al., 2014). Linking mechanisms between these modifiable risk factors and neurodegeneration are neuroinflammation, glycosylation, and glycation alterations, as exemplified by the AGE/RAGE axis discussed earlier (Rungratanawanich et al., 2021; Sparvero et al., 2009; Takuma et al., 2009; Videira and Castro-Caldas, 2018). Whilst pharmacological treatments may provide symptomatic relief, and research into identifying disease biomarkers and more efficacious disease-modifying therapies remains urgent, implementation of interventions to improve modifiable risk factors may in part reduce the burden of neurodegenerative disease and can also be utilized as adjunctive approaches (Table 6.2).

6.3 Conclusion and future perspectives

The human glycome is expansive and dynamic. The glycocalyx is an elegant and complex language made up of an estimated 10^{12} different glycans as "words", and is a reporter of the cellular and physiological state of an organism telling its story through extensive signaling networks (Reily et al., 2019; Smith and Bertozzi, 2021). However, there is still a long way to go until the language is fully understood and translated into effective diagnostics and treatments for NDDs, such as AD and PD. Glycosylation, the glycome, and alterations to it have been studied in many fields, including congenital disorders of glycosylation, cancer, metabolic disease, psychiatric disease, and NDDs, with alterations to glycosylation now established as a hallmark of neurodegeneration.

New areas of research have revealed that small noncoding mammalian RNAs are also modified by specific glycans to form glycoRNAs, which are embedded in plasma membranes and involved in immune regulation (Flynn et al., 2021). This landmark finding has not only expanded the view of the sheer size of the glycome but it has also provided inroads to a deeper appreciation of the significant biological influence exerted by the glycome on all cellular processes that sustain life and drive disease (Flynn et al., 2021). More recently, findings suggest that snoRNA, snRNA, rRNA, tRNA, miRNA, and Y_RNA are also targets of glycosylation and may be found in a range of organelles, including mitochondria (Li et al., 2023a). Further, it has been proposed that glycoRNAs may play a major role in modulating the immune response in the brain, confer neuroprotection in NDDs and, if altered, drive neurodegeneration (Siddiqui et al., 2019). This remains an enigmatic field to be elucidated.

Whilst several glycan-derived pharmacological interventions have been investigated in diverse fields, an increased understanding and appreciation of the substantial role that glycosylation plays in neurodegeneration will allow the future development of targeted therapies and strategies from prevention to treatment for NDDs.

Acknowledgments

This work was supported by the University of Nottingham (JRS). JGS was supported by a Royce Simmons Foundation Mid-Career Research Fellowship through the Dementia Australia Research Foundation. The authors acknowledge the assistance of Riemke Aggio-Bruce for the generation of several figures.

Author contributions

ME wrote the manuscript and generated several of the figures. NV contributed to several sections of the manuscript and generated the protein figures. JGS and JRS contributed to the conceptualization of the manuscript. JRS further contributed to the figures.

References

Abeliovich, A., Gitler, A.D., 2016. Defects in trafficking bridge Parkinson's disease pathology and genetics. Nature 539 (7628), 207–216. Available at https://doi.org/10.1038/nature20414.

Aebi, M., 2013. N-linked protein glycosylation in the ER. Biochimica et Biophysica Acta (BBA) - Molecular Cell Research 1833 (11), 2430–2437. Available at https://doi.org/10.1016/j.bbamcr.2013.04.001.

Aguzzi, A., Heppner, F.L., 2000. Pathogenesis of prion diseases: a progress report. Cell Death & Differentiation 7 (10), 889–902. Available at https://doi.org/10.1038/sj.cdd.4400737.

Ahmed, N., Battah, S., Karachalias, N., Babaei-Jadidi, R., Horányi, M., Baróti, K., et al., 2003. Increased formation of methylglyoxal and protein glycation, oxidation and nitrosation in triosephosphate isomerase deficiency. Biochimica et Biophysica Acta - Molecular Basis of Disease 1639 (2), 121–132. Available at https://doi.org/10.1016/j.bbadis.2003.08.002.

Akasaka-Manya, K., Kawamura, M., Tsumoto, H., Saito, Y., Tachida, Y., Kitazume, S., et al., 2017. Excess APP O-glycosylation by GalNAc-T6 decreases Ab production. Journal of Biochemistry 161 (1), 99–111. Available at https://doi.org/10.1093/jb/mvw056.

Akasaka-Manya, K., Manya, H., 2020. The role of APP O-glycosylation in Alzheimer's disease. Biomolecules 10 (11), 1569. Available at https://doi.org/10.3390/biom10111569.

Alghusen, I.M., Carman, M.S., Wilkins, H.M., Strope, T.A., Gimore, C., Fedosyuk, H., et al., 2024. O-GlcNAc impacts mitophagy via the PINK1-dependent pathway. Frontiers in Aging Neuroscience 16, 1387931. Available at https://doi.org/10.3389/fnagi.2024.1387931.

Alizadeh, M., Kheirouri, S., 2019. Curcumin against advanced glycation end products (AGEs) and AGEs-induced detrimental agents. Critical Reviews in Food Science and Nutrition 59 (7), 1169–1177. Available at https://doi.org/10.1080/10408398.2017.1396200.

Alves, R.N., Iglesia, R.P., Prado, M.B., Escobar, M.I.M., Boccacino, J.M., Fernandes, C.Fd.L., et al., 2020. A new take on prion protein dynamics in cellular trafficking. International Journal of Molecular Sciences 21 (20), 1–19. Available at https://doi.org/10.3390/ijms21207763.

Aminov, R., Aminova, L., 2023. The role of the glycome in symbiotic host-microbe interactions. Glycobiology 33 (12), 1106–1116. Available at https://doi.org/10.1093/glycob/cwad073.

Anis, E., Xie, A., Brundin, L., Brundin, P., 2022. Digesting recent findings: gut alpha-synuclein, microbiome changes in Parkinson's disease. Trends in Endocrinology & Metabolism 33 (2), 147–157. Available at https://doi.org/10.1016/j.tem.2021.11.005.

Aragno, M., Mastrocola, R., 2017. Dietary sugars and endogenous formation of advanced glycation endproducts: emerging mechanisms of disease. Nutrients 9 (4), 385. Available at https://doi.org/10.3390/nu9040385.

Balana, A.T., Levine, P.M., Craven, T.W., Mukherjee, S., Pedowitz, N.J., Moon, S.P., et al., 2021. O-GlcNAc modification of small heat shock proteins enhances their anti-amyloid chaperone activity. Nature Chemistry 13 (5), 441–450. Available at https://doi.org/10.1038/s41557-021-00648-8.

Balana, A.T., Mahul-Mellier, A.-L., Nguyen, B.A., Horvath, M., Javed, A., Hard, E.R., et al., 2024. Author correction: O-GlcNAc forces an α-synuclein amyloid strain with notably diminished seeding and pathology. Nature Chemical Biology 20 (5), 656. Available at https://doi.org/10.1038/s41589-024-01587-4 656.

Barber, E., Houghton, M.J., Visvanathan, R., Williamson, G., 2022. Measuring key human carbohydrate digestive enzyme activities using high-performance anion-exchange chromatography with pulsed amperometric detection. Nature Protocols 17 (12), 2882–2919. Available at https://doi.org/10.1038/s41596-022-00736-0.

Beauchamp, L.C., Dore, V., Villemagne, V.L., Xu, S., Finkelstein, D., Barnham, K.J., et al., 2023. Using 18F-AV-133 VMAT2 PET imaging to monitor progressive nigrostriatal degeneration in Parkinson disease. Neurology 101 (22), E2314. Available at https://doi.org/10.1212/WNL.0000000000207748.

Becker, D.J., Lowe, J.B., 2003. Fucose: biosynthesis and biological function in mammals. Glycobiology 13 (7), 41R. Available at https://doi.org/10.1093/glycob/cwg054.

Bellenguez, C., Küçükali, F., Jansen, I.E., Kleineidam, L., Moreno-Grau, S., Amin, N., et al., 2022. New insights into the genetic etiology of Alzheimer's disease and related dementias. Nature Genetics 54 (4), 412–436. Available at https://doi.org/10.1038/s41588-022-01024-z.

Beringue, V., Mallinson, G., Kaisar, M., Tayebi, M., Sattar, Z., Jackson, G., et al., 2003. S. Hawke, regional heterogeneity of cellular prion protein isoforms in the mouse brain. Brain 126 (9), 2065–2073. Available at https://doi.org/10.1093/brain/awg205.

Bhattacharya, R., Alam, M.R., Kamal, M.A., Seo, K.J., Singh, L.R., 2023. AGE-RAGE axis culminates into multiple pathogenic processes: a central road to neurodegeneration. Frontiers in Molecular Neuroscience 16, 1155175. Available at https://doi.org/10.3389/fnmol.2023.1155175.

Bhattacherjee, A., Daskhan, G.C., Bains, A., Watson, A.E.S., Eskandari-Sedighi, G., St. Laurent, C.D., et al., 2021. Increasing phagocytosis of micoglia by targeting CD33 with liposomes displaying glycan ligands. Journal of Controlled Release 338, 680–693. Available at https://doi.org/10.1016/j.jconrel.2021.09.010.

Birkett, C.R., Hennion, R.M., Bembridge, D.A., Clarke, M.C., Chree, A., Bruce, M.E., et al., 2001. Scrapie strains maintain biological phenotypes on propagation in a cell line in culture. EMBO Journal 20 (13), 3351–3358. Available at https://doi.org/10.1093/emboj/20.13.3351.

Boix, C.P., Lopez-Font, I., Cuchillo-Ibañez, I., Sáez-Valero, J., 2020. Amyloid precursor protein glycosylation is altered in the brain of patients with Alzheimer's disease. Alzheimer's Research and Therapy 12 (1). Available at https://doi.org/10.1186/s13195-020-00664-9.

Bourgognon, J.M., Steinert, J.R., 2019. The metabolome identity: basis for discovery of biomarkers in neurodegeneration. Neural Regeneration Research 14 (3), 387–390. Available at https://doi.org/10.4103/1673-5374.245464.

Bourgognon, J.M., Spiers, J.G., Scheiblich, H., Antonov, A., Bradley, S.J., Tobin, A.B., et al., 2018. Alterations in neuronal metabolism contribute to the pathogenesis of prion disease. Cell Death and Differentiation 25 (8), 1408–1425. Available at https://doi.org/10.1038/s41418-018-0148-x.

Bourgognon, J.M., Spiers, J.G., Robinson, S.W., Scheiblich, H., Glynn, P., Ortori, C., et al., 2021. Inhibition of neuroinflammatory nitric oxide signaling suppresses glycation and prevents neuronal dysfunction in mouse prion disease. Proceedings of the National Academy of Sciences of the United States of America 118 (10). Available at https://doi.org/10.1073/pnas.2009579118.

Braak, H., Tredici, K.D., Rüb, U., De Vos, R.A.I., Jansen Steur, E.N.H., Braak, E., 2003. Staging of brain pathology related to sporadic Parkinson's disease. Neurobiology of Aging 24 (2), 197–211. Available at https://doi.org/10.1016/S0197-4580(02)00065-9.

Bradberry, M.M., Peters-Clarke, T.M., Shishkova, E., Chapman, E.R., Coon, J.J., 2023. N-glycoproteomics of brain synapses and synaptic vesicles. Cell Reports 42 (4). Available at https://doi.org/10.1016/j.celrep.2023.112368.

Brockhausen, I., 1999. Pathways of O-glycan biosynthesis in cancer cells. Biochimica et Biophysica Acta (BBA) - General Subjects 1473 (1), 67–95. Available at https://doi.org/10.1016/s0304-4165(99)00170-1.

Bu, J., Liu, J., Liu, K., Wang, Z., 2019. Diagnostic utility of gut α-synuclein in Parkinson's disease: a systematic review and meta-analysis. Behavioural Brain Research 364, 340–347. Available at https://doi.org/10.1016/j.bbr.2019.02.039.

Burke, C.M., Walsh, D.J., Mark, K.M.K., Deleault, N.R., Nishina, K.A., Agrimi, U., et al., 2020. Cofactor and glycosylation preferences for in vitro prion conversion are predominantly determined by strain conformation. PLoS Pathogens 16 (4). Available at https://doi.org/10.1371/journal.ppat.1008495.

Burnham-Marusich, A.R., Berninsone, P.M., 2012. Multiple proteins with essential mitochondrial functions have glycosylated isoforms. Mitochondrion 12 (4), 423–427. Available at https://doi.org/10.1016/j.mito.2012.04.004.

Callender, J.A., Sevillano, A.M., Soldau, K., Kurt, T.D., Schumann, T., Pizzo, D.P., et al., 2020. Prion protein post-translational modifications modulate heparan sulfate binding and limit aggregate size in prion disease. Neurobiology of Disease 142. Available at https://doi.org/10.1016/j.nbd.2020.104955.

Camacho, M.V., Telling, G., Kong, Q., Gambetti, P., Notari, S., 2019. Role of prion protein glycosylation in replication of human prions by protein misfolding cyclic amplification. Laboratory Investigation 99 (11), 1741–1748. Available at https://doi.org/10.1038/s41374-019-0282-1.

Capellari, S., Zaidi, S.I.A., Urig, C.B., Perry, G., Smith, M.A., Petersen, R.B., 1999. Prion protein glycosylation is sensitive to redox change. Journal of Biological Chemistry 274 (49), 34846–34850. Available at https://doi.org/10.1074/jbc.274.49.34846.

Cavaliere, F., Gülöksüz, S., 2022. Shedding light on the etiology of neurodegenerative diseases and dementia: the exposome paradigm. npj Mental Health Research 1 (1). Available at https://doi.org/10.1038/s44184-022-00018-3.

Cetin, S., Knez, D., Gobec, S., Kos, J., Pišlar, A., 2022. Cell models for Alzheimer's and Parkinson's disease: at the interface of biology and drug discovery. Biomedicine & Pharmacotherapy 149, 112924. Available at https://doi.org/10.1016/j.biopha.2022.112924.

Chai, P., Lebedenko, C.G., Flynn, R.A., 2023. RNA crossing membranes: systems and mechanisms contextualizing extracellular RNA and cell surface GlycoRNAs. Annual Review of Genomics and Human Genetics 24, 85–107. Available at https://doi.org/10.1146/annurev-genom-101722-101224.

Chakraborty, J., Basso, V., Ziviani, E., 2017. Post translational modification of Parkin. Biology Direct 12 (1). Available at https://doi.org/10.1186/s13062-017-0176-3.

Chang, J.J., Kulkarni, S., Pasricha, T.S., 2024. Upper gastrointestinal mucosal damage and subsequent risk of Parkinson disease. JAMA Network Open 7 (9). Available at https://doi.org/10.1001/jamanetworkopen.2024.31949.

Chatterjee, S., Mudher, A., 2018. Alzheimer's disease and type 2 diabetes: a critical assessment of the shared pathological traits. Frontiers in Neuroscience 12, 383. Available at https://doi.org/10.3389/fnins.2018.00383.

Chengcheng, M., Panpan, A., Yalong, Y., Mingyu, S., Wei, X., Jing, C., et al., 2024. GDNF improves the cognitive ability of PD mice by promoting glycosylation and membrane distribution of DAT. Scientific Reports 14 (1). Available at https://doi.org/10.1038/s41598-024-68609-y.

Choi, B.R., Cho, W.H., Kim, J., Lee, H.J., Chung, C., Jeon, W.K., et al., 2014. Increased expression of the receptor for advanced glycation end products in neurons and astrocytes in a triple transgenic mouse model of Alzheimer's disease. Experimental and Molecular Medicine 46 (2). Available at https://doi.org/10.1038/emm.2013.147.

Chun, Y.S., Kwon, O.-H., Chung, S., 2017. O-GlcNAcylation of amyloid-β precursor protein at threonine 576 residue regulates trafficking and processing. Biochemical and Biophysical Research Communications 490 (2), 486–491.

Chung, K.K.K., Zhang, Y., Lim, K.L., Tanaka, Y., Huang, H., Gao, J., et al., 2001. Parkin ubiquitinates the α-synuclein-interacting protein, synphilin-1: implications for Lewy-body formation in Parkinson disease. Nature Medicine 7 (10), 1144–1150. Available at https://doi.org/10.1038/nm1001-1144.

Cole, R.N., Hart, G.W., 2001. Cytosolic O-glycosylation is abundant in nerve terminals. Journal of Neurochemistry 79 (5), 1080–1089. Available at https://doi.org/10.1046/j.1471-4159.2001.00655.x.

Collinge, J., Sidle, K.C.L., Meads, J., Ironside, J., Hill, A.F., 1996. Molecular analysis of prion strain variation and the aetiology of 'new variant' CJD. Nature 383 (6602), 685–690. Available at https://doi.org/10.1038/383685a0.

Conroy, L.R., Hawkinson, T.R., Young, L.E.A., Gentry, M.S., Sun, R.C., 2021. Emerging roles of N-linked glycosylation in brain physiology and disorders. Trends in Endocrinology and Metabolism 32 (12), 980–993. Available at https://doi.org/10.1016/j.tem.2021.09.006.

Cornelissen, L.A.M., Blanas, A., Zaal, A., van der Horst, J.C., Kruijssen, L.J.W., O'Toole, T., et al., 2020. Tn antigen expression contributes to an immune suppressive microenvironment and drives tumor growth in colorectal cancer. Frontiers in Oncology 10, 1622. Available at https://doi.org/10.3389/fonc.2020.01622.

Craig Venter, Adams, Myers, Li, Mural, Sutton, et al., 2001. The sequence of the human genome. Science 291 (5507), 1304–1351. Available at https://doi.org/10.1126/science.1058040.

Deane, R., Yan, S.D., Submamaryan, R.K., LaRue, B., Jovanovic, S., Hogg, E., et al., 2003. RAGE mediates amyloid-β peptide transport across the blood-brain barrier and accumulation in brain. Nature Medicine 9 (7), 907–913. Available at https://doi.org/10.1038/nm890.

Derk, J., MacLean, M., Juranek, J., Schmidt, A.M., 2018. The receptor for advanced glycation endproducts (RAGE) and mediation of inflammatory neurodegeneration. Journal of Alzheimer's Disease & Parkinsonism 08 (01). Available at https://doi.org/10.4172/2161-0460.1000421.

Dobie, C., Skropeta, D., 2021. Insights into the role of sialylation in cancer progression and metastasis. British Journal of Cancer 124 (1), 76–90. Available at https://doi.org/10.1038/s41416-020-01126-7.

Dube, D.H., Bertozzi, C.R., 2005. Glycans in cancer and inflammation - potential for therapeutics and diagnostics. Nature Reviews Drug Discovery 4 (6), 477–488. Available at https://doi.org/10.1038/nrd1751.

Dusoswa, S.A., Verhoeff, J., Abels, E., Méndez-Huergo, S.P., Croci, D.O., Kuijper, L.H., et al., 2020. Glioblastomas exploit truncated O-linked glycans for local and distant immune modulation via the macrophage galactose-type lectin. Proceedings of the National Academy of Sciences of the United States of America 117 (7), 3693–3703. Available at https://doi.org/10.1073/pnas.1907921117.

Eckhardt, M., Bukalo, O., Chazal, G., Wang, L., Goridis, C., Schachner, M., et al., 2000. Mice deficient in the polysialyltransferase ST8SiaIV/PST-1 allow discrimination of the roles of neural cell adhesion molecule protein and polysialic acid in neural development and synaptic plasticity. The Journal of Neuroscience 20 (14), 5234–5244. Available at https://doi.org/10.1523/jneurosci.20-14-05234.2000.

Ellgaard, L., Helenius, A., 2003. Quality control in the endoplasmic reticulum. Nature Reviews Molecular Cell Biology 4 (3), 181–191. Available at https://doi.org/10.1038/nrm1052.

Ermonval, M., Mouillet-Richard, S., Codogno, P., Kellermann, O., Botti, J., 2003. Evolving views in prion glycosylation: functional and pathological implications. Biochimie 85 (1-2), 33–45. Available at https://doi.org/10.1016/S0300-9084(03)00040-3.

Ermonval, M., Petit, D., Le Duc, A., Kellermann, O., Gallet, P.F., 2009. Glycosylation-related genes are variably expressed depending on the differentiation state of a bioaminergic neuronal cell line: implication for the cellular prion protein. Glycoconjugate Journal 26 (4), 477–493. Available at https://doi.org/10.1007/s10719-008-9198-5.

Eskandari-Sedighi, G., Jung, J., Macauley, M.S., 2023. CD33 isoforms in microglia and Alzheimer's disease: friend and foe. Molecular Aspects of Medicine 90.

Esmail, S., Manolson, M.F., 2021. Advances in understanding N-glycosylation structure, function, and regulation in health and disease. European Journal of Cell Biology 100 (7-8), 151186. Available at https://doi.org/10.1016/j.ejcb.2021.151186.

Farag, M.R., Khalil, S.R., Zaglool, A.W., Hendam, B.M., Moustafa, A.A., Cocco, R., et al., 2021. Thiacloprid induced developmental neurotoxicity via ros-oxidative injury and inflammation in chicken embryo: the possible attenuating role of chicoric and rosmarinic acids. Biology 10 (11). Available at https://doi.org/10.3390/biology10111100.

Fisher, P., Spencer, H., Thomas-Oates, J., Wood, A.J., Ungar, D., 2019a. Modeling glycan processing reveals golgi-enzyme homeostasis upon trafficking defects and cellular differentiation. Cell Reports 27 (4), 1231. Available at https://doi.org/10.1016/j.celrep.2019.03.107.

Fisher, P., Thomas-Oates, J., Wood, A.J., Ungar, D., 2019b. The N-glycosylation processing potential of the mammalian golgi apparatus. Frontiers in Cell and Developmental Biology 7, 157. Available at https://doi.org/10.3389/fcell.2019.00157.

Flowers, S.A., Grant, O.C., Woods, R.J., Rebeck, G.W., 2020. O-glycosylation on cerebrospinal fluid and plasma apolipoprotein e differs in the lipid-binding domain. Glycobiology 30 (2), 74–85. Available at https://doi.org/10.1093/glycob/cwz084.

Flynn, R.A., Pedram, K., Malaker, S.A., Batista, P.J., Smith, B.A.H., Johnson, A.G., et al., 2021. Small RNAs are modified with N-glycans and displayed on the surface of living cells. Cell 184 (12), 3109. Available at https://doi.org/10.1016/j.cell.2021.04.023.

Frigerio, L., Lord, J.M., 2000. Glycoprotein degradation: do sugars hold the key? Current Biology 10 (18), R674. Available at https://doi.org/10.1016/S0960-9822(00)00680-1.

Fujita, N., Tamura, A., Higashidani, A., Tonozuka, T., Freeze, H.H., Nishikawa, A., 2008. The relative contribution of mannose salvage pathways to glycosylation in PMI-deficient mouse embryonic fibroblast cells. FEBS Journal 275 (4), 788–798. Available at https://doi.org/10.1111/j.1742-4658.2008.06246.x.

Gao, Y., Luan, X., Melamed, J., Brockhausen, I., 2021. Role of glycans on key cell surface receptors that regulate cell proliferation and cell death. Cells 10 (5), 1252. Available at https://doi.org/10.3390/cells10051252.

Gatta, E., Lefebvre, T., Gaetani, S., Dos Santos, M., Marrocco, J., Mir, A.M., et al., 2016. Evidence for an imbalance between tau O-GlcNAcylation and phosphorylation in the hippocampus of a mouse model of Alzheimer's disease. Pharmacological Research 105, 186–197. Available at https://doi.org/10.1016/j.phrs.2016.01.006.

Gaunitz, S., Tjernberg, L.O., Schedin-Weiss, S., 2021a. The N-glycan profile in cortex and hippocampus is altered in Alzheimer disease. Journal of Neurochemistry 159 (2), 292–304. Available at https://doi.org/10.1111/jnc.15202.

Gaunitz, S., Tjernberg, L.O., Schedin-weiss, S., 2021b. What can n-glycomics and n-glycoproteomics of cerebrospinal fluid tell us about alzheimer disease? Biomolecules 11 (6). Available at https://doi.org/10.3390/biom11060858.

Giasson, B.I., Lee, V.M.Y., 2001. Parkin and the molecular pathways of Parkinson's disease. Neuron 31 (6), 885–888. Available at https://doi.org/10.1016/S0896-6273(01)00439-1.

Goedert, M., Spillantini, M.G., Del Tredici, K., Braak, H., 2013. 100 years of Lewy pathology. Nature Reviews Neurology 9 (1), 13–24. Available at https://doi.org/10.1038/nrneurol.2012.242.

Gonzalez-Gil, A., Porell, R.N., Fernandes, S.M., Maenpaa, E., Li, T.A., Li, T., et al., 2022. Human brain sialoglycan ligand for CD33, a microglial inhibitory Siglec implicated in Alzheimer's disease. Journal of Biological Chemistry 298 (6). Available at https://doi.org/10.1016/j.jbc.2022.101960.

Grassi, D., Howard, S., Zhou, M., Diaz-Perez, N., Urban, N.T., Guerrero-Given, D., et al., 2018. Identification of a highly neurotoxic α-synuclein species inducing mitochondrial damage and mitophagy in Parkinson's disease. Proceedings of the National Academy of Sciences of the United States of America 115 (11). Available at https://doi.org/10.1073/pnas.1713849115 E2634.

Greto, V.L., Cvetko, A., Štambuk, T., Dempster, N.J., Kifer, D., Deriš, H., et al., 2021. Extensive weight loss reduces glycan age by altering IgG N-glycosylation. International Journal of Obesity 45 (7), 1521–1531. Available at https://doi.org/10.1038/s41366-021-00816-3.

Griciuc, A., Serrano-Pozo, A., Parrado, A.R., Lesinski, A.N., Asselin, C.N., Mullin, K., et al., 2013. Alzheimer's disease risk gene cd33 inhibits microglial uptake of amyloid beta. Neuron 78 (4), 631–643. Available at https://doi.org/10.1016/j.neuron.2013.04.014.

Griffith, Mathes, Schmitz, 1995. β-Amyloid precursor protein is modified with O-linked N-acetylglucosamine. Journal of Neuroscience Research 41 (2), 270–278. Available at https://doi.org/10.1002/jnr.490410214.

Guix, F.X., Ill-Raga, G., Bravo, R., Nakaya, T., De Fabritiis, G., Coma, M., et al., 2009. Amyloid-dependent triosephosphate isomerase nitrotyrosination induces glycation and tau fibrillation. Brain 132 (5), 1335–1345. Available at https://doi.org/10.1093/brain/awp023.

Hampel, H., Hardy, J., Blennow, K., Chen, C., Perry, G., Kim, S.H., et al., 2021. The amyloid-β pathway in Alzheimer's disease. Molecular Psychiatry 26 (10), 5481–5503. Available at https://doi.org/10.1038/s41380-021-01249-0.

Hargett, A.A., Renfrow, M.B., 2019. Glycosylation of viral surface proteins probed by mass spectrometry. Current Opinion in Virology 36, 56–66. Available at https://doi.org/10.1016/j.coviro.2019.05.003.

Hart, G.W., Slawson, C., Ramirez-Correa, G., Lagerlof, O., 2011. Cross talk between O-GlcNAcylation and phosphorylation: roles in signaling, transcription, and chronic disease. Annual Review of Biochemistry 80, 825–858. Available at https://doi.org/10.1146/annurev-biochem-060608-102511.

Haukedal, H., Freude, K.K., 2021. Implications of glycosylation in Alzheimer's disease. Frontiers in Neuroscience 14, 625348. Available at https://doi.org/10.3389/fnins.2020.625348.

Hay, A.J.D., Murphy, T.J., Popichak, K.A., Zabel, M.D., Moreno, J.A., 2022. Adipose-derived mesenchymal stromal cells decrease prion-induced glial inflammation in vitro. Scientific Reports 12 (1). Available at https://doi.org/10.1038/s41598-022-26628-7.

Hay, A.J.D., Popichak, K.A., Mumford, G., Bian, J., Shirley, P., Wolfrath, L., et al., 2025. Microglia-specific NF-κB signaling is a critical regulator of prion-induced glial inflammation and neuronal loss. PLOS Pathogens 21 (6). Available at https://doi.org/10.1371/journal.ppat.1012582.

Heneka, M.T., van der Flier, W.M., Jessen, F., Hoozemanns, J., Thal, D.R., Boche, D., et al., 2025. Neuroinflammation in Alzheimer disease. Nature Reviews Immunology 25 (5), 321–352. Available at https://doi.org/10.1038/s41577-024-01104-7.

Jankovic, J., Tan, E.K., 2020. Parkinson's disease: etiopathogenesis and treatment. Journal of Neurology, Neurosurgery and Psychiatry 91 (8), 795–808. Available at https://doi.org/10.1136/jnnp-2019-322338.

Jiang, H., Jiang, Q., Feng, J., 2004. Parkin increases dopamine uptake by enhancing the cell surface expression of dopamine transporter. The Journal of Biological Chemistry 279 (52), 54380–54386.

Jin, K., Peel, A.L., Mao, X.O., Xie, L., Cottrell, B.A., Henshall, D.C., et al., 2004. Increased hippocampal neurogenesis in Alzheimer's disease. Proceedings of the National Academy of Sciences of the United States of America 101 (1), 343–347. Available at https://doi.org/10.1073/pnas.2634794100.

Jin, L.W., di Lucente, J., Ruiz Mendiola, U., Tang, X., Zivkovic, A.M., Lebrilla, C.B., et al., 2023. The role of FUT8-catalyzed core fucosylation in Alzheimer's amyloid-β oligomer-induced activation of human microglia. GLIA 71 (5), 1346–1359. Available at https://doi.org/10.1002/glia.24345.

Joshi, G., Wang, Y., 2015. Golgi defects enhance APP amyloidogenic processing in Alzheimer's disease. BioEssays 37 (3), 240–247. Available at https://doi.org/10.1002/bies.201400116.

Joshi, G., Chi, Y., Huang, Z., Wang, Y., 2014. Aβ-induced Golgi fragmentation in Alzheimer's disease enhances Aβ production. Proceedings of the National Academy of Sciences of the United States of America 111 (13). Available at https://doi.org/10.1073/pnas.1320192111 E1230,.

Kanekiyo, T., Xu, H., Bu, G., 2014. ApoE and Aβ in Alzheimer's disease: accidental encounters or partners? Neuron 81 (4), 740–754. Available at https://doi.org/10.1016/j.neuron.2014.01.045.

Kang, H.E., Bian, J., Kane, S.J., Kim, S., Selwyn, V., Crowell, J., et al., 2020. Incomplete glycosylation during prion infection unmasks a prion protein epitope that facilitates prion detection and strain discrimination. Journal of Biological Chemistry 295 (30), 10420–10433. Available at https://doi.org/10.1074/jbc.ra120.012796.

Kang, Y., Zhang, Q., Xu, S., Yu, Y., 2024. The alteration and role of glycoconjugates in Alzheimer's disease. Frontiers in Aging Neuroscience 16, 1398641. Available at https://doi.org/10.3389/fnagi.2024.1398641.

Katorcha, E., Makarava, N., Savtchenko, R., d'Azzo, A., Baskakov, I.V., McFadden, G., 2014. Sialylation of prion protein controls the rate of prion amplification, the cross-species barrier, the ratio of PrPSc glycoform and prion infectivity. PLoS Pathogens 10 (9), e1004366. Available at https://doi.org/10.1371/journal.ppat.1004366.

Katorcha, E., Makarava, N., Savtchenko, R., Baskakov, I.V., 2015. Sialylation of the prion protein glycans controls prion replication rate and glycoform ratio. Scientific Reports 5. Available at https://doi.org/10.1038/srep16912.

Katorcha, E., Daus, M.L., Gonzalez-Montalban, N., Makarava, N., Lasch, P., Beekes, M., et al., 2016. Reversible off and on switching of prion infectivity via removing and reinstalling prion sialylation. Scientific Reports 6. Available at https://doi.org/10.1038/srep33119.

Khadka, A., Spiers, J.G., Cheng, L., Hill, A.F., 2023. Extracellular vesicles with diagnostic and therapeutic potential for prion diseases. Cell and Tissue Research 392 (1), 247–267. Available at https://doi.org/10.1007/s00441-022-03621-0.

Kim, C., Nam, D.W., Park, S.Y., Song, H., Hong, H.S., Boo, J.H., et al., 2013. O-linked β-N-acetylglucosaminidase inhibitor attenuates β-amyloid plaque and rescues memory impairment. Neurobiology of Aging 34 (1), 275–285. Available at https://doi.org/10.1016/j.neurobiolaging.2012.03.001.

Kim, D.Y., Kim, S.M., Cho, E.J., Kwak, H.B., Han, I.O., 2024. Protective effect of increased O-GlcNAc cycling against 6-OHDA induced Parkinson's disease pathology. Cell Death and Disease 15 (4). Available at https://doi.org/10.1038/s41419-024-06670-1.

Kizuka, Y., Kitazume, S., Fujinawa, R., Saito, T., Iwata, N., Saido, T.C., et al., 2015. An aberrant sugar modification of BACE1 blocks its lysosomal targeting in Alzheimer's disease. EMBO Molecular Medicine 7 (2), 175–189. Available at https://doi.org/10.15252/emmm.201404438.

Kleene, R., Schachner, M., 2004. Glycans and neural cell interactions. Nature Reviews Neuroscience 5 (3), 195–208. Available at https://doi.org/10.1038/nrn1349.

Kobayashi, A., Hirata, T., Shimazaki, T., Munesue, Y., Aoshima, K., Kimura, T., et al., 2023. A point mutation in GPI-attachment signal peptide accelerates the development of prion disease. Acta Neuropathologica 145 (5), 637–650. Available at https://doi.org/10.1007/s00401-023-02553-5.

Koshy, S.M., Kincaid, A.E., Bartz, J.C., 2022. Transport of prions in the peripheral nervous system: pathways, cell types, and mechanisms. Viruses 14 (3). Available at https://doi.org/10.3390/v14030630.

Krieg, J., Hartmann, S., Vicentini, A., Gläsner, W., Hess, D., Hofsteenge, J., 1998. Recognition signal for C-mannosylation of Trp-7 in RNase 2 consists of sequence Trp-x-x-Trp. Molecular Biology of the Cell 9 (2), 301–309. Available at https://doi.org/10.1091/mbc.9.2.301.

Kruger, L.C., Isom, L.L., 2016. Voltage-gated na+ channels: not just for conduction. Cold Spring Harbor Perspectives in Biology 8 (6). Available at https://doi.org/10.1101/cshperspect.a029264.

Lawler, P.E., Bollinger, J.G., Schindler, S.E., Hodge, C.R., Iglesias, N.J., Krishnan, V., et al., 2023. Apolipoprotein E O-glycosylation is associated with amyloid plaques and APOE genotype. Analytical Biochemistry 672. Available at https://doi.org/10.1016/j.ab.2023.115156.

Lederkremer, G.Z., Glickman, M.H., 2005. A window of opportunity: timing protein degradation by trimming of sugars and ubiquitins. Trends in Biochemical Sciences 30 (6), 297–303. Available at https://doi.org/10.1016/j.tibs.2005.04.010.

Lee, J., Scagel, C.F., 2013. Chicoric acid: chemistry, distribution, and production. Frontiers in Chemistry 1, 40. Available at https://doi.org/10.3389/fchem.2013.00040.

Lee, B.E., Kim, H.Y., Kim, H.J., Jeong, H., Kim, B.G., Lee, H.E., et al., 2020. O-GlcNAcylation regulates dopamine neuron function, survival and degeneration in Parkinson disease. Brain 143 (12), 3699–3716. Available at https://doi.org/10.1093/brain/awaa320.

Lee, B.E., Suh, P.-G., Kim, J.-I., 2021. O-GlcNAcylation in health and neurodegenerative diseases. Experimental and Molecular Medicine 53 (11), 1674–1682. Available at https://doi.org/10.1038/s12276-021-00709-5.

Levavasseur, E., Laffont-Proust, I., Morain, E., Faucheux, B.A., Privat, N., Peoc'h, K., et al., 2008. Regulating factors of PrPres glycosylation in Creutzfeldt-Jakob disease - implications for the dissemination and the diagnosis of human prion strains. PLoS ONE 3 (7). Available at https://doi.org/10.1371/journal.pone.0002786.

Levine, P.M., Galesic, A., Balana, A.T., Mahul-Mellier, A.-L., Navarro, M.X., De Leon, C.A., et al., 2019. α-Synuclein O-GlcNAcylation alters aggregation and toxicity, revealing certain residues as potential inhibitors of Parkinson's disease. Proceedings of the National Academy of Sciences of the United States of America 116 (5), 1511–1519. Available at https://doi.org/10.1073/pnas.1808845116.

Lewis, Y.E., Galesic, A., Levine, P.M., De Leon, C.A., Lamiri, N., Brennan, C.K., et al., 2017. O-GlcNAcylation of α-Synuclein at serine 87 reduces aggregation without affecting membrane binding. ACS chemical biology 12 (4), 1020–1027. Available at https://doi.org/10.1021/acschembio.7b00113.

Ley, S.H., Hamdy, O., Mohan, V., Hu, F.B., 2014. Prevention and management of type 2 diabetes: dietary components and nutritional strategies. The Lancet 383 (9933), 1999–2007. Available at https://doi.org/10.1016/S0140-6736(14)60613-9.

Li, F., Ding, J., 2019. Sialylation is involved in cell fate decision during development, reprogramming and cancer progression. Protein & Cell 10 (8), 550–565. Available at https://doi.org/10.1007/s13238-018-0597-5.

Li, X., Lu, F., Wang, J.Z., Gong, C.X., 2006. Concurrent alterations of O-GlcNAcylation and phosphorylation of tau in mouse brains during fasting. European Journal of Neuroscience 23 (8), 2078–2086. Available at https://doi.org/10.1111/j.1460-9568.2006.04735.x.

Li, J., Liu, D., Sun, L., Lu, Y., Zhang, Z., 2012. Advanced glycation end products and neurodegenerative diseases: mechanisms and perspective. Journal of the Neurological Sciences 317 (1-2), 1–5. Available at https://doi.org/10.1016/j.jns.2012.02.018.

Li, J., Hsu, H.C., Mountz, J.D., Allen, J.G., 2018. Unmasking fucosylation: from cell adhesion to immune system regulation and diseases. Cell Chemical Biology 25 (5), 499–512. Available at https://doi.org/10.1016/j.chembiol.2018.02.005.

Li, H., Guglielmetti, C., Sei, Y.J., Zilberter, M., Le Page, L.M., Shields, L., et al., 2023a. Neurons require glucose uptake and glycolysis in vivo. Cell Reports 42 (4). Available at https://doi.org/10.1016/j.celrep.2023.112335.

Li, J., Yue, S., Gao, Z., Hu, W., Liu, Z., Xu, G., et al., 2023b. Novel approach to enriching glycosylated RNAs: specific capture of GlycoRNAs via solid-phase chemistry. Analytical Chemistry 95 (32), 11969–11977. Available at https://doi.org/10.1021/acs.analchem.3c01630.

Liao, H., Winkler, J., Wißfeld, J., Shahraz, A., Klaus, C., Neumann, H., 2021. Low molecular weight polysialic acid prevents lipopolysaccharide-induced inflammatory dopaminergic neurodegeneration in humanized SIGLEC11 transgenic mice. GLIA 69 (12), 2845–2862. Available at https://doi.org/10.1002/glia.24073.

Lin, T., Van Husen, L.S., Yu, Y., Tjernberg, L.O., Schedin-Weiss, S., 2022. Lack of N-glycosylation increases amyloidogenic processing of the amyloid precursor protein. Glycobiology 32 (6), 506–517. Available at https://doi.org/10.1093/glycob/cwac009.

Liu, F., Iqbal, K., Grundke-Iqbal, I., Hart, G.W., Gong, C.X., 2004. O-GlcNAcylation regulates phosphorylation of tau: a mechanism involved in Alzheimer's disease. Proceedings of the National Academy of Sciences of the United States of America 101 (29), 10804–10809. Available at https://doi.org/10.1073/pnas.0400348101.

Liu, F., Shi, J., Tanimukai, H., Gu, J., Gu, J., Grundke-Iqbal, I., et al., 2009. Reduced O-GlcNAcylation links lower brain glucose metabolism and tau pathology in Alzheimer's disease. Brain 132 (7), 1820–1832. Available at https://doi.org/10.1093/brain/awp099.

Long, H., Zhang, S., Zeng, S., Tong, Y., Liu, J., Liu, C., et al., 2022. Interaction of RAGE with α-synuclein fibrils mediates inflammatory response of microglia. Cell Reports 40 (12), 111401. Available at https://doi.org/10.1016/j.celrep.2022.111401.

Losev, Y., Paul, A., Frenkel-Pinter, M., Abu-Hussein, M., Khalaila, I., Gazit, E., et al., 2019. Novel model of secreted human tau protein reveals the impact of the abnormal N-glycosylation of tau on its aggregation propensity. Scientific Reports 9 (1). Available at https://doi.org/10.1038/s41598-019-39218-x.

Losev, Y., Frenkel-Pinter, M., Abu-Hussien, M., Viswanathan, G.K., Elyashiv-Revivo, D., Geries, R., et al., 2021. Differential effects of putative N-glycosylation sites in human Tau on Alzheimer's disease-related neurodegeneration. Cellular and Molecular Life Sciences 78 (5), 2231–2245. Available at https://doi.org/10.1007/s00018-020-03643-3.

Lue, L.F., Walker, D.G., Brachova, L., Beach, T.G., Rogers, J., Schmidt, A.M., et al., 2001. Involvement of microglial receptor for advanced glycation endproducts (RAGE)in Alzheimer's disease: identification of a cellular activation mechanism. Experimental Neurology 171 (1), 29–45. Available at https://doi.org/10.1006/exnr.2001.7732.

Lv, C., Huang, S., Wang, Y., Hu, Z., Zhao, G., Ma, C., et al., 2021. Chicoric acid encapsulated within ferritin inhibits tau phosphorylation by regulating AMPK and GluT1 signaling cascade. Journal of Functional Foods 86, 104681. Available at https://doi.org/10.1016/j.jff.2021.104681.

Maïza, A., Chantepie, S., Vera, C., Fifre, A., Huynh, M.B., Stettler, O., et al., 2018. The role of heparan sulfates in protein aggregation and their potential impact on neurodegeneration. FEBS Letters 592 (23), 3806–3818. Available at https://doi.org/10.1002/1873-3468.13082.

Makarava, N., Baskakov, I.V., 2023. Role of sialylation of N-linked glycans in prion pathogenesis. Cell and Tissue Research 392 (1), 201–214. Available at https://doi.org/10.1007/s00441-022-03584-2.

Makarava, N., Chang, J.C.-Y., Baskakov, I.V., 2020. Region-specific sialylation pattern of prion strains provides novel insight into prion neurotropism. International Journal of Molecular Sciences 21 (3). Available at https://doi.org/10.3390/ijms21030828.

Makarava, N., Mychko, O., Chang, J.C.Y., Molesworth, K., Baskakov, I.V., 2021. The degree of astrocyte activation is predictive of the incubation time to prion disease. Acta Neuropathologica Communications 9 (1). Available at https://doi.org/10.1186/s40478-021-01192-9.

Marotta, N.P., Lin, Y.H., Lewis, Y.E., Ambroso, M.R., Zaro, B.W., Roth, M.T., et al., 2015. O-GlcNAc modification blocks the aggregation and toxicity of the protein α-synuclein associated with Parkinson's disease. Nature Chemistry 7 (11), 913–920. Available at https://doi.org/10.1038/nchem.2361.

Martínez-Menárguez, J.Á., Martínez-Alonso, E., Cara-Esteban, M., Tomás, M., 2021. Focus on the small GTPase Rab1: a key player in the pathogenesis of Parkinson's disease. International Journal of Molecular Sciences 22 (21). Available at https://doi.org/10.3390/ijms222112087.

McDade, E., Voytyuk, I., Aisen, P., Bateman, R.J., Carrillo, M.C., De Strooper, B., et al., 2021. The case for low-level BACE1 inhibition for the prevention of Alzheimer disease. Nature Reviews Neurology 17 (11), 703–714. Available at https://doi.org/10.1038/s41582-021-00545-1.

McEver, R.P., 2015. Selectins: initiators of leucocyte adhesion and signalling at the vascular wall. Cardiovascular Research 107 (3), 331–339. Available at https://doi.org/10.1093/cvr/cvv154.

McFarthing, K., Buff, S., Rafaloff, G., Pitzer, K., Fiske, B., Navangul, A., et al., 2024. Parkinson's disease drug therapies in the clinical trial pipeline: 2024 update. Journal of Parkinson's Disease 14 (5), 899–912. Available at https://doi.org/10.3233/JPD-240272.

Mirabile, I., Jat, P.S., Brandner, S., Collinge, J., 2015. Identification of clinical target areas in the brainstem of prion-infected mice. Neuropathology and Applied Neurobiology 41 (5), 613–630. Available at https://doi.org/10.1111/nan.12189.

Mishra, R., Elgland, M., Begum, A., Fyrner, T., Konradsson, P., Nyström, S., et al., 2019. Impact of N-glycosylation site variants during human PrP aggregation and fibril nucleation. Biochimica et Biophysica Acta (BBA) - Proteins and Proteomics 1867 (10), 909–921. Available at https://doi.org/10.1016/j.bbapap.2019.03.010.

Morales, M.M., Pratt, M.R., 2024. The post-translational modification O-GlcNAc is a sensor and regulator of metabolism. Open Biology 14 (10). Available at https://doi.org/10.1098/rsob.240209.

Moremen, K.W., Tiemeyer, M., Nairn, A.V., 2012. Vertebrate protein glycosylation: diversity, synthesis and function. Nature Reviews. Molecular Cell Biology. 13 (7), 448–462. Available at https://doi.org/10.1038/nrm3383.

Morris, M., Knudsen, G.M., Maeda, S., Trinidad, J.C., Ioanoviciu, A., Burlingame, A.L., et al., 2015. Tau post-translational modifications in wild-type and human amyloid precursor protein transgenic mice. Nature Neuroscience 18 (8), 1183–1189. Available at https://doi.org/10.1038/nn.4067.

Muller, D., Wang, C., Skibo, G., Toni, N., Cremer, H., Calaora, V., et al., 1996. PSA–NCAM Is required for activity-induced synaptic plasticity. Neuron 17 (3), 413–422. Available at https://doi.org/10.1016/s0896-6273(00)80174-9.

Muronets, V.I., Kudryavtseva, S.S., Kurochkina, L.P., Leisi, E.V., Stroylova, Y.Y., Schmalhausen, E.V., 2025. Factors affecting pathological amyloid protein transformation: from post-translational modifications to chaperones. Biochemistry (Moscow) 90 (1), S164. Available at https://doi.org/10.1134/S0006297924604003.

Myers, T.D., Li, Y., Taiclet, S., Cabada-Aguirre, P., Kuti, E., McClure, K., et al., 2024. Neuromuscular dysfunction and pathogenesis in triosephosphate isomerase deficiency. Scientific Reports 14 (1). Available at https://doi.org/10.1038/s41598-024-69618-7.

Narimatsu, Y., Joshi, H.J., Nason, R., Van Coillie, J., Karlsson, R., Sun, L., et al., 2019. An atlas of human glycosylation pathways enables display of the human glycome by gene engineered cells. Molecular Cell 75 (2), 394. Available at https://doi.org/10.1016/j.molcel.2019.05.017.

Nissley, D.A., Jiang, Y., Trovato, F., Sitarik, I., Narayan, K.B., To, P., et al., 2022. Universal protein misfolding intermediates can bypass the proteostasis network and remain soluble and less functional. Nature Communications 13 (1). Available at https://doi.org/10.1038/s41467-022-30548-5.

Oláh, J., Orosz, F., Keserü, G.M., Kovári, Z., Kovács, J., Hollán, S., et al., 2002. Triosephosphate isomerase deficiency: a neurodegenerative misfolding disease. Biochemical Society Transactions 30 (2), 30–38. Available at https://doi.org/10.1042/bst0300030.

Otazu, G.K., Dayyani, M., Badie, B., 2021. Role of RAGE and its ligands on inflammatory responses to brain tumors. Frontiers in Cellular Neuroscience 15, 770472. Available at https://doi.org/10.3389/fncel.2021.770472.

Ott, A., Stolk, R.P., Hofman, A., van Harskamp, F., Grobbee, D.E., Breteler, M.M.B., 1996. Association of diabetes mellitus and dementia: the rotterdam study. Diabetologia 39 (11), 1392–1397. Available at https://doi.org/10.1007/s001250050588.

Ott, A., Stolk, R.P., Van Harskamp, F., Pols, H.A.P., Hofman, A., Breteler, M.M.B., 1999. Diabetes mellitus and the risk of dementia: the rotterdam study. Neurology 53 (9), 1937–1942. Available at https://doi.org/10.1212/wnl.53.9.1937.

Padmaraju, V., Bhaskar, J.J., Prasada Rao, U.J.S., Salimath, P.V., Rao, K.S., 2011. Role of advanced glycation on aggregation and DNA binding properties of α-synuclein. Journal of Alzheimer's Disease 24 (2), 211–221. Available at https://doi.org/10.3233/JAD-2011-101965.

Palmigiano, A., Barone, R., Sturiale, L., Sanfilippo, C., Bua, R.O., Romeo, D.A., et al., 2016. CSF N-glycoproteomics for early diagnosis in Alzheimer's disease. Journal of Proteomics 131, 29–37. Available at https://doi.org/10.1016/j.jprot.2015.10.006.

Panes, J.D., Saavedra, P., Pineda, B., Escobar, K., Cuevas, M.E., Moraga-Cid, G., et al., 2021. PrPC as a transducer of physiological and pathological signals. Frontiers in Molecular Neuroscience 14, 762918. Available at https://doi.org/10.3389/fnmol.2021.762918.

Paquet, S., Langevin, C., Chapuis, J., Jackson, G.S., Laude, H., Vilette, D., 2007. Efficient dissemination of prions through preferential transmission to nearby cells. Journal of General Virology 88 (2), 706–713. Available at https://doi.org/10.1099/vir.0.82336-0.

Park, J.-S., Ji, I.J., Kim, D.-H., An, H.J., Yoon, S.-Y., 2016. The Alzheimer's disease-associated R47H variant of TREM2 has an altered glycosylation pattern and protein stability. Frontiers in neuroscience 10. Available at https://doi.org/10.3389/fnins.2016.00618.

Park, J., Ha, H.J., Chung, E.S., Baek, S.H., Cho, Y., Kim, H.K., et al., 2021. O-GlcNAcylation ameliorates the pathological manifestations of Alzheimer's disease by inhibiting necroptosis. Science Advances 7 (3). Available at https://doi.org/10.1126/sciadv.abd3207.

Pellegrini, C., D'Antongiovanni, V., Miraglia, F., Rota, L., Benvenuti, L., Di Salvo, C., et al., 2022. Enteric α-synuclein impairs intestinal epithelial barrier through caspase-1-inflammasome signaling in Parkinson's disease before brain pathology. npj Parkinson's Disease 8 (1). Available at https://doi.org/10.1038/s41531-021-00263-x.

Perdivara, I., Petrovich, R., Allinquant, B., Deterding, L.J., Tomer, K.B., Przybylski, M., 2009. Elucidation of O-glycosylation structures of the β-amyloid precursor protein by liquid chromatography–mass spectrometry using electron transfer dissociation and collision induced dissociation. Journal of Proteome Research 8 (2), 631–642. Available at https://doi.org/10.1021/pr800758g.

Perrone, A., Giovino, A., Benny, J., Martinelli, F., 2020. Advanced glycation end products (AGEs): biochemistry, signaling, analytical methods, and epigenetic effects. Oxidative Medicine and Cellular Longevity 2020, 3818196.

Pickrell, A.M., Youle, R.J., 2015. The roles of PINK1, Parkin, and mitochondrial fidelity in parkinson's disease. Neuron 85 (2), 257–273. Available at https://doi.org/10.1016/j.neuron.2014.12.007.

Pinho, T.S., Correia, S.C., Perry, G., Ambrósio, A.F., Moreira, P.I., 2019. Diminished O-GlcNAcylation in Alzheimer's disease is strongly correlated with mitochondrial anomalies. Biochimica et Biophysica Acta - Molecular Basis of Disease 1865 (8), 2048–2059. Available at https://doi.org/10.1016/j.bbadis.2018.10.037.

Pradeep, P., Kang, H., Lee, B., 2023. Glycosylation and behavioral symptoms in neurological disorders. Translational Psychiatry 13 (1). Available at https://doi.org/10.1038/s41398-023-02446-x.

Prusiner, S.B., 1989. Creutzfeldt-Jakob disease and scrapie prions. Alzheimer Disease and Associated Disorders 3 (1-2), 52–78. Available at https://doi.org/10.1097/00002093-198903010-00007.

Quartu, M., Serra, M.P., Boi, M., Ibba, V., Melis, T., Del Fiacco, M., 2008. Polysialylated-neural cell adhesion molecule (PSA-NCAM) in the human trigeminal ganglion and brainstem at prenatal and adult ages. BMC Neuroscience 9. Available at https://doi.org/10.1186/1471-2202-9-108.

Rahbar, S., Blumenfeld, O., Ranney, H.M., 1969. Studies of an unusual hemoglobin in patients with diabetes mellitus. Biochemical and Biophysical Research Communications 36 (5), 838–843.

Rawal, P., Zhao, L., 2021. Sialometabolism in brain health and Alzheimer's disease. Frontiers in Neuroscience 15, 648617. Available at https://doi.org/10.3389/fnins.2021.648617.

Rebelo, A.L., Gubinelli, F., Roost, P., Jan, C., Brouillet, E., Van Camp, N., et al., 2021. Complete spatial characterisation of N-glycosylation upon striatal neuroinflammation in the rodent brain. Journal of Neuroinflammation 18 (1). Available at https://doi.org/10.1186/s12974-021-02163-6.

Rebelo, A.L., Drake, R.R., Marchetti-Deschmann, M., Saldova, R., Pandit, A., 2024. Changes in tissue protein N-glycosylation and associated molecular signature occur in the human Parkinsonian brain in a region-specific manner. PNAS Nexus 3 (1). Available at https://doi.org/10.1093/pnasnexus/pgad439.

Reily, C., Stewart, T.J., Renfrow, M.B., Novak, J., 2019. Glycosylation in health and disease. Nature Reviews Nephrology 15 (6), 346–366. Available at https://doi.org/10.1038/s41581-019-0129-4.

Rexach, J.E., Clark, P.M., Mason, D.E., Neve, R.L., Peters, E.C., Hsieh-Wilson, L.C., 2012. Dynamic O-GlcNAc modification regulates CREB-mediated gene expression and memory formation. Nature Chemical Biology 8 (3), 253–261. Available at https://doi.org/10.1038/nchembio.770.

Robinson, S.W., Nugent, M.L., Dinsdale, D., Steinert, J.R., 2014. Prion protein facilitates synaptic vesicle release by enhancing release probability. Human Molecular Genetics 23 (17), 4581–4596. Available at https://doi.org/10.1093/hmg/ddu171.

Roland, B.P., Zeccola, A.M., Larsen, S.B., Amrich, C.G., Talsma, A.D., Stuchul, K.A., et al., 2016. Structural and genetic studies demonstrate neurologic dysfunction in triosephosphate isomerase deficiency is associated with impaired synaptic vesicle dynamics. PLoS Genetics 12 (3). Available at https://doi.org/10.1371/journal.pgen.1005941.

Roller, R.F., Malik, A., Carillo, M.A., Garg, M., Rella, A., Raulf, M.K., et al., 2020. Semisynthesis of functional glycosylphosphatidylinositol-anchored proteins. Angewandte Chemie - International Edition 59 (29), 12035–12040. Available at https://doi.org/10.1002/anie.202002479.

Roth, J., 2002. Protein N-glycosylation along the secretory pathway: relationship to organelle topography and function, protein quality control, and cell interactions. Chemical Reviews 102 (2), 285–303. Available at https://doi.org/10.1021/cr000423j.

Rudd, P.M., Merry, A.H., Wormald, M.R., Dwek, R.A., 2002. Glycosylation and prion protein. Current Opinion in Structural Biology 12 (5), 578–586. Available at https://doi.org/10.1016/S0959-440X(02)00377-9.

Ruiz-Canada, C., Kelleher, D.J., Gilmore, R., 2009. Cotranslational and posttranslational N-glycosylation of polypeptides by distinct mammalian OST isoforms. Cell 136 (2), 272–283. Available at https://doi.org/10.1016/j.cell.2008.11.047.

Rungratanawanich, W., Qu, Y., Wang, X., Essa, M.M., Song, B.J., 2021. Advanced glycation end products (AGEs) and other adducts in aging-related diseases and alcohol-mediated tissue injury. Experimental and Molecular Medicine 53 (2), 168–188. Available at https://doi.org/10.1038/s12276-021-00561-7.

Rush, J.S., 2015. Role of flippases in protein glycosylation in the endoplasmic reticulum. Lipid Insights 2015, 45–53. Available at https://doi.org/10.4137/Lpi.s31784.

Russell, A.C., Šimurina, M., Garcia, M.T., Novokmet, M., Wang, Y., Rudan, I., et al., 2017. The N-glycosylation of immunoglobulin G as a novel biomarker of Parkinson's disease. Glycobiology 27 (5), 501–510. Available at https://doi.org/10.1093/glycob/cwx022.

Ryan, P., Xu, M., Davey, A.K., Danon, J.J., Mellick, G.D., Kassiou, M., et al., 2019. O-GlcNAc modification protects against protein misfolding and aggregation in neurodegenerative disease. ACS Chemical Neuroscience 10 (5), 2209–2221. Available at https://doi.org/10.1021/acschemneuro.9b00143.

Santra, M., Dill, K.A., de Graff, A.M.R., 2019. Proteostasis collapse is a driver of cell aging and death. Proceedings of the National Academy of Sciences of the United States of America 116 (44), 22173–22178. Available at https://doi.org/10.1073/pnas.1906592116.

Sanyal, S., Frank, C.G., Menon, A.K., 2008. Distinct flippases translocate glycerophospholipids and oligosaccharide diphosphate dolichols across the endoplasmic reticulum. Biochemistry 47 (30), 7937–7946. Available at https://doi.org/10.1021/bi800723n.

Sato, Y., Naito, Y., Grundke-Iqbal, I., Iqbal, K., Endo, T., 2001. Analysis of N -glycans of pathological tau: possible occurrence of aberrant processing of tau in Alzheimer's disease. FEBS Letters 496 (2-3), 152–160. Available at https://doi.org/10.1016/s0014-5793(01)02421-8.

Scallon, B.J., Tam, S.H., McCarthy, S.G., Cai, A.N., Raju, T.S., 2007. Higher levels of sialylated Fc glycans in immunoglobulin G molecules can adversely impact functionality. Molecular Immunology 44 (7), 1524–1534. Available at https://doi.org/10.1016/j.molimm.2006.09.005.

Schedin-Weiss, S., Winblad, B., Tjernberg, L.O., 2014. The role of protein glycosylation in Alzheimer disease. FEBS Journal 281 (1), 46–62. Available at https://doi.org/10.1111/febs.12590.

Scheper, A.F., Schofield, J., Bohara, R., Ritter, T., Pandit, A., 2023. Understanding glycosylation: regulation through the metabolic flux of precursor pathways. Biotechnology Advances 67, 108184. Available at https://doi.org/10.1016/j.biotechadv.2023.108184.

Scherer, T., Sakamoto, K., Buettner, C., 2021. Brain insulin signalling in metabolic homeostasis and disease. Nature Reviews Endocrinology 17 (8), 468–483. Available at https://doi.org/10.1038/s41574-021-00498-x.

Schilling, S., Mehr, A., Ludewig, S., j. Stephan, M., Zimmermann, A., August, P., et al., 2017. APLP1 is a synaptic cell adhesion molecule, supporting maintenance of dendritic spines and basal synaptic transmission. Journal of Neuroscience 37 (21), 5345–5365. Available at https://doi.org/10.1523/JNEUROSCI.1875-16.2017.

Schilling, K.M., Tao, L., Wu, B., Kiblen, J.T.M., Ubilla-Rodriguez, N.C., Pushie, M.J., et al., 2020. Both N-terminal and C-terminal histidine residues of the prion protein are essential for copper coordination and neuroprotective self-regulation. Journal of Molecular Biology 432 (16), 4408–4425. Available at https://doi.org/10.1016/j.jmb.2020.05.020.

Schilling, K.M., Jorwal, P., Ubilla-Rodriguez, N.C., Assafa, T.E., Gatdula, J.R.P., Vultaggio, J.S., et al., 2023. N-glycosylation is a potent regulator of prion protein neurotoxicity. Journal of Biological Chemistry 299 (9). Available at https://doi.org/10.1016/j.jbc.2023.105101.

Schmidt, J.W., Catterall, W.A., 1987. Palmitylation, sulfation, and glycosylation of the alpha subunit of the sodium channel. Role of post-translational modifications in channel assembly. The Journal of

Biological Chemistry 262 (28), 13713–13723. Available at https://doi.org/10.1016/s0021-9258(19)76485-0.

Schneider, J.S., Singh, G., 2022. Altered expression of glycobiology-related genes in Parkinson's disease brain. Frontiers in Molecular Neuroscience 15, 1078854. Available at https://doi.org/10.3389/fnmol.2022.1078854.

Sevillano, A.M., Aguilar-Calvo, P., Kurt, T.D., Lawrence, J.A., Soldau, K., Nam, T.H., et al., 2020. Prion protein glycans reduce intracerebral fibril formation and spongiosis in prion disease. Journal of Clinical Investigation 130 (3), 1350–1362. Available at https://doi.org/10.1172/JCI131564.

Shaheen, N., Shaheen, A., Osama, M., Nashwan, A.J., Bharmauria, V., Flouty, O., 2024. MicroRNAs regulation in Parkinson's disease, and their potential role as diagnostic and therapeutic targets. npj Parkinson's Disease 10 (1). Available at https://doi.org/10.1038/s41531-024-00791-2.

Shahmoradian, S.H., Lewis, A.J., Genoud, C., Hench, J., Moors, T.E., Navarro, P.P., et al., 2019. Lewy pathology in Parkinson's disease consists of crowded organelles and lipid membranes. Nature Neuroscience 22 (7), 1099–1109. Available at https://doi.org/10.1038/s41593-019-0423-2.

Shahraz, A., Kopatz, J., Mathy, R., Kappler, J., Winter, D., Kapoor, S., et al., 2015. Anti-inflammatory activity of low molecular weight polysialic acid on human macrophages. Scientific Reports 5. Available at https://doi.org/10.1038/srep16800.

Shaw, B.C., Katsumata, Y., Simpson, J.F., Fardo, D.W., Estus, S., 2021. Analysis of genetic variants associated with levels of immune modulating proteins for impact on alzheimer's disease risk reveal a potential role for siglec14. Genes 12 (7). Available at https://doi.org/10.3390/genes12071008.

Shimura, H., Schlossmacher, M.G., Hattori, N., Frosch, M.P., Trockenbacher, A., Schneider, R., et al., 2001. Ubiquitination of a new form of α-synuclein by parkin from human brain: implications for Parkinson's disease. American Association for the Advancement of Science, United States Science 293 (5528), 263–269. Available at https://doi.org/10.1126/science.1060627.

Siddiqui, S.S., Matar, R., Merheb, M., Hodeify, R., Vazhappilly, C.G., Marton, J., et al., 2019. Siglecs in brain function and neurological disorders. Cells 8 (10). Available at https://doi.org/10.3390/cells8101125.

Siew, J.J., Chern, Y., Khoo, K.H., Angata, T., 2023. Roles of siglecs in neurodegenerative diseases. Molecular Aspects of Medicine 90. Available at https://doi.org/10.1016/j.mam.2022.101141.

Šimunić-Briški, N., Dukarić, V., Očić, M., Madžar, T., Vinicki, M., Frkatović-Hodžić, A., et al., 2024. Regular moderate physical exercise decreases glycan age index of biological age and reduces inflammatory potential of Immunoglobulin G. Glycoconjugate Journal 41 (1), 67–76. Available at https://doi.org/10.1007/s10719-023-10144-5.

Smith, B.A.H., Bertozzi, C.R., 2021. Author correction: the clinical impact of glycobiology: targeting selectins, siglecs and mammalian glycans. Nature Reviews Drug Discovery 20 (3), 244. Available at https://doi.org/10.1038/s41573-021-00160-1.

Somerville, R.A., Hamilton, S., Fernie, K., 2005. Transmissible spongiform encephalopathy strain, PrP genotype and brain region all affect the degree of glycosylation of PrPSc. The Journal of General Virology 86 (Pt 1), 241–246. Available at https://doi.org/10.1099/vir.0.80251-0.

Sparvero, L.J., Asafu-Adjei, D., Kang, R., Tang, D., Amin, N., Im, J., et al., 2009. RAGE (Receptor for advanced glycation endproducts), RAGE ligands, and their role in cancer and inflammation. Journal of Translational Medicine 7. Available at https://doi.org/10.1186/1479-5876-7-17.

Spiers, J.G., Steinert, J.R., 2021. Nitrergic modulation of ion channel function in regulating neuronal excitability. Channels 15 (1), 666–679. Available at https://doi.org/10.1080/19336950.2021.2002594.

Spiers, J.G., Breda, C., Robinson, S., Giorgini, F., Steinert, J.R., 2019. Drosophila Nrf2/Keap1 mediated redox signaling supports synaptic function and longevity and impacts on circadian activity. Frontiers in Molecular Neuroscience 12, 86. Available at https://doi.org/10.3389/fnmol.2019.00086.

Spiers, J.G., Cortina Chen, H.J., Barry, T.L., Bourgognon, J.M., Steinert, J.R., 2022. Redox stress and metal dys-homeostasis appear as hallmarks of early prion disease pathogenesis in mice. Free Radical Biology and Medicine 192, 182–190. Available at https://doi.org/10.1016/j.freeradbiomed.2022.09.025.

Spiers, J.G., Chen, H.-J.C., Steinert, J.R., 2023. Redox mechanisms and their pathological role in prion diseases: the road to ruin. PLoS Pathogens 19 (4). Available at https://doi.org/10.1371/journal.ppat.1011309.

Spiro, R.G., 2002. Protein glycosylation: nature, distribution, enzymatic formation, and disease implications of glycopeptide bonds. Glycobiology 12 (4), 43R. Available at https://doi.org/10.1093/glycob/12.4.43r.

Srivastava, S., Katorcha, E., Makarava, N., Barrett, J.P., Loane, D.J., Baskakov, I.V., 2018. Inflammatory response of microglia to prions is controlled by sialylation of PrPSc. Scientific Reports 8 (1). Available at https://doi.org/10.1038/s41598-018-29720-z.

Stone, A., Cujic, O., Rowlett, A., Aderhold, S., Savage, E., Graham, B., et al., 2023. Triose-phosphate isomerase deficiency is associated with a dysregulation of synaptic vesicle recycling in *Drosophila melanogaster*. Frontiers in Synaptic Neuroscience 15, 1124061. Available at https://doi.org/10.3389/fnsyn.2023.1124061.

Sui, Y.T., Bullock, K.M., Erickson, M.A., Zhang, J., Banks, W.A., 2014. Alpha synuclein is transported into and out of the brain by the blood-brain barrier. Peptides 62, 197–202. Available at https://doi.org/10.1016/j.peptides.2014.09.018.

Sumsuzzman, D.M., Choi, J., Jin, Y., Hong, Y., 2021. Neurocognitive effects of melatonin treatment in healthy adults and individuals with Alzheimer's disease and insomnia: a systematic review and meta-analysis of randomized controlled trials. Neuroscience and Biobehavioral Reviews 127, 459–473. Available at https://doi.org/10.1016/j.neubiorev.2021.04.034.

Sun, R.C., Young, L.E.A., Bruntz, R.C., Markussen, K.H., Zhou, Z., Conroy, L.R., et al., 2021. Brain glycogen serves as a critical glucosamine cache required for protein glycosylation. Cell Metabolism 33 (7), 1404. Available at https://doi.org/10.1016/j.cmet.2021.05.003.

Tajes, M., Guivernau, B., Ramos-Fernández, E., Bosch-Morató, M., Palomer, E., Guix, F.X., et al., 2013. The pathophysiology of triose phosphate isomerase dysfunction in Alzheimer's disease. Histology and Histopathology 28 (1), 43–51. Available at https://doi.org/10.14670/HH-28.43.

Takuma, K., Fang, F., Zhang, W., Yan, S., Fukuzaki, E., Du, H., et al., 2009. RAGE-mediated signaling contributes to intraneuronal transport of amyloid-β and neuronal dysfunction. Proceedings of the National Academy of Sciences of the United States of America 106 (47), 20021–20026. Available at https://doi.org/10.1073/pnas.0905686106.

Taylor, E.W., Wang, K., Nelson, A.R., Bredemann, T.M., Fraser, K.B., Clinton, S.M., et al., 2014. O-GlcNAcylation of AMPA receptor GluA2 is associated with a novel form of long-term depression at hippocampal synapses. Journal of Neuroscience 34 (1), 10–21. Available at https://doi.org/10.1523/JNEUROSCI.4761-12.2014.

Tena, J., Maezawa, I., Barboza, M., Wong, M., Zhu, C., Alvarez, M.R., et al., 2022. Regio-specific N-glycome and N-glycoproteome map of the elderly human brain with and without Alzheimer's disease. Molecular and Cellular Proteomics 21 (11). Available at https://doi.org/10.1016/j.mcpro.2022.100427.

Tijardović, M., Marijančević, D., Bok, D., Kifer, D., Lauc, G., Gornik, O., et al., 2019. Intense physical exercise induces an anti-inflammatory change in IgG N-glycosylation profile. Frontiers in Physiology 10, 1522. Available at https://doi.org/10.3389/fphys.2019.01522.

Tosh, N., Quadrelli, S., Galloway, G., Mountford, C., 2019. Two new fucose-α (1–2)-glycans assigned in the healthy human brain taking the number to seven. Scientific Reports 9 (1). Available at https://doi.org/10.1038/s41598-019-54933-1.

Tseng, P.T., Zeng, B.Y., Chen, Y.W., Yang, C.P., Su, K.P., Chen, T.Y., et al., 2022. The dose and duration-dependent association between melatonin treatment and overall cognition in Alzheimer's dementia: a network metaanalysis of randomized placebo-controlled trials. Current Neuropharmacology 20 (10), 1816–1833. Available at https://doi.org/10.2174/1570159X20666220420122322.

Tuzi, N.L., Cancellotti, E., Baybutt, H., Blackford, L., Bradford, B., Plinston, C., et al., 2008. Host PrP glycosylation: a major factor determining the outcome of prion infection. PLoS Biology 6 (4), 872–882. Available at https://doi.org/10.1371/journal.pbio.0060100.

Underwood, P.C., Zhang, L., Mohr, D.C., Prentice, J.C., Nelson, R.E., Budson, A.E., et al., 2024. Glycated hemoglobin A 1c time in range and dementia in older adults with diabetes. JAMA Network Open 7 (8), e2425354. Available at https://doi.org/10.1001/jamanetworkopen.2024.25354.

Urano, Y., Takahachi, M., Higashiura, R., Fujiwara, H., Funamoto, S., Imai, S., et al., 2020. Curcumin derivative GT863 inhibits amyloid-beta production via inhibition of protein N-glycosylation. Cells 9 (2), 349. Available at https://doi.org/10.3390/cells9020349.

Van Dyck, C.H., Swanson, C.J., Aisen, P., Bateman, R.J., Chen, C., Gee, M., et al., 2023. Lecanemab in early Alzheimer's disease. New England Journal of Medicine 388 (1), 9–21. Available at https://doi.org/10.1056/NEJMoa2212948.

Váradi, C., Nehéz, K., Hornyák, O., Viskolcz, B., Bones, J., 2019. Serum N-glycosylation in Parkinson's disease: a novel approach for potential alterations. Molecules 24 (12). Available at https://doi.org/10.3390/molecules24122220.

Varki, A., 2017. Biological roles of glycans. Glycobiology 27 (1), 3–49. Available at https://doi.org/10.1093/glycob/cww086.

Vassileff, N., Cheng, L., Hill, A.F., 2020. Extracellular vesicles - propagators of neuropathology and sources of potential biomarkers and therapeutics for neurodegenerative diseases. Journal of Cell Science 133 (23). Available at https://doi.org/10.1242/jcs.243139.

Videira, P.A.Q., Castro-Caldas, M., 2018. Linking glycation and glycosylation with inflammation and mitochondrial dysfunction in Parkinson's disease. Frontiers in Neuroscience 12, 381. Available at https://doi.org/10.3389/fnins.2018.00381.

Vitorakis, N., Piperi, C., 2024. Pivotal role of AGE-RAGE axis in brain aging with current interventions. Ageing Research Reviews 100, 102429. Available at https://doi.org/10.1016/j.arr.2024.102429.

Vorberg, I., Priola, S.A., 2002. Molecular basis of scrapie strain glycoform variation. Journal of Biological Chemistry 277 (39), 36775–36781. Available at https://doi.org/10.1074/jbc.M206865200.

Vulevic, J., Juric, A., Tzortzis, G., Gibson, G.R., 2013. A mixture of trans-galactooligosaccharides reduces markers of metabolic syndrome and modulates the fecal microbiota and immune function of overweight adults 1-3. Journal of Nutrition 143 (3), 324–331. Available at https://doi.org/10.3945/jn.112.166132.

Wakabayashi, K., Takahashi, H., Takeda, S., Ohama, E., Ikuta, F., 1988. Parkinson's disease: the presence of Lewy bodies in Auerbach's and Meissner's plexuses. Acta Neuropathologica 76 (3), 217–221. Available at https://doi.org/10.1007/bf00687767.

Walmsley, A.R., Zeng, F., Hooper, N.M., 2001. Membrane topology influences N-glycosylation of the prion protein. The EMBO Journal 20 (4), 703–712.

Wang, W., 2023. Editorial: orthodox vs paradox: the roles of glycomics, genetics and beyond in immunity, immune disorders and glycomedicine. Frontiers in Immunology 14, 1305552. Available at https://doi.org/10.3389/fimmu.2023.1305552.

Wang, Y., Neumann, H., 2010. Alleviation of neurotoxicity by microglial human siglec-11. The Journal of Neuroscience 30 (9), 3482–3488. Available at https://doi.org/10.1523/jneurosci.3940-09.2010.

Wang, Z., Udeshi, N.D., O'Malley, M., Shabanowitz, J., Hunt, D.F., Hart, G.W., 2010. Enrichment and site mapping of O-linked N-acetylglucosamine by a combination of chemical/enzymatic tagging, photochemical cleavage, and electron transfer dissociation mass spectrometry. Molecular and Cellular Proteomics 9 (1), 153–160. Available at https://doi.org/10.1074/mcp.M900268-MCP200.

Wang, Q., Groenendyk, J., Michalak, M., 2015. Glycoprotein quality control and endoplasmic reticulum stress. Molecules 20 (8), 13689–13704. Available at https://doi.org/10.3390/molecules200813689.

Wang, N., Li, R., Feng, B., Cheng, Y., Guo, Y., Qian, H., 2022. Chicoric acid prevents neuroinflammation and neurodegeneration in a mouse Parkinson's disease model: immune response and transcriptome profile of the spleen and colon. International Journal of Molecular Sciences 23 (4), 2031. Available at https://doi.org/10.3390/ijms23042031.

Watson, C.N., Belli, A., Di Pietro, V., 2019. Small non-coding RNAs: new class of biomarkers and potential therapeutic targets in neurodegenerative disease. Frontiers in Genetics 10, 364. Available at https://doi.org/10.3389/fgene.2019.00364.

Wernette-Hammond, M.E., Lauer, S.J., Corsini, A., Walker, D., Taylor, J.M., Rall, S.C., 1989. Glycosylation of human apolipoprotein E. The Carbohydrate Attachment site is Threonine 194. Journal of Biological Chemistry 264 (15), 9094–9101.

Wheatley, E.G., Albarran, E., White, C.W., Bieri, G., Sanchez-Diaz, C., Pratt, K., et al., 2019. Neuronal O-GlcNAcylation improves cognitive function in the aged mouse brain. Current Biology 29 (20), 3359. Available at https://doi.org/10.1016/j.cub.2019.08.003.

Wilkinson, H., Thomsson, K.A., Rebelo, A.L., Hilliard, M., Pandit, A., Rudd, P.M., et al., 2021. The O-glycome of human nigrostriatal tissue and its alteration in Parkinson's disease. Journal of Proteome Research 20 (8), 3913–3924. Available at https://doi.org/10.1021/acs.jproteome.1c00219.

Wilson, C.M., Farmery, M.R., Bulleid, N.J., 2000. Pivotal role of calnexin and mannose trimming in regulating the endoplasmic reticulum-associated degradation of major histocompatibility complex class I heavy chain. The Journal of Biological Chemistry 275 (28), 21224–21232.

Wißfeld, J., Abou Assale, T., Cuevas-Rios, G., Liao, H., Neumann, H., 2024. Therapeutic potential to target sialylation and SIGLECs in neurodegenerative and psychiatric diseases. Frontiers in Neurology 15, 1330874. Available at https://doi.org/10.3389/fneur.2024.1330874.

Wu, E.L., Qi, Y., Park, S., Mallajosyula, S.S., MacKerell, A.D., Klauda, J.B., et al., 2015. Insight into early-stage unfolding of GPI-anchored human prion protein. Biophysical Journal 109 (10), 2090–2100. Available at https://doi.org/10.1016/j.bpj.2015.10.009.

Xu, Q., Bernardo, A., Walker, D., Kanegawa, T., Mahley, R.W., Huang, Y., 2006. Profile and regulation of apolipoprotein E (ApoE) expression in the CNS in mice with targeting of green fluorescent protein gene to the ApoE locus. Journal of Neuroscience 26 (19), 4985–4994. Available at https://doi.org/10.1523/JNEUROSCI.5476-05.2006.

Xu, L., Yu, H., Sun, H., Hu, B., Geng, Y., 2020. Dietary melatonin therapy alleviates the lamina cribrosa damages in patients with mild cognitive impairments: a double-blinded, randomized controlled study. Medical Science Monitor 26. Available at https://doi.org/10.12659/msm.923232.

Xu, M., Jin, H., Wu, Z., Han, Y., Chen, J., Mao, C., et al., 2022. Mass spectrometry-based analysis of serum N-glycosylation changes in patients with Parkinson's disease. ACS Chemical Neuroscience 13 (12), 1719–1726. Available at https://doi.org/10.1021/acschemneuro.2c00264.

Yamagishi, S.I., 2019. Role of advanced glycation endproduct (AGE)-receptor for advanced glycation endproduct (RAGE) axis in cardiovascular disease and its therapeutic intervention. Circulation Journal 83 (9), 1822–1828. Available at https://doi.org/10.1253/circj.CJ-19-0618.

Yamazaki, Y., Zhao, N., Caulfield, T.R., Liu, C.C., Bu, G., 2019. Apolipoprotein E and Alzheimer disease: pathobiology and targeting strategies. Nature Reviews Neurology 15 (9), 501–518. Available at https://doi.org/10.1038/s41582-019-0228-7.

Yang, X., Qian, K., 2017. Protein O-GlcNAcylation: emerging mechanisms and functions. Nature Reviews Molecular Cell Biology 18 (7), 452–465. Available at https://doi.org/10.1038/nrm.2017.22.

Yang, M., Wu, C., Zhang, T., Shi, L., Li, J., Liang, H., et al., 2022. Chicoric acid: natural occurrence, chemical synthesis, biosynthesis, and their bioactive effects. Frontiers in Chemistry 10, 888673. Available at https://doi.org/10.3389/fchem.2022.888673.

Yazaki, M., Tagawa, K., Maruyama, K., Sorimachi, H., Tsuchiya, T., Ishiura, S., et al., 1996. Mutation of potential N-linked glycosylation sites in the Alzheimer's disease amyloid precursor protein (APP). Neuroscience Letters 221 (1), 57–60. Available at https://doi.org/10.1016/s0304-3940(96)13285-7.

Yeini, E., Ofek, P., Pozzi, S., Albeck, N., Ben-Shushan, D., Tiram, G., et al., 2021. P-selectin axis plays a key role in microglia immunophenotype and glioblastoma progression. Nature Communications 12 (1). Available at https://doi.org/10.1038/s41467-021-22186-0.

Yuzwa, S.A., Yadav, A.K., Skorobogatko, Y., Clark, T., Vosseller, K., Vocadlo, D.J., 2011. Mapping O-GlcNAc modification sites on tau and generation of a site-specific O-GlcNAc tau antibody. Amino Acids 40 (3), 857–868. Available at https://doi.org/10.1007/s00726-010-0705-1.

Yuzwa, S.A., Shan, X., MacAuley, M.S., Clark, T., Skorobogatko, Y., Vosseller, K., et al., 2012. Increasing O-GlcNAc slows neurodegeneration and stabilizes tau against aggregation. Nature Chemical Biology 8 (4), 393–399. Available at https://doi.org/10.1038/nchembio.797.

Zhan, X., Stamova, B., Jin, L.W., Decarli, C., Phinney, B., Sharp, F.R., 2016. Gram-negative bacterial molecules associate with Alzheimer disease pathology. Neurology 87 (22), 2324–2332. Available at https://doi.org/10.1212/WNL.0000000000003391.

Zhang, X., Wang, Y., 2016. Glycosylation quality control by the golgi structure. Journal of Molecular Biology 428 (16), 3183–3193. Available at https://doi.org/10.1016/j.jmb.2016.02.030.

Zhang, Q., Ma, C., Chin, L.S., Li, L., 2020. Integrative glycoproteomics reveals protein n-glycosylation aberrations and glycoproteomic network alterations in Alzheimer's disease. Science Advances 6 (40). Available at https://doi.org/10.1126/sciadv.abc5802.

Zhang, Z., Xue, P., Bendlin, B.B., Zetterberg, H., De Felice, F., Tan, X., et al., 2025. Melatonin: a potential nighttime guardian against Alzheimer's. Molecular Psychiatry 30 (1), 237–250. Available at https://doi.org/10.1038/s41380-024-02691-6.

Zhao, J., Lang, M., 2023. New insight into protein glycosylation in the development of Alzheimer's disease. Cell Death Discovery 9 (1). Available at https://doi.org/10.1038/s41420-023-01617-5.

Zhong, Q., Xiao, X., Qiu, Y., Xu, Z., Chen, C., Chong, B., et al., 2023. Protein posttranslational modifications in health and diseases: Functions, regulatory mechanisms, and therapeutic implications. MedComm 4 (3). Available at https://doi.org/10.1002/mco2.261.

Zhou, J.N., Liu, R.Y., Kamphorst, W., Hofman, M.A., Swaab, D.F., 2003. Early neuropathological Alzheimer's changes in aged individuals are accompanied by decreased cerebrospinal fluid melatonin levels. Journal of Pineal Research 35 (2), 125–130. Available at https://doi.org/10.1034/j.1600-079X.2003.00065.x.

Zhu, Y., Shan, X., Yuzwa, S.A., Vocadlo, D.J., 2014. The emerging link between O-GlcNAc and Alzheimer disease. Journal of Biological Chemistry 289 (50), 34472–34481. Available at https://doi.org/10.1074/jbc.R114.601351.

CHAPTER 7

Dysregulated molecular pathways in amyotrophic lateral sclerosis–frontotemporal dementia spectrum disorder

Hui Xu and Weidong Le*
Key Laboratory of Liaoning Province for Research on the Pathogenic Mechanisms of Neurological Diseases, First Affiliated Hospital of Dalian Medical University, Dalian, Liaoning, P.R. China

7.1 Introduction

Amyotrophic lateral sclerosis (ALS) and frontotemporal dementia (FTD) are two devastating neurodegenerative disorders that, despite their distinct clinical manifestations, are now recognized to exist along a common disease spectrum. While ALS is primarily characterized by the progressive degeneration of motor neurons, FTD typically presents as a cognitive-behavioral syndrome driven by the loss of cortical pyramidal neurons and Von Economo neurons within the frontal and temporal lobes. Nonetheless, the ALS–FTD continuum is firmly supported by the significant overlap in clinical, genetic, and pathological features. Clinically, a substantial proportion of patients exhibit symptoms of both motor neuron dysfunction and cognitive or behavioral impairment (Heylen et al., 2025). This overlap is further underscored by shared neuropathological hallmarks, most notably cytoplasmic mislocalization and aggregation of transactive response DNA-binding protein 43 (TDP-43). Furthermore, discoveries of common genetic mutations, such as the hexanucleotide repeat expansion in the chromosome 9 open reading frame 72 (*C9orf72*), and mutations in the TAR DNA-binding protein gene, fused in sarcoma (*FUS*), and others, provide a molecular foundation for this spectrum.

The clinical and pathological heterogeneity of the ALS–FTD spectrum is reflected not only in diverse phenotypes but also in the diversity of molecular pathways implicated in its pathogenesis. With the advent of high-throughput genomic and proteomic technologies, research has begun to unravel the complex molecular underpinnings, revealing a convergent network of core pathological processes. These include proteostasis dysregulation, aberrant RNA metabolism, mitochondrial dysfunction,

* Weidong Le - as contributor.

Molecular Insights into Proteinopathies
DOI: https://doi.org/10.1016/B978-0-443-34176-2.00012-2
© 2027 Elsevier Inc. All rights are reserved, including those for text and data mining, AI training, and similar technologies.

neuroinflammation, impaired axonal transport, and defects in the autophagy–lysosome pathway (Casterton et al., 2020; Rummens and Da Cruz, 2025). For instance, the accumulation of misfolded proteins, such as TDP-43, FUS, and dipeptide repeat proteins (DPRs) produced by the mutant *C9orf72* allele, disrupts cellular proteostasis and serves as a unifying feature across the spectrum. Similarly, the dysfunction of RNA-binding proteins (RBPs) like TDP-43 and FUS leads to widespread disturbances in RNA processing, splicing, and transport (Masrori et al., 2022). Mitochondrial dysfunction is another recurring theme, with evidence for impaired energy metabolism, oxidative stress, and defective lysosome-mediated autophagy of mitochondria in both ALS and FTD. Neuroinflammation, driven by glial activation and innate immune responses, has emerged as both a contributor to disease progression and a potential therapeutic target (Park et al., 2025). The convergence of these disrupted molecular pathways underscores the need for an integrated approach to understanding ALS–FTD pathogenesis. This chapter synthesizes current knowledge of these key molecular mechanisms, explores how their interplay drives neurodegeneration, and highlights the promise of shared therapeutic strategies that target these common mechanisms.

7.2 Protein homeostasis imbalance and abnormalities of transactive response DNA-binding protein 43 and fused in sarcoma

7.2.1 Transactive response DNA-binding protein 43 proteinopathy

TDP-43 proteinopathy is a defining pathological hallmark of both ALS and FTD, as well as their overlapping spectrum disorders. In ALS–FTD patients, TDP-43 is abnormally aggregated within the cytoplasm of neurons and glial cells, forming insoluble inclusions that are ubiquitin- and phosphorylated-TDP-43-positive (Mehta et al., 2023). Under normal physiological conditions, TDP-43 is a predominantly nuclear RNA/DNA-binding protein; however, disease states are characterized by its nuclear depletion and mislocalization to the cytoplasm, where it undergoes aberrant posttranslational modifications, including hyperphosphorylation, ubiquitination, and proteolytic cleavage. These modifications not only promote the formation of pathological cytoplasmic aggregates but also result in the loss of normal nuclear TDP-43 function and the acquisition of toxic properties in the cytoplasm. The loss of nuclear TDP-43 impairs its essential roles in RNA metabolism, including transcriptional regulation, alternative splicing, RNA transport, and stabilization, thereby disrupting the expression and processing of numerous neuronal transcripts critical for cell survival (Moens et al., 2025). Dysregulation of RNA splicing, notably the inclusion of cryptic exons in key neuronal genes encoding proteins such as Stathmin 2 and Unc-13 Homolog A, has been identified as a pathogenic mechanism linking TDP-43 dysfunction to neurodegeneration. Furthermore, the toxic gain-of-function of cytoplasmic TDP-43 aggregates is associated with sequestration of RNA and other RBPs, disturbance of stress granule dynamics, mitochondrial dysfunction, and perturbation of protein homeostasis, all of which exacerbate neuronal (and presumably glial) injury. Experimental models have shown that optogenetically induced TDP-43 aggregation recapitulates neurodegenerative features, including impaired neuronal integrity and behavioral deficits. Collectively, these findings underscore the dual pathogenic roles of TDP-43 in ALS–FTD: loss of its nuclear functions critical for RNA homeostasis and acquisition of cytoplasmic toxicity through aggregate formation, both of which are central to disease progression and represent primary therapeutic targets.

7.2.2 Fused in sarcoma proteinopathy

Mutations in the *FUS* gene, which encodes the RNA-binding protein FUS, represent a significant molecular driver in the pathogenesis of ALS and FTD. These mutations commonly disrupt the proline-tyrosine nuclear localization signal at the C-terminus of FUS, leading to its abnormal mislocalization from the nucleus to the cytoplasm, where it forms pathological inclusions or aggregates mechanistically linked to neurodegeneration (Zhang et al., 2024). The aberrant cytoplasmic localization of FUS is a hallmark feature in both ALS and FTD. Structural investigations have shown that certain mutations, such as P525L and R521C, diminish the binding affinity of FUS for its nuclear import receptor, Transportin-1, thereby promoting cytoplasmic retention and aggregation. Notably, the aggregation process is exacerbated by posttranslational modifications and environmental stressors, including osmotic imbalance and oxidative stress, which can further drive liquid–liquid phase separation (LLPS) of FUS and its transition from dynamic liquid droplets to irreversible, amyloid-like aggregates (Lin et al., 2021). Importantly, while genetic variants significantly accelerate this process, FUS mislocalization can also manifest in the absence of preexisting mutations, suggesting that nongenetic stressors or posttranslational modifications may similarly disrupt its nucleocytoplasmic equilibrium. These cytoplasmic FUS inclusions are not only observed in neurons but also in glia, which are associated with impaired nucleocytoplasmic transport, synaptic dysfunction, and mitochondrial abnormalities (Rezvykh et al., 2023). The pathological inclusions often sequester other RBPs and disrupt normal cellular homeostasis, suggesting a toxic gain-of-function mechanism. Collectively, *FUS* mutations and their downstream effects on protein localization and aggregation constitute a central molecular pathway of neurotoxicity in ALS–FTD, underscoring the importance of understanding FUS biology for developing targeted therapeutics.

Beyond its aberrant localization, FUS proteinopathy profoundly disrupts multiple cellular processes, including DNA repair, RNA metabolism, and stress granule dynamics, that are pivotal to neuronal survival. FUS is normally involved in the regulation of DNA damage responses, particularly in the repair of double-strand breaks, and its nuclear depletion due to cytoplasmic mislocalization impairs these critical repair pathways. Additionally, FUS plays a central role in RNA processing, such as alternative splicing, mRNA transport, and local translation at synapses. Mutant FUS alters the splicing of numerous RNA targets, particularly those involved in synaptic function, leading to early synaptic dysfunction that may initiate neurodegeneration (Zhang et al., 2023). Furthermore, FUS is a key component of stress granules (SGs), membraneless organelles that form via LLPS under cellular stress to sequester and triage RNA and proteins. ALS-associated *FUS* mutations enhance the protein's propensity to undergo aberrant phase transitions, resulting in persistent, compositionally abnormal cytoplasmic granules that differ structurally and functionally from physiological SGs (Tsai et al., 2020). These pathological assemblies can sequester essential factors such as heterogeneous nuclear ribonucleoprotein A3 (hnRNPA3) and disrupt stress responses, proteostasis, and mitochondrial function (Desouky et al., 2024). The interplay between FUS aggregation and its interactions with RNA, other proteins, and cellular chaperones further modulates its toxicity and the progression of neurodegeneration. Ultimately, FUS proteinopathy acts through a multifaceted disruption of DNA repair, RNA metabolism, and SG biology, converging on pathways that promote neuronal injury and the clinical manifestations of ALS–FTD spectrum disorders.

7.2.3 Protein degradation system dysfunction

The disruption of protein degradation systems, specifically the ubiquitin-proteasome system (UPS) and the autophagy–lysosome pathway, is a central pathological hallmark of ALS and FTD, leading to the accumulation of neurotoxic intracellular protein aggregates. The UPS is crucial for the selective degradation of misfolded or damaged proteins, and its impairment results in the buildup of toxic proteins, which is a feature observed in neurodegenerative diseases including ALS and FTD (Nixon and Rubinsztein, 2024). Similarly, defects in the autophagy–lysosome system hinder the clearance of aggregated proteins and damaged organelles, further exacerbating proteostasis imbalance (Chua et al., 2022; Lin et al., 2022). Mutations in genes encoding key regulatory components of these pathways, such as ubiquilin-2 (*UBQLN2*) and optineurin (*OPTN*), intensify the disturbance of protein homeostasis. UBQLN2, a member of the ubiquilin family, acts as a shuttle factor, delivering ubiquitinated substrates to both the proteasome and autophagic machinery for degradation. Mutations in *UBQLN2* have been shown to impair proteasome assembly, disrupt mitochondrial protein degradation, and alter the dynamics of SGs formation, all of which contribute to the pathogenesis of ALS–FTD (Hao et al., 2020). Similarly, *OPTN* mutations compromise selective autophagy, particularly mitophagy, and are linked to the accumulation of toxic protein inclusions. Furthermore, cargo receptors like sequestosome 1 (p62/SQSTM1) bridge ubiquitinated proteins to the autophagy machinery.

Mutations in the *SQSTM1* gene encoding p62 have been identified in both ALS and FTD, and their dysfunction leads to the impaired degradation of protein aggregates and dysregulated signaling pathways of antioxidant response and apoptosis. The convergence of UPS and autophagy–lysosome pathway dysfunction, often exacerbated by genetic mutations affecting UBQLN2, OPTN, and p62, underpins the accumulation of neurotoxic protein aggregates and the progressive neurodegeneration observed in ALS–FTD spectrum disorders. Thus understanding the molecular interplay between these protein degradation systems and their genetic regulators is fundamental to elucidating disease mechanisms and developing targeted therapeutic strategies.

7.3 RNA metabolism abnormalities and chromosome 9 open reading frame 72 expansion

7.3.1 Molecular mechanisms of chromosome 9 open reading frame 72 repeat expansion

The GGGGCC (G4C2) hexanucleotide repeat expansion in the *C9orf72* gene is the most common genetic etiology of ALS and FTD. Its pathogenicity is linked to distinct molecular mechanisms involving both RNA toxicity and the formation of abnormal secondary structures. The expanded G4C2 repeats, often numbering in the hundreds or thousands, are transcribed into sense and antisense RNAs that fold into stable secondary structures, such as G-quadruplexes and hairpins, which are not typically present in healthy individuals. These aberrant RNA structures can sequester various RBPs, forming nuclear RNA foci that disrupt normal RNA metabolism. This sequestration impairs essential cellular processes, including alternative splicing, RNA transport, and translation, thereby contributing to neuronal dysfunction and degeneration (Braems et al., 2022; Feng et al., 2024). The formation of RNA foci has been observed in both patient tissues and model systems. These foci act as molecular sinks, sequestering RBPs such as hnRNPs, including HNRNPK. The consequent mislocalization of these proteins further exacerbates cellular stress and DNA damage. Additionally, the repeat-containing RNAs undergo repeat-associated non-ATG translation, producing DPRs that are neurotoxic and contribute to cellular dysfunction through multiple mechanisms, including nucleocytoplasmic transport impairment and mitochondrial toxicity (Odeh and Shorter, 2020). The accumulation of RNA foci and DPRs is a hallmark of C9orf72-associated ALS–FTD and has been linked to the disruption of protein homeostasis, autophagy, and stress granule dynamics. Thus the expanded G4C2 repeats in C9orf72 trigger a cascade of pathogenic events by inducing RNA toxicity, abnormal secondary structure formation, and sequestration of RBPs, collectively leading to widespread dysregulation of RNA metabolism and neuronal degeneration in the ALS–FTD spectrum (Jiang et al., 2026).

7.3.2 RAN translation and toxic dipeptide repeat proteins

Hexanucleotide repeat expansions in the *C9orf72* gene represent the most prevalent genetic etiology of ALS and FTD. The G4C2 repeats undergo a noncanonical form of translation known as RAN translation, which does not require a conventional AUG start codon. RAN translation of both sense and antisense repeat-containing transcripts gives rise to five distinct DPRs: poly-GA, poly-GR, poly-GP, poly-PA, and poly-PR. The production of these DPRs is central to the pathogenesis of *C9orf72*-mediated ALS–FTD, as their accumulation has been consistently observed in patient brain tissues and multiple animal models. Notably, DPRs, particularly the arginine-rich poly-GR and poly-PR, have been demonstrated to be highly neurotoxic. These peptides interfere with essential cellular processes by forming aggregates that disrupt nucleocytoplasmic transport, impair protein homeostasis, and induce cellular stress responses. For instance, DPR aggregates disrupt the nuclear pore complex and compromise autophagy and proteostasis, leading to neuronal dysfunction and death (Mori et al., 2023). Recent studies have revealed that the accumulation of DPRs correlates with impaired lysosomal function, ER–Golgi trafficking, and the formation of cytoplasmic inclusions, all contributing to the neurodegenerative process in ALS–FTD. Experimental reduction of DPR levels, either by targeting RAN translation or promoting their degradation, ameliorates disease phenotypes in cellular and animal models, underscoring the pathogenic significance of DPRs and the therapeutic potential of modulating RAN translation. Collectively, the dysregulation of RAN translation and subsequent DPR toxicity constitute a key molecular mechanism linking C9orf72 repeat expansions to neurodegeneration in the ALS–FTD spectrum.

7.3.3 RNA-binding protein dysfunction

RBPs such as hnRNPs and serine/arginine-rich splicing factors play pivotal roles in posttranscriptional gene regulation, including RNA splicing, transport, stability, and translation. In the context of ALS and FTD, dysfunction of these RBPs is a central pathogenic mechanism. Disease-associated mutations or pathological conditions can lead to the formation of abnormal RNA species, such as expanded GGGGCC repeats in the *C9orf72* gene, which sequester RBPs like hnRNPs and SRSFs, thereby impairing their normal cellular functions. For instance, in C9orf72-associated ALS–FTD, the expanded repeat RNAs form nuclear foci, aberrantly bind, and trap multiple RBPs, disrupting their availability for physiological RNA processing tasks (Picchiarelli and Dupuis, 2020). This sequestration can lead to widespread defects in RNA metabolism, such as alternative splicing, polyadenylation, and RNA transport, which are critical for neuronal function and survival. Moreover, the pathological aggregation of RBPs, including TDP-43 and FUS, which often colocalize with SGs, further exacerbates RNA processing defects and contributes to neurodegeneration by altering the assembly and dynamics of ribonucleoprotein complexes (Liu et al., 2020). Importantly, the loss of function of these RBPs not only affects their direct RNA targets but can also perturb the broader RBP network, leading to compensatory or deleterious effects on other RBPs and their regulated transcripts. The cumulative consequence is a global disturbance in posttranscriptional gene regulation, which underlies the selective vulnerability of neurons in ALS–FTD spectrum disorders. Understanding how aberrant RNAs sequester hnRNPs, SRSFs, and related RBPs provides vital insights into disease pathogenesis and highlights potential therapeutic avenues targeting these RNA-protein interactions or restoring RBP function.

7.4 Mitochondrial dysfunction and energy metabolism imbalance

7.4.1 Mitochondrial dynamics abnormalities

Mitochondrial dynamics, encompassing the continuous processes of mitochondrial fission and fusion, are essential for maintaining cellular energy homeostasis, mitochondrial quality control, and the metabolic demands of neuronal survival. In the context of ALS and FTD spectrum disorders, increasing evidence points to a critical imbalance between mitochondrial fission and fusion as a key pathological feature, which disrupts energy supply and compromises cell viability. Normally, fusion—mediated by proteins such as mitofusin 2 (MFN2) and optic atrophy 1 (OPA1)—allows mixing of mitochondrial contents, facilitating repair of damaged mitochondria and optimization of ATP production. Conversely, fission, largely regulated by proteins like dynamin-related protein 1, enables the removal of dysfunctional mitochondria via mitophagy and supports mitochondrial distribution along axons. In ALS and FTD models, aberrant expression and function of these key proteins have been documented, leading to excessive mitochondrial fragmentation or hyperfusion, both of which are detrimental to neuronal health. For instance, studies in superoxide dismutase 1 (SOD1)-G93A transgenic mice, a well-established ALS model, reveal reduced expression of MFN2 and OPA1, correlating with pronounced mitochondrial structural abnormalities—such as mitochondrial swelling, disrupted cristae, and increased autophagolysosomes within motor neurons—at symptomatic stages. Furthermore, disruptions to the OPA1-mitofilin complex, often driven by coiled-coil-helix-coiled-coil-helix domain containing 10 (CHCHD10) mutations or TDP-43 pathology, impair mitochondrial fusion and respiration, further linking dysfunctional mitochondrial dynamics to neurodegeneration in FTD (Venkatesan et al., 2023). These molecular alterations not only compromise ATP production but also sensitize neurons to oxidative stress and apoptosis. The critical role of mitochondrial dynamics is further underscored by interventions that restore the fission–fusion balance, which delay disease progression in ALS models and highlight the therapeutic potential of these pathways. Collectively, the dysregulation of mitochondrial dynamics—marked by imbalanced fission and fusion and aberrant expression of MFN2 and OPA1—emerges as a central mechanism underpinning energy failure and neuronal loss in ALS–FTD spectrum disorders.

7.4.2 Oxidative stress and mitochondrial DNA damage

Mitochondrial dysfunction serves as a central feature in the pathogenesis of ALS–FTD spectrum disorders, where it plays a pivotal role in the accumulation of reactive oxygen species (ROS) and the subsequent induction of oxidative stress. Impaired mitochondrial function disrupts ATP production and leads to excessive generation of ROS, which in turn damages mitochondrial DNA (mtDNA), proteins, and lipids, establishing a vicious cycle that exacerbates neuronal injury and accelerates neurodegeneration of motor neurons and glial cells. The accumulation of ROS not only causes direct oxidative lesions to mtDNA but also impairs DNA repair mechanisms, promoting persistent DNA damage and triggering aberrant cell cycle activation in postmitotic neurons. Such mtDNA mutations further compromise the efficiency of the electron transport chain, aggravating energy metabolism deficits and amplifying ROS production (Kim et al., 2023). Studies have demonstrated that oxidative stress-induced mtDNA damage can activate mitophagy as a compensatory mechanism to remove dysfunctional mitochondria, but when this process is overwhelmed or impaired, damaged mtDNA accumulates and is released into the cytoplasm, activating innate immune responses and neuroinflammation through pathways such as cyclic GMP-AMP synthase–stimulator of interferon genes. Moreover, the release of oxidized mtDNA acts as a damage-associated molecular pattern, further amplifying inflammatory cascades and neuronal apoptosis (Gao et al., 2023). In ALS and related neurodegenerative diseases, the convergence of mitochondrial dysfunction, oxidative stress, and mtDNA damage is reinforced by genetic and epigenetic factors, including mutations in key regulatory proteins (such as SOD1 and TDP-43), which fundamentally disrupt mitochondrial homeostasis and redox homeostasis. Collectively, these findings underscore the critical interplay between mitochondrial dysfunction, ROS accumulation, and mtDNA damage. This self-reinforcing cycle drives the energy deficits, oxidative injury, and neuroinflammation that underpin ALS–FTD spectrum disorders, highlighting these molecular pathways as promising therapeutic targets.

7.4.3 Mitochondrial transport impairment

Axonal mitochondrial transport is a critical process for neuronal health, ensuring that energy supply and calcium buffering are maintained at synaptic terminals and throughout the neuron. Within the ALS–FTD clinical spectrum, mounting evidence reveals that disruptions in mitochondrial trafficking are a central pathogenic feature, directly impacting synaptic function and neuronal maintenance. Several genetic mutations linked to ALS–FTD, such as those affecting valosin-containing protein (VCP), FUS, TDP-43, and microtubule-associated protein Tau (MAPT), have been shown to impair the normal movement of mitochondria along axons. For instance, studies in *Drosophila* and mammalian models demonstrate that VCP is crucial for regulating mitochondrial motility, with loss or mutation of VCP leading to enhanced retrograde transport, reduced axonal mitochondrial density, and compromised ATP production, ultimately impairing neuronal function. Similarly, *FUS* mutations disrupt mitochondrial distribution and transport, correlating with synaptic abnormalities and reduced neuronal complexity, likely through altered interactions with mitochondrial tethering proteins such as syntaphilin. *MAPT* mutations, notably V337M and N297K, directly affect microtubule stability and organelle trafficking, resulting in defective mitochondrial and lysosomal movement in patient-derived motor neurons, which further underscores the importance of intact axonal transport mechanisms in neuronal survival. Additionally, TDP-43 pathology, common in ALS–FTD, is associated with defects in mitochondrial morphology, transport, and function, suggesting that mitochondrial mislocalization and dysfunction are key contributors to neurodegeneration. These transport impairments compromise the ability of neurons to meet local energetic demands at synapses, disrupt calcium homeostasis, and promote synaptic degeneration, thereby driving the progressive loss of function characteristic of ALS–FTD spectrum disorders. Collectively, these findings emphasize that mitochondrial transport deficits are not merely a consequence but a driving force in the pathogenesis of ALS–FTD, providing a compelling target for therapeutic intervention.

7.5 Neuroinflammation and glial cell responses

7.5.1 Microglial activation

Microglial activation constitutes a fundamental neuroinflammatory axis in ALS–FTD spectrum disorders. Under physiological conditions, microglia maintain neural homeostasis and support neuronal integrity through surveillance and phagocytosis of neurons. However, in response to pathological stimuli, microglia transition from a resting state to a proinflammatory phenotype, characterized by morphological changes and upregulation of surface markers and inflammatory mediators. This activation is accompanied by the release of pro-inflammatory cytokines such as interleukin-1β (IL-1β) and tumor necrosis factor-alpha (TNF-α), which contribute to a neurotoxic environment. In ALS and FTD, both genetic and sporadic forms show evidence of microglial activation, with studies demonstrating increased expression of inflammatory markers and microglial reactivity in affected brain regions (Rivers-Auty et al., 2024). The pro-inflammatory milieu generated by activated microglia accelerates neuronal death through several mechanisms, including the induction of oxidative stress, mitochondrial dysfunction, and impairment of synaptic function. Notably, the NLR family pyrin domain-containing 3 inflammasome pathway has been implicated in mediating microglial inflammatory responses to neuronal injury and protein aggregates, further amplifying the release of IL-1β and fostering chronic neuroinflammation (Malpetti et al., 2023). Microglial activation is not only a response to neuronal damage but also actively drives disease progression by perpetuating a cycle of neuroinflammation and neurodegeneration (Costa-Pinto et al., 2024). In ALS, for example, microglial activation in the spinal cord correlates with the loss of motor neurons and disease severity. Similarly, in FTD, microglial activation is observed in the frontal and temporal cortices and is associated with cognitive decline. The switch to a pro-inflammatory phenotype is influenced by multiple factors, including genetic mutations (e.g., in granulin (*GRN*), *C9orf72*, triggering receptor expressed on myeloid cells 2 (*TREM2*)), metabolic dysregulation, and the presence of misfolded proteins such as TDP-43 in neurons. Importantly, the chronic pro-inflammatory environment established by activated microglia not only accelerates neuronal loss but also impairs the resolution of inflammation, contributing to the relentless progression of ALS–FTD spectrum disorders. Thus microglial activation and the resultant release of inflammatory cytokines are pivotal in creating a self-sustaining neurotoxic environment that exacerbates neuronal degeneration in ALS–FTD.

7.5.2 Astrocyte dysfunction

Astrocyte dysfunction has emerged as a central contributor to the pathogenesis of ALS and FTD, particularly within the ALS–FTD spectrum. Under physiological conditions, astrocytes provide essential metabolic, trophic, and structural support to neurons, regulate synaptic transmission, maintain blood-brain barrier integrity, and modulate neuroinflammation. However, in the context of ALS and FTD, astrocytes undergo pathological transformation, losing their supportive roles and acquiring neurotoxic properties that exacerbate neuronal degeneration. This pathological transition is primarily defined by impaired glutamate uptake, disrupted metabolic processes, mitochondrial dysfunction, and aberrant release of neurotoxic factors, all of which contribute to excitotoxicity and oxidative stress in surrounding neurons. Mutations in genes such as *SOD1* and *C9orf72* play a pivotal role in driving these astrocytic changes although they are also seen in the absence of mutations. For instance, *SOD1* mutations lead to dysfunctional astrocyte metabolism, reduced glutamate transport, and heightened inflammatory signaling, fostering a hostile microenvironment for motor neurons (Jury et al., 2020). Similarly, *C9orf72* repeat expansions are implicated in both loss- and gain-of-function mechanisms that disrupt astrocyte immune responses and metabolic balance, further promoting neuroinflammation and neurodegeneration (Nguyen et al.,

2021). Additionally, astrocytes with TDP-43 pathology exhibit compromised glutamate uptake, increased cytoplasmic inclusions, and impaired ATP handling, all of which contribute to motor neuron injury through both cell-autonomous and noncell-autonomous mechanisms. Beyond these genetic factors, reactive astrocytes in ALS–FTD exhibit altered calcium homeostasis, mitochondrial abnormalities, and dysregulated lipid and lactate metabolism, collectively undermining their neuroprotective capacity and driving neurotoxicity (Luan et al., 2024). The cumulative effect of these disturbances reprograms astrocytes from supportive cells to active contributors to disease. They not only fail to maintain neuronal health but also propagate neurotoxic cascades, primarily through excitotoxicity, oxidative stress, and chronic inflammation, thereby accelerating disease progression in ALS–FTD spectrum disorders.

7.5.3 Activation of inflammatory signaling pathways

The activation of inflammatory signaling pathways, particularly those involving nuclear factor kappa B (NF-κB) and NLRP3, is a central feature in the pathogenesis and progression of ALS–FTD spectrum disorders. Neuroinflammation, once considered a secondary response to neuronal injury, is now recognized as a driving force that exacerbates neurodegeneration in ALS and FTD. The NLRP3 inflammasome has emerged as a key mediator in this process. Recent evidence demonstrates that DPRs produced from C9orf72 hexanucleotide repeat expansions, especially poly-glycine-arginine, directly activate the NLRP3 inflammasome in microglia and macrophages, leading to the secretion of pro-inflammatory cytokines such as IL-1β (Shu et al., 2023). This inflammasome activation has been observed not only in cellular models but also in organotypic hippocampal slice cultures, highlighting the therapeutic potential of targeting this pathway. In vivo, genetic ablation of *NLRP3* in mouse models of C9orf72-associated ALS–FTD significantly improves survival, preserves behavioral function, and prevents neurodegeneration, underscoring the integral role of inflammasome-mediated innate immunity in disease pathogenesis. Furthermore, transcriptomic analyses across neurodegenerative diseases, including ALS–FTD, reveal a consistent upregulation of neuroinflammatory pathways such as Toll-like receptor, TNF, and NF-κB signaling, supporting a pan-neurodegenerative signature characterized by innate immune activation and phagocytosis. The NF-κB pathway, in particular, has been implicated in promoting neuronal dysfunction and disease progression; pharmacological inhibition of NF-κB in ALS–FTD mouse models ameliorates cognitive and motor deficits, restores synaptic and dendritic integrity, and attenuates hallmark pathological features. Collectively, these findings establish the activation of NF-κB, NLRP3, and related inflammatory signaling cascades as pathogenic mechanisms that not only mark but also drive ALS–FTD, offering promising targets for disease-modifying interventions.

7.6 Axonal transport disorders and cytoskeletal abnormalities

7.6.1 Dysfunction of motor proteins

Motor proteins such as kinesin family member 5A (KIF5A) and dynactin subunit 1 (DCTN1) play essential roles in the maintenance of neuronal homeostasis by mediating axonal transport, which is critical for the delivery of organelles, proteins, and signaling molecules along the extensive axons of motor neurons. Mutations in these motor proteins have been implicated in the pathogenesis of ALS and FTD, as well as their spectrum disorders. Specifically, KIF5A is vital for anterograde transport of mitochondria and other cargoes, while DCTN1 facilitates retrograde transport by linking dynein to its cargo. Mutations in *KIF5A* disrupt microtubule-based transport, leading to axonal transport deficits, which have been observed in several neurodegenerative disease models (Parra Bravo et al., 2024). The dysfunction of motor proteins is not only limited to impaired transport but also affects the homeostasis of other cellular processes, such as mitochondrial dynamics and protein degradation pathways. For instance, the disruption of axonal transport machinery due to *KIF5A* or *DCTN1* mutations exacerbates mitochondrial dysfunction and oxidative stress, further promoting neurodegeneration (Sleigh et al., 2019).

7.6.2 Microtubule instability

Microtubule stability is crucial for neuronal function, particularly within the context of neurodegenerative diseases such as ALS and FTD. Mutations affecting microtubule proteins, notably tubulin alpha 4a (TUBA4A), have been implicated in both familial and sporadic cases of ALS. *In silico* molecular-dynamic modeling studies have demonstrated that *TUBA4A* mutations induce significant structural transitions affecting GTP binding, polypeptide chain stability, tubulin assembly, and increase the propensity for aggregation. These alterations impair the formation and stabilization of the tubulin heterocomplex, ultimately leading to microtubule disintegration and contributing to the vulnerability and degeneration of motor neurons in ALS. Recent multicenter genetic studies have further expanded the phenotypic spectrum of TUBA4A-related disorders, revealing that missense variants can differentially disrupt microtubule dynamics and stability, manifesting as both primary myopathies and multisystem proteinopathies, sometimes with central nervous system involvement. The destabilization of microtubules due to *TUBA4A* and other tubulin mutations accelerates microtubule depolynerization, leading to impaired axonal transport, organelle mislocalization, and subsequent neuronal degeneration. This is consistent with broader findings in neurodegenerative diseases, where microtubule instability is recognized as a key early event that disrupts axonal transport and precedes overt neurodegeneration. Moreover, microtubule destabilization is exacerbated by pathological alterations in microtubule-associated proteins such as Tau, whose aggregation and loss of function further compromise microtubule integrity in ALS-FTD and related tauopathies (Theunissen et al., 2021). Collectively, these insights underscore the pivotal role of microtubule stability in neuronal health and highlight how *TUBA4A* mutations and accelerated microtubule depolymerization act synergistically to drive neurodegenerative processes in the ALS–FTD spectrum.

7.6.3 Dysregulation of cytoskeleton-related signaling pathways

Dysregulation of cytoskeleton-associated signaling pathways, including Ras homolog family member A/rho-associated protein kinase (RhoA/ROCK) and mitogen-activated protein kinase (MAPK), plays a pivotal role in the pathogenesis of ALS–FTD spectrum disorders. The neuronal cytoskeleton, comprising microtubules, actin filaments, and intermediate filaments, is essential for maintaining neuronal morphology, axonal transport, and synaptic integrity. Aberrant regulation of key signaling pathways, such as RhoA/ROCK and MAPK, disrupts the dynamic remodeling of cytoskeletal structures, leading to impaired neuronal function and degeneration. The RhoA/ROCK pathway, for instance, is involved in actin cytoskeleton organization, and its dysregulation can result in altered cell shape, motility, and axonal guidance, all of which are critical for proper neuronal connectivity. Similarly, the MAPK pathway influences cytoskeletal dynamics through phosphorylation events that affect microtubule stability and neuronal survival. Emerging evidence highlights that these signaling cascades are frequently perturbed in neurodegenerative diseases, including ALS and FTD, contributing to the formation of protein aggregates containing cytoskeletal components, mitochondrial dysfunction, and disrupted axonal transport (Hung and Patani, 2024). Furthermore, as noted above, genetic studies have identified mutations affecting cytoskeletal proteins (e.g., TUBA4A) and their regulators, which directly impact microtubule assembly and stability, further implicating cytoskeletal signaling dysregulation in disease pathogenesis. Additionally, transcriptomic analyses of ALS–FTD patient tissues reveal upregulation of genes involved in cytoskeletal organization and signaling, supporting the notion of a "pan-neurodegenerative" gene expression signature shared across related disorders. Therapeutic strategies targeting these pathways, such as ROCK inhibitors or modulators of cytoskeletal dynamics, are under investigation and hold promise for mitigating neurodegeneration by restoring cytoskeletal homeostasis. Collectively, these findings underscore the central role of cytoskeleton-associated signaling pathway dysregulation in ALS–FTD and highlight it as a critical target for future research and therapeutic development.

7.7 Autophagy–lysosome system dysfunction

7.7.1 Decreased autophagic activity

Autophagy is a critical cellular process responsible for the degradation and recycling of misfolded proteins and damaged organelles, thereby maintaining neuronal homeostasis. In the ALS and FTD spectrum, a growing body of evidence implicates decreased autophagic activity as a central pathogenic mechanism. Mutations in autophagy-related genes such as autophagy-related 7 (*ATG7*) and *SQSTM1* (encoding p62) have been shown to disrupt autophagic flux, impeding the efficient clearance of proteotoxic aggregates and dysfunctional cellular organelles (Table 7.1). Specifically, loss-of-function mutations in *ATG7* impair the formation of autophagosomes, while *SQSTM1* mutations compromise the ability of p62 to function as an autophagy receptor, leading to the accumulation of ubiquitinated protein aggregates and defective mitochondria. These defects result in a toxic intracellular environment that promotes neuronal cell death, a hallmark of both ALS and FTD (Deng et al., 2017). Furthermore, other ALS-FTD-associated genes, such as TANK-binding kinase 1 (*TBK1*) and *OPTN*, converge on the autophagic pathway, and their mutations similarly lead to impaired autophagy and enhanced neurodegeneration. Investigations using patient-derived cells and animal models have demonstrated that these genetic disruptions in autophagy-related pathways result in the accumulation of pathological protein inclusions, such as TDP-43, and exacerbate neurodegenerative processes. Importantly, therapeutic strategies aimed at restoring autophagic function, whether by genetic or pharmacological means, have shown promise in ameliorating disease phenotypes in preclinical models, highlighting the centrality of autophagy impairment in ALS–FTD pathogenesis (Li et al., 2025). Thus decreased autophagic activity due to mutations in key autophagy genes not only hinders the clearance of deleterious proteins and organelles but also acts as a fundamental driver in the progression of neurodegeneration within the ALS–FTD disease spectrum (see Fig. 7.1).

7.7.2 Lysosomal dysfunction

Lysosomal dysfunction is increasingly recognized as a convergent pathological feature in the spectrum of ALS and FTD. A hallmark of this dysfunction is the progressive decline in lysosomal enzyme activity, which impairs the cellular degradation system and leads to the accumulation of undigested macromolecules, such as proteins and lipids, within neurons and glial cells. This accumulation overwhelms the intracellular recycling and degradation mechanisms, exacerbating neurodegenerative pathology. The lysosomal–autophagy pathway, in particular, is crucial for the clearance of damaged proteins and organelles, and its impairment has been consistently observed in ALS–FTD, resulting in the buildup of toxic protein aggregates, such as TDP-43 and Tau. Several genetic

Table 7.1 Amyotrophic lateral sclerosis–frontotemporal dementia genes and their disease mutations linked to autophagy.

Gene symbol	Protein	Cellular functions	The effect of the genetic variants	Autophagy involvement
C9ORF72	C9orf72	Proteostasis and vesicle dynamics	Loss/gain of function	Regulates initiation of autophagy
TBK1	TANK-binding kinase 1	Proteostasis and immunity	Loss of function	Phosphorylates autophagy receptors (p62 and OPTN); regulates selective autophagy
OPTN	Optineurin	Proteostasis, vesicle trafficking and axon homeostasis	Loss of function	A substrate of TBK1 and autophagy receptor protein; selective autophagy
SQSTM1	p62	Proteostasis, amino acid sensing, DNA damage response, and oxidative stress	Loss of function	A substrate of TBK1 and Unc-51 Like Autophagy Activating Kinase 1; autophagy receptor; selective autophagy
UBQLN2	Ubiquilin-2	Proteasome, proteostasis, and vesicle trafficking	Gain/loss of function	Potential autophagy receptor
TARDBP	Transactive response DNA-binding protein 43	RNA regulation	Loss/gain of function	Regulates autophagy initiation and autophagosome–lysosome fusion
FUS	Fused in sarcoma	RNA regulation, and DNA damage repair	Loss of function	Amyotrophic lateral sclerosis-linked mutations, P525L and R522G, impair autophagy
VCP	Valosin-containing protein	ER-associated degradation, DNA damage, and membrane dynamics	Loss of function	Regulates the clearance of lysosomes
SOD1	Superoxide dismutase 1	Dismutation reaction	Gain of function	Mutant SOD1 disrupts autophagy

(*Continued*)

Table 7.1 (Continued)

Gene symbol	Protein	Cellular functions	The effect of the genetic variants	Autophagy involvement
ALS2	Alsin	Proteostasis and endosome biogenesis	Loss of function	Pathogenic mutations in ALS2 disrupt the formation of amphisomes
VAPB	Vesicle-associated membrane protein-associated protein B/C	Proteostasis, calcium homeostasis, and proteins trafficking	Loss of function	Regulates ER-mitochondrial contact
SigR1	Sigma receptor-1	Proteostasis, Ca^{2+} signaling, ion channel activity, synaptic plasticity	Loss of function	Regulates autophagosome–lysosome fusion

studies have identified key genes involved in lysosomal function whose mutations are associated with ALS–FTD. Notably, transmembrane protein 106B (TMEM106B) has emerged as a significant genetic modifier in frontotemporal lobar degeneration with TDP-43 pathology and in disease progression of C9orf72-associated ALS–FTD. *TMEM106B* encodes a lysosomal transmembrane protein that regulates lysosomal acidification and trafficking. Loss of TMEM106B function leads to impaired lysosomal acidification, reduced cathepsin activity, and defective autophagosome–lysosome fusion, resulting in the accumulation of neurotoxic DPRs and further disruption of autophagic flux. Other genes, such as *GRN* (encoding progranulin), charged multivesicular body protein 2B, and *C9orf72*, have also been implicated in lysosomal dysfunction, affecting processes such as lysosomal biogenesis, cargo trafficking, and fusion events. The convergence of these genetic and molecular findings underscores the importance of lysosomal integrity for neuronal health and highlights lysosomal dysfunction as a shared mechanism driving neurodegeneration across the ALS–FTD spectrum. Understanding the precise roles of lysosomal genes and their variants, including *TMEM106B*, provides critical insight into disease pathogenesis and offers potential avenues for therapeutic intervention targeting the restoration of lysosomal function.

7.7.3 Interaction between the autophagy–lysosome system and other pathways

The autophagy–lysosome system is intricately interconnected with other cellular pathways, notably those regulating protein homeostasis and mitochondrial function, forming a complex regulatory network critical for neuronal survival. In the ALS and FTD spectrum, mounting evidence highlights that disturbances in autophagy-lysosome dynamics are not isolated events but are deeply entwined with broader cellular dysfunctions. For example, the autophagy-lysosome pathway is essential for the degradation of misfolded proteins and damaged organelles, processes that are tightly coordinated with the UPS (see section 1.3) to maintain proteostasis. Defects in either pathway can lead to the accumulation of toxic protein aggregates, a hallmark of neurodegenerative diseases such as ALS and FTD. Moreover, selective forms of autophagy, such as mitophagy, directly link the autophagy–lysosome system to mitochondrial quality control, ensuring the clearance of dysfunctional mitochondria that would otherwise contribute to cellular stress and degeneration. Recent studies have further elucidated how lysosomal channels and ion homeostasis influence autophagic flux, with lysosomal dysfunction exacerbating autophagy impairment and *vice versa*, thereby amplifying neurodegenerative processes (Giuliano et al., 2023). Genetic mutations associated with ALS–FTD, such as those affecting C9orf72, TBK1, p62, and VCP, often disrupt both autophagy and endolysosomal trafficking, underscoring the

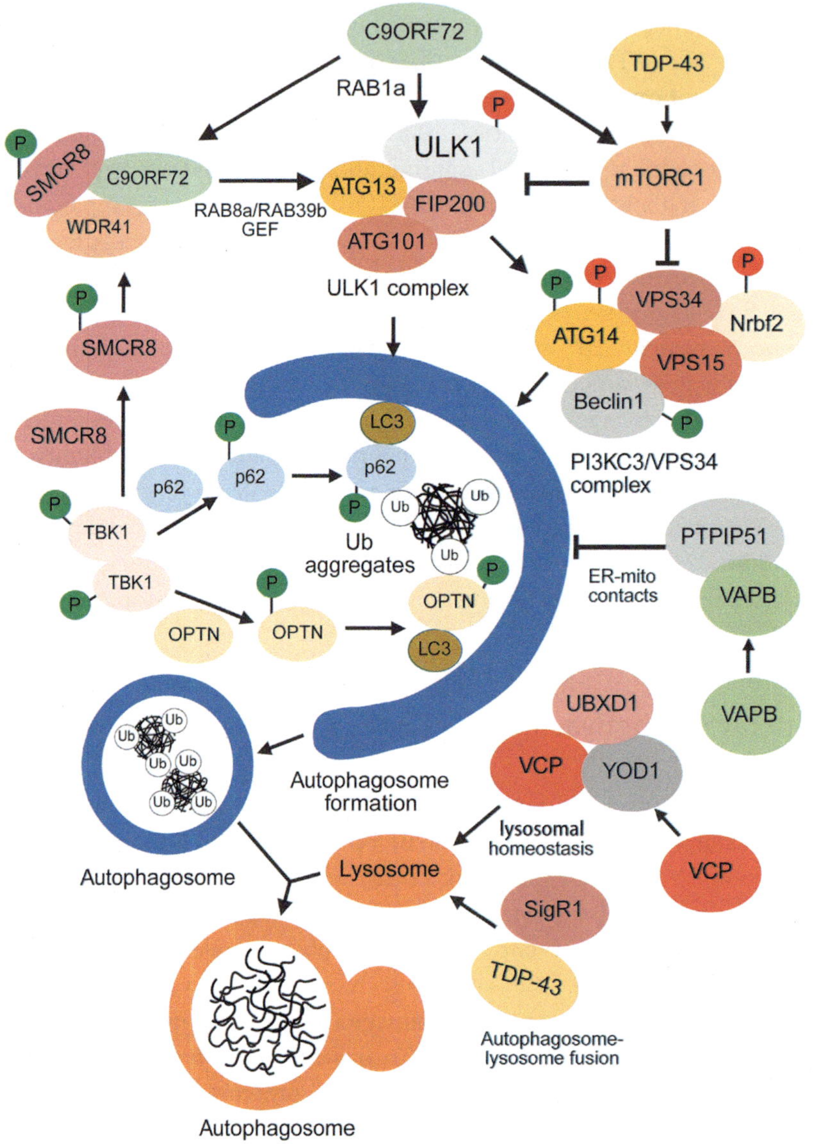

Figure 7.1 Interactions of ALS-FTD associated genes with autophagy pathway. Several ALS-FTD genes take part in the selective autophagy pathway, which controls the clearance of ubiquitinated proteins and damaged organelles. Autophagy kinase TBK1 phosphorylates two known autophagy receptors, p62 and OPTN, to facilitate the clearance of ubiquitinated proteins. TBK1 also phosphorylates SMCR8 to regulate the formation of SMCR8, WDR41 and C9ORF72 complex, which acts as a GEF (GDP/GTP exchange factor) for RAB8a/13b, which is associated with the dynamics of autophagosomes. The C9ORF72 complex potentially regulates the ULK1 kinase activity. C9ORF72 can interact with the Rab1a and ULK1 complex to regulate autophagy initiation. Moreover, C9ORF72 can

(*continued*)

convergence of these molecular pathways in disease pathogenesis (Beckers and Van Damme, 2025). Additionally, cross-talk exists between autophagy and cellular stress responses, such as oxidative stress, whereby autophagy activation can mitigate ROS and maintain redox balance, while impaired autophagy exacerbates oxidative damage and protein aggregation. The critical role of mitophagy exemplifies the interaction between autophagy–lysosome function and mitochondrial homeostasis. When this process is impaired, damaged mitochondria accumulate, increasing neuronal susceptibility to degradation. Collectively, these findings position the autophagy–lysosome system as a central hub, interacting pathways for proteostasis, mitochondrial quality control, and stress response. The dysregulation of this integrated network is now recognized as the core mechanism in the pathogenesis of ALS–FTD spectrum disorders.

7.8 Conclusions

ALS–FTD spectrum disorders represent a complex neurodegenerative continuum characterized by overlapping clinical, genetic, and pathological hallmarks. As synthesized in this chapter, the pathogenesis of ALS–FTD is underpinned by the dysregulation of multiple, interrelated molecular pathways, including proteostasis, RNA metabolism, mitochondrial function, neuroinflammation, axonal transport, and the autophagy–lysosomal system. The intricate cross-talk among these pathways highlights the multifactorial etiology of neuronal degeneration and the clinical heterogeneity observed in ALS–FTD spectrum disorders.

From an expert perspective, the development of ALS–FTD research has evolved from single-gene or single-pathway studies to a more integrated, systems-biology framework. This paradigm shift acknowledges the fact that no single molecular mechanism can solely account for the phenotypic diversity and heterogeneous

negatively regulate autophagy by increasing mTORC1 activity. TDP-43 regulates mTORC1 activity; however, it can enhance autophagosome–lysosome fusion in an mTORC1 independent manner. VAPB, together with PTPIP51, deregulates autophagy flux by controlling the ER-mitochondria contacts. VCP, which cooperates with UBXD1 and YOD1, maintains lysosomal homeostasis. Underlined genes are directly linked to ALS/FTD. Abbreviations: ALS: Amyotrophic lateral sclerosis; FTD: Frontaltemporal dementia; TDP-43: Transactive response DNA-binding protein 43; SMCR8; Smith–Magenis Chromosome Region gene 8; WDR41: WD repeat domain 41; ULK1: Unc-51 Like Autophagy Activating Kinase 1; C9ORF72: Chromosome 9 open reading frame 72; GEF: Guanine nucleotide exchange factor; mTORC1: mechanistic target of rapamycin complex 1; TBK1: TANK-binding kinase 1; OPTN: Optineurin; LC3: microtubule-associated protein1 light chain 3; VAPB: Vesicle-associated membrane protein-associated protein B; VCP: Valosin-containing protein; SigR1: Sigma receptor-1; UBXD1: ubiquitin regulatory X (UBX) domain-containing protein 1; YOD1: yeast OTU deubiquitinating enzyme 1; ER: endoplasmic reticulum; mito: mitochondria; PTPIP51: protein tyrosine phosphatase interacting protein 51.

progression characteristic in ALS–FTD. Instead, it is the convergence and interplay of multiple disrupted cellular processes that ultimately drive neurodegeneration. For instance, protein misfolding and impaired proteostasis not only lead to toxic aggregate formation but also affect RNA metabolism and mitochondrial integrity. Concurrently, defects in the autophagy–lysosomal pathway can exacerbate the accumulation of dysfunctional proteins and organelles, further amplifying cellular stress and neuroinflammatory responses.

Balancing divergent research perspectives is crucial for advancing the field. While reductionist approaches have been invaluable for dissecting specific molecular events, systems biology and multiomics strategies are increasingly essential for unraveling the dynamic relationships between pathways. Integrating high-dimensional data from genomics, transcriptomics, proteomics, and metabolomics, alongside advanced cellular and animal models, provides a more comprehensive view of ALS–FTD pathogenesis. Moreover, cross-disciplinary collaborations among molecular biologists, clinicians, geneticists, and computational scientists are imperative to translate these mechanistic insights into clinically relevant biomarkers and robust therapeutic targets.

The identification of validated biomarkers remains a significant challenge but also a critical opportunity for clinical advancement. Understanding the molecular underpinnings of the ALS–FTD spectrum not only aids in early diagnosis and longitudinal disease monitoring but also facilitates the development of personalied therapeutic interventions. Targeted therapies that can modulate specific pathways—such as enhancing proteostasis, correcting RNA processing defects, restoring mitochondrial function, or dampening neuroinflammation—hold promise. However, given the interconnectedness of these molecular pathways, combination therapies or pharmacological agents with pleiotropic effects may ultimately prove more effective than single-target approaches.

Looking ahead, future research should focus on elucidating the dynamic interactions among these molecular pathways throughout the course of disease. Longitudinal studies employing patient-derived samples, high-resolution imaging, and functional assays will be instrumental in capturing the temporal evolution of molecular alterations in ALS-FTD. In parallel, the application of precision medicine approaches—including genetic profiling and patient stratification—will be essential to tailor interventions to individual disease mechanisms and improve clinical outcomes.

In conclusion, unraveling the molecular complexity of ALS–FTD spectrum disorders presents a formidable challenge and a pivotal opportunity for modern neuroscience. Future progress in this field hinges on our ability to integrate diverse research perspectives, continuously refine our mechanistic models, and translate these insights into tangible benefits for patients. A collaborative and balanced framework will be essential for realizing precision medicine in diagnosing, longitudinal monitoring, and treating these devastating neurodegenerative disorders.

References

Beckers, J., Van Damme, P., 2025. The role of autophagy in the pathogenesis and treatment of amyotrophic lateral sclerosis (ALS) and frontotemporal dementia (FTD). Autophagy Reports 4 (1). Available at https://doi.org/10.1080/27694127.2025.2474796.

Braems, E., Bercier, V., Van Schoor, E., Heeren, K., Beckers, J., Fumagalli, L., et al., 2022. HNRNPK alleviates RNA toxicity by counteracting DNA damage in C9orf72 ALS. Acta Neuropathologica 144 (3), 465–488. Available at https://doi.org/10.1007/s00401-022-02471-y.

Casterton, R.L., Hunt, R.J., Fanto, M., 2020. Pathomechanism heterogeneity in the amyotrophic lateral sclerosis and frontotemporal dementia disease spectrum: providing focus through the lens of autophagy. Journal of Molecular Biology 432 (8), 2692–2713. Available at https://doi.org/10.1016/j.jmb.2020.02.018, https://www.journals.elsevier.com/journal-of-molecular-biology.

Chua, J.P., De Calbiac, H., Kabashi, E., Barmada, S.J., 2022. Autophagy and ALS: mechanistic insights and therapeutic implications. Autophagy 18 (2), 254–282. Available at https://doi.org/10.1080/15548627.2021.1926656, http://www.tandfonline.com/toc/kaup20/current.

Costa-Pinto, S., Gonçalves-Ribeiro, J., Tedim-Moreira, J., 2024. Communication defects with astroglia contribute to early impairments in the motor cortex plasticity of SOD1G93A mice. Neurobiology of Disease 193, 106435.

Deng, Z., Sheehan, P., Chen, S., Yue, Z., 2017. Is amyotrophic lateral sclerosis/frontotemporal dementia an autophagy disease? Molecular Neurodegeneration 12 (1). Available at https://doi.org/10.1186/s13024-017-0232-6.

Desouky, M.A., Michel, H.E., Elsherbiny, D.A., 2024. Recent pharmacological insights on abating toxic protein species burden in neurological disorders: emphasis on 26S proteasome activation. Life Sciences 15 (359), 123206.

Feng, Y., Xu, Z., Jin, H., Chen, Y., Fu, C., Zhang, Y., et al., 2024. Metformin ameliorates mitochondrial damage induced by C9orf72 poly(GR) via upregulating AKT phosphorylation. Journal of Cellular Biochemistry 125 (3). Available at https://doi.org/10.1002/jcb.30526.

Gao, C., Jiang, J., Tan, Y., Chen, S., 2023. Microglia in neurodegenerative diseases: mechanism and potential therapeutic targets. Signal Transduction and Targeted Therapy 8 (1). Available at https://doi.org/10.1038/s41392-023-01588-0.

Giuliano, S., Montemagno, C., Domdom, M.-A., Teisseire, M., Brest, P., Klionsky, D.J., et al., 2023. Should evidence of an autolysosomal de-acidification defect in Alzheimer and Parkinson diseases call for caution in prescribing chronic PPI and DMARD? Autophagy 19 (10), 2800–2806. Available at https://doi.org/10.1080/15548627.2023.2214960.

Hao, Z., Wang, R., Ren, H., Wang, G., 2020. Role of the C9ORF72 gene in the pathogenesis of amyotrophic lateral sclerosis and frontotemporal dementia. Neuroscience Bulletin 36 (9), 1057–1070. Available at https://doi.org/10.1007/s12264-020-00567-7, http://www.springer.com/biomed/neuroscience/journal/12264.

Heylen, A., Vermeiren, Y., De Deyn, P.P., Van Dam, D., 2025. Monoaminergic alterations at the subregional cervical and thoracic spinal cord level of patients within the FTD-ALS continuum and early-onset AD: low thoracic dopaminergic activity in ALS. Journal of Neurochemistry 169 (3). Available at https://doi.org/10.1111/jnc.70046, http://onlinelibrary.wiley.com/journal/10.1111/(ISSN)1471-4159.

Hung, C., Patani, R., 2024. 4R Tau drives endolysosomal and autophagy dysfunction in frontotemporal dementia. Autophagy 20 (5), 1201–1202. Available at https://doi.org/10.1080/15548627.2023.2300917.

Jiang, X., Schaeffer, L., Patni, D., Russo, T., Lee, C.-Z., Aguilar, C., et al., 2026. Blocking RAN translation without altering repeat RNAs rescues C9ORF72-related ALS and FTD phenotypes. Science (New York, N.Y.) 391 (6785). Available at https://doi.org/10.1126/science.adv2600.

Jury, N., Abarzua, S., Diaz, I., Guerra, M.V., Ampuero, E., Cubillos, P., et al., 2020. Widespread loss of the silencing epigenetic mark H3K9me3 in astrocytes and neurons along with hippocampal-dependent cognitive impairment in C9orf72 BAC transgenic mice. Clinical Epigenetics 12 (1). Available at https://doi.org/10.1186/s13148-020-0816-9.

Kim, J., Kim, H.S., Chung, J.H., 2023. Molecular mechanisms of mitochondrial DNA release and activation of the cGAS-STING pathway. Experimental and Molecular Medicine 55 (3), 510–519. Available at https://doi.org/10.1038/s12276-023-00965-7, https://www.nature.com/emm/.

Li, A., Cao, S.-Q., Fang, E.F., Su, H., 2025. Pharmacological activation of mitophagy antagonizes motor neuron degeneration in a cross-species platform of amyotrophic lateral sclerosis. Autophagy 1–3. Available at https://doi.org/10.1080/15548627.2025.2610450.

Lin, B.C., Higgins, N.R., Phung, T.H., Monteiro, M.J., 2022. UBQLN proteins in health and disease with a focus on UBQLN2 in ALS/FTD. FEBS Journal 289 (20), 6132–6153. Available at https://doi.org/10.1111/febs.16129, http://onlinelibrary.wiley.com/journal/10.1111/(ISSN)1742-4658.

Lin, Y.C., Kumar, M.S., Ramesh, N., Anderson, E.N., Nguyen, A.T., Kim, B., et al., 2021. Interactions between ALS-linked FUS and nucleoporins are associated with defects in the nucleocytoplasmic transport pathway. Nature Neuroscience 24 (8), 1077–1088. Available at https://doi.org/10.1038/s41593-021-00859-9, http://www.nature.com/neuro/index.html.

Liu, T., Woo, J.A.A., Bukhari, M.Z., LePochat, P., Chacko, A., Selenica, M.L.B., et al., 2020. CHCHD10-regulated OPA1-mitofilin complex mediates TDP-43-induced mitochondrial phenotypes associated with frontotemporal dementia. FASEB Journal 34 (6), 8493–8509. Available at https://doi.org/10.1096/fj.201903133RR, https://onlinelibrary.wiley.com/journal/15306860.

Luan, T., Li, Q., Huang, Z., Feng, Y., Xu, D., Zhou, Y., et al., 2024. Axonopathy underlying amyotrophic lateral sclerosis: unraveling complex pathways and therapeutic insights. Neuroscience Bulletin 40 (11), 1789–1810. Available at https://doi.org/10.1007/s12264-024-01267-2, https://www.springer.com/journal/12264.

Malpetti, M., Cope, T.E., Street, D., Jones, P.S., Hezemans, F.H., Mak, E., et al., 2023. Microglial activation in the frontal cortex predicts cognitive decline in frontotemporal dementia. Brain 146 (8), 3221–3231. Available at https://doi.org/10.1093/brain/awad078, http://brain.oxfordjournals.org/.

Masrori, P., Beckers, J., Gossye, H., Van Damme, P., 2022. The role of inflammation in neurodegeneration: novel insights into the role of the immune system in C9orf72 HRE-mediated ALS/FTD. Molecular Neurodegeneration 17 (1). Available at https://doi.org/10.1186/s13024-022-00525-z.

Mehta, P.R., Brown, A.L., Ward, M.E., Fratta, P., 2023. The era of cryptic exons: implications for ALS-FTD. Molecular Neurodegeneration 18 (1). Available at https://doi.org/10.1186/s13024-023-00608-5, https://molecularneurodegeneration.biomedcentral.com.

Moens, T.G., Da Cruz, S., Neumann, M., Shelkovnikova, T.A., Shneider, N.A., Van Den Bosch, L., 2025. Amyotrophic lateral sclerosis caused by FUS mutations: advances with broad implications. The Lancet Neurology 24 (2), 166–178. Available at https://doi.org/10.1016/S1474-4422(24)00517-9, https://www.sciencedirect.com/science/journal/14744422.

Mori, G., Uozumi, et al., 2023. RNA dysmetabolism and repeat-associated non-AUG translation in frontotemporal lobar degeneration/amyotrophic lateral sclerosis due to C9orf72 hexanucleotide repeat expansion. JMA Journal 6, 9–15.

Nguyen, H., Zerimech, S., Baltan, S., 2021. Astrocyte mitochondria in white-matter injury. Neurochemical Research 46 (10), 2696–2714. Available at https://doi.org/10.1007/s11064-021-03239-8.

Nixon, R.A., Rubinsztein, D.C., 2024. Mechanisms of autophagy–lysosome dysfunction in neurodegenerative diseases. Nature Reviews. Molecular Cell Biology 25 (11), 926–946. Available at https://doi.org/10.1038/s41580-024-00757-5, https://www.nature.com/molcellbio.

Odeh, H.M., Shorter, J., 2020. Arginine-rich dipeptide-repeat proteins as phase disruptors in C9-ALS/FTD. Emerging Topics in Life Sciences 4 (3), 293–305. Available at https://doi.org/10.1042/ETLS20190167, https://portlandpress.com/emergtoplifesci/article/4/3/293/225720/Argininerich-dipeptiderepeat-proteins-as-phase.

Park, K.H., Yu, E., Choi, S., Kim, S., Park, C., Lee, J.E., et al., 2025. Optogenetic induction of TDP-43 aggregation impairs neuronal integrity and behavior in Caenorhabditis elegans. Translational Neurodegeneration 14 (1). Available at https://doi.org/10.1186/s40035-025-00480-x, https://translationalneurodegeneration.biomedcentral.com/.

Parra Bravo, C., Naguib, S.A., Gan, L., 2024. Cellular and pathological functions of Tau. Nature Reviews. Molecular Cell Biology 25 (11), 845–864. Available at https://doi.org/10.1038/s41580-024-00753-9, https://www.nature.com/molcellbio.

Picchiarelli, G., Dupuis, L., 2020. Role of RNA binding proteins with prion-like domains in muscle and neuromuscular diseases. Cell Stress 4 (4), 76–91. Available at https://doi.org/10.15698/cst2020.04.217.

Rezvykh, A.P., Ustyugov, A.A., Chaprov, K.D., Teterina, E.V., Nebogatikov, V.O., Spasskaya, D.S., et al., 2023. Cytoplasmic aggregation of mutant FUS causes multistep RNA splicing perturbations in the course of motor neuron pathology. Federation Nucleic Acids Research 51 (11), 5810–5830. Available at https://doi.org/10.1093/nar/gkad319, https://academic.oup.com/nar/issue.

Rivers-Auty, J., Hoyle, C., Pointer, A., Lawrence, C., Pickering-Brown, S., Brough, D., et al., 2024. C9orf72 dipeptides activate the NLRP3 inflammasome. Brain Communications 6 (5). Available at https://doi.org/10.1093/braincomms/fcae282.

Rummens, J., Da Cruz, S., 2025. RNA-binding proteins in ALS and FTD: from pathogenic mechanisms to therapeutic insights. Molecular Neurodegeneration 20 (1). Available at https://doi.org/10.1186/s13024-025-00851-y.

Shu, X., Wei, C., Tu, W.-Y., Zhong, K., Qi, S., Wang, A., et al., 2023. Negative regulation of TREM2-mediated C9orf72 poly-GA clearance by the NLRP3 inflammasome. Cell Reports 42 (2).

Sleigh, J.N., Rossor, A.M., Fellows, A.D., Tosolini, A.P., Schiavo, G., 2019. Axonal transport and neurological disease. Nature reviews. Neurology. 15 (12), 691–703. Available at https://doi.org/10.1038/s41582-019-0257-2.

Theunissen, F., West, P.K., Brennan, S., Petrović, B., Hooshmand, K., Akkari, P.A., et al., 2021. New perspectives on cytoskeletal dysregulation and mitochondrial mislocalization in amyotrophic lateral sclerosis. Translational Neurodegeneration 10 (1). Available at https://doi.org/10.1186/s40035-021-00272-z, http://www.translationalneurodegeneration.com/.

Tsai, Y.L., Coady, T.H., Lu, L., Zheng, D., Alland, I., Tian, B., et al., 2020. ALS/FTD-associated protein FUS induces mitochondrial dysfunction by preferentially sequestering respiratory chain complex mRNAs. Genes and Development 34 (11-12), 785–805. Available at https://doi.org/10.1101/gad.335836.119, http://genesdev.cshlp.org/content/34/11-12/785.full.

Venkatesan, D., Iyer, M., Narayanasamy, A., Gopalakrishnan, A.V., Vellingiri, B., 2023. Plausible role of mitochondrial DNA copy number in neurodegeneration—a need for therapeutic approach in Parkinson's disease (PD). Molecular Neurobiology 60 (12), 6992–7008. Available at https://doi.org/10.1007/s12035-023-03500-x.

Zhang, H., Gu, J., Zhang, Y., Guo, H., Zhang, S., Song, J., et al., 2023. Graphene quantum dots modulate stress granule assembly and prevent abnormal phase transition of fused in sarcoma protein. ACS Nano 17 (11), 10129–10141. Available at https://doi.org/10.1021/acsnano.3c00001, http://pubs.acs.org/journal/ancac3.

Zhang, Y., Xu, N., Yan, C., Zhou, X., Qiao, Q., Miao, L., et al., 2024. Live-cell imaging to resolve salt-induced liquid–liquid phase separation of FUS protein by dye self-labeling. Chemical & Biomedical Imaging 2 (1), 70–80. Available at https://doi.org/10.1021/cbmi.3c00094.

CHAPTER 8

New treatments on the horizon for Alzheimer's and other neurodegenerative diseases

Hannah Bailey[1], Robert Layfield[1], and Eva Žerovnik[2,3]
[1]Department of Protein Biochemistry, The University of Nottingham Medical School Queen's Medical Centre Nottingham, Nottingham, United Kingdom,
[2]Department of Biochemistry and Molecular and Structural Biology, Jožef Stefan Institute, Ljubljana, Slovenia,
[3]Jožef Stefan International Postgraduate School (IPS), Ljubljana, Slovenia

8.1 Common processes in neurological disorders

Alzheimer's disease (AD) is a neurodegenerative disorder characterized by progressive impairment of cognitive and behavioral functions such as memory and comprehension, often with the addition of neuropsychiatric symptoms, including agitation and delusions, in the late-stage disease. The strongest risk factor for AD is ageing, with a risk of over 30% for those aged 85 (Alzheimer's Association., 2019). AD is the most common form of dementia, although most cases are likely to be mixed dementias (Barker et al., 2002; DeTure and Dickson, 2019). In postmortem brain tissue analysis, AD is rarely found without neurodegenerative copathologies (DeTure and Dickson, 2019). The main pathological hallmarks of the AD brain are the accumulation of amyloid-β (Aβ) plaques and neurofibrillary tangles, principally composed of hyperphosphorylated tau protein. As with many other neurodegenerative diseases, neuroinflammation involving the activation of microglia and astrocytes occurs, likely as a protective mechanism early in the disease course; however, prolonged neuroinflammation promotes neurodegeneration as the disease progresses (Di Benedetto et al., 2022; Kwon and Koh, 2020).

Understanding the molecular and cellular mechanisms of neurodegeneration and neural repair underpins the search for new treatments. The molecular and cellular mechanisms that underlie AD and neurodegenerative diseases are essentially a response to protein misfolding and aggregation. As evident from genetic cases, the disease often starts with loss of function of the mutated disease-associated protein, accompanied by toxic effects of protein aggregation. Inflammation as a result of innate immunity responses is also important, although the full temporal aspects of the immune component of AD are incompletely understood (Heppner et al., 2015). The modified amyloid cascade hypothesis of AD states that Aβ oligomers—precursors of mature extracellular protein aggregates—are responsible for the neuronal and synaptic damage. These oligomers are believed to impair intracellular membranes (Bucciantini et al., 2012), likely resulting in

Molecular Insights into Proteinopathies
DOI: https://doi.org/10.1016/B978-0-443-34176-2.00003-1
© 2027 Elsevier Inc. All rights are reserved, including those for text and data mining, AI training, and similar technologies.

mitochondrial and lysosomal dysfunction, which subsequently lead to oxidative stress and impairment of protein clearance by autophagy, respectively. These events are presumably upstream of hyperphosphorylation and aggregation of tau protein, for example, as evidenced by tau pathology in the case of AD with amyloid precursor protein (APP) mutations. Pathological consequences of intracellular tau protein aggregation events are (like Aβ oligomers) elevated oxidative stress, impairments in mitochondrial energy levels (Wang et al., 2014), and an impact on autophagic flux leading to accumulation of autophagosomes (Orr and Oddo, 2013). Further, extracellular accumulation of Aβ peptide outside of neurons as amyloid plaques activates microglia, leading to chronic inflammation (Li et al., 2026).

AD shares many features with other neurological disorders. In addition to oxidative stress and neuroinflammation, glutamate-mediated excitotoxicity is a part of a pathogenic "triad" also documented in amyotrophic lateral sclerosis (ALS) and also characterizes the neurobiology of epilepsy and other major mental health disorders (Holton, 2021). The exact sequence of events and mechanisms by which these features emerge during the progression of neurological disorders remain unclear. For example, TREM2 is a cell-surface receptor that has an important role in microglial activation (Ulrich and Holtzman, 2016), playing a critical role in response to inflammation as well as in the clearance of extracellular amyloid plaques. TREM2 activation may be beneficial in the early stages of AD but detrimental in the later stages. Therefore TREM2-targeted treatments for AD must consider both positive effects and potential risks.

Building on a previous review from 2016 (Zerovnik et al., 2016), we provide an update update on developments in early detection, prevention, and disease modification of AD over the past decade. Some new therapies are also reviewed and discussed.

8.2 New *Alzheimer's disease* molecular biomarkers may lead to early detection

Early diagnosis of AD is of utmost importance because disease-modifying treatments are imminent or already available. Until recently, diagnosis of AD was typically obtained when the brains of patients were already severely affected, based on clinical signs of mental decline. Therefore early diagnosis, possibly at a stage of mild cognitive impairment, and the identification of individuals at higher risk of developing the disease, is highly desirable.

Molecular biomarkers from peripheral fluids likely represent the most clinically accessible source. An early event in AD is the overproduction of the more toxic Aβ peptide Aβ42, as opposed to Aβ40. Various imaging techniques enable one to detect and quantify extracellular Aβ plaques in living patients (Adlard, et al., 2014b), offering a less invasive approach to performing a lumbar puncture to obtain CSF, in order to analyze the ratio between Aβ40 and Aβ42 peptides. A more convenient approach would be to detect a biomarker in peripheral blood and follow changes in its level upon disease progression (Khan and Alkon, 2015). Such an approach has been reported by researchers at Rockefeller University, who measured levels of contact system activation (the intrinsic pathway of coagulation) in patient serum. The contact system cascade starts with factor XII, followed by the downstream release of bradykinin. The study showed activation of the inflammation cascade, especially in those AD patients who demonstrated high levels of Aβ42 in their CSF. Importantly, the Tg6799 mouse model of AD reproduced the findings that were observed with human blood samples. Zamolodchikov et al. (2015) proposed that chronic activation of the contact system by Aβ42 results in constant low levels of bradykinin-mediated inflammatory processes in the blood of AD patients (i.e., chronic inflammation). This in turn leads to increased blood–brain barrier permeability and cytokine upregulation (Zamolodchikov et al., 2015).

Several important high-profile human biomarker studies of dementia have been published in the last 5 years (e.g. Guo et al., 2024; Duggan et al., 2025). For example, Walker et al. (2021) performed a large-scale plasma proteomic analysis. Pathway analyses of dementia-associated proteins implicate immune, lipid, metabolic signaling, and hemostasis pathways in dementia pathogenesis. Guo et al. (2024) performed a proteomics study on a large British biobank. From >1400 proteins detected in blood plasma, glial fibrillary acidic protein (GFAP), NEFL, GDF15, and LTBP2 are consistently associated with the incidence of different forms of dementia (vascular, AD, and all-cause dementia). GFAP proved to be an optimal biomarker for dementia prediction, even more than 10 years before the diagnosis. In a previous study by Chatterjee et al. (2021) demonstrated that plasma GFAP levels are elevated in cognitively normal older adults at risk of AD, suggesting astrocytic damage or glia activation begins from the presymptomatic stage of AD and is also associated with brain Aβ load.

A further approach to detect biomarkers in human peripheral blood was reported by researchers at the University of Melbourne (Cheng et al., 2015). They focused on differences in microRNAs (miRNAs), short noncoding regulatory RNAs, between AD patients and healthy controls. The miRNA in serum and other biological fluids can be either free or reside in exosomes (extracellular vesicles). Using next-generation deep sequencing, they profiled exosomal miRNA from serum and validated the results by quantitative reverse transcription PCR. Other risk factors for AD, including clinical, medical, and cognitive factors, and amyloid neuroimaging, were assessed. An AD-specific 16-miRNA signature was then selected, which, together with other risk factors, among them apolipoprotein ε4 allele status, resulted in a sensitivity and specificity of 87% and 77%, respectively, for predicting AD (Cheng et al., 2015). Among recent biomarkers of AD are other circulating RNAs (Riscado et al., 2021). While many studies have focused on identifying miRNA biomarkers, other RNA biotypes such as long noncoding RNAs (lncRNAs) and circRNAs have also been investigated. For example, elevated lncRNA 51A has been linked to alternative splicing of the AD risk gene *SORL1*, leading to impaired APP processing and increased Aβ formation (Ciarlo et al., 2013). Another lncRNA, BACE1-AS, has been suggested as a possible blood-based AD biomarker, including for identifying individuals in the preclinical stages of AD (Fotuhi et al., 2019).

Biomarkers can be defined as a measurable signature of a biological condition or process. Frequently employed and useful biomarkers of oxidative stress and neuroinflammation, including novel promising biomarkers, have been developed for AD (Menzel et al., 2021; Teleanu et al., 2022). Some of them are collated in Table 8.1, which might be taken as a proposal, only, demanding more clinical studies to confirm their relevance.

An important feature of AD pathology is protein aggregation. This can be assessed indirectly by detecting the amount of the aggregating proteins, such as Aβ and tau, in blood plasma or directly by imaging the aggregates by microscopic means, either in prepared brain slices (postmortem) or in the blood plasma of living patients. Protein aggregation in blood samples of neurodegenerative disease, including AD, can be assessed by measuring the Aβ40/Aβ42 ratio (Vogelgsang et al., 2018) and by measuring circulating hyperphosphorylated tau (Thijssen et al., 2021). Studies suggest that p-tau217 is a stronger indicator of amyloid-PET positivity (Thijssen et al., 2021) than total phosphorylated tau.

Table 8.1 Selected biomarkers to trace in biofluids (blood, saliva, and urine) in patients with Alzheimer's disease and possibly other proteinopathies.

Blood	Saliva	Urine
Glial fibrillary acidic protein (De Meyer et al., 2025)	Lactoferrin (Guo et al., 2025)	Cystatin B (Gordin et al., 2024)
Neurofilament light chain (De Meyer et al., 2025)	Melatonin (Guo et al., 2025)	Cystatin C (Vaidya et al., 2008)
Phosphorylated tau, particularly p-tau217) (Verde, 2022; Thijssen et al., 2021)	Cortisol (Guo et al., 2025)	Clusterin (Vaidya et al., 2008; Gordin et al., 2024)
Brain-derived growth factor (Ibrahim et al., 2022)	Cystatin B (Contini et al., 2023)	IL-18 (Vaidya et al., 2008)
Cytokines: IL-2, IL-5, IL-6, IL-8, IL-12, IL-13, IL-16, CCL3, CCL4, CCL17, CXCL10, TNF-α, TNF-β, and VEGF-C (Xu et al., 2023)	Aβ	Kidney injury molecule-1 (Vaidya et al., 2008)
Aβ40/Aβ42 ratio (Vogelgsang et al., 2018)		

Until recently, insoluble protein aggregates could be measured only on postmortem brain samples. By using a novel single-molecule pull-downbased assay (MAPTau), it was possible to characterize individual tau aggregates in AD, both from postmortem brain samples and in biofluids, including blood serum (Böken et al., 2024). Furthermore, using superresolution microscopy, it has recently been demonstrated that aggregate-activated fluorophores (ProteoStat and Amytracker 630 [AT630]) can be used to quantify pathogenic aggregates in tissue, blood, or cells, paving the path for future molecular studies relevant to neurodegeneration (Morten et al., 2022). Especially, AT630 fluorophore has outstanding photophysical properties that enable superresolution imaging of α-synuclein, tau, and Aβ aggregates, achieving ~4 nm precision (Morten et al., 2022). The physics of the superresolution microscopy methods is beyond this review chapter. More detailed description of superresolution techniques such as classical SRMs (STED, PALM, SIM, STORM), their applications in studies of neurological diseases, and comparisions of their respective advantages and disadvantages are available elsewhere (Gong et al., 2023).

8.3 Preventive measures for *Alzheimer's disease*

8.3.1 Diet and exercise

Because the costs of treating individuals living with AD are enormous and the disease places a significant burden on the relatives and those directly affected, preventive measures are certainly welcome. Physical exercise has been proven as one way to reduce the burden of Aβ plaques in patient studies (Yu et al., 2014; Coelho et al., 2014). Nutrition is an obvious source of health-promoting substances. In a recent survey, the effects of some natural compounds on Aβ aggregation and implications for possible new AD treatments were reviewed (Bu et al., 2016; Asthana et al., 2013; Hahnen et al., 2008). Among the best-known plant compounds for their preventive effects on AD are the natural plant-derived products resveratrol and curcumin. It was reported that resveratrol, like many other polyphenols, contributes to Aβ fibril disaggregation (Ladiwala et al., 2010), and importantly, inhibits Aβ oligomerisation, preventing toxicity (Ono et al., 2012). Similar effects of resveratrol were observed on α-synuclein oligomerization, thus potentially preventing synaptic dysfunction in both AD and PD (Hamaguchi et al., 2009; Takahashi et al., 2015). It was previously shown that resveratrol did not inhibit Aβ production by the two (β and γ) secretases; instead, it enhanced proteasome activity (Marambaud et al., 2005). In accordance, the decrease of Aβ was prevented by selective proteasome inhibitors and by siRNA-directed silencing of the proteasome subunit β5 (Marambaud et al., 2005). Curcumin's inhibitory action on Aβ fibril formation is also well documented (Caesar et al., 2012; Yang et al., 2005, 2006). Of note, the same two compounds (resveratrol and curcumin) exert antiinflammatory and antioxidant actions (Bo et al., 2013; Menon and Sudheer, 2007).

There is more and more evidence about the beneficial effects of aerobic exercise for better cognition in people at risk of developing AD. Irisin is an exercise-induced hormone of the muscles. When released in response to exercise, it induces the release of neprilysin (NEP) from astrocytes following the downregulation of ERK-STAT3 signaling (Kim et al., 2023). The cellular and molecular mechanisms by which exercise-induced irisin attenuates Aβ pathology were recently clarified. It has been shown that irisin significantly reduces Aβ burden by increasing the release of the Aβ-degrading enzyme NEP. Integrin αV/β5 acts as the irisin receptor, which then leads to downregulation of ERK-STAT3 signaling and finally leads to release of astrocytic NEP and clearance of Aβ. Therefore this may represent a new target pathway for therapies aimed at the prevention and treatment of AD.

8.3.2 Regulation of cold stress signaling as a novel therapeutic approach for *Alzheimer's disease*

An emerging area relevant to AD therapeutics relates to the manipulation of endogenous cold stress signaling pathways, which comprise the cellular and molecular events that occur in response to low temperatures and are aimed at helping an organism survive the cold. RNA binding motif 3 (RBM3) is a critical protein that orchestrates cellular responses to cold stress, regulating reassembly of synapses in small mammals after hibernation, and has also been shown to preserve synapses and promote neuroprotection in AD and prion disease mouse models (Peretti et al., 2015). RBM3 protein expression is induced by even mild hypothermia in a variety of human cells, regulating cold stress signaling and the translational control of other proteins that mediate hypothermic neuroprotection (Lin et al, 2023).

In the disease context, RBM3 promotes neuroprotection in AD (5xFAD) and prion-infected mouse models, in part through the preservation of synaptic integrity (Peretti et al., 2015). In these models, synapse loss at later stages of disease progression coincides with a reduced capacity to upregulate RBM3 in response to experimental cooling. In both models, RNAi-mediated knockdown of RBM3 at normothermia increased synapse loss, accelerated disease progression, and reduced survival. Enhancing RBM3 expression either by cooling animals early in the disease course to enhance endogenous protein expression or by using direct lentiviral delivery of exogenous RBM3 at normothermia (without cooling) prevented hippocampal synapse loss, alleviated behavioral and memory impairments in both models, and, in prion mice, improved survival (Peretti et al., 2015). More recent work demonstrates that long-lasting elevation of endogenous RBM3 in mice at normothermia in vivo can be achieved via splicing regulation using RNA-targeting antisense oligonucleotides (ASOs) produced with Food and Drug Administration (FDA)-approved chemistry, blunting spongiform change and neuronal loss in the prion-infected mice (Preußner et al., 2023).

Mechanistically, the ASOs prevent inclusion of a "poison exon" in RBM3 mRNA that is otherwise included at normothermia (Lin et al., 2023) and which in turn blocks nonsense-mediated decay of the mRNA and promotes RBM3 protein expression. Collectively, these findings indicate the potential for manipulating cold stress signaling pathways as a therapeutic approach in diverse neurodegenerative disorders, including, but not limited to, AD.

8.3.3 The metabolic syndrome as a risk factor for *Alzheimer's disease*

There is a well-established link between AD and type 2 diabetes mellitus (T2DM) (Nisar et al., 2020). Therefore reversing diabetes might be beneficial in reducing the risk of AD. Metformin is an effective medication for T2DM that has been proposed to decrease the risk of age-related disorders, including AD, and it is currently undergoing phase 3 clinical trials for AD (Swain and Metya, 2025). Metformin is compared to the effect of another drug, pterostilbene (PTS), a stilbenoid (Fig. 8.1). In the study, "Pterostilbene Ameliorates Cognitive Impairment in Polycystic Ovary Syndrome Rat Model through Improving Insulin Resistance via the IRS-1/PI3K/Akt/GSK-3β Pathway: A Comparative Study with Metformin" the authors (Foda et al., 2024) claim that the IRS-1/PI3K/AKT/glycogen synthase kinase (GSK)-3β pathway is very important for clear cognition. Their research (Foda et al., 2024) shows that PTS reduces the activity of the tau kinase GSK-3β, and thus mitigates tau hyperphosphorylation (Foda et al., 2024). PTS has multifaceted effects, also acting on neuroinflammation by reducing the expression of interleukin 1β (IL-1β), tumor necrosis factor alpha (TNF-α), inducible nitric oxide synthase, cyclooxygenases (COX), and nuclear factor kappa B (NFκB). PTS is a naturally derived stilbenoid, structurally related to resveratrol (Fig. 8.1), with potential antioxidant, antiinflammatory, pro-apoptotic, antineoplastic, and cytoprotective activities. Upon

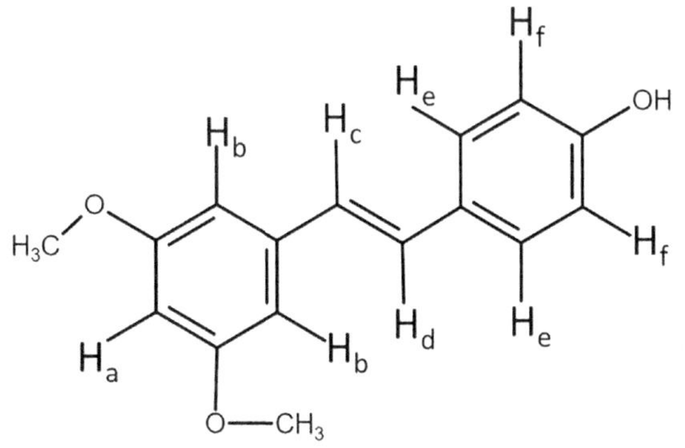

Figure 8.1 Chemical structure of pterostilbene.

administration, PTS exerts its antioxidant activity by scavenging reactive oxygen species (ROS), thereby preventing oxidative stress and ROS-induced cell damage. It may also activate the nuclear factor erythroid 2-related factor 2 (Nrf2)-mediated pathway and increase the expression of various antioxidant enzymes, such as superoxide dismutase (SOD).

As T2DM has a close link to AD risk (Han and Li, 2010) and diabetes causes kidney injury, it may be worthwhile to also measure acute kidney injury (AKI) markers in urine samples from AD patients (as suggested in Table 8.1). AKI is multifactorial, and the exact etiology is often unknown, but it is associated with biomarkers that are sensitive and specific for the detection of tubular injury at an early stage, irrespective of the underlying etiology. Classical biomarkers of AKI were described as early as 2008 (Vaidya et al., 2008). Among them were cystatin C, clusterin, IL-18, and kidney injury molecule-1. The novel biomarkers urinary clusterin and urinary cystatin B could also fulfill these criteria (Gordin et al., 2024); however, they have not yet been evaluated in human clinical trials, and have thus far been studied only in dogs with AKI. However, in 2022, a study was conducted to find an association between AKI and dementia in the atherosclerosis risk in communities (Tung et al., 2022). It was shown that people who were hospitalized with AKI had a significantly higher risk of dementia compared to those without a hospitalization for AKI (Tung et al., 2022).

8.4 New routes towards a successful treatment for *Alzheimer's disease*

By now, despite substantial research investment by the scientific community, we have not achieved a cure for AD. However, new aspects of the disease are being targeted therapeutically, and, when intervention occurs sufficiently early, this can delay the disease course. With a more understanding of the underlying pathology and molecular causes, we can hope to improve the treatment and even hope for a cure.

8.4.1 Decreasing amyloid-β

In accordance with the modified amyloid cascade hypothesis, many new trials searching for treatments for AD are based on reducing the levels of soluble Aβ peptide or clearing amyloid plaques from the brain blood vessels. An innovative approach has been explored in a mouse model of AD by researchers at the University of Queensland using ultrasound. High-frequency sound waves activated the microglial cells, which in turn digested Aβ plaques. As reported by Leinenga and Götz, (2015), memory loss was reversed in these mice. As already described under "Preventive measures," some natural compounds from plants also reduce amyloid plaques and inhibit intracellular Aβ aggregation.

8.4.2 Amyloid-β immunization

Another potential route to reduce toxic Aβ oligomers is by immunization. It is important to differentiate between passive and active immunization, where active means that the body produces endogenous antibodies against the peptide or protein of interest, which is presented to it by the peripheral blood. However, with the potential hazards of Aβ peptides and oligomers, peripheral injection runs the potential risk of promoting "seeding effects" that may drive amyloid pathology. In addition, those with compromised immune systems, which can be the case when living with AD, may fail to raise strong immune responses. Passive immunization with an engineered, preferentially humanized antibodies is therefore safer, even though it may require repeated administration. Passive immunization trials with antibodies directed against Aβ have been performed for some time, as reviewed by Lambracht-Washington and Rosenberg (2013). Several of these earlier studies had to be abandoned due to serious side effects, including aseptic meningoencephalitis (Orgogozo et al., 2003).

A more recent antibody for passive immunization is BAN2401, a monoclonal antibody directed against Aβ that selectively binds and eliminates Aβ protofibrils. BAN2401 was a promising candidate for Aβ immunotherapy in AD patients at an early stage (Lannfelt et al., 2014). The clinical study evaluated its effect on cognition and biomarkers reflecting the progression of the disease. Crenezumab is a further monoclonal antibody used for passive immunization. It recognizes primarily aggregated Aβ, including oligomeric and fibrillar species and amyloid plaques. In order to avoid side effects such as vasogenic edema, this humanized antibody is prepared on an IgG4 backbone, exerting activation of microglia (Adolfsson et al., 2012).

An announcement about a possible "cure" for AD was made when researchers evaluated a high-affinity naturally occurring human antibody against Aβ oligomers. These antibodies were isolated from people aged around 100 years, who were still cognitively intact, assuming they produce powerful auto-antibodies against the endogenous Aβ peptide. The resulting biological, Aducanumab (by Biogen), is a high-affinity monoclonal antibody against Aβ, based on the natural sequence of these auto-antibodies. It recognizes Aβ's N-terminal structural epitope that is present in the aggregated form of Aβ but absent in monomers. First clinical trials (phase 1b) of passive immunization with Aducanumab were seemingly successful, lowering Aβ plaques in the brain as seen by PET imaging and also improving cognitive performance. However, severe side effects were observed, such as brain swelling and headaches, especially at the highest, most effective dose. Medium doses produced fewer side effects and were still moderately efficient. The trials were subsequently repeated with a wider dose range and more patients. The phase 3 clinical trial program for Aducanumab comprised two parallel studies (EMERGE, ENGAGE), which were terminated early after a futility analysis. In both trials, strong amyloid reduction was observed, but the clinical data were inconclusive (Jessen et al., 2024).

Lecanemab is the first antibody to demonstrate clinical efficacy in the phase 3 program (Jessen et al., 2024). While it is fully approved by the United States FDA and by the regulatory bodies of Japan, China, and South Korea, the European Medicines Agency (EMA) recommended against approval in July 2024. The main reason for a long procedure of approval was the claim that its benefits for cognition did not counterbalance the risk of serious side effects, in particular the frequent occurrence of amyloid-related imaging abnormalities, which involved swelling and potential bleeding in the brain of patients. Finally, in April 2025, the EMA approved Lecanemab for the treatment of early AD (Van Dyck et al., 2023), making it the first disease-modifying therapy for AD to be made available in the EU-27 and EEA Member States.

Donanemab, another amyloid-targeting antibody that binds to soluble and insoluble Aβ with proven efficacy, was fully approved by the FDA in July 2024. Very recently, in July 2025, the EMA granted marketing authorization for donanemab, too. The recommendation is based on the results of the phase 3 TRAILBLAZER-ALZ 2 clinical trial. It has been proven that this monoclonal antibody developed by Eli Lilly (under the name Kisunla) can slow the progression of AD in people in the early stages of the disease, with low or moderate levels of tau, that is, mild dementia or mild cognitive decline.

See, https://fpmaragall.org/en/post/the-ema-approves-the-marketing-of-donanemab-for-the-treatment-of-early-stage-alzheimers-disease/

Both biologics, monoclonal antibodies—Lecanemab and Donanemab, are part of a new generation of treatments that act on biological mechanisms underlying AD, in contrast to the therapeutic approaches used to date, which have focused solely on symptomatic relief.

8.4.3 Decreasing inflammation in *Alzheimer's disease*

Microglia are a unique CNS resident myeloid cell population, derived from primitive myeloid progenitors (Ginhoux et al., 2010). When microglial cells become over-activated, they induce significant production of cytotoxic molecules such as superoxide (Colton and Gilbert, 1987), NO (Moss and Bates, 2001), and TNF-α (Lee et al., 1993). In AD, microglia can bind to soluble Aβ oligomers and fibrils via cell-surface receptors, and this process participates in an inflammatory response. The inflammasome sensor NLRP3 is important for mediating neuroinflammation as it can sense a range of aggregated substances, including Aβ aggregates (Halle et al., 2008). Recently, several articles have described the role of inflammation in AD progression (Heneka et al., 2013; Heneka et al., 2015). NLRP3-deficient APP/PS1 mice (a mouse model of AD) have decreased deposition of Aβ (Heneka et al., 2013).

A study on the role of microglia in an AD animal model was conducted at Stanford University in 2015 (Johansson et al., 2015). The deterioration in microglial function with age and in AD is driven, in large part, by increased signaling activity of the prostaglandin receptor protein EP2, which resides both on the surface of microglial and neural cells. Activation of the EP2 receptor by prostaglandin E2 or PGE2 leads to inflammation. In more detail, the authors Johansson et al. (Johansson et al., 2015) examined peritoneal macrophages of young (4 months) and aged (21 months) C57B6/J mice. When exposed to soluble Aβ oligomers, macrophages from young mice produced a rather modest response, not causing inflammation. In contrast, exposure of macrophages from older mice to Aβ initiated a significant increase in EP2 activity, resulting in inflammation and a reduced amount of Aβ-digesting enzymes. A molecule to block the downstream activity of the EP2 receptor, that is, an inhibitor, would be highly desirable. In fact, nonsteroidal antiinflammatory drugs block two enzymes: COX-1 and COX-2, which produce prostaglandin PGE2 that triggers EP2 action. Nonsteroidal antiinflammatory drugs are widely used in the elderly, yet no clear benefits for preventing or ameliorating AD symptoms have been observed in patients to date. Some improvements in cognition in animal models were observed for ibuprofen and mefenamic acid (Meraz-Ríos et al., 2013).

Notably, human clinical trials of antiinflammatory substances (such as aspirin and naproxen) mostly failed (Martin et al., 2008). This seems consistent with recent studies indicating that inflammation is a transient and early phase of AD (Heppner et al., 2015; Meraz-Ríos et al., 2013). Inflammasome formation can also result from impaired processes of Beclin-1-dependent autophagy, which has been reviewed elsewhere (Salminen et al., 2013). A very recent study was performed on the possible use of a nonsteroidal antiinflammatory drug, Montelukast (Wu et al., 2025). It did not ameliorate the inflammatory and cognitive defects in aged mice, in contrast to young mice, where it was efficient in reducing both these defects. Altogether, the study demonstrates that the different inflammatory status of old versus young animals lead to different outcomes.

8.4.4 Metal ions balance

Metal ion balance is critically important for brain physiology and may be disrupted in AD (Adlard and Bush, 2012; Adlard et al., 2011). Cu^{2+} ion homeostasis was reestablished by treating an AD mouse model with mild metal chelators, such as clioquinol. Researchers at the University of Melbourne screened substances based on a clioquinol backbone to be used as possible chelator drugs. Even though the drug PBT2 reversed memory loss in mice, a Phase III trial was not performed. Phase I and II clinical trials showed that PBT2 did not significantly improve the burden of amyloid plaques, and the trial was stopped after Phase II, as reported by Prana. Notably, however, evidence suggests that the same compound could be effective in ameliorating cognitive decline accompanying aging (Adlard et al., 2015; Adlard et al., 2014a) and even in reversing signs of Parkinson's disease (Lei et al., 2015).

8.4.5 Augmenting autophagy

Autophagy is a major route to clear protein aggregates and damaged organelles, and may be impaired or overwhelmed in AD (Orr and Oddo, 2013) as well as other neurodegenerative diseases (Ciechanover and Kwon, 2015) such as PD and ALS (Morimoto et al., 2007; Shen et al., 2013). Therefore it is a viable therapeutic target in AD, as reviewed by Friedman et al. (2015). Autophagy is induced by several different signaling cascades, of which the best known is the mammalian target of rapamycin (mTOR). Multiple mTOR-dependent and mTOR-independent drugs have been tested in AD models. Examples of mTOR-dependent drugs are rapamycin analog temsirolimus (an approved anticancer drug) and an mTOR-independent drug rilmenidine (an antihypertension drug). Another such drug is lithium, which inhibits inositol monophosphate, which, through the downregulation of IP3 signaling, upregulates autophagy. Lithium also inhibits GSK-3β, which ultimately suppresses autophagy (Friedman et al., 2015). Treatment with rapamycin has already been suggested as an approach to ameliorate toxicity caused by protein aggregates in Huntington's disease and other proteinopathies (Sarkar et al., 2009; Sarkar and Rubinsztein, 2008). Rapamycin has been used in longevity studies, and this suggests possible benefits for AD (Richardson et al., 2015). We previously proposed that the process of autophagy may also be relevant for progressive myoclonic epilepsies (such as Lafora disease) and neuropsychiatric disorders (such as depression and bipolar disease) (Polajnar and Žerovnik, 2011; Polajnar and Žerovnik, 2014), for which no clear protein inclusions have been shown as yet, or they are transient. The Wnt signaling pathway is closely connected to autophagy; thus targeting this signaling pathway could also be a great opportunity for both neuroprotection and neurorepair (Serafino et al., 2020).

Inhibition of autophagic flux can cause neural death (Lee et al., 2015). However, it is not certain that enhancing autophagy would be beneficial in all cases of AD. Induction of autophagy seems appropriate at the early stages of AD, whereas later, when extensive accumulation of autophagolysosomes is observed, it might be counter-productive to enhance autophagic flux (Orr and Oddo, 2013). Drugs stabilizing microtubule-trafficking or those promoting lysosomal fusion and lysosomal enzyme function can enhance the later stages of autophagic flux. Inhibitors of histone deacetylase, among them sodium valproate, contribute to tubulin acetylation and stabilize microtubules (Asthana et al., 2013; Hahnen et al., 2008). Fusion of autophagosomes with lysosomes is needed for the final degradation of the aggregated substrates, which takes place at acidic pH. An increase in the activity of lysosomal cathepsins, especially of cathepsin B—one of the Aβ-degrading enzymes (Mueller-Steiner et al., 2006)—can be achieved by deleting its protein inhibitor cystatin B and can reverse autophagy dysfunction in the TgCRND8 mouse model of AD (Yang et al., 2011). However, cystatins C and B may themselves be neuroprotective (Kaur et al., 2010), inducing autophagy by an as yet unknown mechanism (Tizon et al., 2010). Both cystatins are Aβ binding proteins (Škerget et al., 2010; Mi et al., 2007; Sastre et al., 2004). Of interest, stefin B tetramers inhibit Aβ fibril formation but not Aβ monomer formation (Škerget et al., 2010). We suggested that stefin B may act as an amateur chaperone (Taler-Verčič and Žerovnik, 2010), similarly to crystallins. Thus the role of cystatins in AD and neurodegeneration in general remains somewhat controversial and underexplored (Žerovnik, 2009).

8.5 Combination therapies and multitarget agents

Progress beyond amyloid antibody treatments may be achieved by combination therapies, by individualized biomarker-guided patient selection, or by much earlier treatments (which proves so for antiinflammatory drugs) (Jessen et al., 2024). In 2024 the combination therapies were represented in the AD drug development pipeline. The Dominantly Inherited AD-Treatment Unit will probe a combination of Lecanemab, an anti-Aβ antibody, and E2814, an anti-tau agent. Several trials are assessing the combination of dasatinib and quercetin for its senolytic activity in AD, and a combination of insulin and empagliflozin addresses metabolic and bioenergetic pathways (Cummings et al., 2024).

A combination of dronabinol (tetrahydrocannabinol) and palmi-toylethanolamide (an endogenous cannabinoid with putative antiinflammatory properties) is being used in a trial to treat agitation and neuroinflammation in AD (Kwon et al., 2024). KarXT, consisting of a combination of xanomeline and trospium, is also being tested (Kaul et al., 2024). It is an antipsychotic agent, not directed towards D2 dopamine receptors, but rather affecting the muscarinic system, and might be useful against agitation in AD (Nunes et al., 2024).

Acetylcholinesterase (AChE) inhibitors have been used to treat cognition in AD, in particular, memory. However, they present with significant side effects and poor compliance. Pharmacological agents that modulate the activity of type-2 cannabinoid receptors (CB2R), on the other hand, are effective against neuroinflammation. The authors describe the design, synthesis, and pharmacological effects in vitro and in vivo of dual-acting compounds that inhibit AChE and butyrylcholinesterase and target CB2R (Mugnaini et al., 2024). They proved neuroprotective in a model using the SH-SY5Y cell line by preventing oxidative stress-induced cell toxicity and reversing scopolamine-induced amnesia in the Y-maze forced alternation test in mice (Mugnaini et al., 2024).

Other multitarget-directed ligands are broadly explored (Waiker et al., 2023). For example, compound AV-2 acts by binding to and was evaluated against human cholinesterase and human β-secretase. Compound AV-2 also prevented Aβ aggregation in the thioflavin T assay. Additionally, compound AV-2 significantly ameliorated scopolamine- and Aβ-induced cognitive impairments in the in vivo behavioral Y-maze and Morris water maze studies, respectively. The ex vivo and biochemical analysis further revealed good hippocampal AChE inhibition and the antioxidant potential of the compound AV-2 (Verma et al., 2024). It is regarded as the lead candidate and could be explored further for AD therapy.

TREM2 signaling modulators are being explored to treat neurodegenerative disease (Zgorzynska, 2024). As mentioned, TREM2 activation may be beneficial in the early stages of AD but detrimental in the later stages. Recently, innovative nanoparticle dual gene delivery was applied in animal models, showing promise to reverse neurodegeneration (Jiang et al., 2024). The nanoparticles promote RNAi of BACE1 (an APP secretase that promotes Aβ formation) and gene expression of TREM2, decreasing the effect of the former gene target and increasing expression of the latter.

8.6 Lessons from other disorders: amyotrophic lateral sclerosis/ frontotemporal dementia

ALS is a fatal neurodegenerative disease characterized by motor neuron degeneration, resulting in progressive muscle wasting and paralysis. As in AD, disturbance of proteostasis is a key pathogenic mechanism in ALS, with dysregulation of protein folding and clearance by both autophagy and the ubiquitin-proteasome system reported (Cozzi and Ferrari, 2022; Medinas et al., 2017; McAlary et al., 2020). ALS is a highly complex disease, with other pathogenic mechanisms including impairment to mitochondrial function, axonal transport, vesicular transport, and RNA metabolism, in addition to oxidative stress, neuroinflammation, and excitotoxicity (Le Gall et al., 2020). ALS is considered to be on a disease spectrum with frontotemporal dementia (FTD) due to overlapping genetics and symptoms, with 10%–15% of ALS patients also being diagnosed with FTD (Ahmed et al., 2021).

The most common hallmark of ALS observed in 97% of patients is the mislocalization of the RNA-binding protein (RBP) TDP-43 from the nucleus to the cytoplasm in motor neurons, where it forms toxic protein aggregates that may also spread in a prion-like mechanism (Jo et al., 2020; Pongrácová et al., 2024). Up to 57% of AD patients have also been shown to have TDP-43 inclusions, with TDP-43 pathology also reported in several other neurodegenerative diseases such as FTD, Huntington's disease, and Parkinson's disease. Collectively, these diseases are referred to as TDP-43 proteinopathies, which may all benefit from a TDP-43-targeted treatment, or a proteostasis-targeted treatment more generally (De Boer et al., 2021; Meneses et al., 2021). However, the lessons to be learnt from the ALS field in relation to emerging RNA therapies are not limited to targeting TDP-43, as strategies for gene repression in patients can be shared even if the target gene differs.

8.6.1 Tofersen

A recent breakthrough in the ALS field is the development of the ASO tofersen (trade name Qalsody) for treating ALS patients with *SOD1* mutations (SOD1-ALS). Tofersen was approved by the United States FDA in April 2023, and has since been approved by the EMA (Everett and Bucelli, 2024). SOD1 is an antioxidant enzyme that converts superoxide radicals to molecular oxygen and hydrogen peroxide (Bunton-Stasyshyn et al., 2015). *SOD1* mutations are found in approximately 2% of ALS patients, and they are the second most common genetic cause of ALS in Europe, accounting for 14.8% of familial ALS (fALS) cases and 1.2% of sporadic ALS (sALS) cases. In Asia, *SOD1* mutations are the most common cause of ALS, representing 30% of fALS and 1.5% of sALS cases (Zou et al., 2017). Despite reports of mutant SOD1 protein loss of function, it is generally considered that *SOD1* mutations exert most of their neurotoxic effects due to the mutant protein gaining toxic function (Bunton-Stasyshyn et al., 2015; Sau et al., 2007; Saccon et al., 2013). This includes the misfolding and aggregation of mutant SOD1 protein, which can then sequester and misfold wild-type SOD1 in a prion-like mechanism (Bunton-Stasyshyn et al., 2015).

ASOs are short oligonucleotide sequences that bind to complementary RNA to reduce their translation to protein, which can occur by several mechanisms. They are chemically modified to improve stability, cellular uptake, and affinity for their target (Lauffer et al., 2024). Tofersen directly binds to mutant *SOD1* mRNA and induces RNase H-mediated degradation, therefore reducing synthesis of the SOD1 protein (Miller et al., 2022). Intrathecal administration of tofersen in the Phase I–III trial revealed a significant reduction in plasma SOD1 and neurofilament light chain (NfL, or NLC), a marker of neuronal injury (Miller et al., 2022; Gaetani et al., 2019), and also an important AD biomarker (see Table 8.1 and Section 8.2). Clinical endpoints were not met after 28 weeks, and adverse events were reported in some participants, mostly due to lumbar puncture, with serious neurologic adverse events reported in 7% of participants. This did not prevent FDA approval, as tofersen was on an accelerated approval pathway allowing success via a surrogate endpoint (NfL levels) (Everett and Bucelli, 2024). An open-label extension of the study has revealed a possible clinical benefit to starting tofersen treatment earlier (Everett and Bucelli, 2024; Miller et al., 2022).

8.6.2 Other antisense oligonucleotides in amyotrophic lateral sclerosis/frontotemporal dementia clinical trials

While tofersen remains the only FDA-approved ASO for the treatment of ALS as of 2024, its success has encouraged the investigation of ASOs targeting several other ALS genes, which are at different stages of development and clinical trials. Notable examples will now be discussed.

Mutations in the *FUS* gene are associated with early-onset, aggressive forms of ALS and rare FTD cases (Van Langenhove et al., 2010; Korobeynikov et al., 2022). Like TDP-43, FUS is also an RBP, with functions including regulation of transcription, splicing, and mRNA transport (Dormann and Haass, 2013). Pathogenic effects of *FUS* mutations include the mislocalization of FUS protein from the nucleus to the cytoplasm in motor neurons, where it forms toxic protein aggregates, similar to TDP-43 (Nolan et al., 2016). An ASO targeting *FUS* mRNA (ION363, known as jacifusen) has been used to treat 12 *FUS*-ALS patients in an open-label expanded access program. Safety and possible efficacy were demonstrated, with reduction of NfL in the CSF of most participants, reduced FUS protein and pathology in postmortem CNS tissue of 4 participants, and even an unprecedented functional improvement in one participant. Jacifusen is being further evaluated by a clinical trial (Korobeynikov et al., 2022; Shneider et al., 2025; Boros et al., 2022).

As mentioned, TDP-43 cytoplasmic mislocalisation and aggregation are major hallmarks of ALS. In conjunction, there is a loss of TDP-43 nuclear function, which may be harmfully exacerbated by an ASO (Van Daele et al., 2024). However, ASOs may be useful tools to target genes that are impacted by TDP-43 as an ALS treatment. In healthy neurons, TDP-43 regulates splicing by repressing the inclusion of cryptic exons–exons which are usually excluded from mature mRNA (Donde et al., 2019). Therefore when TDP-43 is depleted from the nucleus, such as in ALS, these cryptic exons can be included in mRNA transcripts, which often results in a stop codon or reduced RNA stability. Two examples of genes that have reduced expression due to cryptic exon inclusion upon TDP-43 depletion are *UNC13A* and stathmin-2 (*STMN2*). As a result, reduced STMN2 and UNC13A levels have been reported in ALS patients (Ma et al., 2022; Klim et al., 2019; Melamed et al., 2019; Brown et al., 2022). Furthermore, rare variants in both genes have also been linked to ALS (Van Es et al., 2009; Diekstra et al., 2014; Theunissen et al., 2021). Both genes are important for neuronal function; therefore their depletion in ALS is likely to have neurotoxic effects. For example, STMN2 is a microtubule-binding protein that is vital for axon stability and regeneration, and UNC13A is important for the release of synaptic vesicles during neurotransmission (Klim et al., 2019; Willemse et al., 2023). Since TDP-43 pathology is a feature of 97% of ALS patients, restoration of UNC13A or STMN2 levels may be a promising widespread ALS treatment. Importantly, cryptic splicing of UNC13A and STMN2 has also been demonstrated in TDP-43-associated AD, demonstrating potential for widespread utility of therapies for TDP-43 proteinopathies (Agra Almeida Quadros et al., 2024).

UNC13A and *STMN2* targeted splice-regulating "exon-skipping" ASOs are being developed that block cryptic splice sites with the aim of preventing cryptic exon inclusion, therefore replacing the normal function of TDP-43. Use of the ASO QRL-201 to restore STMN2 by this method is in Phase I clinical trials (Genge et al., 2024), and several other *STMN2* or *UNC13A* targeting ASOs are being investigated in preclinical models (Keuss et al., 2024). Splice-regulating ASOs have already been FDA-approved for the treatment of spinal muscular atrophy (nusinersen/Spinraza) and Duchenne muscular dystrophy (eteplirsen/Exondys 51), highlighting their potential in neurodegenerative diseases (Syed, 2016; Crawford et al., 2023). Since dysregulated splicing has also been implicated in AD, splice-regulating RNA therapies may also be applicable to this disease (Apicco et al., 2019; Biamonti et al., 2021).

ATXN2 (encoding ataxin-2) is a further ALS-associated gene that may be targeted by ASOs as a widespread ALS (or other TDP-43 proteinopathy) treatment. Ataxin-2 is important for neuronal homeostasis due to its roles in calcium regulation, RNA metabolism, translation, and the formation of stress granules—dynamic protein-containing membraneless foci that form upon cell stress. Within the *ATXN2* gene, expansion of a CAG repeat region in exon 1 results in an extended polyglutamine tract, altering protein conformational dynamics, which may result in both a partial loss of function and toxic gain of function (Costa et al., 2024). An intermediate CAG expansion (27–33 CAG repeats) has been shown to increase the risk of ALS (Glass et al., 2022). Ataxin-2 is thought to be disease-modifying, as it interacts with TDP-43 and FUS, and the intermediate expansions have been shown to increase TDP-43 and FUS pathology (Elden et al., 2010; Farg et al., 2013; Hart and Gitler, 2012; Nihei et al., 2012). However, a (now-terminated) Phase I/II trial of BIIB105 (https://clinicaltrials.gov/study/NCT04494256), an *ATXN2*-targeted ASO, was ultimately unsuccessful at reducing NfL levels or improving clinical outcomes, although it did reduce ataxin-2 protein levels in patient CSF.

8.6.3 Small noncoding RNA therapeutics

Therapies utilizing miRNAs as an alternate method to ASOs for reducing protein expression are also being investigated, with several showing promise in preclinical models. A miRNA-based therapy that has been tested in two patients is APB-102 (also called AMT-162), for which the company UniQure acquired the rights for Phase I/II clinical trials, as of 2023 (Tzeplaeff et al., 2023). The therapy consists of an SOD1-targeting miRNA delivered by an adeno-associated virus. The two patients had familial SOD1-ALS and were treated with a single intrathecal injection. While, in theory, this treatment should have similar effects as tofersen, decreasing SOD1 protein levels, one benefit of viral delivery is that only a single dose is required (Mueller et al., 2020).

8.6.4 Proteolysis-targeting chimeras

Protein misfolding and aggregation are hallmarks of neurodegenerative diseases, including ALS and AD. Protein aggregates have long been regarded as "undruggable" using conventional methods. However, recently, proteolysis targeting chimeras (PROTACs) have emerged as a method to induce the degradation of disease-associated proteins. PROTACs consist of two binding ligands: one to recruit E3 ubiquitin ligase, and a second to bind to the target protein. The proximity enables the E3 ligase to ubiquitinate the target protein, labeling it for degradation by the proteasome. Multiple rounds of ubiquitination may occur, therefore ensuring efficient degradation without high drug concentrations (Tseng et al., 2023).

Tseng et al recently developed a PROTAC targeting C-terminal TDP-43 (C-TDP-43), a region that has been shown to form aggregate-prone fragments that are the major component of protein inclusions in ALS patients (Tseng et al., 2023; Neumann et al., 2006). The PROTAC was shown to degrade C-TDP-43 aggregates and oligomers in Neuro-2a cells. This modestly improved Neuro-2a viability and motor neuron defects in *Caenorhabditis elegans*, although the beneficial effects may be limited by the loss of TDP-43 function (Tseng et al., 2023). Further studies in ALS cell and animal models will help to determine the utility of PROTACs for TDP-43 targeting.

A SOD1-targeting PROTAC has also been developed for SOD1-ALS treatment, based on the Dorfin E3 ligase, which naturally degrades mutant SOD1 protein. However, Dorfin has a short half-life; therefore a chimeric protein with improved stability was developed, consisting of the substrate-binding region of Dorfin and the E3 ligase activity (U-box) domain of Hsc70-interacting protein (CHIP). The Dorfin–CHIP protein demonstrated specificity for only the mutant form of the SOD1 protein and reduced formation of mutant SOD1 aggregates in Neuro-2a cells, improving cell viability (Ishigaki et al., 2007). PROTAC approaches have previously been considered in AD and represent an area where more research is warranted (Zhou et al., 2025).

8.6.5 Ribonuclease-targeting chimeras

Ribonuclease-targeting chimeras (RIBOTACs) have a similar mechanism to PROTACs, instead linking target disease-associated RNA molecules to RNases for degradation. RIBOTACs allow the removal of harmful RNA in addition to preventing translation to harmful protein, which is of particular use in certain genetic forms of ALS. This includes the most commonly known genetic cause of ALS, a hexanucleotide repeat expansion (HRE) in an intron of the chromosome 9 open reading frame 72 *(C9orf72)* gene. Translation of the HRE forms toxic dipeptide repeat proteins (DPRs), while the HRE RNA itself is also pathogenic. RNA aggregates or foci form, which can sequester RBPs, therefore affecting their localisation and function (McEachin et al., 2020). A *C9orf72* HRE-targeted RIBOTAC developed by Bush et al., (2021), successfully degraded the mutant RNA with limited off-target effects, leading to a reduction in DPRs. These results were observed in both C9orf72-ALS patient induced pluripotent stem cell-derived spinal neurons and the C9orf72-ALS/FTD bacterial artificial chromosome transgenic mouse model (Bush et al., 2021).

8.7 RNA-based therapeutics for *Alzheimer's disease*—as a new feature

Although there are no RNA-based therapeutics currently approved for the treatment of AD, there are currently three ASO drugs in clinical trials as of June 2025 (Dudzisz and Wandzik, 2025). Two ASOs aim to reduce tau aggregates by targeting the *MAPT* gene, with BIIB080 ($MAPT_{RX}$) targeting *MAPT* premRNA and NIO752 targeting *MAPT* mRNA. BIIB080 is currently in Phase II trials, since a phase 1b trial demonstrated safety and a dose-dependent reduction in total and phosphorylated tau-181 (markers associated with cognitive decline) in the CSF of mild AD patients (Edwards et al., 2023; Mummery et al., 2023; Ziogas et al., 2023). NIO752 is in phase 1b clinical trial, with no current information on its progress. Another ASO being tested in phase 1b clinical trial is ION269, which targets *APP* mRNA. It aims to reduce Aβ and is being tested in patients with Down syndrome with brain amyloid positivity (arising from trisomy 21, which carries the *APP* gene, giving a gene-dosage effect), indicating a risk of AD (Dudzisz and Wandzik, 2025).

8.8 Conclusions

In the last 10 years, several important breakthroughs in the search for the treatment of AD have been made. The aim of this review is, on one hand, to describe these recent developments as illustrated by animal model studies as well as preclinical and clinical trials, and, on the other hand, to highlight the critical pathogenic mechanisms that these studies reveal/confirm.

Undoubtedly, to truly treat AD and similar neurodegenerative diseases (proteinopathies) at their source, a better understanding of the molecular and cellular, including glial, neuronal, and synaptic mechanisms, needs to be understood. We support the argument that the modified amyloid cascade hypothesis presents a good basis for such an understanding, proven by success in reducing soluble Aβ oligomers by antiAβ oligomers antibodies or aggregation inhibitors, among them polyphenols and other antioxidants derived from plants (see Chapter 9 by Pogacnik da Silva L. et al., this book). Diminishing (dissolving) Aβ plaques as well as the intraneural aggregates of tau, and keeping in check accompanying inflammation, proved to be of some help if started early at still mild cognitive impairment. Intensified clearance of the aggregates by autophagy and restoring metal ion balance may also be beneficial in certain (early) stages of disease. For example, the copper compound Cu(ATSM) recently displayed neuroprotective properties in the APP/PS1 mouse model, reducing neurovascular dysfunction and Aβ burden (Pyun et al., 2026).

At least, a better understanding of prion-like templating, seeding, and cell-to-cell spreading of the protein aggregates may be used to stop the routes of spreading (Brettschneider et al., 2015), which is out of scope of this review.

Funding and disclosures

This work was funded by the project J7–4050 (E-Ž) and the program P1-0140 (led by Boris Turk) via the Slovenian Research and Innovation Agency. RL and HB are supported by Stand Against MND and the MND Association (UK), reference 905–792.

References

Adlard, P.A., Bush, A.I., 2012. Metal chaperones: a holistic approach to the treatment of Alzheimer's disease. Frontiers in Psychiatry 3. Available at https://doi.org/10.3389/fpsyt.2012.00015Australia.

Adlard, P.A., Manso, Y., Comes, G., Hidalgo, J., Bush, A.I., 2011. Copper modulation as a therapy for Alzheimer's disease? International Journal of Alzheimer's Disease. Available at https://doi.org/10.4061/2011/370345.

Adlard, P.A., Sedjahtera, A., Gunawan, L., Bray, L., Hare, D., Lear, J., et al., 2014a. A novel approach to rapidly prevent age-related cognitive decline. Aging Cell 13 (2), 351–359. Available at https://doi.org/10.1111/acel.12178.

Adlard, P.A., Tran, B.A., Finkelstein, D.I., Desmond, P.M., Johnston, L.A., Bush, A.I., et al., 2014b. A review of β-amyloid neuroimaging in Alzheimer's disease. Frontiers in Neuroscience 8. Available at https://doi.org/10.3389/fnins.2014.00327.

Adlard, P.A., Parncutt, J., Lal, V., James, S., Hare, D., Doble, P., et al., 2015. Metal chaperones prevent zinc-mediated cognitive decline. Neurobiology of Disease 81, 196–202. Available at https://doi.org/10.1016/j.nbd.2014.12.012.

Adolfsson, O., Pihlgren, M., Toni, N., Varisco, Y., Buccarello, A.L., Antoniello, K., et al., 2012. An effector-reduced anti- -amyloid (A) antibody with unique a binding properties promotes neuroprotection and glial engulfment of A. Journal of Neuroscience 32 (28), 9677–9689. Available at https://doi.org/10.1523/jneurosci.4742-11.2012.

Agra Almeida Quadros, A., Li, Z., Wang, X., Ndayambaje, I.S., Aryal, S., Ramesh, N., et al., 2024. Cryptic splicing of stathmin-2 and UNC13A mRNAs is a pathological hallmark of TDP-43-associated Alzheimer's disease. Acta Neuropathologica 147 (1), 9. Available at 10.1007/s00401-023-02655-0.

Ahmed, R.M., Bocchetta, M., Todd, E.G., Tse, N.Y., Devenney, E.M., Tu, S., et al., 2021. Tackling clinical heterogeneity across the amyotrophic lateral sclerosis-frontotemporal dementia spectrum using a transdiagnostic approach. Brain Communications 3 (4). Available at https://doi.org/10.1093/braincomms/fcab257.

Alzheimer's Association., 2019. Alzheimer's disease facts and figures. Alzheimer's & Dementia 15 (3), 321–387.

Apicco, D.J., Zhang, C., Maziuk, B., Jiang, L., Ballance, H.I., Boudeau, S., et al., 2019. Dysregulation of RNA splicing in tauopathies. Cell Reports 29 (13), 4377. Available at https://doi.org/10.1016/j.celrep.2019.11.093.

Asthana, J., Kapoor, S., Mohan, R., Panda, D., 2013. Inhibition of HDAC6 deacetylase activity increases its binding with microtubules and suppresses microtubule dynamic instability in MCF-7 Cells. Journal of Biological Chemistry 288 (31), 22516–22526. Available at https://doi.org/10.1074/jbc.m113.489328.

Barker, W.W., Luis, C.A., Kashuba, A., Luis, M., Harwood, D.G., Loewenstein, D., et al., 2002. Relative frequencies of Alzheimer disease, Lewy body, vascular and frontotemporal dementia, and hippocampal sclerosis in the State of Florida Brain Bank. Alzheimer Disease and Associated Disorders 16 (4), 203–212. Available at https://doi.org/10.1097/00002093-200210000-00001.

Di Benedetto, G., Burgaletto, C., Bellanca, C.M., Munafò, A., Bernardini, R., Cantarella, G., 2022. Role of microglia and astrocytes in Alzheimer's disease: from neuroinflammation to $Ca^{(2+)}$) homeostasis dysregulation. Cells 11 (17), 2728. Available at https://doi.org/10.3390/cells11172728.

Biamonti, G., Amato, A., Belloni, E., Di Matteo, A., Infantino, L., Pradella, D., et al., 2021. Alternative splicing in Alzheimer's disease. Aging Clinical and Experimental Research 33 (4), 747–758. Available at https://doi.org/10.1007/s40520-019-01360-x.

Bo, S., Ciccone, G., Castiglione, A., Gambino, R., De Michieli, F., Villois, P., et al., 2013. Anti-inflammatory and antioxidant effects of resveratrol in healthy smokers a randomized, double-blind, placebo-controlled, cross-over trial. Current Medicinal Chemistry 20 (10), 1323–1331. Available at https://doi.org/10.2174/0929867311320100009.

De Boer, E.M.J., Orie, V.K., Williams, T., Baker, M.R., De Oliveira, H.M., Polvikoski, T., et al., 2021. TDP-43 proteinopathies: a new wave of neurodegenerative diseases. Journal of Neurology, Neurosurgery and Psychiatry 92 (1), 86–95. Available at https://doi.org/10.1136/jnnp-2020-322983.

Böken, D., Cox, D., Burke, M., Lam, J.Y.L., Katsinelos, T., Danial, J.S.H., et al., 2024. Single-molecule characterization and super-resolution imaging of Alzheimer's disease-relevant Tau aggregates in human samples. Angewandte Chemie - International Edition 63 (21). Available at https://doi.org/10.1002/anie.202317756.

Boros, B.D., Schoch, K.M., Kreple, C.J., Miller, T.M., 2022. Antisense oligonucleotides for the study and treatment of ALS. Neurotherapeutics 19 (4), 1145–1158. Available at https://doi.org/10.1007/s13311-022-01247-2.

Brettschneider, J., Tredici, K.D., Lee, V.M.Y., Trojanowski, J.Q., 2015. Spreading of pathology in neurodegenerative diseases: a focus on human studies. Nature Reviews Neuroscience 16 (2), 109–120. Available at https://doi.org/10.1038/nrn3887.

Brown, A.L., Wilkins, O.G., Keuss, M.J., Hill, S.E., Zanovello, M., Lee, W.C., et al., 2022. TDP-43 loss and ALS-risk SNPs drive mis-splicing and depletion of UNC13A. Nature 603 (7899), 131–137. Available at https://doi.org/10.1038/s41586-022-04436-3.

Bu, X.L., Rao, P.P.N., Wang, Y.J., 2016. Anti-amyloid aggregation activity of natural compounds: implications for Alzheimer's drug discovery. Molecular Neurobiology 53 (6), 3565–3575. Available at https://doi.org/10.1007/s12035-015-9301-4.

Bucciantini, M., Nosi, D., Forzan, M., Russo, E., Calamai, M., Pieri, L., et al., 2012. Toxic effects of amyloid fibrils on cell membranes: the importance of ganglioside GM1. FASEB Journal 26 (2), 818–831. Available at https://doi.org/10.1096/fj.11-189381.

Bunton-Stasyshyn, R.K.A., Saccon, R.A., Fratta, P., Fisher, E.M.C., 2015. SOD1 function and its implications for amyotrophic lateral sclerosis pathology: new and renascent themes. Neuroscientist 21 (5), 519–529. Available at https://doi.org/10.1177/1073858414561795.

Bush, J.A., Aikawa, H., Fuerst, R., Li, Y., Ursu, A., Meyer, S.M., et al., 2021. Ribonuclease recruitment using a small molecule reduced c9ALS/FTD r(G4C2) repeat expansion in vitro and in vivo ALS models. Translational Medicine 13 (617). Available at https://doi.org/10.1126/SCITRANSLMED.ABD5991.

Caesar, I., Jonson, M., Peter, K., Nilsson, R., Thor, S., Hammarström, P., et al., 2012. Curcumin promotes a-beta fibrillation and reduces neurotoxicity in transgenic drosophila. PLoS ONE 7 (2), e31424. Available at https://doi.org/10.1371/journal.pone.0031424.

Chatterjee, P., Pedrini, S., Stoops, E., Goozee, K., Villemagne, V.L., Asih, P.R., et al., 2021. Plasma glial fibrillary acidic protein is elevated in cognitively normal older adults at risk of Alzheimer's disease. Translational Psychiatry 11 (1). Available at https://doi.org/10.1038/s41398-020-01137-1.

Cheng, L., Doecke, J.D., Sharples, R.A., Villemagne, V.L., Fowler, C.J., Rembach, A., et al., 2015. A F Hill, Prognostic serum miRNA biomarkers associated with Alzheimer's disease shows concordance with neuropsychological and neuroimaging assessment. Molecular Psychiatry 20 (10), 1188–1196. Available at https://doi.org/10.1038/mp.2014.127.

Ciarlo, E., Massone, S., Penna, I., Nizzari, M., Gigoni, A., Dieci, G., et al., 2013. An intronic ncRNA-dependent regulation of SORL1 expression affecting Aβ formation is upregulated in post-mortem Alzheimer's disease brain samples. DMM Disease Models and Mechanisms 6 (2), 424–433. Available at https://doi.org/10.1242/dmm.009761.

Ciechanover, A., Kwon, Y.T., 2015. Degradation of misfolded proteins in neurodegenerative diseases: therapeutic targets and strategies. Experimental & Molecular Medicine 47 (3), e147. Available at https://doi.org/10.1038/emm.2014.117.

Coelho, F.G.de M., Vital, T.M., Stein, A.M., Arantes, F.J., Rueda, A.V., Camarini, R., et al., 2014. Acute aerobic exercise increases brain-derived neurotrophic factor levels in elderly with Alzheimer's Disease. Journal of Alzheimer's Disease 39 (2), 401–408. Available at https://doi.org/10.3233/jad-131073.

Colton, C.A., Gilbert, D.L., 1987. Production of superoxide anions by a CNS macrophage, the microglia. FEBS Letters 223 (2), 284–288. Available at https://doi.org/10.1016/0014-5793(87)80305-8.

Contini, C., Serrao, S., Manconi, B., Olianas, A., Iavarone, F., Guadalupi, G., et al., 2023. Characterization of cystatin B interactome in saliva from healthy elderly and Alzheimer's Disease patients. Life 13 (3), 748. Available at https://doi.org/10.3390/life13030748.

Costa, R.G., Conceição, A., Matos, C.A., Nóbrega, C., 2024. The polyglutamine protein ATXN2: from its molecular functions to its involvement in disease. Cell Death and Disease 15 (6). Available at https://doi.org/10.1038/s41419-024-06812-5.

Cozzi, M., Ferrari, V., 2022. Autophagy dysfunction in ALS: from transport to protein degradation. Journal of Molecular Neuroscience 72 (7), 1456–1481. Available at https://doi.org/10.1007/s12031-022-02029-3.

Crawford, T.O., Swoboda, K.J., De Vivo, D.C., Bertini, E., Hwu, W.L., Finkel, R.S., et al., 2023. Continued benefit of nusinersen initiated in the presymptomatic stage of spinal muscular atrophy: 5-year update of the NURTURE study. Muscle and Nerve 68 (2), 157–170. Available at https://doi.org/10.1002/mus.27853.

Cummings, J., Zhou, Y., Lee, G., Zhong, K., Fonseca, J., Cheng, F., 2024. Alzheimer's disease drug development pipeline: 2024. Alzheimer's & Dementia (New York, N. Y.) 10 (2), e12465. Available at 10.1002/trc2.12465.

DeTure, M.A., Dickson, D.W., 2019. The neuropathological diagnosis of Alzheimer's disease. Molecular Neurodegeneration 14 (1). Available at https://doi.org/10.1186/s13024-019-0333-5.

Diekstra, F.P., Van Deerlin, V.M., Van Swieten, J.C., Al-Chalabi, A., Ludolph, A.C., Weishaupt, J.H., et al., 2014. C9orf72 and UNC13A are shared risk loci for amyotrophic lateral sclerosis and frontotemporal dementia: a genome-wide meta-analysis. Annals of Neurology 76 (1), 120–133. Available at https://doi.org/10.1002/ana.24198.

Donde, A., Sun, M., Ling, J.P., Braunstein, K.E., Pang, B., Wen, X., et al., 2019. Splicing repression is a major function of TDP-43 in motor neurons. Acta Neuropathologica 138 (5), 813–826. Available at https://doi.org/10.1007/s00401-019-02042-8.

Dormann, D., Haass, C., 2013. Fused in sarcoma (FUS): an oncogene goes awry in neurodegeneration. Molecular and Cellular Neuroscience 56, 475–486. Available at https://doi.org/10.1016/j.mcn.2013.03.006.

Dudzisz, K., Wandzik, I., 2025. Antisense oligonucleotides: a promising advancement in neurodegenerative disease treatment. European Journal of Pharmacology 999, 177644. Available at https://doi.org/10.1016/j.ejphar.2025.177644.

Duggan, M.R., Yang, S., Gomez, G.T., Cui, Y., Capuano, A.W., Chen, J., et al., 2025. Proteomic signatures of corona and herpes viral antibodies identify IGDCC4 as a mediator of neurodegeneration. Science Advances 11 (22). Available at https://doi.org/10.1126/sciadv.adt7176.

Edwards, A.L., Collins, J.A., Junge, C., Kordasiewicz, H., Mignon, L., Wu, S., et al., 2023. Exploratory Tau biomarker results from a multiple ascending-dose study of biib080 in alzheimer disease: a randomized clinical trial. JAMA Neurology 80 (12), 1344–1352. Available at https://doi.org/10.1001/jamaneurol.2023.3861.

Elden, A.C., Kim, H.J., Hart, M.P., Chen-Plotkin, A.S., Johnson, B.S., Fang, X., et al., 2010. Ataxin-2 intermediate-length polyglutamine expansions are associated with increased risk for ALS. Nature 466 (7310), 1069–1075. Available at https://doi.org/10.1038/nature09320.

Everett, W.H., Bucelli, R.C., 2024. Tofersen for SOD1 ALS. Neurodegenerative Disease Management 14 (5), 149–160. Available at https://doi.org/10.1080/17582024.2024.2402216.

Farg, M.A., Soo, K.Y., Warraich, S.T., Sundaramoorthy, V., Blair, I.P., Atkin, J.D., 2013. Ataxin-2 interacts with FUS and intermediate-length polyglutamine expansions enhance FUS-related pathology in amyotrophic lateral sclerosis. Human Molecular Genetics 22 (4), 717–728. Available at https://doi.org/10.1093/hmg/dds479.

Foda, A.M., Ibrahim, S.S., Ibrahim, S.M., Elbaz, E.M., 2024. Pterostilbene ameliorates cognitive impairment in polycystic ovary syndrome rat model through improving insulin resistance via the IRS-1/PI3K/Akt/GSK-3β Pathway: A Comparative Study with Metformin. ACS Chemical Neuroscience 15 (16), 3064–3077. Available at https://doi.org/10.1021/acschemneuro.4c00352.

Fotuhi, S.N., Khalaj-Kondori, M., Hoseinpour Feizi, M.A., Talebi, M., 2019. Long non-coding RNA BACE1-as may serve as an Alzheimer's disease blood-based biomarker. Journal of Molecular Neuroscience 69 (3), 351–359. Available at https://doi.org/10.1007/s12031-019-01364-2.

Friedman, L.G., Qureshi, Y.H., Yu, W.H., 2015. Promoting autophagic clearance: viable therapeutic targets in Alzheimer's disease. Neurotherapeutics 12 (1), 94–108. Available at https://doi.org/10.1007/s13311-014-0320-z.

Gaetani, L., Blennow, K., Calabresi, P., Di Filippo, M., Parnetti, L., Zetterberg, H., 2019. Neurofilament light chain as a biomarker in neurological disorders. Journal of Neurology, Neurosurgery & Psychiatry 90 (8), 870–881. Available at https://doi.org/10.1136/jnnp-2018-320106.

Le Gall, L., Anakor, E., Connolly, O., Vijayakumar, U.G., Duddy, W.J., Duguez, S., 2020. Molecular and cellular mechanisms affected in ALS. Journal of Personalized Medicine 10 (3), 101. Available at 10.3390/jpm10030101.

Genge, A., Salmon, K., Polzer, J., Martinez, C., Boggs, B., Eon, V., et al., 2024. QRL-201-01—A multicenter, randomized, double-blind, placebo-controlled multiple ascending dose study to evaluate the safety and tolerability of QRL-201 in amyotrophic lateral sclerosis (P6-11.010). Neurology 102 (7_supplement_1), 3505. Available at https://doi.org/10.1212/WNL.0000000000205187.

Ginhoux, F., Greter, M., Leboeuf, M., Nandi, S., See, P., Gokhan, S., et al., 2010. Fate mapping analysis reveals that adult microglia derive from primitive macrophages. Science 330 (6005), 841–845. Available at https://doi.org/10.1126/science.1194637.

Glass, J.D., Dewan, R., Ding, J., Gibbs, J.R., Dalgard, C., Keagle, P.J., et al., 2022. ATXN2 intermediate expansions in amyotrophic lateral sclerosis. Brain 145 (8), 2671–2676. Available at https://doi.org/10.1093/brain/awac167.

Gong, J., Jin, Z., Chen, H., He, J., Zhang, Y., Yang, X., 2023. Super-resolution fluorescence microscopic imaging in pathogenesis and drug treatment of neurological disease. Advanced Drug Delivery Reviews 196, 114791. Available at https://doi.org/10.1016/j.addr.2023.114791.

Gordin, E., Viitanen, S., Gordin, D., Szlosek, D., Peterson, S., Spillmann, T., et al., 2024. A clinical study on urinary clusterin and cystatin b in dogs with spontaneous acute kidney injury. Veterinary Sciences 11 (5), 200. Available at https://doi.org/10.3390/vetsci11050200.

Guo, Y., You, J., Zhang, Y., Liu, W.-S., Huang, Y.-Y., Zhang, Y.-R., et al., 2024. Plasma proteomic profiles predict future dementia in healthy adults. Nature Aging 4 (2), 247–260. Available at https://doi.org/10.1038/s43587-023-00565-0.

Guo, H., Yang, R., Cheng, W., Li, Q., Du, M., 2025. An update of salivary biomarkers for the diagnosis of Alzheimer's disease. International Journal of Molecular Sciences 26 (5), 2059. Available at https://doi.org/10.3390/ijms26052059.

Hahnen, E., Hauke, J., Tränkle, C., Eyüpoglu, I.Y., Wirth, B., Blümcke, I., 2008. Histone deacetylase inhibitors: possible implications for neurodegenerative disorders. Expert Opinion on Investigational Drugs 17 (2), 169–184. Available at https://doi.org/10.1517/13543784.17.2.169.

Halle, A., Hornung, V., Petzold, G.C., Stewart, C.R., Monks, B.G., Reinheckel, T., et al., 2008. The NALP3 inflammasome is involved in the innate immune response to amyloid-β Nature Immunology 9 (8), 857–865. Available at https://doi.org/10.1038/ni.1636.

Hamaguchi, T., Ono, K., Murase, A., Yamada, M., 2009. Phenolic compounds prevent Alzheimer's pathology through different effects on the amyloid-β aggregation pathway. The American Journal of Pathology 175 (6), 2557–2565. Available at https://doi.org/10.2353/ajpath.2009.090417.

Han, W., Li, C., 2010. Linking type 2 diabetes and Alzheimer's disease. Proceedings of the National Academy of Sciences 107 (15), 6557–6558. Available at https://doi.org/10.1073/pnas.1002555107.

Hart, M.P., Gitler, A.D., 2012. ALS-associated ataxin 2 PolyQ expansions enhance stress-induced caspase 3 activation and increase TDP-43 pathological modifications. Journal of Neuroscience 32 (27), 9133–9142. Available at https://doi.org/10.1523/jneurosci.0996-12.2012.

Heneka, M.T., Kummer, M.P., Stutz, A., Delekate, A., Schwartz, S., Vieira-Saecker, A., et al., 2013. NLRP3 is activated in Alzheimer's disease and contributes to pathology in APP/PS1 mice. Nature 493 (7434), 674–678. Available at https://doi.org/10.1038/nature11729.

Heneka, M.T., Carson, M.J., Khoury, J.E., Landreth, G.E., Brosseron, F., Feinstein, D.L., et al., 2015. Neuroinflammation in Alzheimer's disease. The Lancet Neurology 14 (4), 388–405. Available at https://doi.org/10.1016/S1474-4422(15)70016-5.

Heppner, F.L., Ransohoff, R.M., Becher, B., 2015. Immune attack: the role of inflammation in Alzheimer disease. Nature Reviews Neuroscience 16 (6), 358–372. Available at https://doi.org/10.1038/nrn3880.

Holton, K.F., 2021. Micronutrients May Be a Unique Weapon Against the Neurotoxic Triad of Excitotoxicity, Oxidative Stress and Neuroinflammation: A Perspective. Frontiers in Neuroscience 15. Available at 10.3389/fnins.2021.726457.

Ibrahim, A.M., Chauhan, L., Bhardwaj, A., Sharma, A., Fayaz, F., Kumar, B., et al., 2022. Brain-derived neurotropic factor in neurodegenerative disorders. Biomedicines 10 (5). Available at https://doi.org/10.3390/biomedicines10051143.

Ishigaki, S., Niwa, J.-ichi, Yamada, S.-ichi, Takahashi, M., Ito, T., Sone, J., et al., 2007. Dorfin-CHIP chimeric proteins potently ubiquitylate and degrade familial ALS-related mutant SOD1 proteins and reduce their cellular toxicity. Neurobiology of Disease 25 (2), 331–341. Available at https://doi.org/10.1016/j.nbd.2006.09.017.

Jessen, F., Kramberger, M.G., Angioni, D., Aarsland, D., Balasa, M., Bennys, K., et al., 2024. Progress in the treatment of Alzheimer's disease is needed – position statement of European Alzheimer's disease consortium (EADC) Investigators. The Journal of Prevention of Alzheimer's Disease 11 (5), 1212–1218. Available at https://doi.org/10.14283/jpad.2024.153.

Jiang, S., Cai, G., Yang, Z., Shi, H., Zeng, H., Ye, Q., et al., 2024. Biomimetic nanovesicles as a dual gene delivery system for the synergistic gene therapy of Alzheimer's disease. ACS Nano 18 (18), 11753–11768. Available at https://doi.org/10.1021/acsnano.3c13150.

Jo, M., Lee, S., Jeon, Y.-M., Kim, S., Kwon, Y., Kim, H.-J., 2020. The role of TDP-43 propagation in neurodegenerative diseases: integrating insights from clinical and experimental studies. Experimental & Molecular Medicine 52 (10), 1652–1662. Available at https://doi.org/10.1038/s12276-020-00513-7.

Johansson, J.U., Woodling, N.S., Wang, Q., Panchal, M., Liang, X., Trueba-Saiz, A., et al., 2015. Prostaglandin signaling suppresses beneficial microglial function in Alzheimer's disease models. Journal of Clinical Investigation 125 (1), 350–364. Available at https://doi.org/10.1172/JCI77487.

Kaul, I., Sawchak, S., Correll, C.U., Kakar, R., Breier, A., Zhu, H., et al., 2024. Efficacy and safety of the muscarinic receptor agonist KarXT (xanomeline–trospium) in schizophrenia (EMERGENT-2) in the USA: results from a randomised, double-blind, placebo-controlled, flexible-dose phase 3 trial. The Lancet 403 (10422), 160–170. Available at https://doi.org/10.1016/s0140-6736(23)02190-6.

Kaur, G., Mohan, P., Pawlik, M., DeRosa, S., Fajiculay, J., Che, S., et al., 2010. Cystatin C rescues degenerating neurons in a cystatin B-knockout mouse model of progressive myoclonus epilepsy. The American Journal of Pathology 177 (5), 2256–2267. Available at https://doi.org/10.2353/ajpath.2010.100461.

Keuss, M.J., Harley, P., Ryadnov, E., Jackson, R.E., Zanovello, M., Wilkins, O.G., et al., 2024. Loss of TDP-43 induces synaptic dysfunction that is rescued by UNC13A splice-switching ASOs. bioRxiv [Preprint]. 599684. Available at 10.1101/2024.06.20.599684.

Khan, T.K., Alkon, D.L., 2015. Peripheral biomarkers of Alzheimer's disease. Journal of Alzheimer's Disease 44 (3), 729–744. Available at https://doi.org/10.3233/JAD-142262.

Klim, J.R., Williams, L.A., Limone, F., Guerra San Juan, I., Davis-Dusenbery, B.N., Mordes, D.A., et al., 2019. ALS-implicated protein TDP-43 sustains levels of STMN2, a mediator of motor neuron growth and repair. Nature Neuroscience 22 (2), 167–179. Available at https://doi.org/10.1038/s41593-018-0300-4.

Kim, E., Kim, H., Jedrychowski, M.P., Bakiasi, G., Park, J., Kruskop, J., et al., 2023. Irisin reduces amyloid-β by inducing the release of neprilysin from astrocytes following downregulation of ERK-STAT3 signaling. Neuron 111 (22), 3619. Available at https://doi.org/10.1016/j.neuron.2023.08.012.

Korobeynikov, V.A., Lyashchenko, A.K., Blanco-Redondo, B., Jafar-Nejad, P., Shneider, N.A., 2022. Antisense oligonucleotide silencing of FUS expression as a therapeutic approach in amyotrophic lateral sclerosis. Nature Medicine 28 (1), 104–116. Available at https://doi.org/10.1038/s41591-021-01615-z.

Kwon, H.S., Koh, S.H., 2020. Neuroinflammation in neurodegenerative disorders: the roles of microglia and astrocytes. Translational Neurodegeneration 9 (1). Available at https://doi.org/10.1186/s40035-020-00221-2.

Kwon, K.J., Kim, H.Y., Han, S.H., Shin, C.Y., 2024. Future therapeutic strategies for Alzheimer's disease: focus on behavioral and psychological symptoms. International Journal of Molecular Sciences 25 (21). Available at https://doi.org/10.3390/ijms252111338.

Ladiwala, A.R.A., Lin, J.C., Bale, S.S., Marcelino-Cruz, A.M., Bhattacharya, M., Dordick, J.S., et al., 2010. Resveratrol selectively remodels soluble oligomers and fibrils of amyloid Aβ into off-pathway conformers. Journal of Biological Chemistry 285 (31), 24228–24237. Available at https://doi.org/10.1074/jbc.m110.133108.

Lambracht-Washington, D., Rosenberg, R.N., 2013. Advances in the development of vaccines for alzheimer's disease. Discovery Medicine 15 (84), 319–326.

Lannfelt, L., Möller, C., Basun, H., Osswald, G., Sehlin, D., Satlin, A., et al., 2014. Perspectives on future Alzheimer therapies: amyloid-β protofibrils - a new target for immunotherapy with BAN2401 in Alzheimer's disease. Alzheimer's Research & Therapy 6 (2), 16. Available at https://doi.org/10.1186/alzrt246.

Lauffer, M.C., van Roon-Mom, W., Aartsma-Rus, A., 2024. Possibilities and limitations of antisense oligonucleotide therapies for the treatment of monogenic disorders. Communications Medicine 4 (1). Available at https://doi.org/10.1038/s43856-023-00419-1.

Lee, S.C., Liu, W., Dickson, D.W., Brosnan, C.F., Berman, J.W., 1993. Cytokine production by human fetal microglia and astrocytes: differential induction by *Lipopolysaccharide* and IL-1β Journal of Immunology 150 (7), 2659–2667.

Lee, K.I., Kim, M.J., Koh, H., Lee, J.I., Namkoong, S., Oh, W.K., et al., 2015. The anti-hypertensive drug reserpine induces neuronal cell death through inhibition of autophagic flux. Biochemical and Biophysical Research Communications 462 (4), 402–408. Available at https://doi.org/10.1016/j.bbrc.2015.04.145.

Lei, P., Ayton, S., Appukuttan, A.T., Volitakis, I., Adlard, P.A., Finkelstein, D.I., et al., 2015. Clioquinol rescues Parkinsonism and dementia phenotypes of the tau knockout mouse. Neurobiology of Disease 81, 168–175. Available at https://doi.org/10.1016/j.nbd.2015.03.015.

Leinenga, G., Götz, J., 2015. Scanning ultrasound removes amyloid-β and restores memory in an Alzheimer's disease mouse model. Science Translational Medicine 7 (278), 278ra33. Available at https://doi.org/10.1126/scitranslmed.aaa2512.

Li, B.-F., Chen, X.-Y., Xie, R., Mo, Y.-F., Wu, Y.-Z., Meng, Y., Han, X.-L., Chen, M.-H., Peng, Y.-J., 2026. Targeting microglia-mediated neuroinflammation in Alzheimer's disease: mechanisms and therapeutic approaches. Frontiers in Immunology 17. Available at 10.3389/fimmu.2026.1818802.

Lin, J.Q., Khuperkar, D., Pavlou, S., Makarchuk, S., Patikas, N., Lee, F.C.Y., et al., 2023. HNRNPH1 regulates the neuroprotective cold-shock protein RBM3 expression through poison exon exclusion. EMBO Journal 42 (14). Available at https://doi.org/10.15252/embj.2022113168.

Ma, X.R., Prudencio, M., Koike, Y., Vatsavayai, S.C., Kim, G., Harbinski, F., et al., 2022. TDP-43 represses cryptic exon inclusion in the FTD–ALS gene UNC13A. Nature 603 (7899), 124–130. Available at https://doi.org/10.1038/s41586-022-04424-7.

Marambaud, P., Zhao, H., Davies, P., 2005. Resveratrol promotes clearance of Alzheimer's disease amyloid-β peptides. Journal of Biological Chemistry 280 (45), 37377–37382. Available at https://doi.org/10.1074/jbc.m508246200.

Martin, B.K., Szekely, C., Brandt, J., Piantadosi, S., Breitner, J.C.S., Craft, S., et al., 2008. Cognitive function over time in the Alzheimer's disease anti-inflammatory prevention trial (ADAPT): results of a randomized, controlled trial of naproxen and celecoxib. Archives of Neurology 65 (7), 896–905. Available at https://doi.org/10.1001/archneur.2008.65.7.nct70006.

McAlary, L., Chew, Y.L., Lum, J.S., Geraghty, N.J., Yerbury, J.J., Cashman, N.R., 2020. Amyotrophic lateral sclerosis: proteins, proteostasis, prions, and promises. Frontiers in Cellular Neuroscience 14. Available at https://doi.org/10.3389/fncel.2020.581907.

McEachin, Z.T., Parameswaran, J., Raj, N., Bassell, G.J., Jiang, J., 2020. RNA-mediated toxicity in C9orf72 ALS and FTD. Neurobiology of Disease 145. Available at https://doi.org/10.1016/j.nbd.2020.105055.

Medinas, D.B., Valenzuela, V., Hetz, C., 2017. Proteostasis disturbance in amyotrophic lateral sclerosis. Human Molecular Genetics 26 (2), R91. Available at https://doi.org/10.1093/hmg/ddx274.

Melamed, Z.'ev, López-Erauskin, J., Baughn, M.W., Zhang, O., Drenner, K., Sun, Y., et al., 2019. Premature polyadenylation-mediated loss of stathmin-2 is a hallmark of TDP-43-dependent neurodegeneration. Nature Neuroscience 22 (2), 180–190. Available at https://doi.org/10.1038/s41593-018-0293-z.

Meneses, A., Koga, S., O'Leary, J., Dickson, D.W., Bu, G., Zhao, N., 2021. TDP-43 pathology in Alzheimer's disease. Molecular Neurodegeneration 16 (1). Available at https://doi.org/10.1186/s13024-021-00503-x.

Menon, V.P., Sudheer, A.R., Aggarwal, B.B., Surh, YJ., Shishodia, S., 2007. Antioxidant and anti-inflammatory properties of curcumin. In: Aggarwal, B.B., Surh, Y.J., Shishodia, S. (Eds.), The Molecular Targets and Therapeutic Uses of Curcumin in Health and Disease. Advances in Experimental Medicine and Biology, vol 595. Springer, Boston, MA, pp. 105–125. Available at https://doi.org/10.1007/978-0-387-46401-5_3.

Menzel, A., Samouda, H., Dohet, F., Loap, S., Ellulu, M.S., Bohn, T., 2021. Common and novel markers for measuring inflammation and oxidative stress ex vivo in research and clinical practice—which to use regarding disease outcomes? Antioxidants 10 (3), 414. Available at https://doi.org/10.3390/antiox10030414.

Meraz-Ríos, M.A., Toral-Rios, D., Franco-Bocanegra, D., Villeda-Hernández, J., Campos-Peña, V., 2013. Frontiers in Integrative Neuroscience 7. Available at https://doi.org/10.3389/fnint.2013.00059.

De Meyer, S., Blujdea, E.R., Schaeverbeke, J.M., Adamczuk, K., Vandenberghe, R., Poesen, K., et al., 2025. Serum biomarkers as prognostic markers for Alzheimer's disease in a clinical setting. Alzheimer's & Dementia: Diagnosis, Assessment & Disease Monitoring 17 (1), e70071. Available at 10.1002/dad2.70071.

Mi, W., Pawlik, M., Sastre, M., Jung, S.S., Radvinsky, D.S., Klein, A.M., et al., 2007. Cystatin C inhibits amyloid-β deposition in Alzheimer's disease mouse models. Nature Genetics 39 (12), 1440–1442. Available at https://doi.org/10.1038/ng.2007.29.

Miller, T.M., Cudkowicz, M.E., Genge, A., Shaw, P.J., Sobue, G., Bucelli, R.C., et al., 2022. Trial of antisense oligonucleotide tofersen for SOD1 ALS. New England Journal of Medicine 387 (12), 1099–1110. Available at https://doi.org/10.1056/NEJMoa2204705.

Morimoto, N., Nagai, M., Ohta, Y., Miyazaki, K., Kurata, T., Morimoto, M., et al., 2007. Increased autophagy in transgenic mice with a G93A mutant SOD1 gene. Brain Research 1167 (1), 112–117. Available at https://doi.org/10.1016/j.brainres.2007.06.045.

Morten, M.J., Sirvio, L., Rupawala, H., Mee Hayes, E., Franco, A., Radulescu, C., et al., 2022. Quantitative super-resolution imaging of pathological aggregates reveals distinct toxicity profiles in different synucleinopathies. Proceedings of the National Academy of Sciences 119 (41). Available at https://doi.org/10.1073/pnas.2205591119.

Moss, D.W., Bates, T.E., 2001. Activation of murine microglial cell lines by lipopolysaccharide and interferon-γ causes NO-mediated decreases in mitochondrial and cellular function. European Journal of Neuroscience 13 (3), 529–538. Available at https://doi.org/10.1046/j.1460-9568.2001.01418.x.

Mueller, C., Berry, J.D., McKenna-Yasek, D.M., Gernoux, G., Owegi, M.A., Pothier, L.M., et al., 2020. SOD1 Suppression with adeno-associated virus and MicroRNA in familial ALS. New England Journal of Medicine 383 (2), 151–158. Available at https://doi.org/10.1056/nejmoa2005056.

Mueller-Steiner, S., Zhou, Y., Arai, H., Roberson, E.D., Sun, B., Chen, J., et al., 2006. Antiamyloidogenic and neuroprotective functions of cathepsin B: implications for Alzheimer's disease. Neuron 51 (6), 703–714. Available at https://doi.org/10.1016/j.neuron.2006.07.027.

Mugnaini, C., Brizzi, A., Paolino, M., Scarselli, E., Castelli, R., de Candia, M., et al., 2024. Novel dual-acting hybrids targeting type-2 cannabinoid receptors and cholinesterase activity show neuroprotective effects in vitro and amelioration of cognitive impairment in vivo. ACS Chemical Neuroscience 15 (5), 955–971. Available at https://doi.org/10.1021/acschemneuro.3c00656.

Mummery, C.J., Börjesson-Hanson, A., Blackburn, D.J., Vijverberg, E.G.B., De Deyn, P.P., Ducharme, S., et al., 2023. Tau-targeting antisense oligonucleotide MAPTRx in mild Alzheimer's disease: a phase 1b, randomized, placebo-controlled trial. Nature Medicine 29 (6), 1437–1447. Available at https://doi.org/10.1038/s41591-023-02326-3.

Neumann, M., Sampathu, D.M., Kwong, L.K., Truax, A.C., Micsenyi, M.C., Chou, T.T., et al., 2006. Ubiquitinated TDP-43 in frontotemporal lobar degeneration and amyotrophic lateral sclerosis. Science 314 (5796), 130–133. Available at https://doi.org/10.1126/science.1134108.

Nihei, Y., Ito, D., Suzuki, N., 2012. Roles of ataxin-2 in pathological cascades mediated by TAR DNA-binding Protein 43 (TDP-43) and fused in sarcoma (FUS). Journal of Biological Chemistry 287 (49), 41310–41323. Available at https://doi.org/10.1074/jbc.m112.398099.

Nisar, O., Pervez, H., Mandalia, B., Waqas, M., Sra, H.K., 2020. Type 3 diabetes mellitus: a link between alzheimer's disease and type 2 diabetes mellitus. Cureus 12 (11), e11703. Available at https://doi.org/10.7759/cureus.11703.

Nolan, M., Talbot, K., Ansorge, O., 2016. Pathogenesis of FUS-associated ALS and FTD: insights from rodent models. Acta Neuropathologica Communications 4 (1), 99. Available at https://doi.org/10.1186/s40478-016-0358-8.

Nunes, E.J., Addy, N.A., Conn, P.J., Foster, D.J., 2024. Targeting the actions of muscarinic receptors on dopamine systems: new strategies for treating neuropsychiatric disorders. Annual Review of Pharmacology and Toxicology 64, 277–289. Available at https://doi.org/10.1146/annurev-pharmtox-051921-023858.

Ono, K., Li, L., Takamura, Y., Yoshiike, Y., Zhu, L., Han, F., et al., 2012. Phenolic compounds prevent amyloid β-protein oligomerization and synaptic dysfunction by site-specific binding. Journal of Biological Chemistry 287 (18), 14631–14643. Available at https://doi.org/10.1074/jbc.m111.325456.

Orgogozo, J.-M., Gilman, S., Dartigues, J.-F., Laurent, B., Puel, M., Kirby, L.C., et al., 2003. Subacute meningoencephalitis in a subset of patients with AD after A 42 immunization. Neurology 61 (1), 46–54. Available at https://doi.org/10.1212/01.WNL.0000073623.84147.A8.

Orr, M.E., Oddo, S., 2013. Autophagic/lysosomal dysfunction in Alzheimer's disease. Alzheimer's Research and Therapy 5, 53. Available at https://doi.org/10.1186/alzrt217.

Peretti, D., Bastide, A., Radford, H., Verity, N., Molloy, C., Martin, M.G., et al., 2015. RBM3 mediates structural plasticity and protective effects of cooling in neurodegeneration. Nature 518 (7538), 236–239. Available at https://doi.org/10.1038/nature14142.

Polajnar, M., Žerovnik, E., 2011. Impaired autophagy: a link between neurodegenerative diseases and progressive myoclonus epilepsies. Trends in Molecular Medicine 17 (6), 293–300. Available at https://doi.org/10.1016/j.molmed.2011.02.005.

Polajnar, M., Žerovnik, E., 2014. Impaired autophagy: a link between neurodegenerative and neuropsychiatric diseases. Journal of Cellular and Molecular Medicine 18 (9), 1705–1711. Available at https://doi.org/10.1111/jcmm.12349.

Pongrácová, E., Buratti, E., Romano, M., 2024. Prion-like spreading of disease in TDP-43 proteinopathies. Brain Sciences 14 (11), 1132. Available at https://doi.org/10.3390/brainsci14111132.

Preußner, M., Smith, H.L., Hughes, D., Zhang, M., Emmerichs, A.K., Scalzitti, S., et al., 2023. ASO targeting RBM3 temperature-controlled poison exon splicing prevents neurodegeneration in vivo. EMBO Molecular Medicine 15 (5). Available at https://doi.org/10.15252/emmm.202217157.

Pyun, J., Noor, A., Runwal, P., Mawal, C., Fuller, O.K., Egan, C.L., Febbraio, M.A., Donnelly, P.S., Short, J.L., Bush, A.I., Nicolazzo, J.A., 2026. Cu(ATSM) Restores Blood–Brain Barrier Abundance of P-Glycoprotein and Improves Cognitive Function in the APP/PS1 Mouse Model of Alzheimer's Disease. ACS Chemical Neuroscience 17, 2389–2405. Available at 10.1021/acschemneuro.6c00252.

Richardson, A., Galvan, V., Lin, A.L., Oddo, S., 2015. How longevity research can lead to therapies for Alzheimer's disease: the rapamycin story. Experimental Gerontology 68, 51–58. Available at https://doi.org/10.1016/j.exger.2014.12.002.

Riscado, M., Baptista, B., Sousa, F., 2021. New RNA-based breakthroughs in Alzheimer's disease diagnosis and therapeutics. Pharmaceutics 13 (9), 1397. Available at https://doi.org/10.3390/pharmaceutics13091397.

Saccon, R.A., Bunton-Stasyshyn, R.K.A., Fisher, E.M.C., Fratta, P., 2013. Is SOD1 loss of function involved in amyotrophic lateral sclerosis? Brain 136 (8), 2342–2358. Available at https://doi.org/10.1093/brain/awt097.

Salminen, A., Kaarniranta, K., Kauppinen, A., 2013. Beclin 1 interactome controls the crosstalk between apoptosis, autophagy and inflammasome activation: Impact on the aging process. Ageing Research Reviews 12 (2), 520–534. Available at https://doi.org/10.1016/j.arr.2012.11.004.

Sarkar, S., Rubinsztein, D.C., 2008. Huntington's disease: degradation of mutant huntingtin by autophagy. FEBS Journal 275 (17), 4263–4270. Available at https://doi.org/10.1111/j.1742-4658.2008.06562.x.

Sarkar, S., Ravikumar, B., Floto, R.A., Rubinsztein, D.C., 2009. Rapamycin and mTOR-independent autophagy inducers ameliorate toxicity of polyglutamine-expanded huntingtin and related proteinopathies. Cell Death and Differentiation 16 (1), 46–56. Available at https://doi.org/10.1038/cdd.2008.110.

Sastre, M., Calero, M., Pawlik, M., Mathews, P.M., Kumar, A., Danilov, V., et al., 2004. Binding of cystatin C to Alzheimer's amyloid β inhibits in vitro amyloid fibril formation. Neurobiology of Aging 25 (8), 1033–1043. Available at https://doi.org/10.1016/j.neurobiolaging.2003.11.006.

Sau, D., De Biasi, S., Vitellaro-Zuccarello, L., Riso, P., Guarnieri, S., Porrini, M., et al., 2007. Mutation of SOD1 in ALS: a gain of a loss of function. Human Molecular Genetics 16 (13), 1604–1618. Available at https://doi.org/10.1093/hmg/ddm110.

Serafino, A., Giovannini, D., Rossi, S., Cozzolino, M., 2020. Targeting the Wnt/β-catenin pathway in neurodegenerative diseases: recent approaches and current challenges. Expert Opinion on Drug Discovery 15 (7), 803–822. Available at https://doi.org/10.1080/17460441.2020.1746266.

Shen, Y.F., Tang, Y., Zhang, X.J., Huang, K.X., Le, W.D., 2013. Adaptive changes in autophagy after UPS impairment in Parkinson's disease. Acta Pharmacologica Sinica 34 (5), 667–673. Available at https://doi.org/10.1038/aps.2012.203.

Shneider, N.A., Harms, M.B., Korobeynikov, V.A., Rifai, O.M., Hoover, B.N., Harrington, E.A., et al., 2025. Antisense oligonucleotide jacifusen for FUS-ALS: an investigator-initiated, multicentre, open-label case series. The Lancet 405 (10494), 2075–2086. Available at https://doi.org/10.1016/S0140-6736(25)00513-6.

Škerget, K., Taler-Verčič, A., Bavdek, A., Hodnik, V., Čeru, S., Tušek-Žnidarič, M., et al., 2010. Interaction between oligomers of stefin B and amyloid-β in vitro and in cells. Journal of Biological Chemistry 285 (5), 3201–3210. Available at https://doi.org/10.1074/jbc.m109.024620.

Swain, S., Metya, A.K., 2025. Exploring metformin's therapeutic potential for Alzheimer's disease: an in-silico perspective using well-tempered funnel metadynamics. Journal of Chemical Information and Modeling 65 (8), 4163–4172. Available at https://doi.org/10.1021/acs.jcim.5c00106.

Syed, Y.Y., 2016. Eteplirsen: first global approval. Drugs 76 (17), 1699–1704. Available at https://doi.org/10.1007/s40265-016-0657-1.

Takahashi, R., Ono, K., Takamura, Y., Mizuguchi, M., Ikeda, T., Nishijo, H., et al., 2015. Phenolic compounds prevent the oligomerization of α-synuclein and reduce synaptic toxicity. Journal of Neurochemistry 134 (5), 943–955. Available at https://doi.org/10.1111/jnc.13180.

Taler-Verčič, A., Žerovnik, E., 2010. Binding of amyloid peptides to domain-swapped dimers of other amyloid-forming proteins may prevent their neurotoxicity. BioEssays 32 (12), 1020–1024. Available at https://doi.org/10.1002/bies.201000079.

Teleanu, D.M., Niculescu, A.G., Lungu, I.I., Radu, C.I., Vladâcenco, O., Roza, E., et al., 2022. An overview of oxidative stress, neuroinflammation and neurodegenerative diseases. International Journal of Molecular Sciences 23 (11). Available at https://doi.org/10.3390/ijms23115938.

Theunissen, F., Anderton, R.S., Mastaglia, F.L., Flynn, L.L., Winter, S.J., James, I., et al., 2021. Novel STMN2 variant linked to amyotrophic lateral sclerosis risk and clinical phenotype. Frontiers in Aging Neuroscience 13. Available at https://doi.org/10.3389/fnagi.2021.658226.

Thijssen, E.H., La Joie, R., Strom, A., Fonseca, C., Iaccarino, L., Wolf, A., et al., 2021. Plasma phosphorylated tau 217 and phosphorylated tau 181 as biomarkers in Alzheimer's disease and frontotemporal lobar degeneration: a retrospective diagnostic performance study. The Lancet Neurology 20 (9), 739–752. Available at https://doi.org/10.1016/S1474-4422(21)00214-3.

Tizon, B., Sahoo, S., Yu, H., Gauthier, S., Kumar, A.R., Mohan, P., et al., 2010. Induction of autophagy by cystatin C: a mechanism that protects murine primary cortical neurons and neuronal cell lines. PLoS ONE 5 (3), e9819. Available at https://doi.org/10.1371/journal.pone.0009819.

Tseng, Y.L., Lu, P.C., Lee, C.C., He, R.Y., Huang, Y.A., Tseng, Y.C., et al., 2023. Degradation of neurodegenerative disease-associated TDP-43 aggregates and oligomers via a proteolysis-targeting chimera. Journal of Biomedical Science 30, 27. Available at https://doi.org/10.1186/s12929-023-00921-7.

Tung, S., Kendrick, J., Surapaneni, A., Scheppach, J.B., Coresh, J., Gottesman, R., et al., 2022. Association between acute kidney injury and dementia in the atherosclerosis risk in communities (ARIC) study. American Journal of Kidney Diseases 80 (4), 495–501. Available at https://doi.org/10.1053/j.ajkd.2022.02.015.

Tzeplaeff, L., Wilfling, S., Requardt, M.V., Herdick, M., 2023. Current state and future directions in the therapy of ALS. Cells 12 (11), 1523. Available at https://doi.org/10.3390/cells12111523.

Ulrich, J.D., Holtzman, D.M., 2016. TREM2 function in Alzheimer's disease and neurodegeneration. ACS Chemical Neuroscience 7 (4), 420–427. Available at https://doi.org/10.1021/acschemneuro.5b00313.

Vaidya, V.S., Ferguson, M.A., Bonventre, J.V., 2008. Biomarkers of acute kidney injury. Annual Review of Pharmacology and Toxicology 48, 463–493. Available at https://doi.org/10.1146/annurev.pharmtox.48.113006.094615.

Van Daele, S.H., Masrori, P., Van Damme, P., Van Den Bosch, L., 2024. The sense of antisense therapies in ALS. Trends in Molecular Medicine 30 (3), 252–262. Available at https://doi.org/10.1016/j.molmed.2023.12.003.

Van Dyck, C.H., Swanson, C.J., Aisen, P., Bateman, R.J., Chen, C., Gee, M., et al., 2023. Lecanemab in early Alzheimer's disease. New England Journal of Medicine 388 (1), 9–21. Available at https://doi.org/10.1056/NEJMoa2212948.

Van Es, M.A., Veldink, J.H., Saris, C.G.J., Blauw, H.M., Van Vught, P.W.J., Birve, A., et al., 2009. Genome-wide association study identifies 19p13.3 (UNC13A) and 9p21.2 as susceptibility loci for sporadic amyotrophic lateral sclerosis. Nature Genetics 41 (10), 1083–1087. Available at https://doi.org/10.1038/ng.442.

Van Langenhove, T., van der Zee, J., Sleegers, K., Engelborghs, S., Vandenberghe, R., Gijselinck, I., et al., 2010. Genetic contribution of FUS to frontotemporal lobar degeneration. Neurology 74 (5), 366–371. Available at https://doi.org/10.1212/WNL.0b013e3181ccc732.

Verde, F., 2022. Tau proteins in blood as biomarkers of Alzheimer's disease and other proteinopathies. Journal of Neural Transmission 129 (2), 239–259. Available at https://doi.org/10.1007/s00702-022-02471-y.

Verma, A., Waiker, D.K., Singh, N., Roy, A., Singh, N., Saraf, P., et al., 2024. Design, synthesis, and biological investigation of quinazoline derivatives as multitargeting therapeutics in Alzheimer's disease therapy. ACS Chemical Neuroscience 15 (4), 745–771. Available at https://doi.org/10.1021/acschemneuro.3c00653.

Vogelgsang, J., Shahpasand-Kroner, H., Vogelgsang, R., Streit, F., Vukovich, R., Wiltfang, J., 2018. Multiplex immunoassay measurement of amyloid-β42 to amyloid-β40 ratio in plasma discriminates between dementia due to Alzheimer's disease and dementia not due to Alzheimer's disease. Experimental Brain Research 236 (5), 1241–1250. Available at https://doi.org/10.1007/s00221-018-5210-x.

Waiker, D.K., Verma, A., Akhilesh, N., Gajendra, T.A., Singh, N., Roy, A., et al., 2023. Design, synthesis, and biological evaluation of piperazine and N-benzylpiperidine hybrids of 5-phenyl-1,3,4-oxadiazol-2-thiol as potential multitargeted ligands for Alzheimer's disease therapy. ACS Chemical Neuroscience 14 (11), 2217–2242. Available at https://doi.org/10.1021/acschemneuro.3c00245.

Walker, K.A., Chen, J., Zhang, J., Fornage, M., Yang, Y., Zhou, L., et al., 2021. Large-scale plasma proteomic analysis identifies proteins and pathways associated with dementia risk. Nature Aging 1 (5), 473–489. Available at https://doi.org/10.1038/s43587-021-00064-0.

Wang, X., Wang, W., Li, L., Perry, G., Lee, H.-gon, Zhu, X., 2014. Oxidative stress and mitochondrial dysfunction in Alzheimer's disease. Biochimica et Biophysica Acta (BBA) - Molecular Basis of Disease 1842 (8), 1240–1247. Available at https://doi.org/10.1016/j.bbadis.2013.10.015.

Willemse, S.W., Harley, P., van Eijk, R.P.A., Demaegd, K.C., Zelina, P., Pasterkamp, R.J., et al., 2023. UNC13A in amyotrophic lateral sclerosis: from genetic association to therapeutic target. Journal of Neurology, Neurosurgery & Psychiatry 94 (8), 649–656. Available at 10.1136/jnnp-2022-330504.

Wu, M., Chen, Y.-F., Yao, W., Zhou, S., Xie, Z., Tao, Y., et al., 2025. The anti-inflammatory drug Montelukast ameliorates cognitive deficits by rescuing the inflammatory levels in young AD animal models. Scientific Reports 15 (1). Available at https://doi.org/10.1038/s41598-025-91785-4.

Xu, Y., Liang, J., Sun, Y., Zhang, Y., Shan, F., Ge, J., et al., 2023. Serum cytokines-based biomarkers in the diagnosis and monitoring of therapeutic response in patients with major depressive disorder. International Immunopharmacology 118, 110108. Available at https://doi.org/10.1016/j.intimp.2023.110108.

Yang, F., Lim, G.P., Begum, A.N., Ubeda, O.J., Simmons, M.R., Ambegaokar, S.S., et al., 2005. Curcumin inhibits formation of amyloid β Oligomers and fibrils, binds plaques, and reduces amyloid in vivo. Journal of Biological Chemistry 280 (7), 5892–5901. Available at https://doi.org/10.1074/jbc.m404751200.

Yang, X., Thomas, D.P., Zhang, X., Culver, B.W., Alexander, B.M., Murdoch, W.J., et al., 2006. Curcumin inhibits platelet-derived growth factor–stimulated vascular smooth muscle cell function and injury-induced neointima formation. Arteriosclerosis, Thrombosis, and Vascular Biology 26 (1), 85–90. Available at https://doi.org/10.1161/01.atv.0000191635.00744.b6.

Yang, D.S., Stavrides, P., Mohan, P.S., Kaushik, S., Kumar, A., Ohno, M., et al., 2011. Reversal of autophagy dysfunction in the TgCRND8 mouse model of Alzheimer's disease ameliorates amyloid pathologies and memory deficits. Brain 134 (1), 258–277. Available at https://doi.org/10.1093/brain/awq341.

Yu, F., Bronas, U.G., Konety, S., Nelson, N.W., Dysken, M., Jack, C., et al., 2014. Effects of aerobic exercise on cognition and hippocampal volume in Alzheimer's disease: study protocol of a randomized controlled trial (The FIT-AD trial). Trials 15, 394. Available at https://doi.org/10.1186/1745-6215-15-394.

Zamolodchikov, D., Chen, Z.L., Conti, B.A., Renné, T., Stricklan, S., 2015. Activation of the factor XII-driven contact system in Alzheimer's disease patient and mouse model plasma. Proceedings of the National Academy of Sciences of the United States of America 112 (13), 4068–4073. Available at https://doi.org/10.1073/pnas.1423764112.

Žerovnik, E., 2009. The emerging role of cystatins in Alzheimer's disease. BioEssays 31 (6), 597–599. Available at https://doi.org/10.1002/bies.200900012.

Zerovnik, E., Jerala, N.K., Layfield, R., 2016. Recent developments in treating Alzheimer's disease. Journal of Alzheimers Disease & Parkinsonism 6, 220. Available at https://doi.org/10.4172/2161-0460.1000220.

Zgorzynska, E., 2024. TREM2 in Alzheimer's disease: structure, function, therapeutic prospects, and activation challenges. Molecular and Cellular Neuroscience 128, 103917. Available at https://doi.org/10.1016/j.mcn.2024.103917.

Zhou, Q., Wang, W., Deng, C., 2025. Advancements in Proteolysis Targeting Chimeras for Targeted Therapeutic Strategies in Alzheimer's Disease. Molecular Neurobiology 62, 9686–9709. Available at 10.1007/s12035-025-04838-0.

Ziogas, N., Mummery, C.J., Rovira, M.B., Cummings, J.L., Fox, N.C., Hansson, O., et al., 2023. Design of a phase 2, randomized, double-blind, placebo-controlled, parallel-group study to assess the efficacy, safety, and tolerability of BIIB080 in subjects with mild cognitive impairment due to Alzheimer's disease or mild Alzheimer's disease dementia. Alzheimer's & Dementia 19 (S21), e077322. Available at https://doi.org/10.1002/alz.077322.

Zou, Z.Y., Zhou, Z.R., Che, C.H., Liu, C.Y., He, R.L., Huang, H.P., 2017. Genetic epidemiology of amyotrophic lateral sclerosis: a systematic review and meta-analysis. Journal of Neurology, Neurosurgery and Psychiatry 88 (7), 540–549. Available at https://doi.org/10.1136/jnnp-2016-315018.

Further reading

del Campo, M., Quesada, C., Vermunt, L., Peeters, C.F.W., Hok-A-Hin, Y.S., Trieu, C., et al., 2024. CSF proteins of inflammation, proteolysis and lipid transport define preclinical AD and progression to AD dementia in cognitively unimpaired individuals. Molecular Neurodegeneration 19 (1). Available at https://doi.org/10.1186/s13024-024-00767-z.

Jack, C.R., Andrews, J.S., Beach, T.G., Buracchio, T., Dunn, B., Graf, A., et al., 2024. Revised criteria for diagnosis and staging of Alzheimer's disease: Alzheimer's association workgroup. Alzheimer's & Dementia 20 (8), 5143–5169. Available at https://doi.org/10.1002/alz.13859.

Noh, M.Y., Kwon, H.S., Kwon, M.S., Nahm, M., Jin, H.K., Bae, J.S., et al., 2025. Biomarkers and therapeutic strategies targeting microglia in neurodegenerative diseases: current status and future directions. Molecular Neurodegeneration 20 (1). Available at https://doi.org/10.1186/s13024-025-00867-4.

Wu, C.K., Fuh, J.L., 2025. A 2025 update on treatment strategies for the Alzheimer's disease spectrum. Journal of the Chinese Medical Association 88 (7), 495–502. Available at https://doi.org/10.1097/JCMA.0000000000001252.

PART 3

Small molecules as modulators of the key processes in proteinopathies

CHAPTER 9

Small molecule modulators of protein aggregation and toxicity: antioxidants and polyphenol compounds

Lea Pogačnik da Silva[1], Nataša Poklar Ulrih[1], and Eva Žerovnik[2,3]
[1]Department of Food Science and Technology, Biotechnical Faculty, University of Ljubljana, Ljubljana, Slovenia,
[2]Department of Biochemistry and Molecular and Structural Biology, Jožef Stefan Institute, Ljubljana, Slovenia,
[3]Jožef Stefan International Postgraduate School (IPS), Ljubljana, Slovenia

9.1 Introduction

Aberrant protein aggregation may be part of a vicious circle: protein aggregation impairs metal balance, oxidative stress causes more oxidatively modified proteins, these accumulate and sequester other proteins, and finally their presence in the cytoplasm can overwhelm the two intracellular protein degradation systems—the ubiquitin–proteasome system and the lysosomal system—which normally work to prevent the accumulation of protein aggregates. One cannot say what the initial trigger for protein aggregation is; many processes in neurons are impacted, and mitochondria, the energy factories of the cell, are consistently affected. Partially, oxidative stress may be due to the Fenton reaction induced by protein aggregates interacting with metals, where, among other free radicals, H_2O_2 (hydrogen peroxide) is formed (Lévy et al., 2019). Free radicals are characterized by the presence of unpaired electrons. These highly reactive molecules are generated through endogenous processes, such as mitochondrial respiration and immune cell activation, or exogenous sources, such as UV light, air pollution, and smoking. When the delicate balance between oxidants and antioxidants is disrupted, this leads to oxidative stress. Oxidative stress further damages DNA, proteins, and lipids, contributing to the pathogenesis of various diseases, including neurodegenerative disorders (see Fig. 9.1).

These cellular processes lead to neuroinflammation and, ultimately, neuronal death. Natural products interacting with these processes are beneficial with respect to preventing neuronal damage (Silla et al., 2025).

9.2 Small molecule modulators of protein aggregation towards amyloid formation

Stefani and Rigacci, as early as 2013, studied the influence of natural phenolic compounds on protein aggregation (Stefani and Rigacci, 2013). They classified the

Molecular Insights into Proteinopathies
DOI: https://doi.org/10.1016/B978-0-443-34176-2.00008-0
© 2027 Elsevier Inc. All rights are reserved, including those for text and data mining, AI training, and similar technologies.

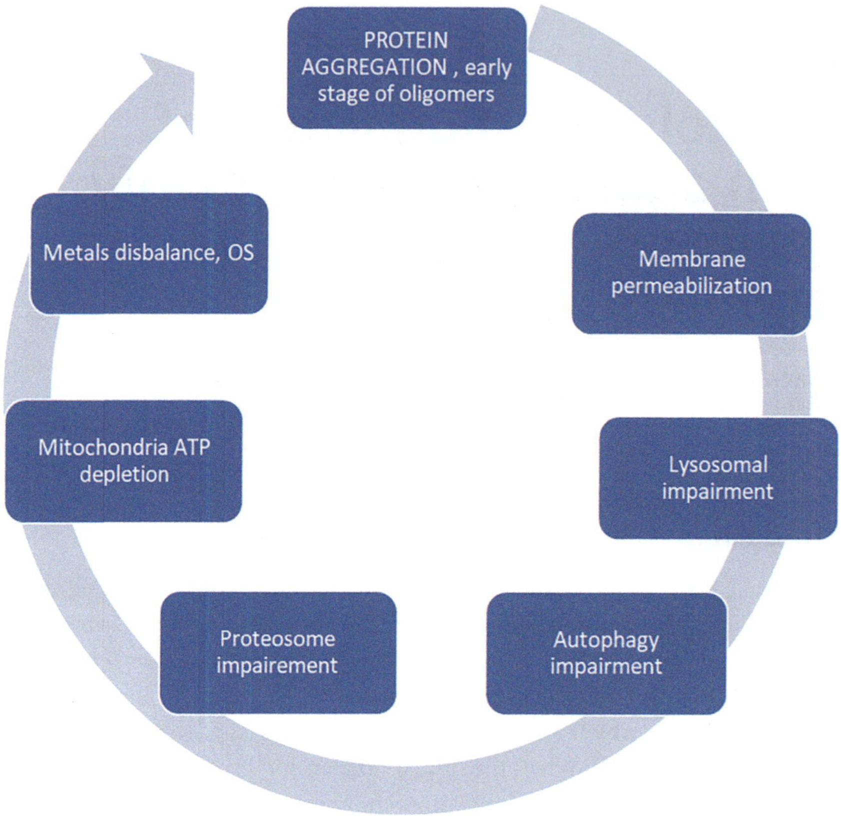

Figure 9.1 Schematic of protein aggregation as part of a vicious circle accompanying and perpetuating neurodegeneration.

molecules into three classes, according to Necula et al. (2007), namely, Class I molecules, inhibiting amyloid oligomerization but not fibrillization; Class II molecules, inhibiting both oligomerization and fibrillization; and Class III molecules, inhibiting fibrillization but not oligomerization.

Some mechanistic in vitro studies have been performed. By using pulsed hydrogen/deuterium exchange mass spectrometry, circular dichroism, and transmission electron microscopy, the influence of resveratrol on α-synuclein (αS) aggregation (relevant to synucleinopathies such as Parkinson's disease [PD]) has been explored (Alaziqi et al., 2024). After using complex modeling for the stages of amyloid fibril nucleation, elongation, secondary nucleation, and fragmentation, the results indicated a different route of protein aggregation in the presence of resveratrol. The main conclusion was that resveratrol either destabilizes aggregation intermediates or interacts with monomeric αS to bypass various aggregation pathways (Alaziqi et al., 2024).

Epigallocatechin-3-gallate (EGCG), the most abundant of the flavanols in green tea, is one of the most powerful polyphenols, exerting neuroprotective effects in PD and Alzheimer's disease (AD), and also acting beneficially in type 2 diabetes (Meng et al., 2010), likely due to its effect on the destabilization of protein fibrils (Bieschke et al., 2010). EGCG prevents the formation of toxic aggregates of αS in vitro (Bieschke et al., 2010; Camilleri et al., 2013; Caruana et al., 2012; Pogačnik et al., 2016). It is believed that EGCG promotes "off-pathway" oligomer formation, which is nontoxic in mammalian cell lines, including human embryonic kidney-293 and rat PC12 cells (Bieschke et al., 2010). Additionally, EGCG protected cell membranes from being destabilized by preformed fibrils (Camilleri et al., 2013).

Similar effects on αS oligomer formation were observed for 14 naturally occurring polyphenolic compounds and for black tea extract. All of the polyphenols inhibited αS aggregation in a dose-dependent manner. Additionally, they promoted disaggregation of preformed oligomers (Caruana et al., 2011). The precise mechanisms of EGCG action remain unknown, although it is believed that EGCG has protective effects through the production of off-pathway "compact" oligomers and by facilitating the conversion of "active" oligomers into amyloid fibrils (Yang et al., 2017).

To test the EGCG fibril-destabilizing effects in vivo, mice underwent oral administration of EGCG followed by MPTP treatment; here, EGCG prevented MPTP-induced αS accumulation (Mandel et al., 2004). Further, a mixture of polyphenols from green tea that included various catechins and, in particular, EGCG, was shown to reduce motor deficits in MPTP-lesioned cynomolgus monkeys, associated with increased numbers of substantia nigra tyrosine hydroxylase-positive neurons and increased DA levels in the striatum (Chen et al., 2015). Strong neuroprotection by green tea was also observed in an Alzheimer's-like rat model, with a reduction in memory deficits, reactive oxygen species (ROS) levels, and thiobarbituric-acid-reactive substances in the hippocampus. Based on these data, green tea supplements have been proposed to reduce the effects of AD due to their high EGCG concentration (Hoensch and Oertel, 2015). In parallel, it was shown that polyphenols from tea lowered the levels of protein oligomers while enhancing cell viability in the dopaminergic substantia nigra/neuroblastoma hybrid cell line MES23.5 treated with a combination of MPP+ and αS monomers or oligomers.

Olive oil is rich in polyphenolic substances, with the main olive oil polyphenol, oleuropein aglycone, being the subject of many studies due to its antiamyloidogenic activity in vitro and in cells. As part of a Mediterranean diet, it is also a potential protective strategy against AD. Extracts of extra virgin olive oil have been explored for their anti-protein aggregation properties. It has been found that at relatively low concentrations (<20 μg/mL), these extracts reduced the rate of Aβ40 aggregation and fibril formation in vitro. However, the polyphenolic components were also shown to bind to preformed Aβ40 fibrils and release soluble Aβ oligomers, which are the more toxic species in vivo. Furthermore, it was shown that oleuropein aglycone stabilized αS monomers and prevented the multiplication of on-pathway oligomers, which favored the growth of very stable and harmless aggregates that did not evolve into cytotoxic amyloids. Oleuropein aglycone prevented αS aggregates from binding to cell membrane components, which resulted in the prevention of oxidative damage to cells and a reduction of cytotoxicity (Martinez-Huelamo et al., 2017; Palazzi et al., 2018).

Other polyphenols have also shown fibril-destabilizing properties through inhibition of initiation and/or propagation of stefin B and αS fibrils in vitro, including quercetin (Hasanbasic et al., 2018; Zhu et al., 2013), curcumin (Hasanbasic et al., 2018; Jiang et al., 2013a,b), and resveratrol (see above) (Hasanbasic et al., 2018). Interestingly, the oxidized form of quercetin was more efficient at destablization of human wild-type αS fibril formation and growth (Zhu et al., 2013). Although quercetin and its oxidized species disaggregated preformed fibrils, it is likely that oxidized quercetin is more potent because of its higher hydrophilicity and polarity. *Drosophila melanogaster* expressing wild-type human αS is a good model for investigating the synucleinopathic aspect of PD and for studying promising neuroprotective strategies (Renaud and Martinoli, 2014). Curcumin and EGCG added to the *Drosophila* diet promoted increased life span, recovered the loss of climbing ability, reduced protein carbonyl content as well as lipid peroxidation, and mitigated the death of brain cells (Siddique et al., 2014a,b,c,d).

Our groups (Žerovnik and coworkers) performed studies on the effect of polyphenols and some other antioxidant substances on the aggregation of a model protein, human stefin B (cystatin B), and another amyloidogenic protein causing amyloid angiopathy, human cystatin C (CysC). First, we measured the influence of several substances on human CysC aggregation and amyloid fibril formation (Jahić et al., 2020), induced at pH 4 in vitro. The effect of three polyphenols—resveratrol, quercetin, and curcumin—and of two antioxidants—vitamin C and N-acetyl-L-cysteine (NAC)—was explored, as well as the effect of sulforaphane and α-lipoic acid. Surprisingly, NAC, vitamin C, and sulforaphane exhibited the largest inhibitory effects on amyloid fibril growth of this protein. The effects of the polyphenols were not significant, apart from resveratrol, which partly inhibited amyloid fibril growth of human CysC. However, with the other amyloidogenic protein, human stefin B, vitamin C and NAC showed less inhibitory effects on its fibrillation. Specifically, they inhibited the process only at low concentrations. Among the polyphenols, curcumin showed a significant inhibitory effect, prolonging the lag phase and reducing the amount of mature fibrils (Hasanbasic et al., 2018), whereas no comparative effect was observed with resveratrol and quercetin. More unexpectedly, a synergistic inhibitory action of polyphenols plus vitamin C on the fibrillation of human stefin B was observed (Jahić Mujkić et al., 2021). Namely, the strongest effect was observed in the case of curcumin (1, 4, 10, and 20 μM) plus 0.5 mM vitamin C. Similarly, various concentrations of quercetin (1, 4, 10, 20, and 50 μM) plus

0.5 mM vitamin C showed stronger inhibition by prolongation of the lag phase than quercetin or vitamin C alone, thus indicating synergy.

From the recent literature, it has been confirmed that polyphenolic compounds are prone to inhibit amyloid fibril formation of proteins and exert a dual role as nutraceuticals and pharmaceuticals (Zhytniakivska et al., 2025). Namely, polyphenolic compounds, such as EGCG and curcumin, have shown significant inhibitory effects on the fibrillation of the prion fragment PrP (106–128) (Regmi et al., 2025) and on Aβ fibrillation (Li et al., 2025). Another example is the extract from the plant *Heliotropium bacciferum*, which demonstrated a dose-dependent inhibition of insulin fibrillation (Al-Bagmi et al., 2025).

Unfortunately, polyphenols have limited bioavailability and cannot always easily cross the blood–brain barrier (BBB). Even though selected polyphenols can cross the BBB (Campos-Bedolla et al., 2014; Pogačnik et al., 2016), to address this difficulty, nanoparticle-based delivery systems have been designed. Polymer nanoparticles with attached polyphenolic compounds facilitate the delivery of polyphenols across the BBB (Fang et al., 2024) (see Fig. 9.2).

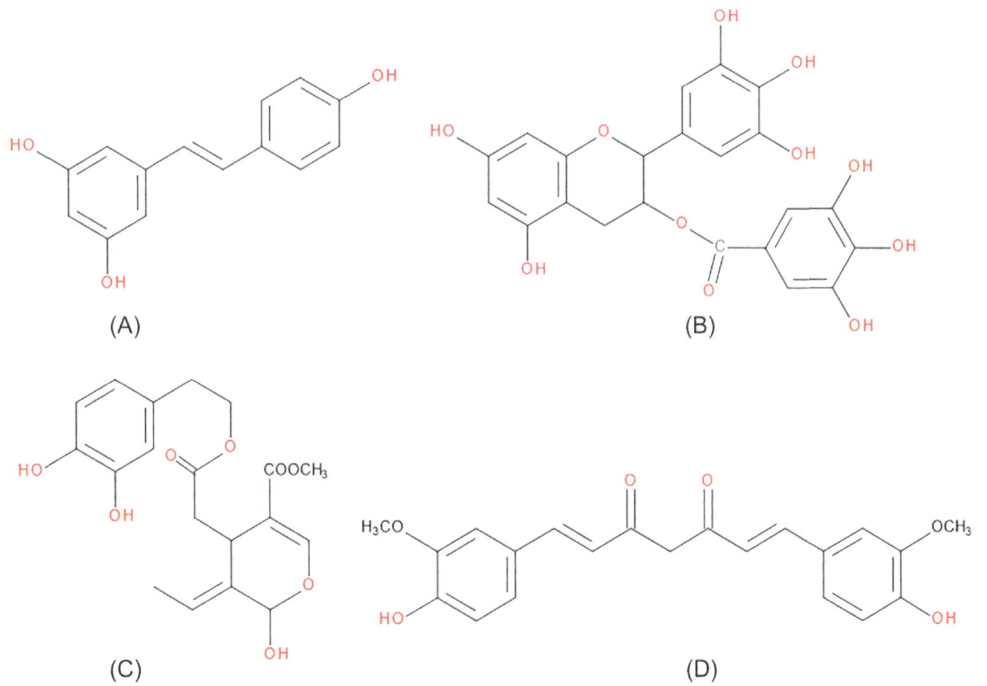

Figure 9.2 Structures of the most studied polyphenols as modulators of protein aggregation and toxicity. (A) Resveratrol, (B) epigallocatechin-3-gallate, (C) oleuropein aglycone, and (D) curcumin.

9.3 Inhibition of membrane permeabilization by small molecules from the groups of antioxidants and polyphenols

Understanding the mechanism by which different conformers and oligomers of Aβ interact with membranes and cause membrane permeability should help in the search for new therapeutics to alleviate Aβ cytotoxicity (Viles, 2023). Similar to Necula et al. (2007), we posit that the modes of action of small-molecule modulators of protein aggregation with lipid membranes can be threefold. Either they prevent oligomerization by binding to and stabilizing protein monomers, they bind to prefibrillar oligomers and prevent their interaction with membranes, or they block amyloid pore formation by changing the conformation of oligomers on the route from α-helical to β-barrel at the stage of the prepore formation. For further reference, see figure 1.1 in Chapter 1: "Amyloid-pore hypothesis: prefibrillar oligomers interactions with membranes; a generic cause of neurodegeneration?" by E. Žerovnik, this book.

Vassallo's group in 2012 used a lipid vesicle permeabilization assay to show the inhibitory action of certain polyphenolic substances toward membrane permeabilization by αS oligomers (Caruana et al., 2012). They screened several natural polyphenols and synthetic N′-benzylidene-benzohydrazide substances, as well as black tea extract. It was found that apigenin, baicalein, morin, nordihydroguaiaretic acid, and black tea extract strongly inhibited membrane damage by αS soluble protofibrils, both wild-type and synucleinopathy-associated mutants (A30P, A53T). Structure–activity analysis further suggested that a 5,7-dihydroxychromen-4-one moiety was crucial for the inhibition. In conclusion, we have identified a group of polyphenols that can effectively hinder membrane damage by αS aggregates. These may serve as a viable source of lead substances for the development and design of novel therapeutic agents for PD and related protein aggregation disorders.

A study of bacoside-A, an Indian traditional medicine substance, was performed by Malishev et al. (2017). The authors showed that this substance can inhibit fibril formation as well as membrane permeabilization by Aβ42, whereas the formation of oligomers was not affected. Preincubation of cells with bacoside-A reduced Aβ42 toxicity in solution and when associated with membrane vesicles (Malishev et al., 2017).

The effect of GT863, a derivative of curcumin, on Aβ-induced plasma membrane damage was assessed by looking for phospholipid peroxidation, membrane fluidity, membrane phase state, membrane potential, membrane resistance, and changes in intracellular Ca^{2+} concentration (Momma et al., 2023). The authors demonstrated that GT863's neuroprotective effects against Aβ oligomer-mediated membrane damage may be due to decreased phospholipid peroxidation, decreased membrane fluidity and resistance, and decreased excessive Ca^{2+} ion influx.

The interactions of EGCG with 1,2-dipalmitoyl-*sn*-glycero-3-phosphatidylcholine liposomes have been addressed in a recent study by the group of Poklar Ulrih

(Šturm and et al., 2022). Incorporation of EGCG either decreased or increased the lipid ordering parameter, depending on the initial membrane order. Significantly, the liposomes with incorporated EGCG were shown to be highly protected against UV light.

9.4 Small molecules from the group of antioxidants and polyphenols repairing mitochondrial damage

Genes commonly implicated in PD are closely linked to mitochondrial homeostasis. Proteins such as DJ-1, leucine-rich repeat kinase 2 (LRRK2), PTEN-induced kinase 1, and Parkin are involved in key mitochondrial processes, including fission and fusion dynamics, lysosomal-autophagic degradation, mitophagy, and ROS sensing (Chaturvedi and Beal, 2013).

The flavone baicalein has been shown to upregulate DJ-1 expression in SH-SY5Y human neuroblastoma cells treated with 6-hydroxydopamine, a neurotoxin used to model PD (Wang et al., 2013). Similarly, EGCG demonstrated neuroprotective effects in *Drosophila* models of PD. In flies expressing the G2019S LRRK2 mutation or lacking Parkin, EGCG improved dopaminergic neuron survival and enhanced climbing ability (Ng et al., 2012). However, these protective effects were significantly reduced when AMP-activated protein kinase (AMPK) was genetically inactivated, highlighting the importance of its functional integrity.

Resveratrol has been extensively studied for its impact on AMPK signaling. In primary skin fibroblast cultures derived from two patients with early-onset PD due to Parkin mutations, resveratrol improved mitochondrial homeostasis (Ferretta et al., 2014). Treatment enhanced mitochondrial function by increasing complex I activity, ATP production, and oxygen consumption while simultaneously reducing ROS levels and lactate accumulation.

Beyond these effects, resveratrol also elevated the NAD^+/NADH ratio and activated both AMPK and SIRT1 pathways. It modulated the transcriptional activity of PGC-1α, as evidenced by increased mRNA expression of its downstream target genes. Furthermore, in fibroblast cultures, resveratrol promoted macroautophagy through a microtubule-associated protein light chain 3-independent mechanism, contributing to the clearance of αS aggregates.

9.5 Effect of small molecules from the group of antioxidants and polyphenols on macroautophagy

The major two processes removing intracellular protein aggregates are macroautophagy (autophagy - in short) and ubiquitin proteasomal system (proteasome - in short). It is well known that autophagic pathways are impaired in PD (Ghavami et al., 2014). The treatment of differentiated dopaminergic cells (PC12) with inhibitors of the proteasome (e.g., MG132 and MG115) resulted in extensive mitochondrial damage and cell death, including apoptosis. The flavone baicalein inhibited these effects following proteasome inhibition (Jung and Lee, 2014). Furthermore, it was shown that resveratrol protected human neuroblastoma cells (SH-SY5Y) against cell death when challenged with rotenone (Lin et al., 2014). A role for autophagy in this process was suggested since, when bafilomycin A1, an autophagosome–lysosome fusion inhibitor, was administered, it abolished these effects. Additionally, resveratrol-mediated autophagy and neuroprotection can be abolished by pharmacological inhibition of heme oxygenase-1. Similarly, it was shown that puerarin (a bioactive isoflavone glycoside) increased the expression of lysosome-associated membrane protein 2a (LAMP2a), a lysosomal membrane receptor involved in the regulation of cytosolic proteins through chaperone-mediated autophagy (Zhu et al., 2014). Here, the levels of LAMP2a present at the lysosomal membrane were directly associated with the proteolytic pathway. Even if there was no direct evidence for the removal of protein oligomers, it was found that quercetin inhibited Aβ-induced paralysis in *Caenorhabditis elegans* through activation of protein degradation pathways (Regitz et al., 2014). Evidently, a role for modulation of protein misfolding and clearance underlies the neuroprotective actions of some polyphenols (Freyssin et al., 2018; Reglodi et al., 2017).

In an MPTP-induced αS mouse model of PD, naringenin significantly downregulated αS and upregulated DA transporter and tyrosine hydroxylase protein expression (Mani et al., 2018). Additionally, it downregulated tumor necrosis factor alpha and interleukin 1β mRNA expression and improved levels of superoxide dismutase. Naringenin also lowered levels of GSH (glutathione) compared to those in the vehicle-treated mice. Upregulation of tyrosine hydroxylase with naringenin treatment corresponded to increased homovanillic acid and 3,4-dihydroxyphenylacetic acid turnover and improved motor function. It is likely that naringenin can counteract MPTP-induced dopaminergic degeneration through the regulation of neuroinflammation, αS fibrillation, and oxidative stress (Yang et al., 2017).

Some of the most recent studies show that polyphenols, such as resveratrol and nobiletin, enhance autophagy, thereby improving cellular detoxification and mitigating damage from stressors such as nanoplastics (Okamoto et al., 2025; Yu et al., 2025).

9.6 Clinical studies of the beneficial effects of polyphenols

9.6.1 Exogenous substances from plant sources

9.6.1.1 Curcumin

Curcumin (diferuloylmethane) is a phytopolyphenol pigment derived from *Curcuma longa*, a perennial plant in the ginger family (*Zingiberaceae*), commonly known as turmeric. Used for centuries in Indian and Chinese medicine (Monroy et al., 2013), curcumin exhibits antioxidant, antiinflammatory, and anti-protein aggregation properties, making it a promising neuroprotective agent (Cole et al., 2007).

Epidemiological data from the Indo-US Cross-National Dementia Epidemiology Study suggest that a curcumin-rich diet may contribute to the lower prevalence of AD in India compared with the USA, although a direct causal link has not been established (Ganguli et al., 2000). Curcumin's pleiotropic effects in the central nervous system (CNS), including its preferential binding to amyloid proteins, support its potential in treating age-related neurodegenerative diseases (NDs) (Maiti and Dunbar, 2018).

In vitro studies have shown that curcumin and its analog rosmarinic acid inhibit Aβ fibril formation in a dose-dependent manner, destabilize existing fibrils, and regenerate Aβ monomers (Ono et al., 2004). It also scavenges nitric oxide and hydroxyl radicals, binds redox-active metals, and suppresses inflammation by inhibiting transcription of inflammatory cytokines, inducible nitric oxide synthase, and cyclooxygenase-2 (Cole et al., 2007; Ono et al., 2004).

Animal studies in murine and rat models of AD have demonstrated curcumin's neuroprotective effects, supporting further clinical research into its use as a dietary supplement for AD prevention and treatment (Frautschy et al., 2001; Lim et al., 2001; Garcia-Alloza et al., 2007). However, its clinical application is limited by poor bioavailability, as noted in several studies (Appleton and Day, 2019; Hewlings and Kalman, 2017; Kotha and Luthria, 2019; Lopresti, 2018).

9.6.1.2 Resveratrol

Resveratrol (3,5,4-trihydroxystilbene) is a phytoalexin produced by plants in response to stressors such as infection, UV radiation, and injury (Park and Pezzuto, 2015). It is found in grape skins, berries, Japanese knotweed, red wine, rhubarb roots, and peanuts (Kim et al., 2016; Ali et al., 2010).

Both in vitro and in vivo studies show that resveratrol crosses the BBB and may benefit neurodegenerative conditions, including PD, AD, cerebral ischemia, prion disease, epilepsy, Huntington's disease, and amyotrophic lateral sclerosis (Martinez-Huelamo et al., 2017; Albani et al., 2010; Mokni et al., 2007). Mechanistically, it reduces secreted and intracellular Aβ levels (Marambaud et al., 2005), scavenges free radicals, and enhances cellular antioxidant defenses (Candelario-Jalil et al., 2007; Cao and Li, 2004).

Animal models have shown that trans-resveratrol can prevent cognitive decline and spatial memory deficits (Kumar et al., 2007; Sharma et al., 2005). However, its therapeutic use in humans is hindered by its low oral bioavailability due to rapid metabolism and excretion (Walle, 2011). Despite this, the extensive evidence of its neuroprotective effects supports continued interest in resveratrol as a potential treatment for NDs (Andrade et al., 2018; Navarro et al., 2018; Renaud and Martinoli, 2019).

9.6.1.3 Green tea catechins

Tea is one of the most widely consumed beverages globally, and interest in its health benefits has grown significantly (Cooper et al., 2005). The neuroprotective effects of green tea (*Camellia sinensis*) are largely attributed to catechins—phenolic compounds with strong antioxidant, iron-chelating, and antiinflammatory properties (Mandel et al., 2012). Major catechins include (–)-epicatechin, (–)-epigallocatechin, (–)-epicatechin-3-gallate, and (−)-epigallocatechin-3-gallate (EGCG) (Mak, 2012).

EGCG, the most abundant catechin (50%–80% of total catechins), exerts its effects through various interactions with proteins and receptors, modulation of signaling pathways, mitochondrial function, DNA methylation, and autophagy (Hoensch and Oertel, 2015; Charo and Ransohoff, 2006; Kim et al., 2014). In vivo studies in mice show that EGCG diminishes Aβ-induced cognitive dysfunction by reducing Aβ production through modulating amyloid precursor protein cleavage by γ-secretase (Shaham-Niv et al., 2018; Rezai-Zadeh et al., 2005; Lee, 1987).

EGCG also inhibits nuclear factor kappa-light-chain-enhancer of activated B cells activation and provides neuroprotection in experimental autoimmune encephalomyelitis (EAE), a primate model of multiple sclerosis (Aktas et al., 2004). EAE is an animal model of autoimmune inflammatory diseases of the CNS, and it resembles MS in many respects (Constantinescu et al., 2012). Epidemiological studies suggest a link between tea consumption and reduced AD risk, with antiamyloid effects summarized by Polito et al. (2018).

Evidence also supports a protective role of green and black tea against PD (Mandel et al., 2012). A randomized, double-blind, placebo-controlled study funded by The Michael J. Fox Foundation and conducted by the Chinese Parkinson Study Group evaluated green tea catechins in 400 untreated PD patients. After six months, all dose groups showed significant improvement, with no adverse effects reported (Chan et al., 2009). However, after 12 months, the improvements were no longer significant compared with placebo, suggesting that while green tea polyphenols may offer symptomatic relief in early PD, they do not appear to modify disease progression. Determining optimal dosage and administration frequency remains a key challenge (Reglodi et al., 2017) (see Fig. 9.3).

9.6.1.4 Structures of gingolides from Ginkgo biloba

Extracts from *Ginkgo biloba* leaves, containing ginkgolides A and B (see Fig. 9.3), have demonstrated protective effects against Aβ-induced neurotoxicity by inhibiting the formation of Aβ-derived diffusible neurotoxic soluble ligands, which are believed to play a key role in the development of AD (Yao et al., 2001). Multiple clinical studies have shown that *G. biloba* leaf extract can improve both cognitive and noncognitive symptoms in patients with dementia (Napryeyenko and Borzenko, 2007;

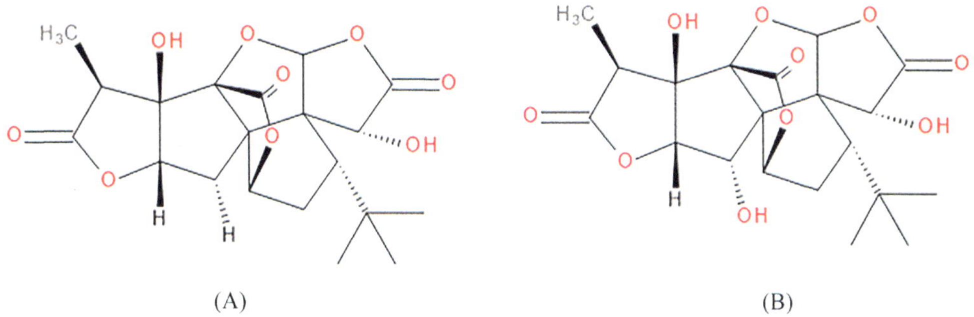

Figure 9.3 Ginkgolides (A) and (B).

Ihl et al., 2012; Hashiguchi et al., 2015; Gavrilova and et al., 2014). These therapeutic effects are thought to be largely due to the extract's antioxidant properties and its ability to enhance the activity of antioxidant enzymes (Mahadevan and Park, 2008).

Preclinical studies, both in vitro and in vivo, involving models of AD and PD, have confirmed the neuroprotective and antioxidant effects of the standardized *G. biloba* extract EGb761 (Rojas et al., 2016). Although EGb761 has not yet been directly evaluated as an antioxidant in clinical trials, indirect evidence supports its potential use in treating NDs. Further clinical research is needed to fully assess its therapeutic value in these conditions.

9.6.2 Endogenous substances

9.6.2.1 Acetyl-L-carnitine

Acetyl-L-carnitine plays a vital role in intermediary metabolism, functioning as an acetyl group donor and facilitating the transport of fatty acids across the mitochondrial membrane during β-oxidation (Traina, 2016). Dietary supplementation with exogenous acetyl-L-carnitine has shown neuroprotective, neurotrophic, antidepressant, and analgesic effects, particularly in the treatment of painful neuropathies (Aktas et al., 2004).

Its primary mechanism involves enhancing mitochondrial respiration, which enables neurons to produce sufficient ATP to maintain normal membrane potential (Owen and Sunram-Lea, 2011). In vitro studies have demonstrated that acetyl-L-carnitine provides dose-dependent neuroprotection against Aβ-induced toxicity, protein oxidation, lipid peroxidation, and apoptosis (Abdul et al., 2006; Forloni et al., 1994). Additionally, it has shown neurotrophic effects in primary cultures of rat embryo motor neurons (Bigini et al., 2002).

Double-blind, controlled clinical trials have reported beneficial effects of acetyl-L-carnitine in the treatment of major depressive disorders and AD, both of which are prevalent in the elderly population (Pettegrew et al., 2000). A metaanalysis by Montgomery et al. (2003) concluded that acetyl-L-carnitine can improve symptoms in patients with mild cognitive impairment and early-stage AD. Overall, the accumulated evidence from in vitro and in vivo studies suggests that acetyl-L-carnitine, as an endogenous compound, holds significant promise for the treatment of AD and other neurodegenerative disorders (Traina, 2016).

9.6.2.2 Coenzyme Q10

Coenzyme Q10 (ubidecarenone) is a crucial component of the mitochondrial respiratory chain, acting as an electron carrier and a powerful antioxidant within mitochondria. Its role in supporting mitochondrial function and combating oxidative stress makes it a compelling candidate for treating NDs characterized by mitochondrial dysfunction, such as PD, Huntington's disease, and Friedreich's ataxia (Shults, 2003; Niedzielska et al., 2016).

Animal studies have shown that coenzyme Q10 can reduce neuronal damage in various models of neurodegeneration (Beal, 1999). Recent findings suggest that it may influence the expression of genes involved in cell signaling, metabolism, and transport, which could contribute to its therapeutic effects (Garrido-Maraver et al., 2014). Clinical trials have indicated that coenzyme Q10 supplementation may reduce inflammatory markers in multiple sclerosis and slow functional decline in PD (Shults et al., 2002; Sanoobar et al., 2015).

Given its good safety profile and the development of new formulations that enhance its bioavailability, coenzyme Q10 is considered a promising agent for the treatment of NDs. Additionally, other lipophilic antioxidants—such as vitamins A and E, carotenoids, and n-3 polyunsaturated fatty acids (PUFAs)—are gaining attention for their potential roles in preventing and treating these conditions. A comprehenive review explored the relationship between lipophilic antioxidants and neurodegeneration, based on findings from cellular, animal, and biomarker studies (Chang et al., 2018).

9.6.2.3 n-3 polyunsaturated fatty acids

n-3 PUFAs (Fig. 9.4), primarily found in fish and fish oil, possess several biological properties that may be beneficial in NDs (Denis et al., 2015). The brain has limited capacity to synthesize key n-3 PUFAs such as eicosapentaenoic acid (EPA), docosahexaenoic acid (DHA), and docosapentaenoic acid, relying instead on uptake from plasma, which is influenced by dietary and hepatic sources (Dyall, 2015).

Studies in mice and rats have shown that supplementation with n-3 PUFAs enhances neurogenesis, executive function, and learning, while deficiency is associated with memory impairment and reduced hippocampal plasticity (Cutuli et al., 2016). Improvements in mitochondrial function have also been observed in both in vivo and ex vivo models of aging and neurodegeneration following PUFA treatment (Eckert et al., 2013).

Figure 9.4 Structures of the long-chain n-3 polyunsaturated fatty acids: (A) eicosapentaenoic acid, (B) docosahexaenoic acid, and (C) docosapentaenoic acid.

EPA and DHA exhibit antiinflammatory effects, reduce age-related microglial activation and oxidative stress, and modulate cytokine levels (Calder, 2011; Trépanier et al., 2016; Maher et al., 2004). Although clinical data are somewhat inconsistent, the overall evidence supports the neuroprotective potential of n-3 PUFAs in a range of neurodegenerative and neurological disorders (see Fig. 9.4).

9.7 Conclusions

The growing body of evidence underscores the significant potential of small molecules—particularly antioxidants and polyphenols—in modulating protein aggregation and mitigating the cellular toxicity associated with NDs. These compounds exhibit multifaceted mechanisms of action, including inhibition of amyloid fibril formation, prevention of membrane permeabilization, restoration of mitochondrial function, and enhancement of autophagic clearance. While in vitro and in vivo studies provide compelling support for their neuroprotective roles, clinical trials have yielded promising, yet sometimes inconclusive, results due to challenges such as poor bioavailability and limited BBB permeability.

Nonetheless, the integration of polyphenol-rich foods and supplements into the diet, especially those characteristic of the Mediterranean diet, emerges as a viable preventive strategy. Advances in delivery systems, such as nanoparticle-based carriers, offer new avenues to overcome pharmacokinetic limitations and enhance therapeutic efficacy.

Given the increasing burden of neurodegenerative disorders in aging populations, further well-designed, double-blind clinical studies are warranted to validate the efficacy of these compounds in both diseased and healthy individuals. Continued interdisciplinary research will be essential to translate these findings into effective, accessible interventions for neuroprotection and cognitive health.

Acknowledgments

The authors thank the Slovenian Research and Innovation Agency for the financial support through the programs P1–0140 (E.Ž contributing) and P4–0121 (N.P.U. as PI).

Copyright

The contents of this chapter are, in some parts, similar to those of a paper by Pogačnik L, Ota A, Ulrih NP. An Overview of Crucial Dietary Substances and Their Modes of Action for Prevention of NDs. Cells. 2020 Feb 28;9(3):576. https://doi.org/10.3390/cells9030576.

Creative Commons Attribution (CC BY) license (http://creativecommons.org/licenses/by/4.0/).

References

Abdul, H.M., Calabrese, V., Calvani, M., Butterfield, D.A., 2006. Acetyl-L-carnitine-induced up-regulation of heat shock proteins protects cortical neurons against amyloid-beta peptide 1-42-mediated oxidative stress and neurotoxicity: implications for Alzheimer's disease. Journal of Neuroscience Research 84 (2), 398–408. Available at https://doi.org/10.1002/jnr.20877.

Aktas, O., Prozorovski, T., Smorodchenko, A., Savaskan, N.E., Lauster, R., Kloetzel, P.M., et al., 2004. Green tea epigallocatechin-3-gallate mediates T cellular NF-κB inhibition and exerts neuroprotection in autoimmune encephalomyelitis. Journal of Immunology 173 (9), 5794–5800. Available at https://doi.org/10.4049/jimmunol.173.9.5794.

Alaziqi, B., Beckitt, L., Townsend, D.J., Morgan, J., Price, R., Maerivoet, A., et al., 2024. Characterization of olive oil phenolic extracts and their effects on the aggregation of the Alzheimer's amyloid-β peptide and tau. ACS Omega 9 (30). Available at https://doi.org/10.1021/acsomega.4c01281.

Al-Bagmi, M.S., Alokail, M.S., Al-Daghri, N.M., Altwaijry, N., Aziz, I.M., Almajhdi, F.N., et al., 2025. Phenolic fractions of *Heliotropium Bacciferum* inhibit human insulin aggregation and protect against amyloid-induced cytotoxicity: structural and biophysical analysis. Biophysical Chemistry 322, 107437. Available at https://doi.org/10.1016/j.bpc.2025.107437.

Albani, D., Polito, L., Signorini, A., Forloni, G., 2010. Neuroprotective properties of resveratrol in different neurodegenerative disorders. BioFactors 36 (5), 370–376. Available at https://doi.org/10.1002/biof.118.

Ali, K., Maltese, F., Choi, Y.H., Verpoorte, R., 2010. Metabolic constituents of grapevine and grape-derived products. Phytochemistry Reviews 9 (3), 357–378. Available at https://doi.org/10.1007/s11101-009-9158-0.

Andrade, S., Ramalho, M.J., Pereira, M.D.C., Loureiro, J.A., 2018. Resveratrol brain delivery for neurological disorders prevention and treatment. Frontiers in Pharmacology 9. Available at https://doi.org/10.3389/fphar.2018.01261.

Appleton, L., Day, A.S., 2019. Re: a meta-analysis of the clinical use of curcumin for irritable bowel syndrome. Journal of Clinical Medicine 8 (11). Available at https://doi.org/10.3390/jcm8111885.

Beal, M.F., 1999. Coenzyme Q10 administration and its potential for treatment of neurodegenerative diseases. BioFactors 9 (2-4), 261–266. Available at https://doi.org/10.1002/biof.5520090222.

Bieschke, J., Russ, J., Friedrich, R.P., Ehrnhoefer, D.E., Wobst, H., Neugebauer, K., et al., 2010. EGCG remodels mature α-synuclein and amyloid-β fibrils and reduces cellular toxicity. Proceedings of the National Academy of Sciences of the United States of America 107 (17), 7710–7715. Available at https://doi.org/10.1073/pnas.0910723107.

Bigini, P., Larini, S., Pasquali, C., Muzio, V., Mennini, T., 2002. Acetyl-l-carnitine shows neuroprotective and neurotrophic activity in primary culture of rat embryo motoneurons. Neuroscience Letters 329 (3), 334–338. Available at https://doi.org/10.1016/s0304-3940(02)00667-5.

Calder, P.C., 2011. Fatty acids and inflammation: the cutting edge between food and pharma. European Journal of Pharmacology 668 (1), S50–S58S50. Available at https://doi.org/10.1016/j.ejphar.2011.05.085.

Camilleri, A., Zarb, C., Caruana, M., Ostermeier, U., Ghio, S., Högen, T., et al., 2013. Mitochondrial membrane permeabilisation by amyloid aggregates and protection by polyphenols. Biochimica et Biophysica Acta (BBA) - Biomembranes 1828 (11), 2532–2543. Available at https://doi.org/10.1016/j.bbamem.2013.06.026.

Campos-Bedolla, P., Walter, F.R., Veszelka, S., Deli, M.A., 2014. Role of the blood-brain barrier in the nutrition of the central nervous system. Archives of Medical Research 45 (8), 610–638. Available at https://doi.org/10.1016/j.arcmed.2014.11.018.

Candelario-Jalil, E., de Oliveira, A.C.P., Gräf, S., Bhatia, H.S., Hüll, M., Muñoz, E., et al., 2007. Resveratrol potently reduces prostaglandin E2production and free radical formation in lipopolysaccharide-activated primary rat microglia. Journal of Neuroinflammation 4 (1). Available at https://doi.org/10.1186/1742-2094-4-25.

Cao, Z., Li, Y., 2004. Potent induction of cellular antioxidants and phase 2 enzymes by resveratrol in cardiomyocytes: protection against oxidative and electrophilic injury. European Journal of Pharmacology 489 (1-2), 39–48. Available at https://doi.org/10.1016/j.ejphar.2004.02.031.

Caruana, M., Högen, T., Levin, J., Hillmer, A., Giese, A., Vassallo, N., 2011. Inhibition and disaggregation of α-synuclein oligomers by natural polyphenolic compounds. FEBS Letters 585 (8), 1113–1120. Available at https://doi.org/10.1016/j.febslet.2011.03.046.

Caruana, M., Neuner, J., Högen, T., Schmidt, F., Kamp, F., Scerri, C., et al., 2012. Polyphenolic compounds are novel protective agents against lipid membrane damage by α-synuclein aggregates in vitro. Biochimica et Biophysica Acta (BBA) - Biomembranes 1818 (11), 2502–2510. Available at https://doi.org/10.1016/j.bbamem.2012.05.019.

Chan, P., Qin, Z., Zheng, Z., Zhang, L., Fang, X., Sun, F., et al., 2009. P2.204 A randomized, double-blind, placebo-controlled, delayed start study to assess safety, tolerability and efficacy of green tea polyphenols in Parkinson's disease. Parkinsonism & Related Disorders 15 (2), S145. Available at https://doi.org/10.1016/S1353-8020(09)70555-3.

Chang, K.H., Cheng, M.L., Chiang, M.C., Chen, C.M., 2018. Lipophilic antioxidants in neurodegenerative diseases. Clinica Chimica Acta 485, 79–87. Available at https://doi.org/10.1016/j.cca.2018.06.031.

Charo, I.F., Ransohoff, R.M., 2006. The many roles of chemokines and chemokine receptors in inflammation. New England Journal of Medicine 354 (6), 610–621. Available at https://doi.org/10.1056/nejmra052723.

Chaturvedi, R.K., Beal, M.F., 2013. Mitochondria targeted therapeutic approaches in Parkinson's and Huntington's diseases. Molecular and Cellular Neuroscience 55, 101–114. Available at https://doi.org/10.1016/j.mcn.2012.11.011.

Chen, M., Wang, T., Yue, F., Li, X., Wang, P., Li, Y., et al., 2015. Tea polyphenols alleviate motor impairments, dopaminergic neuronal injury, and cerebral α-synuclein aggregation in MPTP-intoxicated parkinsonian monkeys. Neuroscience 286, 383–392. Available at https://doi.org/10.1016/j.neuroscience.2014.12.003.

Cole, G.M., Teter, B., Frautschy, S.A., 2007. Neuroprotective effects of curcumin. In: Aggarwal, B.B., Surh, Y.J., Shishodia, S. (Eds.), The Molecular Targets and Therapeutic Uses of Curcumin in Health and Disease. ADVANCES IN EXPERIMENTAL MEDICINE AND BIOLOGY, vol 595. Springer, Boston, MA, pp. 197–212. Available at https://doi.org/10.1007/978-0-387-46401-5_8.

Constantinescu, C.S., Farooqi, N., O'Brien, K., & Gran, B. (2011). Experimental autoimmune encephalomyelitis (EAE) as a model for multiple sclerosis (MS). British Journal of Pharmacology, 164(4), 1079–1106. Available at https://doi.org/10.1111/j.1476-5381.2011.01302.x.

Cooper, R., Morré, D.J., Morré, D.M., 2005. Medicinal benefits of green tea: Part I. Review of noncancer health benefits. Journal of Alternative and Complementary Medicine 11 (3), 521–528. Available at https://doi.org/10.1089/acm.2005.11.521.

Cutuli, D., Pagani, M., Caporali, P., Galbusera, A., Laricchiuta, D., Foti, F., et al., 2016. Effects of omega-3 fatty acid supplementation on cognitive functions and neural substrates: a voxel-based morphometry study in aged mice. Frontiers in Aging Neuroscience 8. Available at https://doi.org/10.3389/fnagi.2016.00038.

Denis, I., Potier, B., Heberden, C., Vancassel, S., 2015. Omega-3 polyunsaturated fatty acids and brain aging. Current Opinion in Clinical Nutrition and Metabolic Care 18 (2), 139–146. Available at https://doi.org/10.1097/mco.0000000000000141.

Dyall, S.C., 2015. Long-chain omega-3 fatty acids and the brain: a review of the independent and shared effects of EPA, DPA and DHA. Frontiers in Aging Neuroscience 7. Available at https://doi.org/10.3389/fnagi.2015.00052.

Eckert, G.P., Lipka, U., Muller, W.E., 2013. Omega-3 fatty acids in neurodegenerative diseases: focus on mitochondria. Prostaglandins Leukotrienes and Essential Fatty Acids 88 (1), 105–114. Available at https://doi.org/10.1016/j.plefa.2012.05.006.

Fang, S., Zhang, K., Liu, D., Yang, Y., Xi, H., Xie, W., et al., 2024. Polyphenol-based polymer nanoparticles for inhibiting amyloid protein aggregation: recent advances and perspectives. Frontiers in Nutrition 11. Available at https://doi.org/10.3389/fnut.2024.1408620.

Ferretta, A., Gaballo, A., Tanzarella, P., Piccoli, C., Capitanio, N., Nico, B., et al., 2014. Effect of resveratrol on mitochondrial function: implications in parkin-associated familiar Parkinson's disease. Biochimica et Biophysica Acta (BBA) - Molecular Basis of Disease 1842 (7), 902–915. Available at https://doi.org/10.1016/j.bbadis.2014.02.010.

Forloni, G., Angeretti, N., Smiroldo, S., 1994. Neuroprotective activity of acetyl-L-carnitine: studies in vitro. Journal of Neuroscience Research 37 (1), 92–96. Available at https://doi.org/10.1002/jnr.490370112.

Frautschy, S.A., Hu, W., Kim, P., Miller, S.A., Chu, T., Harris-White, M.E., et al., 2001. Phenolic anti-inflammatory antioxidant reversal of Aβ-induced cognitive deficits and neuropathology. Neurobiology of Aging 22 (6), 993–1005. Available at https://doi.org/10.1016/S0197-4580(01)00300-1.

Freyssin, A., Page, G., Fauconneau, B., Rioux Bilan, A., 2018. Natural polyphenols effects on protein aggregates in Alzheimer's and Parkinson's prion-like diseases. Neural Regeneration Research 13 (6), 955–961. Available at https://doi.org/10.4103/1673-5374.233432.

Ganguli, M., Chandra, V., Kamboh, M.I., Johnston, J.M., Dodge, H.H., Thelma, B.K., et al., 2000. Apolipoprotein E polymorphism and Alzheimer disease: the Indo-US cross- national dementia study. Archives of Neurology 57 (6), 824–830. Available at https://doi.org/10.1001/archneur.57.6.824.

Garcia-Alloza, M., Borrelli, L.A., Rozkalne, A., Hyman, B.T., Bacskai, B.J., 2007. Curcumin labels amyloid pathology in vivo, disrupts existing plaques, and partially restores distorted neurites in an Alzheimer mouse model. Journal of Neurochemistry 102 (4), 1095–1104. Available at https://doi.org/10.1111/j.1471-4159.2007.04613.x.

Garrido-Maraver, J., Cordero, M.D., Oropesa-Avila, M., Vega, A.F., De La Mata, M., Pavon, A.D., et al., 2014. Clinical applications of coenzyme Q10. Frontiers in Bioscience - Landmark 19 (4), 619–633. Available at https://doi.org/10.2741/4231.

Gavrilova, S.I., Preuss, U.W., Wong, J.W., Hoerr, R., Kaschel, R., Bachinskaya, N., et al., 2014. Efficacy and safety of Ginkgo biloba extract EGb 761 in mild cognitive impairment with neuropsychiatric symptoms: a randomized, placebo-controlled, double-blind, multi-center trial. International Journal of Geriatric Psychiatry 29 (10), 1087–1095. Available at 10.1002/gps.4103.

Ghavami, S., Shojaei, S., Yeganeh, B., Ande, S.R., Jangamreddy, J.R., Mehrpour, M., et al., 2014. Autophagy and apoptosis dysfunction in neurodegenerative disorders. Progress in Neurobiology 112, 24–49. Available at https://doi.org/10.1016/j.pneurobio.2013.10.004.

Hasanbasic, S., Jahic, A., Berbic, S., Znidaric, M.T., Zerovnik, E., 2018. Inhibition of protein aggregation by several antioxidants. Oxidative Medicine and Cellular Longevity 2018, 8613209. Available at https://doi.org/10.1155/2018/8613209.

Hashiguchi, M., Ohta, Y., Shimizu, M., Maruyama, J., Mochizuki, M., 2015. Meta-analysis of the efficacy and safety of *Ginkgo biloba* extract for the treatment of dementia. Journal of Pharmaceutical Health Care and Sciences 1, 14. Available at https://doi.org/10.1186/s40780-015-0014-7.

Hewlings, S.J., Kalman, D.S., 2017. Curcumin: A Review of Its' Effects on Human Health. Foods 6 (10), 92. Available at 10.3390/foods6100092.

Hoensch, H.P., Oertel, R., 2015. The value of flavonoids for the human nutrition: short review and perspectives. Clinical Nutrition Experimental 3, 8–14. Available at https://doi.org/10.1016/j.yclnex.2015.09.001.

Ihl, R., Tribanek, M., Bachinskaya, N., 2012. Efficacy and tolerability of a once daily formulation of *Ginkgo biloba* extract EGb 761® in Alzheimer's disease and vascular dementia: results from a randomised controlled trial. Pharmacopsychiatry 45 (02), 41–46. Available at https://doi.org/10.1055/s-0031-1291217.

Jahić, A., Tušek Žnidarič, M., Pintar, S., Berbić, S., Žerovnik, E., 2020. The effect of three polyphenols and some other antioxidant substances on amyloid fibril formation by Human cystatin C. Neurochemistry International 140, 104806. Available at https://doi.org/10.1016/j.neuint.2020.104806.

Jahić Mujkić, A., Žnidarič, M.T., Berbić, S., Žerovnik, E., 2021. Synergy of the inhibitory action of polyphenols plus vitamin C on amyloid fibril formation: case study of human stefin B. Antioxidants 10 (9), 1471. Available at https://doi.org/10.3390/antiox10091471.

Jiang, J., Jiang, J., Zuo, Y., Gu, Z., 2013a. Rapamycin protects the mitochondria against oxidative stress and apoptosis in a rat model of Parkinson's disease. International Journal of Molecular Medicine 31 (4), 825–832. Available at https://doi.org/10.3892/ijmm.2013.1280.

Jiang, T.F., Zhang, Y.J., Zhou, H.Y., Wang, H.M., Tian, L.P., Liu, J., et al., 2013b. Curcumin ameliorates the neurodegenerative pathology in A53T α-synuclein cell model of Parkinson's disease through the downregulation of mTOR/p70S6K signaling and the recovery of macroautophagy. Journal of Neuroimmune Pharmacology 8 (1), 356–369. Available at https://doi.org/10.1007/s11481-012-9431-7.

Jung, E.B., Lee, C.S., 2014. Baicalein attenuates proteasome inhibition-induced apoptosis by suppressing the activation of the mitochondrial pathway and the caspase-8- and Bid-dependent pathways. European Journal of Pharmacology 730 (1), 116–124. Available at https://doi.org/10.1016/j.ejphar.2014.02.039.

Kim, H.S., Quon, M.J., Kim, Ja, 2014. New insights into the mechanisms of polyphenols beyond antioxidant properties; lessons from the green tea polyphenol, epigallocatechin 3-gallate. Redox Biology 2 (1), 187–195. Available at https://doi.org/10.1016/j.redox.2013.12.022.

Kim, Y., Keogh, J.B., Clifton, P.M., 2016. Polyphenols and glycemic control. Nutrients 8 (1), 17. Available at 10.3390/nu8010017.

Kotha, R.R., Luthria, D.L., 2019. Curcumin: biological, pharmaceutical, nutraceutical, and analytical aspects. Molecules 24 (16). Available at https://doi.org/10.3390/molecules24162930.

Kumar, A., Naidu, P.S., Seghal, N., Padi, S.S.V., 2007. Neuroprotective effects of resveratrol against intracerebroventricular colchicine-induced cognitive impairment and oxidative stress in rats. Pharmacology 79 (1), 17–26. Available at https://doi.org/10.1159/000097511.

Lee, J.W., 1987. Green tea (-)-epigallocatechin-3-gallate inhibits β-amyloid-induced cognitive dysfunction through modification of secretase activity via inhibition of ERK and NF-κB pathways in mice. The Journal of Nutrition 139 (10), 1987–1993. Available at https://doi.org/10.3945/jn.109.109785.

Lévy, E., El Banna, N., Baïlle, D., Heneman-Masurel, A., Truchet, S., Rezaei, H., et al., 2019. Causative links between protein aggregation and oxidative stress: a review. International Journal of Molecular Sciences 20 (16), 3896. Available at https://doi.org/10.3390/ijms20163896.

Li, H., Zheng, C., Wang, Z., Wen, K., Zhang, Y., 2025. Neuroprotective effects of catechins by differentially affecting the binding of beta-amyloid and its aggregates to the target cells. Molecular Neurobiology 62, 9861–9880. Available at https://doi.org/10.1007/s12035-025-04870-0.

Lim, G.P., Chu, T., Yang, F., Beech, W., Frautschy, S.A., Cole, G.M., 2001. The curry spice curcumin reduces oxidative damage and amyloid pathology in an Alzheimer transgenic mouse. Journal of Neuroscience 21 (21), 8370–8377. Available at https://doi.org/10.1523/jneurosci.21-21-08370.2001.

Lin, T.K., Chen, S.D., Chuang, Y.C., Lin, H.Y., Huang, C.R., Chuang, J.H., et al., 2014. Resveratrol partially prevents rotenone-induced neurotoxicity in dopaminergic SH-SY5Y cells through induction of heme oxygenase-1 dependent autophagy. International Journal of Molecular Sciences 15 (1), 1625–1646. Available at https://doi.org/10.3390/ijms15011625.

Lopresti, A.L., 2018. The problem of curcumin and its bioavailability: could its gastrointestinal influence contribute to its overall health-enhancing effects? Advances in Nutrition 9 (1), 41–50. Available at https://doi.org/10.1093/advances/nmx011.

Mahadevan, S., Park, Y., 2008. Multifaceted therapeutic benefits of *Ginkgo biloba* L.: chemistry, efficacy, safety, and uses. Journal of Food Science 73 (1), R14. Available at https://doi.org/10.1111/j.1750-3841.2007.00597.x.

Maher, F.O., Martin, D.S.D., Lynch, M.A., 2004. Increased IL-1β in cortex of aged rats is accompanied by downregulation of ERK and PI-3 kinase. Neurobiology of Aging 25 (6), 795–806. Available at https://doi.org/10.1016/j.neurobiolaging.2003.08.007.

Maiti, P., Dunbar, G.L., 2018. Use of curcumin, a natural polyphenol for targeting molecular pathways in treating age-related neurodegenerative diseases. International Journal of Molecular Sciences 19 (6). Available at https://doi.org/10.3390/ijms19061637.

Mak, J.C.W., 2012. Potential role of green tea catechins in various disease therapies: progress and promise. Clinical and Experimental Pharmacology and Physiology 39 (3), 265–273. Available at https://doi.org/10.1111/j.1440-1681.2012.05673.x.

Malishev, R., Shaham-Niv, S., Nandi, S., Kolusheva, S., Gazit, E., Jelinek, R., 2017. Bacoside-A, an Indian traditional-medicine substance, inhibits β-amyloid cytotoxicity, fibrillation, and membrane interactions. ACS Chemical Neuroscience 8 (4), 884–891. Available at https://doi.org/10.1021/acschemneuro.6b00438.

Mandel, S., Weinreb, O., Amit, T., Youdim, M.B.H., 2004. Cell signaling pathways in the neuroprotective actions of the green tea polyphenol (-)-epigallocatechin-3-gallate: implications for neurodegenerative diseases. Journal of Neurochemistry 88 (6), 1555–1569. Available at https://doi.org/10.1046/j.1471-4159.2003.02291.x.

Mandel, S.A., Weinreb, O., Amit, T., Youdim, M.B.H., 2012. Molecular mechanisms of the neuroprotective/neurorescue action of multi-target green tea polyphenols. Frontiers in Bioscience - Scholar 4 (2), 581–598. Available at https://doi.org/10.2741/s286.

Mani, S., Sekar, S., Barathidasan, R., Manivasagam, T., Thenmozhi, A.J., Sevanan, M., et al., 2018. Naringenin decreases α-synuclein expression and neuroinflammation in MPTP-induced Parkinson's disease model in mice. Neurotoxicity Research 33 (3), 656–670. Available at https://doi.org/10.1007/s12640-018-9869-3.

Marambaud, P., Zhao, H., Davies, P., 2005. Resveratrol promotes clearance of Alzheimer's disease amyloid-β peptides. Journal of Biological Chemistry 280 (45), 37377–37382. Available at https://doi.org/10.1074/jbc.m508246200.

Martinez-Huelamo, M., Rodriguez-Morato, J., Boronat, A., Torre, 2017. Modulation of Nrf2 by olive oil and wine polyphenols and neuroprotection. Antioxidants 6 (4), 73. Available at https://doi.org/10.3390/antiox6040073.

Meng, F., Abedini, A., Plesner, A., Verchere, C.B., Raleigh, D.P., 2010. The flavanol (-)-epigallocatechin 3-gallate inhibits amyloid formation by islet amyloid polypeptide, disaggregates amyloid fibrils, and protects cultured cells against IAPP-induced toxicity. Biochemistry 49 (37), 8127–8133. Available at https://doi.org/10.1021/bi100939a.

Mokni, M., Elkahoui, S., Limam, F., Amri, M., Aouani, E., 2007. Effect of resveratrol on antioxidant enzyme activities in the brain of healthy rat. Neurochemical Research 32 (6), 981–987. Available at https://doi.org/10.1007/s11064-006-9255-z.

Momma, Y., Tsuji, M., Oguchi, T., Ohashi, H., Nohara, T., Ito, N., et al., 2023. The curcumin derivative GT863 protects cell membranes in cytotoxicity by Aβ oligomers. International Journal of Molecular Sciences 24 (4), 3089. Available at https://doi.org/10.3390/ijms24043089.

Monroy, A., Lithgow, G.J., Alavez, S., 2013. Curcumin and neurodegenerative diseases. BioFactors 39 (1), 122–132. Available at https://doi.org/10.1002/biof.1063.

Montgomery, S.A., Thal, L.J., Amrein, R., 2003. Meta-analysis of double blind randomized controlled clinical trials of acetyl-L-carnitine versus placebo in the treatment of mild cognitive impairment and mild Alzheimer's disease. International Clinical Psychopharmacology 18 (2), 61–71. Available at https://doi.org/10.1097/00004850-200303000-00001.

Napryeyenko, O., Borzenko, I., 2007. *Ginkgo biloba* special extract in dementia with neuropsychiatric features: a randomised, placebo-controlled, double-blind clinical trial. Arzneimittel-Forschung/Drug Research. 57 (1), 4–11. Available at https://doi.org/10.1055/s-0031-1296579.

Navarro, G., Martínez -Pinilla, E., Ortiz, R., Noé, V., Ciudad, C.J., Franco, R., 2018. Resveratrol and related stilbenoids, nutraceutical/dietary complements with health-promoting actions: industrial production, safety, and the search for mode of action. Comprehensive Reviews in Food Science and Food Safety 17 (4), 808–826. Available at https://doi.org/10.1111/1541-4337.12359.

Necula, M., Kayed, R., Milton, S., Glabe, C.G., 2007. Small molecule inhibitors of aggregation indicate that amyloid β oligomerization and fibrillization pathways are independent and distinct. Journal of Biological Chemistry 282 (14), 10311–10324. Available at https://doi.org/10.1074/jbc.M608207200.

Ng, C.-H., Guan, M.S.H., Koh, C., Ouyang, X., Yu, F., Tan, E.-K., et al., 2012. AMP kinase activation mitigates dopaminergic dysfunction and mitochondrial abnormalities in drosophila models of Parkinson's disease. The Journal of Neuroscience 32 (41), 14311–14317. Available at https://doi.org/10.1523/jneurosci.0499-12.2012.

Niedzielska, E., Smaga, I., Gawlik, M., Moniczewski, A., Stankowicz, P., Pera, J., et al., 2016. Oxidative stress in neurodegenerative diseases. Molecular Neurobiology 53 (6), 4094–4125. Available at https://doi.org/10.1007/s12035-015-9337-5.

Okamoto, S., Kakimaru, S., Koreishi, M., Sakamoto, M., Nakamura, Y., Ando, H., et al., 2025. Resveratrol, a food-derived polyphenol, promotes melanosomal degradation in skin fibroblasts through coordinated activation of autophagy, lysosomal, and antioxidant pathways. Journal of Functional Foods 125, 106672. Available at https://doi.org/10.1016/j.jff.2025.106672.

Ono, K., Hasegawa, K., Naiki, H., Yamada, M., 2004. Curcumin has potent anti-amyloidogenic effects for Alzheimer's β-amyloid fibrils in vitro. Journal of Neuroscience Research 75 (6), 742–750. Available at https://doi.org/10.1002/jnr.20025.

Owen, L., Sunram-Lea, S.I., 2011. Metabolic agents that enhance ATP can improve cognitive functioning: a review of the evidence for glucose, oxygen, pyruvate, creatine, and L-carnitine. Nutrients 3 (8), 735–755. Available at https://doi.org/10.3390/nu3080735.

Palazzi, L., Bruzzone, E., Bisello, G., Leri, M., Stefani, M., Bucciantini, M., et al., 2018. Oleuropein aglycone stabilizes the monomeric α-synuclein and favours the growth of non-toxic aggregates. Scientific Reports 8 (1). Available at https://doi.org/10.1038/s41598-018-26645-5.

Park, E.J., Pezzuto, J.M., 2015. The pharmacology of resveratrol in animals and humans. Biochimica et Biophysica Acta - Molecular Basis of Disease 1852 (6), 1071–1113. Available at https://doi.org/10.1016/j.bbadis.2015.01.014.

Pettegrew, J.W., Levine, J., McClure, R.J., 2000. Acetyl-L-carnitine physical-chemical, metabolic, and therapeutic properties: relevance for its mode of action in Alzheimer's disease and geriatric depression. Molecular Psychiatry 5 (6), 616–632. Available at https://doi.org/10.1038/sj.mp.4000805.

Pogačnik, L., Pirc, K., Palmela, I., Skrt, M., Kim, K.S., Brites, D., et al., 2016. Potential for brain accessibility and analysis of stability of selected flavonoids in relation to neuroprotection in vitro. Brain Research 1651, 17–26. Available at https://doi.org/10.1016/j.brainres.2016.09.020.

Pogačnik, L., Pirc, K., Palmela, I., Skrt, M., Kim, K.S., Brites, D., et al., 2016. Potential for brain accessibility and analysis of stability of selected flavonoids in relation to neuroprotection in vitro. Brain Research 1651, 17–26. Available at https://doi.org/10.1016/j.brainres.2016.09.020.

Polito, C.A., Cai, Z.Y., Shi, Y.L., Li, X.M., Yang, R., Shi, M., et al., 2018. Association of tea consumption with risk of Alzheimer's disease and anti-beta-amyloid effects of tea. Nutrients 10 (5). Available at https://doi.org/10.3390/nu10050655.

Regitz, C., Marie Dußling, L., Wenzel, U., 2014. Amyloid-beta (A β 1–42)-induced paralysis in C aenorhabditis elegans is inhibited by the polyphenol quercetin through activation of protein degradation pathways. Molecular Nutrition & Food Research 58 (10), 1931–1940. Available at https://doi.org/10.1002/mnfr.201400014.

Reglodi, D., Renaud, J., Tamas, A., Tizabi, Y., Socías, S.B., Del-Bel, E., et al., 2017. Novel tactics for neuroprotection in Parkinson's disease: role of antibiotics, polyphenols and neuropeptides. Progress in Neurobiology 155, 120–148. Available at https://doi.org/10.1016/j.pneurobio.2015.10.004.

Regmi, D., Haque, S., Ul Karim, M.R., Stanic, A., Du, D., 2025. Inhibition of amyloid formation of prion fragment (106–128) by polyphenolic compounds. Biochimica et Biophysica Acta (BBA) - General Subjects 1869 (5), 130778. Available at https://doi.org/10.1016/j.bbagen.2025.130778.

Renaud, J., Martinoli, M.G., 2014. Resveratrol as a protective molecule for neuroinflammation: a review of mechanisms. Current Pharmaceutical Biotechnology 15 (4), 318–329. Available at https://doi.org/10.2174/1389201015666140617101332.

Renaud, J., Martinoli, M.G., 2019. Considerations for the use of polyphenols as therapies in neurodegenerative diseases. International Journal of Molecular Sciences 20 (8). Available at https://doi.org/10.3390/ijms20081883.

Rezai-Zadeh, K., Shytle, D., Sun, N., Mori, T., Hou, H., Jeanniton, D., et al., 2005. Green tea epigallocatechin-3-gallate (EGCG) modulates amyloid precursor protein cleavage and reduces cerebral amyloidosis in Alzheimer transgenic mice. Journal of Neuroscience 25 (38), 8807–8814. Available at https://doi.org/10.1523/JNEUROSCI.1521-05.2005.

Rojas, C., Rojas-Castañeda, J., Rojas, P., 2016. Antioxidant properties of a *Ginkgo biloba* leaf extract (EGb 761) in animal models of Alzheimer's and Parkinson's diseases. Current Topics in Nutraceutical Research 14 (1), 1–16.

Sanoobar, M., Eghtesadi, S., Azimi, A., Khalili, M., Khodadadi, B., Jazayeri, S., et al., 2015. Coenzyme Q10 supplementation ameliorates inflammatory markers in patients with multiple sclerosis: a double blind, placebo, controlled randomized clinical trial. Nutritional Neuroscience 18 (4), 169–176. Available at https://doi.org/10.1179/1476830513Y.0000000106.

Shaham-Niv, S., Rehak, P., Zaguri, D., Levin, A., Adler-Abramovich, L., Vuković, L., et al., 2018. Differential inhibition of metabolite amyloid formation by generic fibrillation-modifying polyphenols. Communications Chemistry 1 (1). Available at https://doi.org/10.1038/s42004-018-0025-z.

Sharma, M., Briyal, S., Gupta, Y.K., 2005. Effect of alpha lipoic acid, melatonin and trans resveratrol on intracerebroventricular streptozotocin induced spatial memory deficit in rats. Indian Journal of Physiology and Pharmacology 49 (4), 395–402.

Shults, C.W., 2003. Coenzyme Q10 in neurodegenerative diseases. Current Medicinal Chemistry 10 (19), 1917–1921. Available at https://doi.org/10.2174/0929867033456882.

Shults, C.W., Oakes, D., Kieburtz, K., Beal, M.F., Haas, R., Plumb, S., et al., 2002. Effects of coenzyme Q 10 in early Parkinson disease: evidence of slowing of the functional decline. Archives of Neurology 59 (10), 1541–1550. Available at https://doi.org/10.1001/archneur.59.10.1541.

Siddique, Y.H., Naz, F., Jyoti, S., Fatima, A., Khanam, S., Rahul, et al., 2014a. Effect of centella asiatica leaf extract on the dietary supplementation in transgenic drosophila model of parkinson's disease. Parkinson's Disease 2014. Available at https://doi.org/10.1155/2014/262058.

Siddique, Y.H., Faisal, M., Naz, F., Jyoti, S., Rahul, 2014b. Role of *Ocimum sanctum* leaf extract on dietary supplementation in the transgenic Drosophila model of Parkinson's disease. Chinese Journal of Natural Medicines 12 (10), 777–781. Available at https://doi.org/10.1016/S1875-5364(14)60118-7.

Siddique, Y.H., Mujtaba, S.F., Faisal, M., Jyoti, S., Naz, F., 2014c. The effect of *Bacopa monnieri* leaf extract on dietary supplementation in transgenic *Drosophila* model of Parkinson's disease. European Journal of Integrative Medicine 6 (5), 571–580. Available at https://doi.org/10.1016/j.eujim.2014.05.007.

Siddique, Y.H., Naz, F., Jyoti, S., 2014d. Effect of curcumin on lifespan, activity pattern, oxidative stress, and apoptosis in the brains of transgenic *Drosophila* model of Parkinson's disease. BioMed Research International 2014. Available at https://doi.org/10.1155/2014/606928.

Silla, A., Punzo, A., Caliceti, C., Barbalace, M.C., Hrelia, S., Malaguti, M., 2025. The role of antioxidant compounds from citrus waste in modulating neuroinflammation: a sustainable solution. Antioxidants 14 (5), 581. Available at https://doi.org/10.3390/antiox14050581.

Stefani, M., Rigacci, S., 2013. Protein folding and aggregation into amyloid: the interference by natural phenolic compounds. International Journal of Molecular Sciences 14 (6), 12411–12457. Available at https://doi.org/10.3390/ijms140612411.

Šturm, L., et al., 2022. Interactions of (−)-epigallocatechin-3-gallate with model lipid membranes. Biochimica et Biophysica Acta (BBA) - Biomembranes 1864 (10), 183999. Available at https://doi.org/10.1016/j.bbamem.2022.183999.

Traina, G., 2016. The neurobiology of acetyl-L-carnitine. Frontiers in Bioscience - Landmark 21 (7), 1314–1329. Available at https://doi.org/10.2741/4459.

Trépanier, M.O., Hopperton, K.E., Orr, S.K., Bazinet, R.P., 2016. N-3 polyunsaturated fatty acids in animal models with neuroinflammation: an update. European Journal of Pharmacology 785, 187–206. Available at https://doi.org/10.1016/j.ejphar.2015.05.045.

Viles, J.H., 2023. Imaging amyloid-β membrane interactions: Ion-channel pores and lipid-bilayer permeability in Alzheimer's disease. Angewandte Chemie - International Edition 62 (25). Available at https://doi.org/10.1002/anie.202215785.

Walle, T., 2011. Bioavailability of resveratrol. Annals of the New York Academy of Sciences 1215 (1), 9–15. Available at https://doi.org/10.1111/j.1749-6632.2010.05842.x.

Wang, Y.H., Yu, H.T., Pu, X.P., Du, G.H., 2013. Baicalein prevents 6-hydroxydopamine-induced mitochondrial dysfunction in SH-SY5Y cells via inhibition of mitochondrial oxidation and up-regulation of DJ-1 protein expression. Molecules 18 (12), 14726–14738. Available at https://doi.org/10.3390/molecules181214726.

Yang, J.E., Rhoo, K.Y., Lee, S., Lee, J.T., Park, J.H., Bhak, G., et al., 2017. EGCG-mediated protection of the membrane disruption and cytotoxicity caused by the 'active oligomer' of α-synuclein. Scientific Reports 7 (1). Available at https://doi.org/10.1038/s41598-017-18349-z.

Yao, Z.-xing, Drieu, K., Papadopoulos, V., 2001. The *Ginkgo biloba* extract EGb 761 rescues the PC12 neuronal cells from β-amyloid-induced cell death by inhibiting the formation of β-amyloid-derived diffusible neurotoxic ligands. Brain Research 889 (1-2), 181–190. Available at https://doi.org/10.1016/s0006-8993(00)03131-0.

Yu, J., Yoon, J.H., Park, M., Lee, H.J., 2025. Nobiletin-mediated autophagy mitigates nanoplastic-induced toxicity in human intestinal Caco-2 cells. FASEB Journal 39 (6). Available at https://doi.org/10.1096/fj.202402761R.

Zhu, M., Han, S., Fink, A.L., 2013. Oxidized quercetin inhibits α-synuclein fibrillization. Biochimica et Biophysica Acta - General Subjects 1830 (4), 2872–2881. Available at https://doi.org/10.1016/j.bbagen.2012.12.027.

Zhu, G., Wang, X., Wu, S., Li, X., Li, Q., 2014. Neuroprotective effects of puerarin on 1-methyl-4-phenyl-1,2,3,6-tetrahydropyridine induced Parkinson's disease model in mice. Phytotherapy Research 28 (2), 179–186. Available at https://doi.org/10.1002/ptr.4975.

Zhytniakivska, O., Chaturvedi, T., Thomsen, M.H., 2025. Plant-based inhibitors of protein aggregation. Biomolecules 15 (4), 481. Available at https://doi.org/10.3390/biom15040481.

CHAPTER 10

Involvement of CB1 and CB2 receptors in neuroprotective effects of cannabinoids

Eva De Lago[1,2,3], Javier Fernández-Ruiz[1,2,3], and Carmen Rodríguez-Cueto[1,2,3]
[1]Departamento de Bioquímica y Biología Molecular, Facultad de Medicina, Instituto Universitario de Investigación en Neuroquímica, Universidad Complutense, Madrid, Spain,
[2]Instituto Ramón y Cajal de Investigación Sanitaria (IRYCIS), Madrid, Spain,
[3]Centro de Investigación Biomédica en Red de Enfermedades Neurodegenerativas (CIBERNED), Madrid, Spain

10.1 Introduction: overview of the endogenous cannabinoid system

The endocannabinoid system was discovered in the late 1980s and early 1990s because of research aimed at identifying the mechanisms of action of Δ^9-tetrahydrocannabinol (Δ^9-THC) and other phytocannabinoids present in the cannabis plant (Mechoulam et al., 2014; Gertsch et al., 2010). It is a modulatory system constituted by several signaling lipids, mainly derivatives of the arachidonic acid, such as arachidonoylethanolamide anandamide (AEA) and 2-arachidonoylglycerol (2-AG), but also some other minor endocannabinoids (e.g., noladin ether, virodhamine) (Maccarrone et al., 2023). Beyond these canonical mediators, the concept of an "expanded endocannabinoid system" has been proposed to encompass structurally related *N*-acylethanolamines, such as palmitoylethanolamide or oleoylethanolamide, as well as monoacylglycerols such as 2-oleylglycerol, which share biosynthetic and degradative pathways with AEA and 2-AG (see below) and signal through some of the endocannabinoid-related targets (see below). This broader framework considerably extends the functional scope of the endocannabinoid system beyond the classical AEA/2-AG axis (Cristino et al., 2020).

These signaling lipids play a role as intercellular messengers in the communication between cells that are in a close vicinity (e.g., paracrine interaction) (Maccarrone et al., 2015; Winters and Vaughan, 2021). The cell that elaborates the molecular message expresses those enzymes that synthesize these endocannabinoids (and related *N*-acylethanolamines and monoacylglycerols), which include *N*-acylethanolamine phospholipase D for the synthesis of anandamide and other *N*-acylethanolamines (Mock et al., 2023) and diacylglycerol lipase for the synthesis of 2-AG and other 2-acylglycerols (Briand-Mésange et al., 2024). However, the synthesis process is highly redundant, with the possibility of alternative pathways involving other lipid-related enzymes specific to certain cells and/or physiological contexts (Simard et al., 2022). This enzymatic plasticity becomes particularly relevant under disease conditions, where shifts in cellular

Molecular Insights into Proteinopathies
DOI: https://doi.org/10.1016/B978-0-443-34176-2.00011-0
© 2027 Elsevier Inc. All rights are reserved, including those for text and data mining, AI training, and similar technologies.

composition (neurons, astrocytes, microglia), activation state, and subcellular compartmentalization can reshape the local "endocannabinoid tone" and thereby influence signaling outcomes (Maccarrone et al., 2015). In addition, a feature of this neuromodulatory system is that the synthesis and release of these endocannabinoids into the extracellular compartment are on demand and do not require their previous compartmentalization in intracellular vesicles. On the contrary, it occurs through free traffic across the plasma membrane, facilitated by their hydrophobic properties or using specific carriers, despite this having raised certain controversy (Nicolussi and Gertsch, 2015; Kaczocha and Haj-Dahmane, 2022). This activity-dependent, transient mode of signaling provides a mechanistic basis for the repeated recruitment of the endocannabinoid system as an endogenous homeostatic buffer under conditions of cell stress and potential damage.

Once released to the extracellular compartment, endocannabinoids will have access to different canonical (e.g., cannabinoid type-1 (CB_1) and cannabinoid type-2 (CB_2) receptors) and atypical receptors (Guo et al., 2026). CB_1 and CB_2 receptors are GPCRs that function by modulating classic intracellular signaling pathways such as PKA/cAMP inhibition, PI3K-Akt signaling, and ERK-JNK-p38 signaling (for both receptors), as well as metabotropic effects on calcium and potassium channels (predominantly associated with the CB_1 receptor) (Ehrlich et al., 2026). In neurons, CB_1 receptor activation typically suppresses presynaptic Ca^{2+} influx, mainly through N- and P/Q-type voltage-gated Ca^{2+} channels, while enhancing K^+ conductance, thereby reducing the release of different neurotransmitters and contributing to synaptic homeostasis (Gómez-Cañas et al., 2023). As regards to atypical receptors, they were discovered and/or associated with the endocannabinoid system after the identification of the canonical receptors and include: (1) a recently deorphanized GPCR, G-protein receptor 55 (GPR55), which functions through modulating the signaling for PLC/PKC, RhoA-ROCK and MAPK and whose endogenous ligand is another signaling lipid, lysophosphatidyl-inositol (Brown and Robin Hiley, 2009); (2) additional GPCRs such as GPR119 and GPR18, which can be engaged by endocannabinoids or phytocannabinoids in a context- and cell-type-dependent manner, contributing to diverse functional outcomes (Cristino et al., 2020); (3) the transient receptor potential vanilloid-1 ($TRPV_1$) or other members of the TRP channel family, which are considered the ionotropic receptor type of the endocannabinoid system (Gómez-Cañas et al., 2023), with the particularity that their binding site is intracellular, which demands the previous cell internalization of their endocannabinoid (e.g., anandamide, oleylethanolamide) or endovanilloid (e.g., *N*-arachidonoyldopamine, *N*-oleyldopamine) ligands; and (4) the nuclear receptors of the peroxisome proliferator activated receptor (PPAR) family, which also demand such previous ligand cell internalization and enable endocannabinoids to access the nucleus, thereby directly modulating gene expression (O'Sullivan, 2016; Iannotti and Vitale, 2021; Gómez-Cañas et al., 2023).

The last key element of the endocannabinoid system is the mechanism(s) for endocannabinoid inactivation, which may occur in a cell substrate other than the cells that generate and receive the endocannabinoid message, although these cells could also be part of the inactivation process. The inactivation mainly involves hydrolyzing enzymes, such as fatty acid amide hydrolase (FAAH), which degrades predominantly anandamide and other *N*-acylethanolamines, and monoacylglycerol lipase (MAGL), which degrades 2-AG and other 2-acylglycerols (Gil-Ordóñez et al., 2018). However, endocannabinoid inactivation is also highly redundant with other hydrolases, such as the family of α/β-domain hydrolases, which may participate in endocannabinoid inactivation in specific cell substrates and contexts (Bononi et al., 2021). This redundancy makes processes such as the pharmacological inhibition of endocannabinoid inactivation for elevating endocannabinoid tone extremely complex. In addition, it is important to note that endocannabinoid-inactivating enzymes are all intracellular and located in different cytosolic compartments, so that their activity requires the previous internalization of the endocannabinoid into the cell. This process is currently controversial, as indicated before, in relation to the occurrence or not of a specific membrane carrier for endocannabinoids, or with the assumption that this would be a simple diffusion process promoted by the fast degradation of the endocannabinoid once incorporated within the cell (Nicolussi and Gertsch, 2015).

The process of endocannabinoid inactivation has sometimes been associated with the possibility that both anandamide and 2-AG may serve as substrates of cyclooxygenase-2 (COX-2) or other eicosanoid-related enzymes (e.g., lipoxygenases) for the generation of prostaglandin derivatives such as prostamides and prostaglandin-glyceryl-esters, respectively (Kingsley et al., 2019; Urquhart et al., 2015). These new signaling lipids have the particularity of losing their ability to bind and activate cannabinoid receptors, which explains why these pathways are frequently interpreted as degradative ones, but of gaining activity at certain prostanoid-related receptors (Burstein, 2019). This enables these derivatives to exert new physiological and/or pathophysiological roles (Stanley and O'Sullivan, 2014). Therefore these COX-2-dependent pathways should be assumed to represent a precursor role of endocannabinoids rather than an additional and simple degradative option.

All major components of the endocannabinoid system are schematically illustrated in Fig. 10.1. As indicated before, they play an important pro-homeostatic role in cell–cell communication in numerous body tissues and organs, which explains their relevance in the regulation of cell integrity and survival (Fernández-Ruiz et al., 2026). For example, in the central nervous system (CNS), endocannabinoids and their receptors work as a retrograde signaling system to modulate synaptic homeostasis as well as to regulate neuron–glia communication and adult neurogenesis (Piette et al., 2020; Winters and Vaughan, 2021; Oddi et al., 2023). Such an assumption has led to interest in the endocannabinoid signaling system as a potential endogenous neuroprotective system,

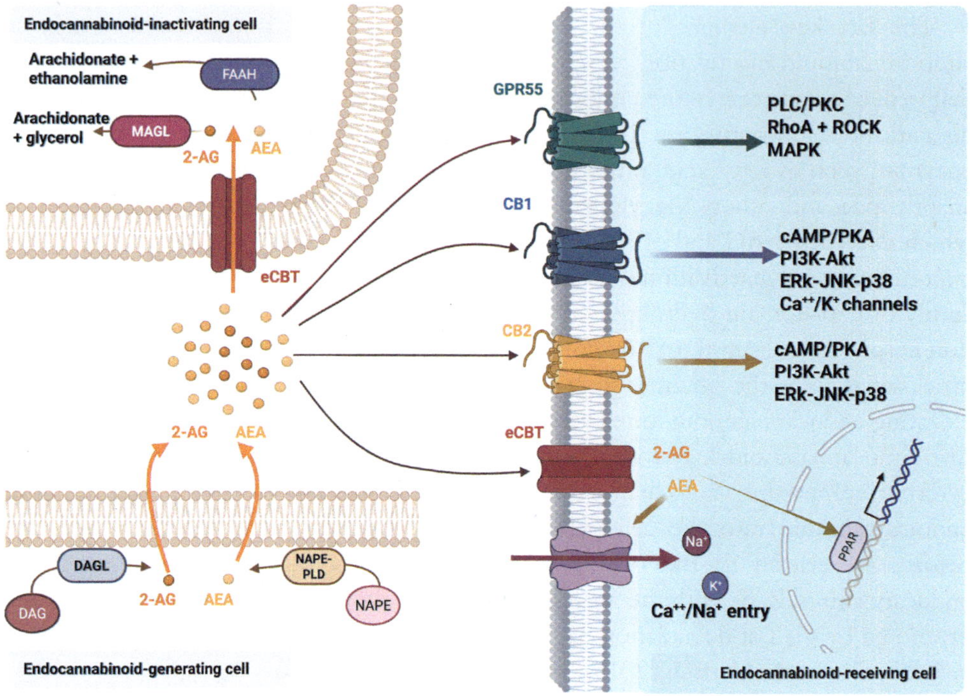

Figure 10.1 Major elements of the endocannabinoid signaling system. Graphical description of the most representative elements of the endocannabinoid signaling system with their more frequent cellular location and functioning in a paracrine interaction. *Created in BioRender. Fernandez-Ruiz, J., 2026. https://BioRender.com/y8xvnci.*

something that several studies have investigated for some years (Fernández-Ruiz, 2019, 2026) and that will be described in more detail below.

10.2 Cannabinoids and neuroprotection

In this chapter, we will address the experimental evidence that supports the neuroprotective role that the modulation of the endocannabinoid system, in particular the modulation of the two canonical cannabinoid receptors, CB_1 and CB_2, may exert in conditions of brain damage and neurodegeneration. Such experimental evidence suggests that compounds active at these receptors (and also at the atypical ones) may serve as promising new neuroprotective therapies, based on four characteristics of these compounds and their mechanisms that represent important advantages compared with more classic types of neuroprotective agents investigated so far. Their broad-spectrum neuroprotective profile is the first advantage, as it enables cannabinoids to synergistically face the numerous neurotoxic stimuli affecting neuronal homeostasis and integrity. Such broad-spectrum activity may be extended, as a second advantage, to their neurogenic activity (generation of new neurons and other neural cells), which is important in most chronic neurodegenerative disorders, as they frequently have long prodromal phases before diagnosis. These two advantages are possible because the endogenous elements of the endocannabinoid system targeted by cannabinoids are located in critical cellular and tissue substrates for the regulation of neural cell homeostasis, integrity, and survival. Lastly, such particular cell and tissue distribution of endocannabinoids and their receptors presumably responds to the fact that they play a pro-homeostatic role addressed to serve as an endogenous protector of neural cell homeostasis against excitotoxicity, oxidative stress, glia-driven inflammation, traumatic injury, protein dysregulation and aggregation, and other neurotoxic insults (Pacher and Mechoulam, 2011; Fernández-Ruiz, 2019; Cristino et al., 2020; Oddi et al., 2023; Fernández-Ruiz et al., 2026). These insults are clearly different from those (e.g., pathogen infection) classically controlled by the immune system.

The first of the particular characteristics of cannabinoids as neuroprotective agents is their pleiotropic profile, which enables these compounds to exert a broad-spectrum activity covering the majority of the different neurotoxic stimuli that kill neurons and other neural cells in neurodegenerative conditions (Fernández-Ruiz et al., 2015a,b, 2017; Rodríguez-Cueto et al., 2021). This includes their antioxidant profile shown by most of the phytocannabinoids (e.g., Δ^9-THC, cannabinol, cannabigerol, cannabidiol (CBD), cannabidiolic acid, cannabidivarin, Δ^9-THCV, etc (Marsicano et al., 2002; Pertwee, 2008; Blebea et al., 2024)). This is also the case of nabilone, levonantradol, dexanabinol, or other synthetic cannabinoid derivatives (Marsicano et al., 2002; Blebea et al., 2024). All of them exert cannabinoid receptor-independent antioxidant properties by acting as mere scavengers of reactive oxygen species (ROS) generated in neurodegenerative conditions, properties that simply derive from their chemical structure (Marsicano et al., 2002; Blebea et al., 2024). This characteristic of phytocannabinoids and some synthetic cannabinoids was, in fact, the first experimental evidence supporting the neuroprotective action of cannabinoids. However, further studies explored in more detail these antioxidant properties, thereby identifying a possible contribution of endogenous mechanisms, for example, the activation of nuclear receptors of the PPAR family or the role of the transcription factor NRF2 (Gugliandolo et al., 2023; Fernández-Ruiz, 2019).

In addition to the antioxidant effects of cannabinoids, neuroprotection with these compounds may also be afforded by normalizing glutamate homeostasis and reducing the overactivation of ionotropic glutamate receptors, thus reducing excitotoxic damage (see Fig. 10.2 for a schematic description). Such an effect can be achieved through several strategies, predominantly by controlling the excess glutamate release via presynaptic CB_1 receptors in glutamatergic terminals (Marsicano et al., 2003). Additional cannabinoid-mediated antiexcitotoxic effects may derive from the blockade of glutamate receptors (e.g., NMDA receptors) located in the postsynaptic neuron

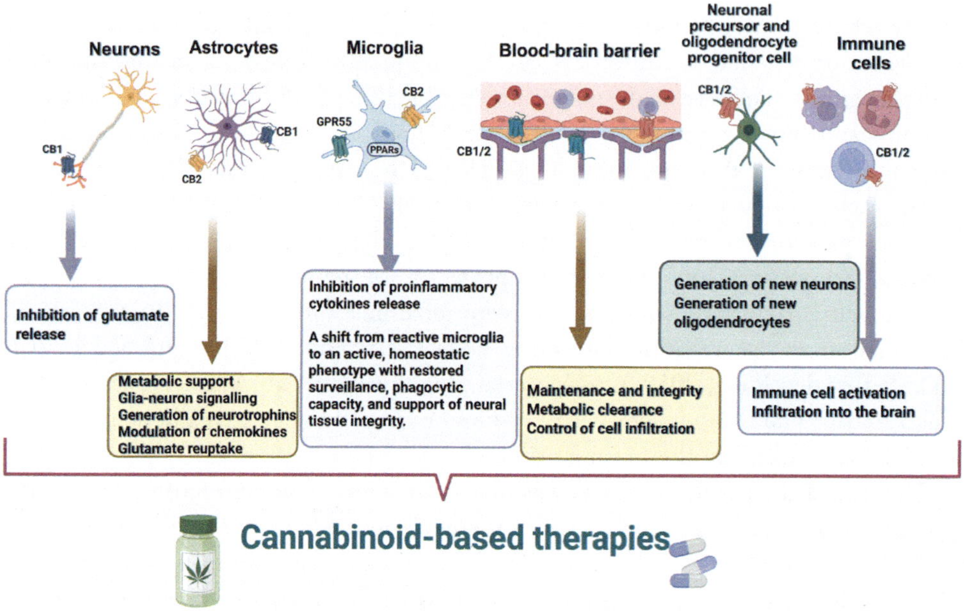

Figure 10.2 Mechanisms proposed for the classic neuroprotective and neurorepair effects of cannabinoids. Graphical description of the cellular and molecular mechanisms underlying the classic neuroprotective and neurorepair effects of cannabinoids in neurodegenerative disorders. *Created in BioRender. Fernandez-Ruiz, J., 2026. https://BioRender.com/y8xvnci.*

within the glutamatergic synapse (Sánchez-Blázquez et al., 2013). This may close voltage-dependent calcium channels, resulting in a reduction of intracellular calcium levels and inhibition of calcium-dependent destructive pathways, although this possibility has been controversial (Fernández-Ruiz et al., 2015a,b; Fernández-Ruiz, 2019). Another option would be to increase the clearance of extracellular glutamate levels by potentiating glutamate reuptake mainly by glial cells, something that cannabinoids could facilitate, given the abundant presence of CB_1 receptors in astrocytes in tripartite glutamatergic synapses and certain preliminary evidence suggesting this possibility (Covelo et al., 2021). Therefore the maintenance of a correct glutamatergic synapse homeostasis by limiting the consequences of excessive excitatory transmission, irrespective of the pre and postsynaptic neuronal or astroglial location of the CB_1 receptor, may serve to preserve the integrity and survival of those neurons under the control of this excitatory transmission (see Fig. 10.2 for a schematic description).

Cannabinoids can also preserve neuronal homeostasis and integrity by reducing glia-driven neuroinflammatory events that result in neuronal toxicity (see Fig. 10.2 for a schematic description). These effects are exerted by targeting the CB_2 receptor, as well as some of the atypical receptors, such as GPR55 or the PPAR-γ nuclear receptors

(Fernández-Ruiz et al., 2015a,b; Aymerich et al., 2018; Rathod et al., 2023; Abd-Nikfarjam et al., 2023). Some studies have also involved the activation of the CB_1 receptor in these antiinflammatory effects (Fernández-Moncada et al., 2023). The contribution of both astrocytes and microglial cells is critical in these effects, as they are the most important cells in which CB_2, GPR55, or PPAR-γ are located, in particular when these cells become activated (Duffy et al., 2021; Fernández-Ruiz et al., 2026; Fernández-Ruiz, 2019). Even, some of these receptors (e.g., CB_2 receptor) become upregulated with the glial cell activation, something observed both in vitro and in vivo in most neurodegenerative conditions (Benito et al., 2003; Cabral and Griffin-Thomas, 2009; Cassano et al., 2017; Fernández-Ruiz et al., 2007; Duffy et al., 2021). As regards astrocytes, they express not only CB_2 but also CB_1 receptors (Stella, 2010). It is well known that both receptors cooperate to regulate several astrocyte functions that are addressed to maintain energy metabolism in neurons and to improve their homeostasis, integrity, and survival, which contribute to potentially rescuing deteriorated neurons (Stella, 2010; Fernández-Ruiz et al., 2015a,b; Molina-Holgado et al., 2003; Fernández-Ruiz et al., 2017; Fernández-Moncada et al., 2024) (see specific studies in Arévalo-Martín et al. (2008)). For example, astrocytes may play an important role in glutamate clearance, as indicated before, in glutamatergic synapses (Dembitskaya and Popov, 2025). They also supply lactate, ketone bodies, and other metabolic substrates to neurons (Dembitskaya and Popov, 2025). They can also generate transforming growth factor-β, neurotrophins, interleukin-10, interleukin-1 receptor antagonist, and other factors that have an important prosurvival function for neurons (Mohammad et al., 2025). Lastly, they may reduce the production of chemokines or other harmful factors (Vida et al., 2025). Cannabinoids acting at CB_1, CB_2, or atypical receptors may potentiate these beneficial effects (Stella, 2010; Hempel and Xi, 2022; Fernández-Moncada et al., 2023). Microglial cells, particularly when they are activated in neuroinflammatory and neurodegenerative conditions, also have an abundant expression of elements of the endocannabinoid system, particularly the CB_2 receptor (Fernández-Ruiz et al., 2007, 2026). This receptor participates in the regulation of microglial cell proliferation and migration to lesion sites (Cabral et al., 2015). It can also facilitate the transformation of those microglial cell populations with pro-inflammatory phenotypes towards antiinflammatory ones (Mecha et al., 2016; Tanaka et al., 2020), a process that also appears to occur in astrocytes (Lawrence et al., 2023). In both cases, this process is susceptible to being regulated by endocannabinoids and their receptors, among other signals (Moreno-García et al., 2020). It is important to mention that these CB_2 receptor-mediated effects appear to involve an inhibition in nuclear factor kappa-light-chain-enhancer of activated B cells (NF-κB) signaling, resulting in changes in the expression of some pro-inflammatory genes under the control of this transcription factor. This includes genes encoding for inducible nitric oxide synthase, COX-2, metalloproteases, tumor necrosis factor-α, or other pro-inflammatory cytokines (Stella,

2010; Zoppi et al., 2014; Baradaran Rahimi and Askari, 2022). Similar responses may also be elicited by the pharmacological modulation with cannabinoids of GPR55 and/or PPAR-γ receptors in glial cells in concordance with the broad-spectrum profile of cannabinoid compounds (Fernández-Ruiz et al., 2017, 2026).

As an extension of the above-described potential of cannabinoids against the pathogenic role exerted by glia-driven inflammatory events in neurodegenerative pathologies, cannabinoids may also be active, through a presumably peripherally located mechanism, against the activation and infiltration of immune cells into the CNS (Ranieri et al., 2016; Sirbu et al., 2023). Such activation/infiltration has been found to operate not only in demyelinating disorders (e.g., multiple sclerosis [MS]) (Rida Zainab et al., 2025), but also in other more classic chronic progressive neurodegenerative pathologies such as Alzheimer's disease (AD), Parkinson's disease (PD) and amyotrophic lateral sclerosis (Shafi et al., 2025). Such an event is possibly facilitated by a certain degree of blood–brain barrier (BBB) deterioration occurring in these disorders, which may facilitate the infiltration of macrophages, neutrophils, lymphocytes, and other immune cells with consequences on the brain homeostasis and integrity that are well documented (Cai et al., 2017). Cannabinoids acting at the immune sites against immune cell activation and at the BBB against cell infiltration have been found to be able to diminish the pathological consequences of such processes, involving the activation of both CB_1 and CB_2 receptors (see Fig. 10.2 for a schematic description).

As it has just been indicated, both CB_1 and CB_2 receptors have also been found to play a protective role against immune cell infiltration at the BBB level (see above). Still, such activity is the consequence of a more general activity of these receptors in the preservation of BBB integrity and function (Fujii et al., 2014) and in the blood supply to the CNS (Hagan et al., 2022). Both are important mechanisms to preserve brain homeostasis and integrity in neurodegenerative conditions (see Fig. 10.2 for a schematic description), and several studies have demonstrated that cannabinoids may contribute to this objective (Hagan et al., 2022). This includes certain cannabinoids, such as CBD, which, acting at some serotonin receptor types, may improve BBB function (Haj-Dahmane and Shen, 2011), while additional cannabinoids may also contribute to these benefits by acting at CB_1 and CB_2 receptors (Vendel and de Lange, 2014). Such nonendocannabinoid targets (serotonin receptor types) and also CB_1 and CB_2 receptors have been found to be present in astrocytes, perivascular microglial cells, and endothelial cells in the BBB (Hagan et al., 2022). Cannabinoids, acting at these receptors and at these cell substrates, may contribute to reducing endothelin-1-induced vasoconstriction, restoring blood supply to the injured brain, inhibiting the above-mentioned leukocyte infiltration by maintaining the integrity of tight junctions, and facilitating the elimination of β-amyloid and other toxic proteins.

Lastly, another important piece of evidence supporting the role of cannabinoids as disease-modifying agents in neurodegenerative disorders is their potential in promoting the replacement of neurons and other neural cells lost in neurodegenerative disorders (Valeri and Mazzon, 2021). This neurorepair activity is essential in pathologies with long prodromal periods, in which the loss of neural cells initiates early but without showing its face, so clinical signs do not appear, then eliminating the option of an early diagnosis and an anticipation in the onset of neuroprotective treatments. This is the case of PD in which the occurrence of the first signs of rigidity, postural instability, and akinesia appears when more than 60% of dopaminergic neurons have already degenerated, so that the possibilities of preserving surviving neurons, although necessary, are limited (Rees et al., 2024). This means that it is mandatory to activate mechanisms of neuroreplacement, for example, by activating neurogenic niches to generate new neurons that may migrate to the atrophic sites and differentiate into dopaminergic neurons (Verkerke et al., 2025). Similar situations exist for AD (Sugaya, 2005) and other neurodegenerative disorders (Ziemka-Nalecz and Zalewska, 2012). This is also the case of demyelinating disorders, in particular MS, with the remyelinating process that may be elicited from oligodendrocyte progenitor cells (Lalmani and Deshmukh, 2026). In both cases, the activation of specific elements of the endocannabinoid system may be useful in such neuroreplacement events, given that elements of this system have been found to be expressed in both neural precursor cells and oligodendrocyte progenitor cells (Ilyasov et al., 2018; Oddi et al., 2023). In addition, a few studies have provided the first evidence that targeting these elements may allow these cells to proliferate, migrate, and differentiate into the cell phenotypes expected for reaching such neuroreplacement (Prenderville et al., 2015), although this possibility requires additional research before it can progress to the clinical scenario (see Fig. 10.2 for a schematic description).

10.3 Cannabinoid effects promoting the elimination of protein aggregates

Adding to the above-described neuroprotective and neurorepair effects derived from targeting specific endocannabinoid elements in neurodegenerative conditions, new experimental evidence suggests that cannabinoids may also preserve neural cells, in particular neurons, through the activation of cell mechanisms (e.g., autophagy signaling and other lysosomal mechanisms) that may eliminate protein aggregates in degenerating neural cells (Hamano et al., 2021). Autophagy is a lysosomal catabolic process responsible for degrading cytosolic components, which is essential for cellular homeostasis and survival (Yang and Klionsky, 2010). In neurodegenerative disorders, protein aggregation, which reflects the proteinopathy feature of most neurodegenerative pathologies, is an essential key pathogenic mechanism in these disorders as it produces waste materials that provoke cellular stress, finally activating cell death pathways and thus eliciting neurodegeneration (Fontaine et al., 2015). Such protein aggregation occurs in the cell nucleus (e.g., nuclear inclusions in Huntington's disease and other polyglutaminopathies; reviewed in van der Wel (2024)), cytoplasm (e.g., Lewy bodies in PD; reviewed in Barbosa et al. (2025); Tau aggregates in the form of neurofibrillary tangles in AD; reviewed in Chinnathambi (2025)); TDP43 aggregates in ALS/frontotemporal dementia; reviewed in Verde et al. (2025)), and/or at extracellular sites (e.g., amyloid plaques in AD; reviewed in Ahmadi et al. (2025)). Intracellular protein aggregates are primarily cleared through autophagy and related lysosomal degradation pathways (Nixon and Rubinsztein, 2024), whereas extracellular aggregates can predominantly be removed via phagocytic mechanisms, largely executed by reactive microglial cells despite certain uncertainties and controversies (Spangenberg et al., 2019).

The interest of cannabinoids as potential modulators of these degradative pathways originally derives from previous studies carried out mainly in the oncology field, which demonstrated that cannabinoids active at CB_1 and/or CB_2 receptors may activate autophagy to kill cancer cells. This represents an important mechanism for the antitumoral properties of certain cannabinoids (Ellert-Miklaszewska et al., 2021). A derived question from these pioneering studies was whether such cannabinoid-mediated effects may also promote the autophagic response, in a noncytotoxic manner, to facilitate the elimination of protein aggregates in degenerating cells (Nixon, 2024). As indicated before, the rationale for this question is that autophagy is not only a mechanism to promote cell death in certain conditions (e.g., transformed cells), but also a lysosomal catabolic process that may serve to degrade cytosolic components, including protein aggregates, thereby providing a key benefit for cell homeostasis and integrity. Importantly, protein aggregates are not only intrinsically toxic, but they can also act as extracellular danger-associated molecular pattern-like signals that sustain maladaptive microglial activation. Such chronic inflammatory states may paradoxically impair efficient aggregate clearance while exacerbating neurotoxicity (Nixon and Rubinsztein, 2024). Therefore interventions addressed to target autophagy signaling as well as increasing the phagocytic activity of microglial cells, without amplifying inflammatory damage, have been proposed as potential therapeutic strategies in neurodegenerative proteinopathies (Spangenberg et al., 2019). Although the responses to these questions are still incipient, progressively increased evidence appears to confirm that the activation of classic and atypical cannabinoid receptors may serve for such a purpose (see Fig. 10.3 for a schematic description). Such potential adds to the other neuroprotective effects exerted by these receptors, which are presented in schematic form in Fig. 10.2, thereby extending the broad-spectrum profile of cannabinoids as agents active against excitotoxicity, oxidative stress, glia-driven neuroinflammation, immune cell infiltration, and BBB deterioration with an additional and important activity against protein aggregation and its consequences on neural cell homeostasis and integrity (Hiebel et al., 2014; Hiebel and Behl, 2015; Gugliandolo et al., 2020). In line

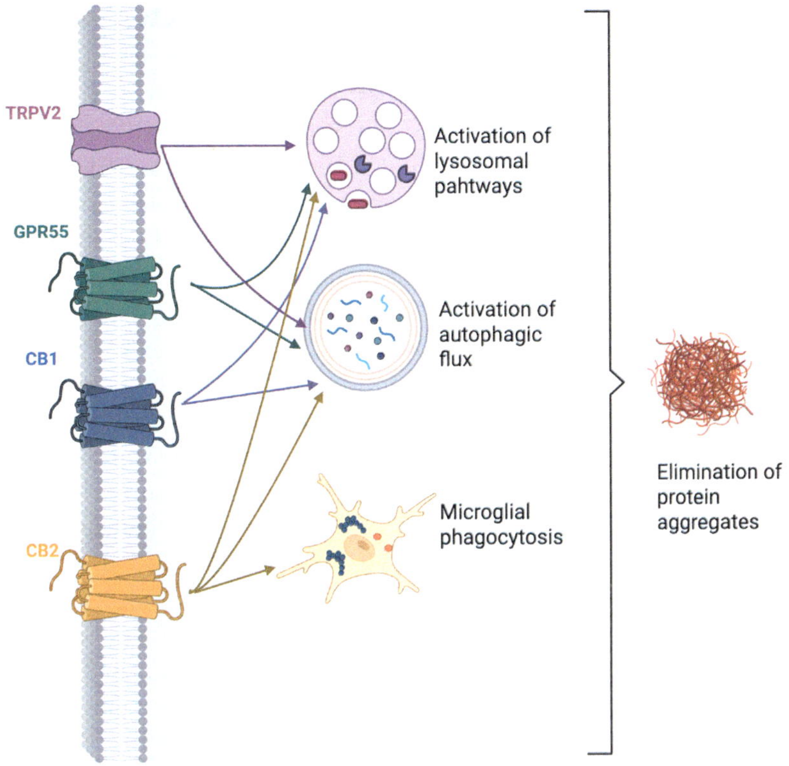

Figure 10.3 Cannabinoid-elicited elimination of protein aggregates in neurodegenerative pathologies. Graphical description of the potential cellular and molecular mechanisms that could underlie cannabinoid-elicited elimination of protein aggregates in neurodegenerative pathologies. *Created in BioRender. Fernandez-Ruiz, J., 2026. https://BioRender.com/y8xvnci.*

with this integrative view, recent work indicates that specific cannabinoid-based interventions may simultaneously enhance neuronal autophagic flux, facilitating the clearance of intracellular aggregates (e.g., Tau, α-synuclein), as well as potentiate microglial phagocytosis and lysosomal processing of extracellular aggregates (e.g., β-amyloid), thus addressing multiple aggregation compartments within the same disease entity (Jiang et al., 2022; Erustes et al., 2025; Vrechi et al., 2025). This includes work carried out by Julián Romero and coworkers after genetic ablation of the endocannabinoid-inactivating enzyme FAAH in the 5xFAD mouse model of AD. This work revealed important upregulation of microglial factors related to phagocytic activity, β-amyloid clearance, and neuron–glia crosstalk (e.g., TREM2, CTSD, C3, C3AR, etc.) compatible with the benefits of modulating, rather than suppressing, neuroinflammation in AD, which results in augmented microglial activation and β-amyloid phagocytosis.

10.3.1 Effects of cannabinoid type-1 receptor on autophagy and other lysosomal pathways

Although it is one of their major properties, the contribution of CB_1 receptors to cannabinoid-induced neuroprotection is not exclusively related to the control of excitotoxicity. Cannabinoids can also target these receptors, as indicated above, to induce autophagy or to modulate the lysosomal function, which, in the context of neurodegeneration, may serve to eliminate damaged proteins that are deposited and aggregated in different subcellular compartments of neurons and other neural cells (Hiebel et al., 2014; Gugliandolo et al., 2023). Such effects may contribute significantly to improving neuronal homeostasis and integrity. This is supported by the location of this receptor in different intracellular compartments (e.g., lysosomes, endosomes, endoplasmic reticulum, mitochondria) that are significantly involved in these processes (Rozenfeld and Devi, 2008; Brailoiu et al., 2011; Fernández-Moncada et al., 2024; Rodríguez et al., 2001). This intracellular topology is conceptually relevant because it allows CB_1 receptors to shape organelle-specific signaling microdomains, such as those regulating mitochondrial bioenergetics or lysosomal stability, which directly intersect with autophagic and proteostatic pathways.

In support of the role exerted by the CB_1 receptor in this process, it is well known that CB_1 receptor-deficient mice exhibit early signs of dysregulated autophagy and lysosomal functions, resulting in pathological accumulation of proteins (Piyanova et al., 2013). In the context of brain ageing, this may support the idea that the reduced CB_1 receptor signaling found in the aged brain may be associated with autophagy dysregulation and an elevated risk to provoke neurodegenerative events (Bilkei-Gorzo, 2012). Further studies from the same group demonstrated that the CB_1 receptor also plays an important role in mitochondrial autophagy (mitophagy), in part due to the presence of this receptor in the mitochondrial membrane (Kataoka et al., 2020). Using several methodologies, these authors found that serine 65-phosphorylated ubiquitin, a key marker for mitophagy, was reduced in the hippocampus of adult CB_1-deficient mice, which was accompanied by reduced mitophagy-like events and altered mitochondrial morphology (Kataoka et al., 2020). They concluded that these mitochondrial changes might be due to impairments in the mitochondrial quality control system, which links age-related decline in CB_1 receptor activity to impaired memory (Kataoka et al., 2020). In the same direction, a further study demonstrated that the activation of the CB_1 receptor with ACEA induced mitophagy by promoting the formation of mitophagosomes and by regulating several autophagy-related markers. This served to maintain normal mitochondrial morphology in neurons and to alleviate oxidative stress and neurological dysfunction in an experimental model of subarachnoid hemorrhage in rats (Liu et al., 2022). In a more recent study in PD, Gugliandolo and coworkers used retinoic acid-differentiated SH-SY5Y neuroblastoma cells exposed to

MPP^+, whose cell viability was preserved with cannabinol (presumably acting at the CB_1 receptor), to carry out a transcriptomic analysis, which proved that this phytocannabinoid was able to attenuate MPP^+-induced elevation in the expression of genes involved in mitochondrial transport, localization and protein targeting (e.g., PINK1/Parkin-mediated autophagy) (Gugliandolo et al., 2023). Collectively, these observations are consistent with the notion that CB_1 receptor-linked mechanisms may influence aggregate-prone proteomes both indirectly, by stabilizing mitochondrial function and limiting ROS-driven protein misfolding, and directly, by facilitating organelle-selective autophagy that sustains lysosomal degradative capacity.

The role of the CB_1 receptor in this context goes beyond the activation of the autophagic flux and involves other lysosomal mechanisms. In an in vitro study using rat cortical neurons, Noonan et al. (2010) demonstrated that the activation of this receptor plays a positive role in the stabilization of lysosomes against β-amyloid-induced lysosomal membrane permeabilisation, an event associated with its toxicity (Gowran and Campbell, 2008). These authors found that this positive role was exerted by preventing the upregulation elicited by β-amyloid of p53, which prevents the interaction of this tumor suppressor protein with the lysosomal membrane (Noonan et al., 2010). Given that lysosomal membrane permeabilisation represents a critical failure point in the collapse of cellular proteostasis, CB_1 receptor-mediated lysosomal stabilization provides a plausible mechanistic link between cannabinoid signaling and protection against aggregate-driven lysosomal dysfunction.

10.3.2 Effects of cannabinoid type-2 receptor on autophagy-lysosomal pathways and microglial phagocytosis

The experimental evidence supporting the contribution of CB_2 receptor activation to the beneficial effects of cannabinoid-induced autophagy (or other lysosomal pathways) in the elimination of protein aggregates in neurons is significantly more recent than that for the CB_1 receptor. For example, a very recent study (Vrechi et al., 2025) used the CB_2 receptor agonist GW405,833 to investigate its effects on the autophagy flux in human neuroblastoma SH-SY5Y cells overexpressing wild-type Tau. These authors found increased Tau degradation mediated by the autophagy response (reversed by the autophagy blocker chloroquine). Tau degradation in this experimental cell model was also found with the selective CB_1 receptor agonist ACEA, but this response was not reversed by chloroquine (Vrechi et al., 2025). This pharmacological dissociation indicates that CB_1- and CB_2-driven signaling pathways may converge on Tau lowering through partially distinct mechanisms, reinforcing the rationale for receptor-selective or combinatorial strategies tailored to disease stage and to the dominant aggregate species involved. In another study, also using activation of the CB_2 receptor but in cortical neurons exposed to β-amyloid (Zhou et al., 2022), the authors described elevated expression of autophagy-related proteins LC3-II, Beclin-1, ATG5, and ATG13. These authors had previously shown a similar induction of autophagy by CB_2 receptor activation in the APP/PS1 transgenic mouse model of AD (Long et al., 2021). In an older study carried out in an experimental model of autoimmune encephalomyelitis in mice, the deletion of the CB_2 receptor impaired the accumulation of the autophagy-related protein LC3-II in the spinal cord, whereas the activation of CB_2 receptors reversed such effects (Shao et al., 2014). Also, in an older study, our group demonstrated improved autophagy markers in LRRK2(G2019S) transgenic mice, a genetic model of PD, with the CB2 receptor agonist HU-308, but we could not confirm whether this response occurred in neurons or in other cell substrates. In fact, looking at neural cells other than neurons, the available information appears to be a little bit more abundant, although it is more circumscribed to the CB_2 receptor-mediated induction of the autophagy flux for eliminating proteins associated with an excessive neuroinflammation (e.g., NLRP3 inflammasome) rather than protein aggregates, which, however, have been investigated more in relation with their CB_2 receptor-mediated phagocytosis by microglial cells. This distinction is especially relevant for proteinopathies, as glial autophagy can indirectly modulate aggregate burden by resolving inflammatory signaling that otherwise compromises phagocytic efficiency and lysosomal maturation. In this sense, a recent study demonstrated that activating the CB_2 receptor in microglial cells in a murine model of spinal cord injury (SCI) resulted in elevated ubiquitination of the NLRP3 inflammasome associated with interactions with some autophagy-related proteins (e.g., p62, microtubule-associated protein 1B light chain 3), which attenuated neuroinflammation and facilitated the functional recovery of SCI mice (Jiang et al., 2022). A similar CB_2 receptor-mediated degradation of the NLRP3 inflammasome in astrocytes, exerted by activating the autophagy-lysosomal pathway, was described in another study using an experimental model of PD, as MPTP-lesioned mice (Zhu et al., 2023). Thus activation of the CB_2 receptor with JWH-133 alleviated motor impairments in mice and reduced NLRP3 deposition on astrocytes, whereas CB_2 receptor knockdown aggravated the pathology and induced the NLRP3/caspase-1/interleukin-1β pathway (Zhu et al., 2023). Subsequent in vitro studies using cultured astrocytes exposed to MPP^+ or LPS confirmed the in vivo results, with the CB_2 receptor activation attenuating the activation of the NLRP3/caspase-1/interleukin-1β pathway, increasing autophagy flux, and facilitating NLRP3 degradation (Zhu et al., 2023).

As regards the CB_2 receptor-mediated cell phagocytosis of protein aggregates, CBD in combination with a very low amount of Δ^9-THC (99:1) was also found to be active against β-amyloid accumulation in a murine model of AD (Ruiz de Martín Esteban et al., 2026). These authors found elevated microglial phagocytic activity on amyloid plaques, resulting in a decrease in plaque protein density analyzed by in vivo imaging procedures (multiphoton microscopy), which reflects a lower toxicity. However, it was difficult to identify the possible molecular mechanisms involved in these effects due to the combination of both phytocannabinoids in an unusual ratio (99:1). The authors assumed that this effect implied the participation of endocannabinoid-based mechanisms (e.g., CB_1 and CB_2 receptors) but also nonendocannabinoid mechanisms presumably located on neighboring cells around plaques (Ruiz de Martín Esteban et al., 2026). Previous studies have revealed the interaction of both Δ^9-THC (Eubanks et al., 2006) and CBD (Chrobak et al., 2021) with β-amyloid-constituent of senile plaques, as well as their ability to enhance microglial phagocytic activity (Hassan et al., 2014; Yang et al., 2022). Previous studies by the same group indicated an important role of the microglial CB_2 receptors activated with the selective Roche agonist RO-6866945 in the metabolism of insoluble forms of β-amyloid protein (Ruiz de Martín Esteban et al., 2022). In addition, the combination of Δ^9-THC and CBD was also investigated, using in this case the most frequent Sativex-like 1:1 ratio, in a mouse model of tauopathy (Casarejos et al., 2013), proving its efficacy in reducing Tau and β-amyloid accumulation. However, this study did not explore whether this reduction was caused by an elevated microglial phagocytosis or by an activation of lysosomal pathways.

Another recent study investigated this removal process in α-synuclein-based models by pro-phagocytic cells, proving the protection of vulnerable neurons (Joers et al., 2024). However, such an effect was found in this study with a CB_2 receptor inverse agonist (SMM-189), which was able to elevate the phagocytic profile of activated microglial cells and infiltrated peripheral immune cells, then contributing to reducing the toxicity of α-synuclein aggregates by facilitating their clearance (Joers et al., 2024). This observation is concordant with the fact that genetic ablation of the CB_2 receptor may promote neuroinflammation-related phagocytosis (Feng et al., 2023), which may adjust the effects derived from the CB_2 receptor inhibition with the classical antiinflammatory response elicited by CB_2 receptor activation (Fernández-Ruiz et al., 2026). Taken together, these apparently divergent findings support a nuanced, state-dependent model in which CB_2 receptor modulation shapes phagocytosis and lysosomal processing according to microglial activation status, receptor coupling, and the equilibrium between pro-inflammatory and pro-resolving programs.

Certain benefits at the autophagy level have also been described with the CB_2 receptor activation in pathological conditions that do not imply the occurrence of protein aggregates. This is the case of bone cancer pain, in which the selective CB_2 agonist JWH-015 ameliorated the impaired autophagic flux in spinal neurons caused by glia-derived inflammatory mediators in a murine model of this disease (Mao et al., 2019). This effect involved a reduction in the LC3β-II/LC3β-I ratio and in the expression of p62 in spinal neurons, as well as a parallel reduction in glia-derived inflammatory IL-1β and IL-6 cytokines (Mao et al., 2019). A similar situation was observed with status epilepticus, which causes neuronal loss and gliosis, which run in parallel to changes in the CB_2 receptor in the hippocampus in developing rats (Wu et al., 2020). The activation of this receptor exerted anticonvulsant effects, which were paralleled by increased neuronal autophagy and elevated phosphorylated-mTOR/mTOR, Beclin-1, and LC3-II/LC3-I, while decreasing p62 and other related proteins (Wu et al., 2020). The opposite situation was found with the selective CB_2 receptor antagonist AM-630. Collectively, these data reinforce the view that CB_2 receptor-linked modulation of autophagy and lysosomal pathways is not restricted to a single disease entity but may represent a more general mechanism by which cannabinoid signaling limits inflammation-driven disruptions of cellular proteostasis.

10.3.3 Effects of atypical cannabinoid receptors on autophagy-lysosomal pathways and microglial phagocytosis

The cannabinoid-mediated elevation in the autophagic flux to eliminate protein aggregates has also been investigated in relation to the possible contribution of some of the atypical cannabinoid receptors, for example, the $TRPV_2$ receptor, which is part of the same family as the $TRPV_1$ receptor, the most extensively investigated endocannabinoid-linked receptor of this family. $TRPV_2$ has been found to be reduced in the *postmortem* hippocampus and cortex of AD mice and patients (Yang et al., 2022). Given this reduction, these authors used the phytocannabinoid CBD, which has been found to activate $TRPV_2$ (Etemad et al., 2022), to investigate the issue using both in vitro (neuronal and microglial cultures) and in vivo (AD mice) approaches (Yang et al., 2022). These authors observed an enhanced microglial phagocytosis and clearance of β-amyloid peptide with this phytocannabinoid, as well as an attenuation of these beneficial effects by knockdown of $TRPV_2$ or TREM2, which confirmed the participation of these vanilloid receptors located in microglial cells (Yang et al., 2022). In a more recent study also using CBD (Vrechi et al., 2025), this phytocannabinoid demonstrated to be active against Tau deposition in cultures of the SH-SY5Y neuronal line, which had been transfected to overexpress wild-type Tau protein. The authors described that the addition of CBD to these cells reduced AT8 and Tau protein expression in parallel to an elevation of LC3-II levels and autophagosome formation, indicating activation of autophagy and enhanced Tau degradation (Vrechi et al., 2025). Coaddition of chloroquine, which, as mentioned above, blocks autophagy, reversed CBD effects, confirming the contribution of autophagic mechanisms. However, these authors did not explore in this study the specific mechanism(s) underlying these CBD effects (Vrechi et al., 2025). They could be related to the activation of vanilloid receptors, for example, $TRPV_1$, as similar CBD-induced effects were previously associated with these receptors (and also with CB_1/CB_2 receptors) (Vrechi et al., 2021) or potentially to TRPV2, which other authors also linked to CBD in a parallel study (Yang et al., 2022). Even CBD effects could be associated with other endocannabinoid-related targets, such as GPR55 (see below), for which CBD is an antagonist (Miao et al., 2025) or even nonendocannabinoid pathways (Fernández-Ruiz et al., 2013). In this regard, it is plausible that CBD engages parallel pro-autophagic signaling nodes (e.g., ERK/Akt–mTOR crosstalk) while receptor-level contributions may differ by cell type (neuronal vs microglial) and by aggregate species (Tau *vs* β-amyloid), an issue that remains to be resolved experimentally. This question was also investigated by another research group using CBD in the same neuroblastoma cell line, SH-SY5Y cells, but subjected to MPP^+ exposure to recapitulate features of PD (Gugliandolo et al., 2020). These authors observed that CBD was able to counteract the loss of viability of these cells while reducing apoptosis (reduction in Bax and caspase 3). These benefits were presumably mediated by the activation of ERK and Akt/mTOR pathways, as blockade of these pathways abolished CBD benefits (Gugliandolo et al., 2020). In parallel, authors analyzed some autophagy-related proteins, which were found to be elevated by the MPP^+ exposure in the case of LC3 (not with Beclin-1) while reduced by CBD, through effects that were apparently mediated by the activation of the $TRPV_1$ but also the CB_2 receptor (Gugliandolo et al., 2020).

CBD has also been found to be protective in AD, but its precise mechanism(s) of action remain to be elucidated. Hao and Feng (2021) carried out a transcriptomic study in the hippocampus of the APP/PS1 murine model chronically treated with CBD with the purpose of identifying possible pathways associated with CBD-mediated benefits in this disease. Immunostaining, transmission electron microscopy, and Western blot analyses confirmed the presence of β-amyloid plaques, autophagosomes, and autophagy-related protein expression, respectively, in the hippocampus of APP/PS1 mice (Hao and Feng, 2021). An RNA-seq analysis in these mice after the chronic treatment with CBD revealed significant upregulation of genes related to immune system response, as well as to the autophagy pathway. The enhancement of the autophagy response in hippocampal neurons by CBD was confirmed with further analysis by transmission electron microscopy (Hao and Feng, 2021). These transcriptomic signatures are consistent with a coordinated "immuno-proteostatic" response, in which CBD modulates immune-related gene networks while also promoting an autophagy-linked program that could support aggregate clearance (Hao and Feng, 2021). This CBD-dependent induction of autophagy was also found in relation to natural ageing. In a recent study, CBD was found to increase lifespan and rescue physiological ageing-associated declines in *C. elegans* by promoting autophagic flux, an effect that was dependent on the autophagy genes *bec-1*, *vps-34*, and *sqst-1* (Wang et al., 2022). CBD also slowed the occurrence of several aberrant neuronal morphologies typical of ageing in *C. elegans*. Such improved neuronal health effect was further demonstrated to involve sirtuin-1 signaling, which acted upstream of the autophagy machinery to promote stress resistance, maintain proteostasis, and preserve neuronal morphology during aging (Wang et al., 2022). These findings indicate that CBD can engage conserved pro-autophagic and proteostatic programs across disease-related and physiological ageing contexts, although the multiplicity of its molecular targets complicates the assignment of a single upstream receptor mechanism.

Beyond the above-described links of CBD with TRP channels, the orphan G-protein-coupled receptor GPR55, as indicated above, has been proposed as an indirect modulator of proteostasis-related processes, particularly in inflammatory and glial contexts, and may contribute to the pleiotropic actions of CBD. Although direct evidence linking GPR55 to autophagy-dependent degradation of pathogenic protein aggregates, such as Tau, β-amyloid, or α-synuclein, is still lacking, several studies in nonneuronal systems suggest that inhibition or antagonism of GPR55 promotes pro-autophagic and lysosome-engaging cellular profiles, likely through relief of ERK/Akt–mTOR signaling and modulation of cytoskeletal dynamics (Whyte et al., 2009; Shrivastava et al., 2011). In microglia and other immune cells, GPR55 signaling influences activation state, migration, and phagocytic competence, thereby conditioning the efficiency of extracellular aggregate clearance and lysosomal maturation. In this framework, the fact that CBD acts as a functional antagonist of GPR55 raises the possibility that suppression of GPR55 signaling may operate as a permissive mechanism facilitating autophagic and phagocytic programs relevant for proteostatic control, in a context- and cell-type-dependent manner.

10.4 Concluding remarks and future perspectives

The studies reviewed here are all concordant with the view that cannabinoids may serve as a novel therapy to delay/arrest neuronal deterioration in chronic progressive neurodegenerative pathologies and in episodic brain damage. This proposed neuroprotective potential derives from their capability to normalize glutamate homeostasis to reduce excitotoxicity, to eliminate ROS to reduce oxidative injury, to attenuate glial reactivity to limit local neuroinflammatory events, and, as the special topic reviewed here, to eliminate protein aggregates by activating autophagic and lysosomal pathways and/or elevating phagocytosis by microglial and/or infiltrated peripheral cells. Importantly, the proteinopathy-oriented cannabinoid effects are mechanistically plausible because they engage both neuronal (autophagy/mitophagy/lysosomal stability) and glial (phagocytosis/lysosome maturation/inflammation resolution) determinants of aggregate burden, potentially addressing intracellular and extracellular aggregate pools within the same disorder (Yang et al., 2022; Vrechi et al., 2025). Most of the experimental evidence supporting this neuroprotective potential has been collected in preclinical models (laboratory animals and/or cell-based models), with a general lack of clinical validation of these potential therapies, so that this objective remains the major challenge for the future. A key near-term translational need is the identification of pharmacodynamic biomarkers (e.g., imaging or biofluid markers reflecting microglial activation states, autophagic flux-related signatures, or aggregate load) that can bridge preclinical mechanisms to human trials in a feasible manner. The progression towards clinical validation of preclinical data may be facilitated by the recent approval of the first cannabinoid-based medicines (e.g., Sativex®, Epidiolex®), as well as by the regulation of standardized preparations of medicinal cannabis in different countries. These already existing formulations have shown a reasonable safety (Gottschling et al., 2020; Bilbao and Spanagel, 2022), as well as a pleiotropic and broad-range profile that appears to be adequate for diseases in which different cytotoxic mechanisms cooperate to damage specific neuronal subpopulations (Fernández-Ruiz, 2019; Fernández-Ruiz et al., 2026). New phytocannabinoid- or synthetic cannabinoid-based formulations that may emerge in the coming years are expected to reproduce these two advantages. In this regard, receptor-selective strategies (e.g., CB_2-preferring agonists or biased ligands) and indirect approaches (e.g., FAAH/MAGL inhibitors) may offer improved tolerability and disease-context specificity, while multitarget phytocannabinoids (e.g., CBD) may provide broader coverage at the cost of mechanistic complexity; comparing these strategies head-to-head in proteinopathy-relevant models will be essential. Therefore, it is expected and desirable that the issue will recruit an important amount of clinical research in the coming years, which will allow the promising expectations generated by these molecules to progress from the present preclinical evidence to genuine clinical exploitation. This will also imply further investigating possible side effects of cannabinoid therapies in this context, although the information collected to date indicates they appear to show an adequate safety profile.

Acknowledgments

Authors research is supported by grants from the MCINN "Programa Nacional de Biomedicina" (RTI-2018–098885-B-100 and PID2021–128906OB-I00) with cofunding from the European Regional Development Fund, the Michael J. Fox Foundation (MJFF-022552), La Caixa and Luzón Foundations (HR21–00931), and the "Instituto de Salud Carlos III" (CIBERNED, CB06/05/0089). The authors are indebted to all colleagues who contributed to the studies mentioned in the chapter.

References

Abd-Nikfarjam, B., Dolati-Somarin, A., Baradaran Rahimi, V., Askari, V.R., 2023. Cannabinoids in neuroinflammatory disorders: focusing on multiple sclerosis, Parkinsons, and Alzheimers diseases. BioFactors 49 (3), 560–583. Available at https://doi.org/10.1002/biof.1936.

Ahmadi, S., Khaledi, S., Ahmadi, K., Hassanzadeh, K., 2025. Amyloid beta in Alzheimer's Disease: mechanisms, biomarker potential, and therapeutic targets. CNS Neuroscience & Therapeutics 31 (12). Available at https://doi.org/10.1002/cns.70688.

Arévalo-Martín, A., García-Ovejero, D., Gómez, O., Rubio-Araiz, A., Navarro-Galve, B., Guaza, C., et al., 2008. CB 2 cannabinoid receptors as an emerging target for demyelinating diseases: from neuroimmune interactions to cell replacement strategies. British Journal of Pharmacology 153 (2), 216–225. Available at https://doi.org/10.1038/sj.bjp.0707466.

Aymerich, M.S., Aso, E., Abellanas, M.A., Tolon, R.M., Ramos, J.A., Ferrer, I., et al., 2018. Cannabinoid pharmacology/therapeutics in chronic degenerative disorders affecting the central nervous system. Biochemical Pharmacology 157, 67–84. Available at https://doi.org/10.1016/j.bcp.2018.08.016.

Baradaran Rahimi, V., Askari, V.R., 2022. A mechanistic review on immunomodulatory effects of selective type two cannabinoid receptor β-caryophyllene. BioFactors 48 (4), 857–882. Available at https://doi.org/10.1002/biof.1869.

Barbosa, P.M., Parmera, J.B., Warner, M.J., 2025. Neuropathology in genetic Parkinson's disease: a focused review of pathological and clinical findings. Journal of Neural Transmission (Vienna, Austria: 1996). Available at 10.1007/s00702-025-03088-7.

Benito, C., Núñez, E., Tolón, R.M., Carrier, E.J., Rábano, A., Hillard, C.J., et al., 2003. Cannabinoid CB2 receptors and fatty acid amide hydrolase are selectively overexpressed in neuritic plaque-associated glia in Alzheimer's disease brains. Journal of Neuroscience 23 (35), 11136–11141. Available at https://doi.org/10.1523/jneurosci.23-35-11136.2003.

Bilbao, A., Spanagel, R., 2022. Medical cannabinoids: a pharmacology-based systematic review and meta-analysis for all relevant medical indications. BMC Medicine 20 (1). Available at https://doi.org/10.1186/s12916-022-02459-1.

Bilkei-Gorzo, A., 2012. The endocannabinoid system in normal and pathological brain ageing. Philosophical Transactions of the Royal Society B: Biological Sciences 367 (1607), 3326–3341. Available at https://doi.org/10.1098/rstb.2011.0388.

Blebea, N.M., Pricopie, A.I., Vlad, R.A., Hancu, G., 2024. Phytocannabinoids: exploring pharmacological profiles and their impact on therapeutical use. International Journal of Molecular Sciences 25 (8), 4204. Available at https://doi.org/10.3390/ijms25084204.

Bononi, G., Tuccinardi, T., Rizzolio, F., Granchi, C., 2021. α/β-Hydrolase domain (ABHD) inhibitors as new potential therapeutic options against lipid-related diseases. Journal of Medicinal Chemistry 64 (14), 9759–9785. Available at https://doi.org/10.1021/acs.jmedchem.1c00624.

Brailoiu, G.C., Oprea, T.I., Zhao, P., Abood, M.E., Brailoiu, E., 2011. Intracellular cannabinoid type 1 (CB1) receptors are activated by anandamide. Journal of Biological Chemistry 286 (33), 29166–29174. Available at https://doi.org/10.1074/jbc.M110.217463.

Briand-Mésange, F., Gennero, I., Salles, J., Trudel, S., Dahan, L., Ausseil, J., et al., 2024. From classical to alternative pathways of 2-arachidonoylglycerol synthesis: alterAGs at the crossroad of endocannabinoid and lysophospholipid signaling. Molecules 29 (15), 3694. Available at https://doi.org/10.3390/molecules29153694.

Brown, C., Robin Hiley, 2009. Is GPR55 an anandamide receptor? Vitamins and Hormones 81, 111–137.

Burstein, S.H., 2019. Eicosanoid mediation of cannabinoid actions. Bioorganic and Medicinal Chemistry 27 (13), 2718–2728. Available at https://doi.org/10.1016/j.bmc.2019.05.018.

Cabral, G.A., Griffin-Thomas, L.T., 2009. Emerging role of the cannabinoid receptor CB2 in immune regulation: therapeutic prospects for neuroinflammation. Expert Reviews in Molecular Medicine 11. Available at https://doi.org/10.1017/S1462399409000957.

Cabral, G.A., Ferreira, G.A., Jamerson, M.J., 2015. Endocannabinoids and the immune system in health and disease. In: Pertwee, R. (Ed.), Endocannabinoids. Handbook of Experimental Pharmacology, vol 231. Springer, Cham, pp. 185–211. Available at https://doi.org/10.1007/978-3-319-20825-1_6.

Cai, W., Zhang, K., Li, P., Zhu, L., Xu, J., Yang, B., et al., 2017. Dysfunction of the neurovascular unit in ischemic stroke and neurodegenerative diseases: an aging effect. Ageing Research Reviews 34, 77–87. Available at https://doi.org/10.1016/j.arr.2016.09.006.

Casarejos, M.J., Perucho, J., Gomez, A., Muñoz, M.P., Fernandez-Estevez, M., Sagredo, O., et al., 2013. Natural cannabinoids improve dopamine neurotransmission and tau and amyloid pathology in a mouse model of tauopathy. Journal of Alzheimer's Disease 35 (3), 525–539. Available at https://doi.org/10.3233/JAD-130050.

Cassano, T., Calcagnini, S., Pace, L., De Marco, F., Romano, A., Gaetani, S., 2017. Cannabinoid receptor 2 signaling in neurodegenerative disorders: from pathogenesis to a promising therapeutic target. Frontiers in Neuroscience 11. Available at https://doi.org/10.3389/fnins.2017.00030.

Chinnathambi, S., 2025. Tau proteotasis in Alzheimer's disease. Advances in Protein Chemistry and Structural Biology 147, 333–353. Available at https://doi.org/10.1016/bs.apcsb.2024.09.003.

Chrobak, W., Pacut, D.W., Blomgren, F., Rodin, A., Swenson, J., Ermilova, I., 2021. Component of cannabis, cannabidiol, as a possible drug against the cytotoxicity of Aβ(31-35) and Aβ(25-35) peptides: an investigation by molecular dynamics and well-tempered metadynamics simulations. ACS Chemical Neuroscience 12 (4), 660–674. Available at https://doi.org/10.1021/acschemneuro.0c00692.

Covelo, A., Eraso-Pichot, A., Fernández-Moncada, I., Serrat, R., Marsicano, G., 2021. CB1R-dependent regulation of astrocyte physiology and astrocyte-neuron interactions. Neuropharmacology 195, 108678. Available at https://doi.org/10.1016/j.neuropharm.2021.108678.

Cristino, L., Bisogno, T., Di Marzo, V., 2020. Cannabinoids and the expanded endocannabinoid system in neurological disorders. Nature Reviews Neurology 16 (1), 9–29. Available at https://doi.org/10.1038/s41582-019-0284-z.

Dembitskaya, Y., Popov, A., 2025. Astrocytes in maintaining neuronal health and brain function: interplay of aging, diet, and environment. Metabolic Brain Disease 40 (8). Available at https://doi.org/10.1007/s11011-025-01706-7.

Duffy, S.S., Hayes, J.P., Fiore, N.T., Moalem-Taylor, G., 2021. The cannabinoid system and microglia in health and disease. Neuropharmacology 190. Available at https://doi.org/10.1016/j.neuropharm.2021.108555.

Ehrlich, B., Patel, M., Zheng, X., Glass, M., 2026. Canonical cannabinoid receptors. In: Cooper, Z.D., Patel, S., Alexander, S.P.H. (Eds.), Behavioral Impact of Cannabinoids. Current Topics in Behavioral Neurosciences, vol 76. Springer, Cham, pp. 37–68. Available at https://doi.org/10.1007/7854_2024_556.

Ellert-Miklaszewska, A., Ciechomska, I.A., Kaminska, B., 2021. Synthetic cannabinoids induce autophagy and mitochondrial apoptotic pathways in human glioblastoma cells independently of deficiency in TP53 or PTEN tumor suppressors. Cancers 13 (3), 419. Available at https://doi.org/10.3390/cancers13030419.

Erustes, A.G., Abílio, V.C., Bincoletto, C., Piacentini, M., Pereira, G.J.S., Smaili, S.S., 2025. Cannabidiol induces autophagy via CB1 receptor and reduces α-synuclein cytosolic levels. Brain Research 1850. Available at https://doi.org/10.1016/j.brainres.2024.149414.

Etemad, L., Karimi, G., Alavi, M.S., Roohbakhsh, A., 2022. Pharmacological effects of cannabidiol by transient receptor potential channels. Life Sciences 300, 120582. Available at https://doi.org/10.1016/j.lfs.2022.120582.

Eubanks, L.M., Rogers, C.J., Beuscher IV, A.E., Koob, G.F., Olson, A.J., Dickerson, T.J., et al., 2006. A molecular link between the active component of marijuana and Alzheimer's disease pathology. Molecular Pharmaceutics 3 (6), 773–777. Available at https://doi.org/10.1021/mp060066m.

Feng, L., Lo, H., You, H., Wu, W., Cheng, X., Xin, J., et al., 2023. Loss of cannabinoid receptor 2 promotes α-synuclein-induced microglial synaptic pruning in nucleus accumbens by modulating the pCREB-c-Fos signaling pathway and complement system. Experimental Neurology 359, 114230. Available at https://doi.org/10.1016/j.expneurol.2022.114230.

Fernández-Moncada, I., Eraso-Pichot, A., Dalla Tor, T., Fortunato-Marsol, B., Marsicano, G., 2023. An enquiry to the role of CB1 receptors in neurodegeneration. Neurobiology of Disease 184, 106235. Available at https://doi.org/10.1016/j.nbd.2023.106235.

Fernández-Moncada, I., Rodrigues, R.S., Fundazuri, U.B., Bellocchio, L., Marsicano, G., 2024. Type-1 cannabinoid receptors and their ever-expanding roles in brain energy processes. Journal of Neurochemistry 168 (5), 693–703. Available at https://doi.org/10.1111/jnc.15922.

Fernández-Ruiz, J., Moro, M.A., Martínez-Orgado, J., 2015a. Cannabinoids in neurodegenerative disorders and stroke/brain trauma: from preclinical models to clinical applications. Neurotherapeutics 12 (4), 793–806. Available at https://doi.org/10.1007/s13311-015-0381-7.

Fernández-Ruiz, J., Romero, J., Ramos, J.A., 2015b. Endocannabinoids and neurodegenerative disorders: Parkinson's disease, Huntington's chorea, Alzheimer's disease, and others. Endocannabinoids. Springer International Publishing, Spain, pp. 233–259.

Fernández-Ruiz, J., Sagredo, O., Gómez-Ruiz, M., de Lago, E., 2026. Ageing, neurodegeneration and the endocannabinoid system. In: Cooper, Z.D., Patel, S., Alexander, S.P.H. (Eds.), Behavioral Impact of Cannabinoids. Current Topics in Behavioral Neurosciences, vol 76. Springer, Cham, pp. 209–248. Available at https://doi.org/10.1007/7854_2025_597.

Fernández-Ruiz, J., Gómez-Ruiz, M., García, C., Hernández, M., Ramos, J.A., 2017. Modeling neurodegenerative disorders for developing cannabinoid-based neuroprotective therapies. Methods in Enzymology 593, 175–198. Available at https://doi.org/10.1016/bs.mie.2017.06.021.

Fernández-Ruiz, J., Romero, J., Velasco, G., Tolón, R.M., Ramos, J.A., Guzmán, M., 2007. Cannabinoid CB2 receptor: a new target for controlling neural cell survival? Trends in Pharmacological Sciences 28 (1), 39–45. Available at https://doi.org/10.1016/j.tips.2006.11.001.

Fernández-Ruiz, J., Sagredo, O., Pazos, M.R., García, C., Pertwee, R., Mechoulam, R., et al., 2013. Cannabidiol for neurodegenerative disorders: important new clinical applications for this phytocannabinoid? British Journal of Clinical Pharmacology 75 (2), 323–333. Available at https://doi.org/10.1111/j.1365-2125.2012.04341.x.

Fernández-Ruiz, J., 2019. The biomedical challenge of neurodegenerative disorders: an opportunity for cannabinoid-based therapies to improve on the poor current therapeutic outcomes. British Journal of Pharmacology 176 (10), 1370–1383. Available at https://doi.org/10.1111/bph.14382.

Fontaine, S.N., Sabbagh, J.J., Baker, J., Martinez-Licha, C.R., Darling, A., Dickey, C.A., 2015. Cellular factors modulating the mechanism of tau protein aggregation. Cellular and Molecular Life Sciences 72 (10), 1863–1879. Available at https://doi.org/10.1007/s00018-015-1839-9.

Fujii, M., Sherchan, P., Krafft, P.R., Rolland, W.B., Soejima, Y., Zhang, J.H., 2014. Cannabinoid type 2 receptor stimulation attenuates brain edema by reducing cerebral leukocyte infiltration following subarachnoid hemorrhage in rats. Journal of the Neurological Sciences 342 (1-2), 101–106. Available at https://doi.org/10.1016/j.jns.2014.04.034.

Gertsch, J., Pertwee, R.G., Di Marzo, V., 2010. Phytocannabinoids beyond the Cannabis plant - Do they exist? British Journal of Pharmacology 160 (3), 523–529. Available at https://doi.org/10.1111/j.1476-5381.2010.00745.x.

Gil-Ordóñez, A., Martín-Fontecha, M., Ortega-Gutiérrez, S., López-Rodríguez, M.L., 2018. Monoacylglycerol lipase (MAGL) as a promising therapeutic target. Biochemical Pharmacology 157, 18–32. Available at https://doi.org/10.1016/j.bcp.2018.07.036.

Gómez-Cañas, M., Rodríguez-Cueto, C., Satta, V., Hernández-Fisac, I., Navarro, E., Fernández-Ruiz, J., 2023. Endocannabinoid-binding receptors as drug targets. In: Maccarrone, M. (Ed.), Endocannabinoid Signaling. Methods in Molecular Biology, vol 2576. Humana, New York, pp. 67–94. Available at https://doi.org/10.1007/978-1-0716-2728-0_6.

Gottschling, S., Ayonrinde, O., Bhaskar, A., Blockman, M., D'agnone, O., Schecter, D., et al., 2020. Safety considerations in cannabinoid-based medicine. International Journal of General Medicine 13, 1317–1333. Available at https://doi.org/10.2147/IJGM.S275049.

Gowran, A., Campbell, V.A., 2008. A role for p53 in the regulation of lysosomal permeability by Δ9-tetrahydrocannabinol in rat cortical neurones: implications for neurodegeneration. Journal of Neurochemistry 105 (4), 1513–1524. Available at https://doi.org/10.1111/j.1471-4159.2008.05278.x.

Gugliandolo, A., Pollastro, F., Bramanti, P., Mazzon, E., 2020. Cannabidiol exerts protective effects in an in vitro model of Parkinson's disease activating AKT/mTOR pathway. Fitoterapia 143, 104553. Available at https://doi.org/10.1016/j.fitote.2020.104553.

Gugliandolo, A., Blando, S., Salamone, S., Pollastro, F., Mazzon, E., D'Angiolini, S., 2023. Transcriptome highlights cannabinol modulation of mitophagy in a Parkinson's disease in vitro model. Biomolecules 13 (8), 1163. Available at https://doi.org/10.3390/biom13081163.

Guo, X., Li, F., Zhang, F., 2026. Structural and dynamic mechanisms of cannabinoid receptors. Biochemical Pharmacology 244, 117568. Available at https://doi.org/10.1016/j.bcp.2025.117568.

Hagan, K., Varelas, P., Zheng, H.A., 2022. Endocannabinoid system of the blood–brain barrier: current understandings and therapeutic potentials. Cannabis and Cannabinoid Research 7 (5), 561–568. Available at https://doi.org/10.1089/can.2021.0101.

Haj-Dahmane, S., Shen, R.Y., 2011. Modulation of the serotonin system by endocannabinoid signaling. Neuropharmacology 61 (3), 414–420. Available at https://doi.org/10.1016/j.neuropharm.2011.02.016.

Hamano, T., Enomoto, S., Shirafuji, N., Ikawa, M., Yamamura, O., Yen, S.-H., et al., 2021. Autophagy and Tau protein. International Journal of Molecular Sciences 22 (14), 7475. Available at https://doi.org/10.3390/ijms22147475.

Hao, F., Feng, Y., 2021. Cannabidiol (CBD) enhanced the hippocampal immune response and autophagy of APP/PS1 Alzheimer's mice uncovered by RNA-seq. Life Sciences 264, 118624. Available at https://doi.org/10.1016/j.lfs.2020.118624.

Hassan, S., Eldeeb, K., Millns, P.J., Bennett, A.J., Alexander, S.P.H., Kendall, D.A., 2014. Cannabidiol enhances microglial phagocytosis via transient receptor potential (TRP) channel activation. British Journal of Pharmacology 171 (9), 2426–2439. Available at https://doi.org/10.1111/bph.12615.

Hempel, B., Xi, Z.X., 2022. Receptor mechanisms underlying the CNS effects of cannabinoids: CB1 receptor and beyond. Advances in Pharmacology 93, 275–333. Available at https://doi.org/10.1016/bs.apha.2021.10.006.

Hiebel, C., Behl, C., 2015. The complex modulation of lysosomal degradation pathways by cannabinoid receptors 1 and 2. Life Sciences 138, 3–7. Available at https://doi.org/10.1016/j.lfs.2015.03.020.

Hiebel, C., Kromm, T., Stark, M., Behl, C., 2014. Cannabinoid receptor 1 modulates the autophagic flux independent of mTOR- and BECLIN1-complex. Journal of Neurochemistry 131 (4), 484–497. Available at https://doi.org/10.1111/jnc.12839.

Iannotti, F.A., Vitale, R.M., 2021. The endocannabinoid system and ppars: focus on their signalling crosstalk, action and transcriptional regulation. Cells 10 (3), 586. Available at https://doi.org/10.3390/cells10030586.

Ilyasov, A.A., Milligan, C.E., Pharr, E.P., Howlett, A.C., 2018. The endocannabinoid system and oligodendrocytes in health and disease. Frontiers in Neuroscience 12. Available at https://doi.org/10.3389/fnins.2018.00733.

Jiang, F., Xia, M., Zhang, Y., Chang, J., Cao, J., Zhang, Z., et al., 2022. Cannabinoid receptor-2 attenuates neuroinflammation by promoting autophagy-mediated degradation of the NLRP3 inflammasome post spinal cord injury. Frontiers in Immunology 13. Available at https://doi.org/10.3389/fimmu.2022.993168.

Joers, V., Murray, B.C., McLaughlin, C., Oliver, D., Staley, H.E., Coronado, J., et al., 2024. Modulation of cannabinoid receptor 2 alters neuroinflammation and reduces formation of alpha-synuclein aggregates in a rat model of nigral synucleinopathy. Journal of Neuroinflammation 21, 240. Available at https://doi.org/10.1186/s12974-024-03221-5.

Kaczocha, M., Haj-Dahmane, S., 2022. Mechanisms of endocannabinoid transport in the brain. British Journal of Pharmacology 179 (17), 4300–4310. Available at https://doi.org/10.1111/bph.15469.

Kataoka, K., Bilkei-Gorzo, A., Nozaki, C., Togo, A., Nakamura, K., Ohta, K., et al., 2020. Age-dependent alteration in mitochondrial dynamics and autophagy in hippocampal neuron of cannabinoid CB1 receptor-deficient Mice. Brain Research Bulletin 160, 40–49. Available at https://doi.org/10.1016/j.brainresbull.2020.03.014.

Kingsley, P.J., Rouzer, C.A., Morgan, A.J., Patel, S., Marnett, L.J., 2019. Aspects of prostaglandin glycerol ester biology. In: Honn, K., Zeldin, D. (Eds.), The Role of Bioactive Lipids in Cancer, Inflammation and Related Diseases. Advances in Experimental Medicine and Biology, vol 1161. Springer, Cham, pp. 77–88. Available at https://doi.org/10.1007/978-3-030-21735-8_8.

Lalmani, Y.S., Deshmukh, R., 2026. Understanding glial cells: implications for multiple sclerosis pathogenesis and therapeutics. Ageing Research Reviews 114, 102961. Available at https://doi.org/10.1016/j.arr.2025.102961.

Lawrence, J.M., Schardien, K., Wigdahl, B., Nonnemacher, M.R., 2023. Roles of neuropathology-associated reactive astrocytes: a systematic review. Acta Neuropathologica Communications 11 (1). Available at https://doi.org/10.1186/s40478-023-01526-9.

Liu, B., Tian, Y., Li, Y., Wu, P., Zhang, Y., Zheng, J., et al., 2022. ACEA attenuates oxidative stress by promoting mitophagy via CB1R/Nrf1/PINK1 pathway after subarachnoid hemorrhage in rats. Oxidative Medicine and Cellular Longevity 2022 (1). Available at https://doi.org/10.1155/2022/1024279.

Long, Cm, Zheng, Qx, Zhou, Y., Liu, Yt, Gong, Lp, Zeng, Yc, et al., 2021. N-linoleyltyrosine exerts neuroprotective effects in APP/PS1 transgenic mice via cannabinoid receptor-mediated autophagy. Journal of Pharmacological Sciences 147 (4), 315–324. Available at https://doi.org/10.1016/j.jphs.2021.08.008.

Maccarrone, M., Bab, I., Bíró, T., Cabral, G.A., Dey, S.K., Di Marzo, V., et al., 2015. Endocannabinoid signaling at the periphery: 50 years after THC. Trends in Pharmacological Sciences 36 (5), 277–296. Available at https://doi.org/10.1016/j.tips.2015.02.008.

Maccarrone, M., Di Marzo, V., Gertsch, J., Grether, U., Howlett, A.C., Hua, T., et al., 2023. Goods and bads of the endocannabinoid system as a therapeutic target: lessons learned after 30 Years. Pharmacological Reviews 75 (5), 885–958. Available at https://doi.org/10.1124/pharmrev.122.000600.

Mao, Y., Huang, Y., Zhang, Y., Wang, C., Wu, H., Tian, X., et al., 2019. Cannabinoid receptor 2-selective agonist JWH015 attenuates bone cancer pain through the amelioration of impaired autophagy flux induced by inflammatory mediators in the spinal cord. Molecular Medicine Reports 20 (6), 5100–5110. Available at https://doi.org/10.3892/mmr.2019.10772.

Marsicano, G., Moosmann, B., Hermann, H., Lutz, B., Behl, C., 2002. Neuroprotective properties of cannabinoids against oxidative stress: role of the cannabinoid receptor CB1. Journal of Neurochemistry 80 (3), 448–456. Available at https://doi.org/10.1046/j.0022-3042.2001.00716.x.

Marsicano, G., Goodenough, S., Monory, K., Hermann, H., Eder, M., Cannich, A., et al., 2003. CB1 Cannabinoid receptors and on-demand defense against excitotoxicity. Science 302 (5642), 84–88. Available at https://doi.org/10.1126/science.1088208.

Mecha, M., Carrillo-Salinas, F.J., Feliú, A., Mestre, L., Guaza, C., 2016. Microglia activation states and cannabinoid system: therapeutic implications. Pharmacology & Therapeutics 166, 40–55. Available at https://doi.org/10.1016/j.pharmthera.2016.06.011.

Mechoulam, R., Hanuš, L.O., Pertwee, R., Howlett, A.C., 2014. Early phytocannabinoid chemistry to endocannabinoids and beyond. Nature Reviews Neuroscience 15 (11), 757–764. Available at https://doi.org/10.1038/nrn3811.

Miao, Y., Zhao, F., Guan, W., 2025. A novel insight into the antidepressant effect of cannabidiol: possible involvement of the 5-HT1A, CB1, GPR55, and PPARγ receptors. International Journal of Neuropsychopharmacology 28 (2). Available at https://doi.org/10.1093/ijnp/pyae064.

Mock, E.D., Gagestein, B., van der Stelt, M., 2023. Anandamide and other N-acylethanolamines: a class of signaling lipids with therapeutic opportunities. Progress in Lipid Research 89, 101194. Available at https://doi.org/10.1016/j.plipres.2022.101194.

Mohammad, Z.B., Yudin, S.C.Y., Goldberg, B.J., Serra, K.L., Klegeris, A., 2025. Exploring neuroglial signaling: diversity of molecules implicated in microglia-to-astrocyte neuroimmune communication. Reviews in the Neurosciences 36 (1), 91–117. Available at https://doi.org/10.1515/revneuro-2024-0081.

Molina-Holgado, F., Pinteaux, E., Moore, J.D., Molina-Holgado, E., Guaza, C., Gibson, R.M., et al., 2003. Endogenous interleukin-1 receptor antagonist mediates anti-inflammatory and neuroprotective actions of cannabinoids in neurons and glia. Journal of Neuroscience 23 (16), 6470–6474. Available at https://doi.org/10.1523/jneurosci.23-16-06470.2003.

Moreno-García, Á., Bernal-Chico, A., Colomer, T., Rodríguez-Antigüedad, A., Matute, C., Mato, S., 2020. Gene expression analysis of astrocyte and microglia endocannabinoid signaling during autoimmune demyelination. Biomolecules 10 (9), 1228. Available at https://doi.org/10.3390/biom10091228.

Nicolussi, S., Gertsch, J., 2015. Endocannabinoid transport revisited. Vitamins and Hormones 98, 441–485. Available at https://doi.org/10.1016/bs.vh.2014.12.011.

Nixon, R.A., 2024. Autophagy-lysosomal-associated neuronal death in neurodegenerative disease. Acta Neuropathologica 148 (1), 142. Available at 10.1007/s00401-024-02799-7.

Nixon, R.A., Rubinsztein, D.C., 2024. Mechanisms of autophagy–lysosome dysfunction in neurodegenerative diseases. Nature Reviews Molecular Cell Biology 25 (11), 926–946. Available at https://doi.org/10.1038/s41580-024-00757-5.

Noonan, J., Tanveer, R., Klompas, A., Gowran, A., McKiernan, J., Campbell, V.A., 2010. Endocannabinoids prevent β-amyloid-mediated lysosomal destabilization in cultured neurons. Journal of Biological Chemistry 285 (49), 38543–38554. Available at https://doi.org/10.1074/jbc.M110.162040.

Oddi, S., Fiorenza, M.T., Maccarrone, M., 2023. Endocannabinoid signaling in adult hippocampal neurogenesis: a mechanistic and integrated perspective. Progress in Lipid Research 91, 101239. Available at https://doi.org/10.1016/j.plipres.2023.101239.

O'Sullivan, S.E., 2016. An update on PPAR activation by cannabinoids. British Journal of Pharmacology 173 (12), 1899–1910. Available at https://doi.org/10.1111/bph.13497.

Pacher, P., Mechoulam, R., 2011. Is lipid signaling through cannabinoid 2 receptors part of a protective system? Progress in Lipid Research 50 (2), 193–211. Available at https://doi.org/10.1016/j.plipres.2011.01.001.

Pertwee, R.G., 2008. The diverse CB 1 and CB 2 receptor pharmacology of three plant cannabinoids: Δ 9-tetrahydrocannabinol, cannabidiol and Δ 9-tetrahydrocannabivarin. British Journal of Pharmacology 153 (2), 199–215. Available at https://doi.org/10.1038/sj.bjp.0707442.

Piette, C., Cui, Y., Gervasi, N., Venance, L., 2020. Lights on endocannabinoid-mediated synaptic potentiation. Frontiers in Molecular Neuroscience 13. Available at https://doi.org/10.3389/fnmol.2020.00132.

Piyanova, A., Albayram, O., Rossi, C.A., Farwanah, H., Michel, K., Nicotera, P., et al., 2013. Loss of CB1 receptors leads to decreased cathepsin D levels and accelerated lipofuscin accumulation in the hippocampus. Mechanisms of Ageing and Development 134 (9), 391–399. Available at https://doi.org/10.1016/j.mad.2013.08.001.

Prenderville, J.A., Kelly, Á.M., Downer, E.J., 2015. The role of cannabinoids in adult neurogenesis. British Journal of Pharmacology 172 (16), 3950–3963. Available at https://doi.org/10.1111/bph.13186.

Ranieri, R., Laezza, C., Bifulco, M., Marasco, D., Malfitano, A.M., 2016. Endocannabinoid system in neurological disorders. Recent Patents on CNS Drug Discovery 10 (2), 90–112. Available at https://doi.org/10.2174/1574889810999160719105433.

Rathod, S.S., Agrawal, Y.O., Nakhate, K.T., Meeran, M.F.N., Ojha, S., Goyal, S.N., 2023. Neuroinflammation in the central nervous system: exploring the evolving influence of endocannabinoid system. Biomedicines 11 (10), 2642. Available at https://doi.org/10.3390/biomedicines11102642.

Rees, R.N., Noyce, A.J., Schrag, A.E., 2024. Identification of prodromal parkinson disease we may be able to but should we? Neurology 102 (11), e209394. Available at https://doi.org/10.1212/WNL.0000000000209394.

Rida Zainab, S., Zeb Khan, J., Khalid Tipu, M., Jahan, F., Irshad, N., 2025. A review on multiple sclerosis: unravelling the complexities of pathogenesis, progression, mechanisms and therapeutic innovations. Neuroscience 567, 133–149. Available at https://doi.org/10.1016/j.neuroscience.2024.12.029.

Rodríguez, J.J., Mackie, K., Pickel, V.M., 2001. Ultrastructural localization of the CB1 cannabinoid receptor in μ-opioid receptor patches of the rat caudate putamen nucleus. Journal of Neuroscience 21 (3), 823–833. Available at https://doi.org/10.1523/jneurosci.21-03-00823.2001.

Rodríguez-Cueto, C., García-Toscano, L., Santos-García, I., Gómez-Almería, M., Gonzalo-Consuegra, C., Espejo-Porras, F., et al., 2021. Targeting the CB2 receptor and other endocannabinoid elements to delay disease progression in amyotrophic lateral sclerosis. British Journal of Pharmacology 178 (6), 1373–1387. Available at https://doi.org/10.1111/bph.15386.

Rozenfeld, R., Devi, L.A., 2008. Regulation of CB1 cannabinoid receptor trafficking by the adaptor protein AP-3. FASEB Journal 22 (7), 2311–2322. Available at https://doi.org/10.1096/fj.07-102731.

Ruiz de Martín Esteban, S., Benito-Cuesta, I., Terradillos, I., Martínez-Relimpio, A.M., Arnanz, M.A., Ruiz-Pérez, G., et al., 2022. Cannabinoid CB2 receptors modulate microglia function and amyloid dynamics in a mouse model of Alzheimer's Disease. Frontiers in Pharmacology 13. Available at https://doi.org/10.3389/fphar.2022.841766.

Ruiz de Martín Esteban, S., Grande, M.T., Martínez-Relimpio, A.M., Herráez-Aguilar, D., Mostany, R., Hillard, C.J., et al., 2026. Treatment with a botanical mixture of cannabidiol:Δ9-tetrahydrocannabinol enhances microglial phagocytosis and shapes amyloid plaques in a mouse model of Alzheimer's disease. Biomedicine & Pharmacotherapy 194, 118902. Available at https://doi.org/10.1016/j.biopha.2025.118902.

Sánchez-Blázquez, P., Rodríguez-Muñoz, M., Vicente-Sánchez, A., Garzón, J., 2013. Cannabinoid receptors couple to nmda receptors to reduce the production of NO and the mobilization of zinc induced by glutamate. Antioxidants & Redox Signaling 19 (15), 1766–1782. Available at https://doi.org/10.1089/ars.2012.5100.

Shafi, A., Akmal, M., Sethi, A., Chauhdary, Z., 2025. A mechanistic insight of neuro-inflammation signaling pathways and implication in neurodegenerative disorders. Inflammopharmacology 34, 309–318. Available at https://doi.org/10.1007/s10787-025-02041-0.

Shao, B.Z., Wei, W., Ke, P., Xu, Z.Q., Zhou, J.X., Liu, C., 2014. Activating cannabinoid receptor 2 alleviates pathogenesis of experimental autoimmune encephalomyelitis via activation of autophagy and inhibiting NLRP3 inflammasome. CNS Neuroscience and Therapeutics 20 (12), 1021–1028. Available at https://doi.org/10.1111/cns.12349.

Shrivastava, A., Kuzontkoski, P.M., Groopman, J.E., Prasad, A., 2011. Cannabidiol induces programmed cell death in breast cancer cells by coordinating the cross-talk between apoptosis and autophagy. Molecular Cancer Therapeutics 10 (7), 1161–1172. Available at https://doi.org/10.1158/1535-7163.MCT-10-1100.

Simard, M., Archambault, A.-S., Lavoie, J.-P.C., Dumais, É., Di Marzo, V., Flamand, N., 2022. Biosynthesis and metabolism of endocannabinoids and their congeners from the monoacylglycerol and N-acyl-ethanolamine families. Biochemical Pharmacology 205, 115261. Available at https://doi.org/10.1016/j.bcp.2022.115261.

Sirbu, C.A., Georgescu, R., Pleşa, F.C., Paunescu, A., Marilena Ţânţu, M., Nicolae, A.C., et al., 2023. Cannabis and cannabinoids in multiple sclerosis: from experimental models to clinical practice - a review. American Journal of Therapeutics 30 (3). Available at https://doi.org/10.1097/MJT.0000000000001568 E220.

Spangenberg, E., Severson, P.L., Hohsfield, L.A., Crapser, J., Zhang, J., Burton, E.A., et al., 2019. Sustained microglial depletion with CSF1R inhibitor impairs parenchymal plaque development in an Alzheimer's disease model. Nature Communications 10 (1). Available at https://doi.org/10.1038/s41467-019-11674-z.

Stanley, C., O'Sullivan, S.E., 2014. Vascular targets for cannabinoids: animal and human studies. British Journal of Pharmacology 171 (6), 1361–1378. Available at https://doi.org/10.1111/bph.12560.

Stella, N., 2010. Cannabinoid and cannabinoid-like receptors in microglia, astrocytes, and astrocytomas. Glia 58 (9), 1017–1030. Available at https://doi.org/10.1002/glia.20983.

Sugaya, K., 2005. Possible use of autologous stem cell therapies for Alzheimers disease. Current Alzheimer Research 2 (3), 367–376. Available at https://doi.org/10.2174/1567205054367919.

Tanaka, M., Sackett, S., Zhang, Y., 2020. Endocannabinoid modulation of microglial phenotypes in neuropathology. Frontiers in Neurology 11. Available at https://doi.org/10.3389/fneur.2020.00087.

Urquhart, P., Nicolaou, A., Woodward, D.F., 2015. Endocannabinoids and their oxygenation by cyclo-oxygenases, lipoxygenases and other oxygenases. Biochimica et Biophysica Acta (BBA) - Molecular and Cell Biology of Lipids 1851 (4), 366–376. Available at https://doi.org/10.1016/j.bbalip.2014.12.015.

Valeri, A., Mazzon, E., 2021. Cannabinoids and neurogenesis: the promised solution for neurodegeneration? Molecules 26 (20), 6313. Available at https://doi.org/10.3390/molecules26206313.

van der Wel, P.C.A., 2024. Solid-state nuclear magnetic resonance in the structural study of polyglutamine aggregation. Biochemical Society Transactions 52 (2), 719–731. Available at https://doi.org/10.1042/BST20230731.

Vendel, E., de Lange, E.C.M., 2014. Functions of the CB1 and CB2 receptors in neuroprotection at the level of the blood–brain barrier. NeuroMolecular Medicine 16 (3), 620–642. Available at https://doi.org/10.1007/s12017-014-8314-x.

Verde, E.M., Secco, V., Ghezzi, A., Mandrioli, J., Carra, S., 2025. Molecular mechanisms of protein aggregation in ALS-FTD: focus on TDP-43 and cellular protective responses. Cells 14 (10), 680. Available at https://doi.org/10.3390/cells14100680.

Verkerke, M., Werkman, M.H., Donega, V., 2025. Neural stem cells of the subventricular zone: a potential stem cell pool for brain repair in Parkinson's disease. Stem Cell Reports 20 (9), 102533. Available at https://doi.org/10.1016/j.stemcr.2025.102533.

Vida, H., Sahar, M., Nikdouz, A., Arezoo, H., 2025. Chemokines in neurodegenerative diseases. Immunology & Cell Biology 103 (3), 275–292. Available at https://doi.org/10.1111/imcb.12843.

Vrechi, T.A.M., Leão, A.H.F.F., Morais, I.B.M., Abílio, V.C., Zuardi, A.W., Hallak, J.E.C., et al., 2021. Cannabidiol induces autophagy via ERK1/2 activation in neural cells. Scientific reports 11 (1), 5434. Available at https://doi.org/10.1038/s41598-021-84879-2.

Vrechi, T.A.M., Guarache, G.C., Oliveira, R.B., Guedes, Ed.C., Erustes, A.G., Leão, A.H.F.F., et al., 2025. Cannabidiol-induced autophagy ameliorates tau protein clearance. Neurotoxicity Research 43 (1). Available at https://doi.org/10.1007/s12640-025-00729-3.

Wang, Z., Zheng, P., Chen, X., Xie, Y., Weston-Green, K., Solowij, N., et al., 2022. Cannabidiol induces autophagy and improves neuronal health associated with SIRT1 mediated longevity. GeroScience 44 (3), 1505–1524. Available at https://doi.org/10.1007/s11357-022-00559-7.

Whyte, L.S., Ryberg, E., Sims, N.A., Ridge, S.A., Mackie, K., Greasley, P.J., et al., 2009. The putative cannabinoid receptor GPR55 affects osteoclast function in vitro and bone mass in vivo. Proceedings of the National Academy of Sciences of the United States of America 106 (38), 16511–16516. Available at https://doi.org/10.1073/pnas.0902743106.

Winters, B.L., Vaughan, C.W., 2021. Mechanisms of endocannabinoid control of synaptic plasticity. Neuropharmacology 197, 108736. Available at https://doi.org/10.1016/j.neuropharm.2021.108736.

Wu, Q., Zhang, M., Liu, X., Zhang, J., Wang, H., 2020. CB2R orchestrates neuronal autophagy through regulation of the mTOR signaling pathway in the hippocampus of developing rats with status epilepticus. International Journal of Molecular Medicine 45 (2), 475–484. Available at https://doi.org/10.3892/ijmm.2019.4439.

Yang, S., Du, Y., Zhao, X., Tang, Q., Su, W., Hu, Y., et al., 2022. Cannabidiol enhances microglial beta-amyloid peptide phagocytosis and clearance via vanilloid family type 2 channel activation. International Journal of Molecular Sciences 23 (10), 5367. Available at https://doi.org/10.3390/ijms23105367.

Yang, Z., Klionsky, D.J., 2010. Mammalian autophagy: core molecular machinery and signaling regulation. Current Opinion in Cell Biology 22 (2), 124–131. Available at https://doi.org/10.1016/j.ceb.2009.11.014.

Zhou, Y., Li, Zx, Liu, Yt, Xu, Z.C., Hu, Y., Lv, W., et al., 2022. N-linoleyltyrosine protects neurons against Aβ1–40-induced cell toxicity via autophagy involving the CB2/AMPK/mTOR/ULK1 pathway. Brain Research Bulletin 188, 203–213. Available at https://doi.org/10.1016/j.brainresbull.2022.08.002.

Zhu, H., Xiao, F., Xiao, Y., Guo, Y., Shan, X., Zhang, Z., et al., 2023. Targeting CB2R in astrocytes for Parkinson's disease therapy: unraveling the Foxg1-mediated neuroprotective mechanism through autophagy-mediated NLRP3 degradation. Journal of Neuroinflammation 20 (1). Available at https://doi.org/10.1186/s12974-023-02989-2.

Ziemka-Nalecz, M., Zalewska, T., 2012. Endogenous neurogenesis induced by ischemic brain injury or neurodegenerative diseases in adults. Acta Neurobiologiae Experimentalis 72 (4), 309–324. Available at https://doi.org/10.55782/ane-2012-1904.

Zoppi, S., Madrigal, J.L., Caso, J.R., García-Gutiérrez, M.S., Manzanares, J., Leza, J.C., et al., 2014. Regulatory role of the cannabinoid CB2 receptor in stress-induced neuroinflammation in mice. British Journal of Pharmacology 171 (11), 2814–2826. Available at https://doi.org/10.1111/bph.12607.

CHAPTER 11

Role of autophagy in aging and aging-related neurodegenerative diseases

Weidong Le and Panpan Wang
Key Laboratory of Liaoning Province for Research on the Pathogenic Mechanisms of Neurological Diseases, First Affiliated Hospital of Dalian Medical University, Dalian, Liaoning, P.R. China

11.1 Introduction

A key initial step in aging research was the observation in 1939 by McCay CM that caloric restriction increased the lifespan of mice and rats. In 1952, Peter Medawar's "mutation accumulation theory" proposed that aging results from the declining force of natural selection after reproduction. This theory spurred biologists to actively research genetic factors by testing early- and late-reproducing fly populations. A landmark study in 1988 discovered that mutations in a single gene, *age-1*, could extend lifespan in *C. elegans*, granting the biology of aging unprecedented scientific credibility. To date, over 800 genes that modulate lifespan have been identified in *C. elegans*, with additional genes continually being discovered under various environmental conditions. As research in the genetics of aging has advanced, a complex network of interacting intracellular signaling pathways has been revealed. These include the insulin signaling pathway, the target of rapamycin (TOR) signaling, sirtuins and NAD+, the circadian clock, mitochondrial function, oxidative stress, chronic inflammation, and proteostasis.

As early as the mid-1900s, it was recognized that the incidence of chronic diseases, such as cardiovascular diseases, neurodegenerative diseases, and macular degeneration, was age-related, though credible evidence was lacking. Over the past 30 years, the development and application of genetic technologies have led to the discovery of more regulatory mechanisms involving intracellular signaling molecules. Notably, many pathways that regulate aging overlap with those involved in disease pathology. For instance, the TOR and insulin pathways have been shown to regulate aging and age-related diseases. Inhibiting either the TOR or insulin pathway can extend the lifespan of model animals, and simultaneous inhibition of both pathways has been shown to extend lifespan by nearly fivefold (Chen et al., 2013). Rapamycin, a TOR inhibitor originally discovered for its potent antifungal properties and later known to inhibit cell growth and act as an immune modulator, is now in clinical trials for lifespan extension. Similarly, metformin, a classic antidiabetic drug, has been shown to significantly reduce the incidence of cancer and neurodegenerative diseases.

Molecular Insights into Proteinopathies
DOI: https://doi.org/10.1016/B978-0-443-34176-2.00004-3
© 2027 Elsevier Inc. All rights are reserved, including those for text and data mining, AI training, and similar technologies.

In Mitchener et al. (1976) discovered that deprivation of serum and amino acids in vitro induces autophagy, a catabolic mechanism that delivers cytoplasmic contents to lysosomes for degradation to maintain energy homeostasis and protect cells against stress, while feeding inhibits this process in vivo (Mortimore et al., 1983). These findings recalled the 1939 discovery by McCay CM that dietary restriction extends the lifespan of rats, suggesting a potential link between autophagy activation and longevity. In 1995, rapamycin was found to activate autophagy by inhibiting mechanistic TOR (mTOR), solidifying the close connection between autophagy and aging. Since then, mTOR and insulin signaling have emerged as key, parallel, yet interacting, conserved nutrient-sensing pathways, making them attractive candidates for mediating the switch from growth to maintenance triggered by dietary restriction across species. Autophagy serves as an energy compensation mechanism that appears to explain an evolutionary hypothesis, which proposes that dietary restriction prompts cells to switch from a growth mode to a survival maintenance mode, thereby extending lifespan.

With the expansion and deepening of autophagy research, it became clear that autophagy is a highly evolutionarily conserved process essential for cellular life. It plays a crucial role in maintaining intracellular homeostasis, especially during nutrient deprivation, and is closely related to various human pathophysiological states, including neurodegenerative diseases, diabetes, cancer, cardiovascular diseases, and immune responses. In this chapter, we will focus on the role of autophagy in aging and in neurodegenerative diseases, whose incidence increases with age. Autophagy-targeted treatment strategies will also be discussed.

11.2 Autophagy and aging

11.2.1 The definition of autophagy

Autophagy is a fundamental catabolic process that delivers cytoplasmic contents to lysosomes for degradation and recycling. There are three primary types of autophagy, classified based on the mechanism of delivery to the lysosome: microautophagy, chaperone-mediated autophagy (CMA), and macroautophagy (commonly referred to simply as autophagy). In microautophagy, cytosolic components are directly engulfed by invagination or protrusions of the lysosomal membrane. CMA involves the selective recognition and translocation of cytoplasmic proteins bearing a KFERQ peptide motif across the lysosomal membrane for degradation. In macroautophagy, the cytoplasmic cargo, including proteins and damaged organelles, is sequestered by double-membrane, cup-shaped structures called phagophores. These phagophores expand and enclose the cargo to form autophagosomes, which subsequently fuse with lysosomes to form autolysosomes, where the enclosed material is degraded (Fig. 11.1). Among these three types, macroautophagy (hereafter referred to as autophagy) is the most extensively

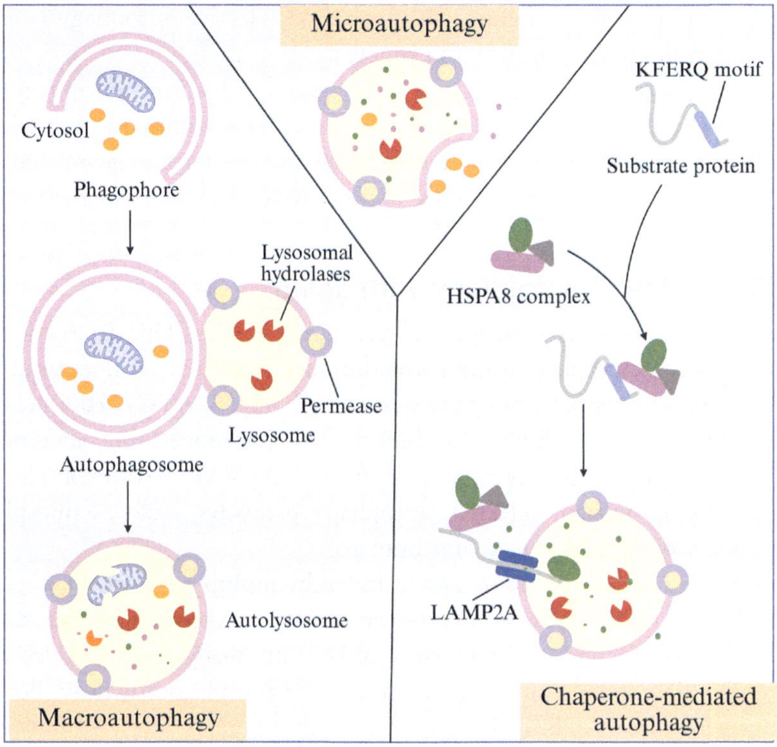

Figure 11.1 The three types of autophagy in mammalian cells. Macroautophagy relies on *de novo* formation of cytosolic double-membrane vesicles, autophagosomes, to sequester and transport cargo to the lysosome. Chaperone-mediated autophagy transports individual unfolded proteins directly across the lysosomal membrane. Microautophagy involves the direct uptake of cargo through invagination of the lysosomal membrane. All three types of autophagy lead to degradation of cargo and release of the breakdown products back into the cytosol for reuse by the cell.

studied. Autophagy occurs at a basal level under normal conditions and is further induced by various stresses, such as nutrient or energy deprivation. This induction allows cells to degrade cytoplasmic material into metabolites for biosynthetic processes or energy production, thereby promoting survival. Under homeostatic conditions, autophagy contributes to cellular maintenance by selectively degrading damaged or superfluous organelles. Thus autophagy is primarily a cytoprotective mechanism.

Based on the selectivity to substrates, autophagy can be categorized into selective and nonselective forms of autophagy. Nonselective autophagy (often referred to as bulk autophagy), induced by stimuli like nutrient starvation, lacks cargo specificity. In contrast, selective autophagy targets specific cargoes for degradation, such as aggregate-prone proteins (aggrephagy), intracellular pathogens (xenophagy), damaged mitochondria (mitophagy) and lysosomes (lysophagy), excess peroxisomes (pexophagy), and dysfunctional endoplasmic reticulum ([ER]-phagy). Selective autophagy is particularly crucial for organelle quality control and renewal.

Since its initial discovery in yeast as a survival mechanism, autophagy has been recognized as critical for diverse biological processes. Perturbations in both selective and nonselective autophagy have been linked to aging and numerous age-related disorders, including inflammation, cancer, diabetes, and neurodegenerative diseases.

11.2.2 The decline of autophagy with aging

Autophagic capacity declines with age across multiple species. This decline is evidenced by a reduced rate of autophagosome formation, accompanied by decreased transcript and protein levels of essential autophagy genes (e.g., ATG5, ATG7, BECN1) in tissues such as the brain, muscle, liver, and blood cells. Moreover, the accumulation of autophagic vacuoles in aged organisms indicates impaired lysosomal clearance. In addition to bulk autophagy, selective autophagy pathways, such as mitophagy, also decline in aged worms, flies, mice, and humans.

The age-related decline in autophagy is driven by multiple factors. First, genetic and epigenetic changes lead to reduced expression of autophagy-related genes, diminishing autophagic capacity. Second, the expression of key autophagy regulators decreases with age, weakening the initiation and maintenance of the autophagy pathway. For example, the autophagy regulator TP53INP2 decreases with age in mouse and human skeletal muscle, inhibiting the pathway and contributing to age-related muscle atrophy (Sebastián et al., 2024). Third, increased oxidative stress during aging, largely due to mitochondrial dysfunction, disrupts redox balance (e.g., the glutathione system) and impairs autophagy, a phenomenon particularly observed in T cells (Bharath and Nikolajczyk, 2020). Oxidative stress can also interfere with autophagy by affecting signaling molecules like FoxO transcription factors (Penniman et al., 2023). Fourth, defects in intracellular trafficking and vesicle positioning prevent the perinuclear relocation of autophagosomes and lysosomes in aged cells, inhibiting their fusion and subsequent substrate degradation. Finally, a vicious cycle involving mitochondrial dysfunction exacerbates autophagy decline. Aging leads to mitochondrial deterioration, increasing reactive oxygen species (ROS) production, causing oxidative stress, which further impairs autophagy. This failure in mitochondrial quality control is especially detrimental in postmitotic tissues like the brain and the heart.

11.2.3 Positive autophagy activity extends the lifespan

A strong correlation exists between robust autophagy and longevity. Impaired autophagy shortens lifespan, as demonstrated by studies where deletion or knockdown of core autophagy genes (e.g., ATG1, ATG5, ATG7, ATG12, Beclin1, ATG18) reduces lifespan in model organisms. This shortened lifespan is often associated with the accumulation of toxic protein aggregates, particularly in neurons, leading to neurodegeneration.

Conversely, enhancing autophagy promotes longevity. Overexpression of autophagy genes such as ATG5 extends lifespan in flies and mice (Pyo et al., 2013). Similarly, upregulating mitophagy, either genetically or pharmacologically, improves lifespan in worms. Overexpression of the TFEB orthologue HLH-30 in *C. elegans* extends lifespan in an autophagy-dependent manner (Lapierre et al., 2013). In mice, disrupting the inhibitory interaction between Beclin1 and BCL2 enhances autophagy, increases lifespan, and reduces age-related pathology in the heart and kidneys (Fernández et al., 2018b). A study published in Nature highlighted the importance of circadian regulation, showing that night-specific, but not day-specific, induction of autophagy is necessary and sufficient to extend the lifespan of *Drosophila* on an *ad libitum* diet (Ulgherait et al., 2021).

11.2.4 Therapeutic potential of autophagy in aging

Enhanced autophagy is a promising strategy to counteract age-related physiological decline and extend lifespan and healthspan. Genetic interventions that boost autophagy confer an "anti-aging" phenotype. For instance, mice overexpressing ATG5 exhibit improved insulin sensitivity and a 17% extension in median lifespan compared to wild-type mice (Pyo et al., 2013). Similarly, mice with a Beclin1 mutation (Beclin1$^{F121A/F121A}$) that disrupts its interaction with the negative regulator BCL2 display elevated basal autophagic flux, reduced age-related pathologies (e.g., in the heart and kidneys), and decreased spontaneous tumorigenesis, resulting in a significantly longer lifespan (Fernández et al., 2018a). Furthermore, this Beclin1 mutation rescues premature aging phenotypes, including premature lethality and infertility, in klotho-deficient mice (Fernández et al., 2018a).

Pharmacological activation of autophagy also extends lifespan. The National Institute on Aging's Intervention Testing Program has identified several compounds that reproducibly increase lifespan in genetically heterogeneous mice, including rapamycin, acarbose, nordihydroguaiaretic acid, 17-α-estradiol, and aspirin. Other studies have found that metformin, drugs targeting the renin-angiotensin-aldosterone system, and sirtuin activators (SRT2104, SRT1720) extend rodent lifespan. Notably, several drug classes with geroprotective potential are now being tested in humans, including metformin, rapamycin, and its analogs (rapalogs), senolytics (which clear senescent cells), sirtuin activators, and NAD+ precursors. Among these, metformin, rapamycin, and rapalogs are known to activate autophagy, suggesting this mechanism contributes to lifespan extension.

11.3 Autophagy in aging-related neurodegenerative diseases

Intracellular protein misfolding and aggregation are hallmark features of neurodegenerative diseases, which are consequently classified as proteinopathies. The postmitotic nature of neurons, combined with the age-related decline in autophagic capacity, makes them particularly vulnerable to accumulating damaged organelles and toxic protein aggregates. Consequently, autophagic dysfunction is implicated in the pathogenesis of many neurodegenerative disorders, making the modulation of autophagy a promising therapeutic avenue. This chapter will describe the evidence for autophagy impairment in major age-related neurodegenerative diseases, including Alzheimer's disease (AD), Parkinson's disease (PD), Huntington's disease (HD), and amyotrophic lateral sclerosis (ALS), and discuss the potential of autophagy upregulation as a treatment strategy.

11.3.1 Autophagy and Alzheimer's disease

AD is the most common neurodegenerative disorder, appearing sporadically in the elderly population above 80 years of age. Familial cases have much earlier onset, even below the age of 60. Its core clinical features are progressive cognitive dysfunction and behavioral impairment. The risk of developing AD increases significantly with age; prevalence doubles approximately every 6.1 years, reaching 20%–30% in individuals over 85. The defining neuropathological hallmarks of AD are the accumulation of intracellular neurofibrillary tangles (NFTs) (composed of hyperphosphorylated tau) and extracellular amyloid plaques. AD remains difficult to treat. Conventional treatments such as cholinesterase inhibitors (e.g., tacrine, donepezil, and galantamine) can alleviate symptoms but do not slow disease progression. More recently, monoclonal antibodies such as Aduhelm and Lecanemab have been developed to treat mild cognitive impairment in AD by clearing amyloid plaques, representing a breakthrough by targeting a putative etiological source. The clinical use of these antibodies validates the clearance of amyloid plaques as a viable therapeutic strategy. Given that autophagy is the primary pathway for degrading misfolded proteins and large protein aggregates within neurons, it plays a crucial role in the pathogenesis, progression, and potential treatment of AD. Therefore regulating autophagy may represent one of the most promising therapeutic avenues for AD.

11.3.1.1 The role of autophagy in amyloid-β metabolism

The two primary pathological hallmarks of AD are the extracellular deposition of amyloid-β (Aβ) plaques and the intracellular accumulation of hyperphosphorylated tau tangles. Aβ is generated by sequential cleavage of type I transmembrane amyloid precursor protein (APP) by β- and γ-secretases. Cleavage by β-site APP-cleaving enzyme produces a soluble APP fragment and an intracellular C-terminal fragment (βCTF). Subsequently, γ-secretase cleaves βCTF to generate Aβ peptides, primarily Aβ40 and the more neurotoxic Aβ42. Intracellular generation and release of Aβ are a major source of the diffuse Aβ and fibrils found extracellularly. Intracellular Aβ is neurotoxic prior to secretion, and after release, soluble extracellular Aβ can bind cell surface receptors, disrupt membrane integrity, or impair lysosomal function following cellular uptake. Aβ levels are increased in AD mouse models, and mutations in APP or the γ-secretase subunits presenilin 1 (PS1) and PS2 cause early-onset familial AD (fAD).

A complex interplay exists between Aβ and autophagy, as autophagy controls both the generation and clearance of Aβ. On one hand, upregulation of autophagy can reduce Aβ accumulation and rescue neurodegeneration. On the other hand, autophagosomes may also contribute to Aβ production under pathological conditions. In aging or disease states, autophagosomes have been observed to contain APP, its cleavage intermediate βCTF, γ-secretase (including PS1), and newly generated Aβ. For example, one study reported enhanced autophagy and increased γ-secretase-mediated APP cleavage in the brains of hypoxia-treated APP/PS1 transgenic mice (Li et al., 2009). The increase in autophagic vacuoles provides additional compartments for Aβ generation, especially as the induction of autophagy can traffic γ-secretase complexes from endosomes and the ER to these acidic vacuoles, where Aβ readily oligomerizes and forms fibrils.

The autophagy-lysosome pathway is also critical for the secretion of Aβ peptides. Quantitative analysis shows that extracellular Aβ secretion is reduced by 90% in autophagy-deficient mice. Conditional knockout of the essential autophagy gene ATG7 in APP transgenic mice resulted in less extracellular Aβ secretion and fewer plaques, but was associated with toxic intracellular Aβ aggregation, neurodegeneration, and memory deficits (Nilsson et al., 2013). Conversely, in CRND8 transgenic AD mice with impaired autophagosome-lysosome fusion, knockout of cystatin B (an endogenous inhibitor of lysosomal proteases) restored autolysosomal function and significantly reduced the deposition of Aβ40, Aβ42, and extracellular senile plaques.

11.3.1.2 The role of autophagy in tau metabolism

The other major pathological feature of AD is the formation of NFTs primarily composed of hyperphosphorylated tau protein. NFTs are toxic to neurons. Tau is a microtubule-associated protein involved in the assembly and stabilization of microtubules. Hyperphosphorylated tau proteins assemble into double-stranded filaments that are 820 nm wide and have a perimeter of 80 nm, and they are highly insoluble and unable to bind to microtubules due to conformational changes. The deposition of Aβ, activation of Ca^{2+}/calmodulin-dependent protein kinase II, cyclin-dependent kinase 5, and glycogen synthase kinase 3β (GSK-3β) can induce the hyperphosphorylation of tau.

A strong link exists between hyperphosphorylated tau and autophagy. Although tau can be degraded by the ubiquitin-proteasome system, the autophagy-lysosome pathway is a major clearance route for tau oligomers, as autophagy impairment leads to the accumulation of tau oligomers and insoluble species. Autophagy-inducing drugs like rapamycin, verapamil, or felodipine induce the clearance of tau in organisms. Furthermore, autophagy impacts tau phosphorylation. Autophagy-deficient mice display hyperphosphorylated tau, while autophagy restoration reduces tau phosphorylation. However, some evidence suggests the autophagy-lysosome pathway may not be essential for basal tau metabolism, as inhibiting autophagy with 3-methyladenine (3-MA) does not always lead to tau accumulation. Additionally, tau itself can modulate autophagy, with tau accumulation disturbing lysosome membrane integrity and retarding autophagosome maturation.

11.3.1.3 The interplay of autophagy with Alzheimer's disease pathological proteins

In the early stage of AD, the autophagy pathway is activated to degrade APP-rich metabolic substrates. However, in later stages of AD, excessive APP and Aβ deposition generally leads to the dysfunction of autophagy. A key impairment is the inhibition of autophagosome fusion with lysosomes, leading to an accumulation of autophagic vacuoles in the neurites of AD patients and mouse models. AD-associated mutations have been found to affect the three main steps of the autophagy pathways: (1) autophagosome synthesis, trafficking, and maturation/degradation. Beclin1 activity is reduced in AD brains and leads to a decrease in autophagosome synthesis. (2) Mutations of PS1 (known to cause fAD) impair autophagosome maturation and degradation by disrupting the lysosomal V-ATPase, whose N-glycosylation is regulated by PS1 and is essential for lysosomal acidification and subsequent degradation of autophagic contents. (3) Tau is known to be critical for retrograde vesicle trafficking as it regulates the assembly and stabilization of neuronal microtubules. The hyperphosphorylation of tau negatively impacts neuronal vesicle trafficking. Chaperonin CCT/TRiC complex, an important chaperone that assists the proper folding of tau, was found to be down-regulated in the AD brain and inhibits the trafficking of autophagosomes and their ultimate fusion with lysosomes (Pavel et al., 2016). Additional genetic risk factors for AD, such as the APOE4 variant, mutant APP, or APP duplication, also damage lysosomal function.

11.3.1.4 Autophagy modulation as therapy in Alzheimer's disease

Enhancing autophagic flux is a promising therapeutic strategy for AD. This can be achieved pharmacologically using mTOR-dependent or independent enhancers, or via gene therapy approaches (Table 11.1).

The first group of compounds are mTOR inhibitors (the negative autophagy regulator), such as rapamycin and its analog (e.g., temsirolimus). These compounds activate neuronal autophagy to clear hyperhosphorylated tau and Aβ in multiple AD model mice. Rapamycin also reduces tau toxicity and extends lifespan in *Drosophila* expressing wild-type or mutant R406W tau by enhancing autophagic clearance of soluble and insoluble tau. While rapamycin is an FDA-approved drug that has been used extensively in renal transplant patients, it is a less than optimal choice for therapeutic induction of autophagy as it is associated with serious adverse effects, most of which are related to long-term immunosuppression. Consequently, novel rapamycin analogs, such as temsirolimus, are desirable to manipulate autophagy in human patients.

The second group of compounds comprises the class of small-molecule enhancers of autophagy that signal through activation of the AMPK pathway. These molecules are either direct AMPK modulators (such as metformin, bosutinib, nilotinib) or indirect upstream activators acting by lowering mitochondrial ATP production and/or intracytoplasmic Ca^{2+} levels. The direct AMPK modulators, metformin, bosutinib, and nilotinib, show beneficial effects in multiple transgenic APP mouse models. Of these, metformin has progressed to pilot clinical studies for AD, showing promising early results with improved memory and executive function. The indirect modulators mainly include the disaccharide trehalose, lithium, imidazoline receptor agonizts (such as clonidine and rilmenidine), L-type Ca^{2+} channel antagonists (felodipine and verapamil), and the calpain inhibitor A-705253. These compounds have been shown to enhance the clearance of tau and Aβ in preclinical trials. While these described chemical compounds that enhance autophagy offer clear beneficial effects in multiple AD models, it is important to acknowledge that many of these compounds may also exert beneficial effects via autophagy-independent mechanisms.

11.3.2 Autophagy and Parkinson's disease

PD is the second most common neurodegenerative disorder, primarily affecting the elderly and characterized by motor system dysfunction. Its core clinical manifestations include resting tremor, bradykinesia, muscle rigidity, and postural instability, often accompanied by nonmotor symptoms such as olfactory loss, depression, constipation, and cognitive impairment. The primary etiology of PD is the degeneration of dopaminergic (DA) neurons in the substantia nigra (SN), although the precise cause of this neuronal loss remains unclear. PD pathogenesis is considered multifactorial, involving a complex interplay of aging, genetic predisposition, and environmental factors. While specific genetic mutations underlie familial forms of PD, the pathogenesis of the more common sporadic form is not well-defined. Aging is the most significant risk factor for PD. DA neurons in the SN begin to undergo degenerative changes after age 30; however, normal aging alone does not produce clinical PD symptoms until DA

Table 11.1 Therapeutic strategies to regulate autophagy and their effects on disease models.

Mechanism to induce autophagy	Small molecules	Application in disease models	Examples
Mechanistic TOR (mTOR)-dependent mTOR-inhibition	Rapamycin	Alzheimer's disease (AD)	Reduces the levels of hyperphosphorylated tau and amyloid-β (Aβ) in AD mice and *Drosophila*
		Parkinson's disease (PD)	Increases the clearance of mutant α-synuclein (α-syn) and attenuates toxicity in PC12 cells, dopaminergic (DA) neurons, and PD animal models
		Amyotrophic lateral sclerosis (ALS)	Decreases TAR DNA binding protein 43 (TDP-43) inclusions. Reduces neuronal TBPH aggregates in ALS *Drosophila* model, partially rescuing lifespan and locomotor defects in these animals
		Huntington's disease (HD)	Facilitates clearance of polyglutamine (PolyQ) aggregates and reduces toxicity in HD *Drosophila* and mice
		Aging	Decreases mortality in mice and life extension in *Drosophila* and mice. Its trial for lifespan extension has entered the clinical stage
mTOR-independent	Small-molecule enhancers of rapamycin (SMERs)	PD	SMERs induce autophagic clearance of A53T α-syn in vitro independently of mTOR
		HD	SMERs trigger the elimination of mutant huntingtin protein (mHtt) aggregates and ameliorate toxicity in cell culture and in HD *Drosophila* models
AMP-activated protein kinase activation	Metformin, bosutinib, nilotinib	AD	Prevent the degeneration of DA neurons, the loss of which affects cognitive and noncognitive functions in Tg2576 AD mice

(*Continued*)

Table 11.1 (Continued)

Mechanism to induce autophagy	Small molecules	Application in disease models	Examples
		PD	Reduce levels of phosphorylated-Ser-129 α-syn and rescue mitochondrial dysfunction in SH-SY5Y cells and *Drosophila*, mouse models. Induce α-syn clearance in vitro. Rescue disease phenotypes in both A53T and PD mice and improve cognitive and motor skills in PD patients
		HD	Significantly clear PolyQ aggregates and improve behavioral dysfunction in the PolyQ disease mouse model
	Trehalose	AD	Induces tau clearance and neuronal survival in AD tauP301S mice
		ALS	Ameliorates early motor phenotypes in the G93A superoxide dismutase 1 ALS mice, but fails to delay end-stage motor phenotypes
		HD	Enhances the degradation of mHtt aggregates in HD mice
IP3-Ca^{2+}-AMPK axis	Lithium, clonidine, rilmenidine	AD	Lithium reduces tau phosphorylation and the formation of Aβ plaques in prepathological AD mice. Clonidine and rilmenidine induce mutant tau clearance and ameliorate motor defects in A152T tau zebrafish
		PD	Lithium increases the clearance of A53T and A30P mutant α-syn in inducible PC12 cell models and ameliorates phenotypes associated with PD in aged mutant Parkin mice
		HD	Clonidine enhances the removal of aberrant mHtt aggregates in HD models

(*Continued*)

Table 11.1 (Continued)

Mechanism to induce autophagy	Small molecules	Application in disease models	Examples
Ca^{2+}-AMPK axis	Felodipine, verapamil	AD	Improve the clearance of insoluble tau and ameliorate the morphological defects seen in tauopathy zebrafish models
		PD	Decrease the levels of α-syn in A53T α-syn transgenic PD mouse model and ameliorates DA neurodegeneration
		HD	Decrease mHtt in an HD mouse model and ameliorate signs of disease
Calpain inhibition	A-705253	AD	Improves clearance of hyperphosphorylated tau and Aβ and has beneficial cognitive effects in 3xtg AD mice
		HD	Downregulation of Calpain decreases mHtt aggregates and is neuroprotective in HD mice and *Drosophila*
FOXO3-mediated induction	Genistein	HD	Stimulates autophagy and promotes the degradation of mHtt

neuron loss exceeds 50% and striatal dopamine content falls below 80% of normal levels. Patients with older age of onset often present with more severe motor symptoms and pathological markers. A key neuropathological hallmark of PD, alongside DA neuron degeneration, is the presence of Lewy bodies and Lewy neurites. These intracellular inclusions are composed primarily of fibrillar α-synuclein (α-syn), and their accumulation contributes to neuronal dysfunction and death. Critically, α-syn levels correlate with disease severity. Evidence from postmortem studies of PD patients and animal models strongly implicates defective autophagy as a central mechanism in the disease's etiology and pathogenesis. The following sections will detail the interplay between autophagy and the development of PD.

11.3.2.1 The role of autophagy in the pathogenesis of Parkinson's disease

Evidence for the role of autophagy in PD progression first emerged from postmortem studies, which revealed an accumulation of abnormal autophagic structures within DA neurons of the SN in PD patients. Subsequent investigations have consistently identified aberrant levels of autophagy-related proteins in the SN and peripheral blood cells of both PD patients and animal models. A key observation is the accumulation of p62-positive inclusions within DA neurons. As p62 is a selective autophagy substrate, its accumulation signifies impaired autophagic flux and disrupted protein clearance in PD.

This autophagy dysfunction is a direct contributor to DA neurodegeneration. The most compelling evidence comes from transgenic animal models; specific knockout of essential autophagy genes (e.g., ATG5, ATG7, or Vmp1) in DA neurons triggers their rapid degeneration, leading to PD-like symptoms in mice. However, while these models robustly demonstrate the consequence of autophagic failure, the rapid pathology may essentially mimic the end-stage of the disease. Consequently, they fail to fully recapitulate the slow, multifactorial progression of human PD, which involves complex interactions where the causal relationship between α-syn pathology and further pathogenesis remains particularly difficult to model.

Importantly, there is genetic evidence suggesting that an impaired autophagy-lysosomal pathway may be a significant pathogenic mechanism of PD, as the gene mutations causing PD often significantly affect the autophagy-lysosome pathway, such as autophagosome biogenesis, fusion with lysosomes, or affect lysosomal function. We have summarized several common PD causative genes as follows:

11.3.2.1.1 SNCA

α-syn is encoded by the *SNCA* gene. Multiplications of the *SNCA* locus cause autosomal dominant forms of PD, as the increased α-syn protein level accelerates the formation of inclusion bodies. These α-syn inclusions were further proved to affect autophagosome maturation and fusion with lysosomes, resulting in decreased protein degradation. α-Syn is also closely related to CMA, another form of autophagy, in which the substrates are transported by a chaperone protein. Wild-type α-syn can be degraded by CMA, while A53T mutant α-syn shows a strong binding ability to LAMP2, a lysosomal membrane protein, inhibiting CMA with impaired lysosome degradation.

11.3.2.1.2 Leucine-rich repeat kinase 2

The leucine-rich repeat kinase 2 (LRRK2) gene mutation is currently the most important genetic factor in PD pathogenesis. The clinical phenotype of PD patients associated with *LRRK2* mutations resembles that of sporadic PD. To date, more than 50 *LRRK2* mutations have been identified, and the G2019S mutation is the most important mutation site that is found in about 2% of total PD patients. LRRK2 is a substrate for CMA, and its PD-associated mutants inhibit CMA function, leading to the accumulation of other CMA substrates, such as α-syn. LRRK2 G2019S mutation increases LRRK2 kinase activity, leading to defective lysosomal positioning mediated by Rab7 and enlarged lysosomes containing nondegraded content.

11.3.2.1.3 Glucocerebrosidase

Homozygous mutations in the gene encoding the *lysosomal enzyme glucocerebrosidase* (*GBA*) can cause lysosomal storage disease and Gaucher disease. Heterozygous *GBA* mutations are among the most common genetic risk factors for PD. The *GBA* gene encodes a lysosomal GBA, whose primary function is to hydrolyze glucosylceramide into glucose and ceramide, while mutant GBA exhibits reduced enzymatic activity, leading to accumulation of glucosylceramide in neurons and inhibition of lysosomal degradation. PD patients with *GBA* mutations show increased α-syn pathology, as well as autophagic and lysosomal defects.

11.3.2.1.4 ATP13A2

ATP13A2 mutation can cause an autosomal recessive form of DA-responsive PD syndrome (Kufor-Rakeb syndrome, KRS). *ATP13A2* encodes a lysosomal ATPase involved in the selective transport of cations on the inner or outer side of the membrane. The clinical manifestations of KRS vary widely and are dependent on the mutation site. Reportedly, ATP13A2 protects cells against a variety of cellular stress-induced injuries, such as mitochondrial respiratory chain complex I inhibition, oxidative damage, metal ion toxicity, and proteasome stress. Deletion of ATP13A2 in vitro has been shown to impair lysosomal membrane stability, lysosomal acidification process, lysosomal substrate, including autophagosome degradation. All these phenotypes can be reversed by the overexpression of wild-type ATP13A2. Depletion of ATP13A2 in vitro has been shown to decrease SYT11 levels, another novel PD-related protein. Reversely, SYT11 overexpression in ATP13A2 knockdown cells was able to rescue the autophagy defects and lysosomal dysfunction in ATP13A2 deficiency cells.

11.3.2.1.5 VPS35

VPS35 mutations can cause autosomal dominant inheritance and DA-responsive PD. The pathogenic 1858G > A (protein D620N) was found in multiple PD families. VPS35 encodes a subunit of the retromer complex whose primary role is to facilitate the reverse transport of endosomes to the Golgi complex, thereby promoting the recycling of certain membrane proteins or SNARES complexes. The D620N mutation of VPS35 reduces interaction with the Wiskott-Aldrich syndrome protein and SCAR homolog (WASH) complex, which significantly inhibits autophagy. While the mechanism is not fully elucidated, it may involve disruptions to the SNARE complex or ATG9 function.

11.3.2.1.6 FBXO7

FBXO7 gene mutations were first discovered in the early-onset PD patients with autosomal recessive inheritance. FBXO7 protein is localized in mitochondria and interacts with Parkin and PINK1 to maintain mitochondrial function and promote Parkin-mediated mitophagy.

11.3.2.1.7 PINK1 and Parkin

Autosomal recessive mutations in *PINK1* or *Parkin*, two genes encoding proteins that are key regulators of a form of mitophagy, cause early-onset Parkinsonism. Parkin is a cytosolic E3 ubiquitin ligase involved in the clearance of damaged mitochondria via the ubiquitin-proteasomal system (UPS). It is recruited to damaged mitochondria by the mitochondrial outer membrane kinase PINK1, whereupon Parkin ubiquitinates mitochondrial substrates, leading to mitophagy. Mutant *PINK1* or *Parkin* inhibits mitochondrial quality control through impairing mitophagy.

11.3.2.2 Autophagy modulation as therapy in Parkinson's disease

Autophagy activation by small-molecules modulators (see 2.1.4, the case of AD) or molecules to inhibit the aggregation of α-syn or disaggregate α-syn polymers, as well as promoting mitochondrial homeostasis, are promising therapeutic strategies for PD, as many compounds have been proven to promote the clearance of α-syn and protect DA neurons in PD models (Table 11.1).

11.3.2.2.1 Mechanistic TOR-dependent autophagy inducers

There is ample evidence indicating that rapamycin exhibits satisfactory antiPD effects in both in vivo and in vitro experiments. Inhibition of mTOR with rapamycin treatment increases the clearance of WT and PD-related mutant forms (A30P and A53T) of α-syn, attenuating toxicity both in vitro and PD animal models. In a wild-type SNCA transgenic mouse model, intracerebral injection of rapamycin reduced aggregation of α-syn in both soma and axonal terminals of DA neurons. Feeding rapamycin ameliorated the toxic effects of paraquat in *Drosophila* and improved the phenotype of Parkin or PINK1 gene deletion in *Drosophila*. However, in human trials, one should be aware of possible serious adverse effects of rapamycin related to long-term immunosuppression.

11.3.2.2.2 Mechanistic TOR-independent autophagy inducers

mTOR-independent autophagy inducers have shown beneficial effects in PD animal models. Several novel small-molecule enhancers of rapamycin (SMERs), including SMER10, SMER18, and SMER28, were found to induce autophagic clearance of α-syn in vitro independently of mTOR (Sarkar et al., 2007b).

11.3.2.2.3 AMPK activators

Several AMPK activators have demonstrated potential in modulating autophagy and mitigating PD pathology in preclinical models: Metformin has been shown to reduce levels of Ser-129-phosphorylated α-syn in SH-SY5Y cells and to partially rescue mitochondrial dysfunction in genetic *Drosophila* models of PD (Dulovic et al., 2014). Nilotinib, a tyrosine kinase inhibitor, induces autophagy and partially rescues disease phenotypes in mouse models of PD (A53T and lentiviral α-syn overexpression). Clinical trials in PD patients have reported improvements in cognitive and motor skills. Resveratrol, a natural plant phenol, induces AMPK-dependent autophagy and promotes α-syn clearance in vitro, an effect potentially mediated by SIRT1, which activates autophagy through the deacetylation of autophagy-related proteins (Park et al., 2012). Lithium, an indirect AMPK activator, increases the clearance of α-syn aggregates in A53T and A30P mutant cell models and reduces PD-associated phenotypes in an aged Parkin mutant transgenic mouse model (Lieu et al., 2014). Other mood-stabilizing drugs, such as sodium valproate and carbamazepine, also induce autophagy in cells under rotenone-induced toxicity (Xiong et al., 2011). Trehalose provides significant relief in various neurodegenerative disease models, including PD, HD, and ALS (Sarkar et al., 2007a). Treatment with trehalose induces autophagy and enhances α-syn clearance both in vitro and in vivo. Felodipine decreases the levels of α-syn in A53T mouse model and ameliorates DA neurodegeneration (Siddiqi et al., 2019).

11.3.3 Autophagy and amyotrophic lateral sclerosis

ALS is a fatal neurodegenerative disease characterized by the selective degeneration of both upper motor neurons in the motor cortex and lower motor neurons in the brainstem and spinal cord. This results in progressive muscle weakness, atrophy, paralysis, and eventual multisystem dysfunction. The disease etiology remains largely unknown, its presentation is highly heterogeneous, and no effective cure exists. ALS typically progresses rapidly, leading to death within a few years of diagnosis. It is categorized into two types: familial ALS (fALS), which accounts for about 10% of cases and is driven by known genetic mutations, and sporadic ALS (sALS), which comprises the remaining 90% of cases with no clear hereditary component.

In 1993, the first genetic mutations were identified in the *superoxide dismutase 1 (SOD1)* gene in individuals with fALS, and so far, more than 100 genetic variants throughout all coding regions have been associated with a risk of ALS, including point mutations, insertions, and deletion mutations. SOD1 mutations account for about 12% of fALS cases and about 1% of sALS cases. With the application of whole-exome sequencing, more than 30 disease-causing genes in ALS have been identified. The nuclear TAR DNA binding protein 43 (TDP-43) accounts for 4%–5% cases of fALS. A hexanucleotide G4C2 repeat expansion in the chromosome 9 open reading frame 72 gene (C9orf72) is the most well-known genetic reason for ALS, which accounts for 30%–40% of fALS. Additionally, many ALS-related genes have been discovered, including fused in sarcoma (FUS), sequestosome 1 (SQSTM1)/p62, optineurin (OPTN), TANK-binding kinase 1 (TBK1), VAMP-associated protein B, valosin-containing protein (VCP), buiquilin2 (UBQLN2), charged multivesicular body protein 2B (CHMP2B), dynactin1, and factor-induced gene 4 (FIG4).

The etiology of sALS is unknown and is considered multifactorial. Proposed mechanisms include free radical-induced oxidative damage, abnormal protein aggregation, excitatory amino acid toxicity, mitochondrial dysfunction, aberrant apoptosis, and immune system defects. A key pathological hallmark of ALS, shared with other neurodegenerative disorders, is the presence of cytoplasmic protein aggregates, such as ubiquitinated inclusions, skein-like inclusions, and Bunina bodies, in motor neurons. These inclusions suggest a critical failure in protein homeostasis (proteostasis). Consequently, strategies aimed at accelerating the clearance of aggregated proteins, particularly through the modulation of autophagy, are emerging as attractive therapeutic avenues for ALS.

11.3.3.1 The role of autophagy in the pathogenesis of amyotrophic lateral sclerosis

ALS progression is age-related, typically manifesting in the fifth or sixth decade of life, and importantly, aging is a prerequisite for the ALS neurodegenerative disease process, as evidence showing that ALS transgenic mouse models exhibit abnormalities in axonal transport, sensorimotor development, and neuronal excitability as early as the neonatal period, not exhibiting neurodegenerative phenotypes until mature adulthood. With aging, autophagy activity is decreased in both motor neurons and skeletal muscle. Cytosolic protein aggregates, which are autophagy substrates, are found almost universally in ALS and typically comprise TDP-43, SOD1, FUS, and dipeptide repeat (DPR) proteins in various combinations. Tau, p62, and ubiquitin can also be found in these cytosolic inclusions. Another group of ALS-causing genes, including p62/SQSTM1, OPTN, TBK1, and VCP, converges on the autophagy pathway.

Models based on disease-causing mutations have implicated dysfunctions in the autophagy-lysosome pathway in ALS pathogenesis. The human genetic data suggest that the formation of ALS-associated protein aggregates is countered by autophagy and that mutations that adversely affect aggregate formation or removal can lead to ALS. Here, we focus on the three most common ALS pathogenic genes, *SOD1*, *TDP-43*, and *C9orf72*, as well as several ALS-related genes, and summarize their association with autophagy and the impact of their mutations on the autophagy pathway, to elucidate the potential role of autophagy in the pathogenesis of ALS.

11.3.3.1.1 Superoxide dismutase 1

SOD1 is a highly conserved cytoplasmic and mitochondrial antioxidizing enzyme that catalyzes the dismutation of superoxide ($O_2\bullet^-$), generating hydrogen peroxide (H_2O_2) and O_2. By its activity on superoxide radical, SOD1 controls the concentrations of reactive oxygen and nitrogen species: ROS and RNS.

Mutations in SOD1 result in reduced stability of the enzyme, which results in its loss of function and consequently to superoxide accumulation in neurons, mitochondrial dysfunctions, oxidative stress, calcium overload, protein aggregation, ER stress, etc. Many studies support the view that misfolding and aggregation of mutant SOD1 (mSOD1) are key events in the pathogenic process of fALS. mSOD1 can be degraded by autophagy and by the UPS. Inhibition of autophagy increased both SOD1 and mSOD1 levels in detergent-soluble and -insoluble fractions, and increased autophagy could ameliorate the mSOD1-mediated cell death, indicating the essential role of autophagy in the clearance of toxic mSOD1 protein oligomers and aggregates. However, excessive accumulation of toxic protein aggregates can itself affect autophagic flux. mSOD1 induces hyperactive autophagy, manifested as an increase of LC3- II and Beclin1 levels, and autophagic vacuoles are consistently found in the spinal cord and brainstem fractions of $SOD1^{G93A}$ mice. Nevertheless, the downstream events, autophagosome maturation and subsequent lysosomal fusion, are inhibited by mSOD1, leading to the accumulation of autophagosomes and ubiquitinated inclusions. In addition, it appears that autophagy in motor neurons plays counteracting roles in the $SOD1^{G93A}$ transgenic mice, as evidenced by a study showing that inhibition of autophagy in motor neurons accelerated disease onset but prolonged lifespan in the $Atg7\ ^{cKO}$-$SOD1^{G93A}$ double-mutant mice (Rudnick et al., 2017).

11.3.3.1.2 TAR DNA binding protein 43

Mutations in the RNA and DNA binding protein TDP-43 contribute to 4%–5% of fALS and nearly 1% of sALS cases, and most of the mutations are concentrated in the C terminus of the protein. TDP-43 aggregates accumulate in ubiquitinated or p62-positive inclusions and are found in almost all ALS patients, suggesting the potential core role of TDP-43 in the progression of ALS disease. Both UPS and autophagy pathways contribute to the clearance of aberrant TDP-43 proteins. Further studies show that soluble (monomeric) TDP-43 is predominantly degraded by the UPS, whereas clearance of aggregated (oligomeric) TDP-43 requires autophagy. In the case of ALS, TDP-43 is often observed in the cytoplasm, and cytoplasmic TDP-43 aggregates are also positive for p62, indicating the autophagy is mainly responsible for the cytoplasmic TDP-43. TDP-43 dysfunction was reported to inhibit the autophagosome-lysosome fusion (Xia et al., 2016).

11.3.3.1.3 Chromosome 9 open reading frame 72 gene

A hexanucleotide (GGGGCC) repeat expansion in a noncoding region of the C9orf72 transcript is the most common genetic abnormality identified in fALS and sALS. The majority (>95%) of healthy controls have ≤11 hexanucleotide repeats in the C9orf72 gene, while much longer expansions are observed in patients with C9-ALS, ranging from hundreds to thousands of repeats. Three possible mechanisms have been proposed: loss of function caused by a decreased expression of the C9orf72 protein, production of toxic C9orf72 repeat RNAs, and accumulation of DPR proteins produced by repeat-associated nonATG translation. Although C9orf72 mRNA was reduced in the postmortem tissue of C9-ALS patients, this reduction does not seem to have an inevitable causal link with the onset of C9-ALS, as evidenced by a zebrafish model with knockeddown of C9orf72. No o ALS phenotype was observed in this model, indicating that the latter two hypotheses regarding the accumulation of toxic substances seem more plausible. P62-positive puncta were accumulated in the spinal cord and brain tissues from C9-ALS patients, further supporting those toxic substances deposit in the cells. There is also some evidence indicating that the absence of C9orf72 can impair autophagic flux, possibly by affecting vesicular trafficking, late endosome maturation, and endosome recycling through interaction with Rab GTPases.

11.3.3.1.4 Ubiquilin-2

UBQLN2 encodes a 624-amino acid multidomain adapter protein critical for maintaining protein homeostasis. More than ten mutations in the UBQLN2 gene have been found in ALS or frontotemporal dementia (FTD) patients. UBQLN2 is involved in UPS by it delivering poly-ubiquitinated proteins to the proteasome for degradation. UBQLN2 also devolves in autophagy-lysosome pathway and participates in delivering cargo to autophagosomes. Studies show that mice with UBQLN2 mutants, but not wild-type UBQLN2, develop pathological characteristics of ALS, indicating the production of toxic mutant UBQLN2 forms. Yet, it is still to be determined if the ALS-associated mutations in UBQLN2 affect protein degradation predominantly through UPS or autophagy or both.

11.3.3.1.5 Valosin-containing protein

VCP was originally discovered in an Italian family whose R191Q mutation resulted in autosomal dominant ALS. VCP is an essential AAA^+-ATPase involved in DNA replication and repair, cell cycle regulation, and protein clearance. Loss of VCP activity results in autophagosome accumulation. VCP is an autophagy inducer; it activates autophagy through negatively regulating mTOR. In VCP mutant models, autophagosomes fail to fuse with lysosomes. Extensive evidence suggests that the roles of VCP in autophagy are varied, and further studies are needed to investigate its role in autophagy regulation and in the pathogenesis of ALS.

11.3.3.1.6 Sequestosome/P62

P62, also known as *SQSTM1*, encodes an adapter protein that is involved in autophagy and UPS. Mutations in the *p62* gene have been found to cause both ALS and FTD. Limited evidence indicates that disease-associated mutations may cause loss of function of p62 in autophagy. Expression of the p62 P392L mutant, an ALS-associated domain mutation, impairs basal autophagy in murine osteoclasts, implicating this mutation in impaired autophagic clearance. Further studies show that the ALS-associated mutation p62 LC3-related domain (LIR) mutant L341V fails to recognize of LC3-II, hindering its functional involvement with autophagic vacuoles. p62 selectively binds mSOD1 in the ALS models. Knockdown of p62 was reported to shorten the lifespan of SOD1H46R and SOD1G93A ALS model mice and to aggravate ALS symptoms, accelerating weight loss, motor dysfunction, and motor neuron degeneration.

11.3.3.1.7 Charged multivesicular body protein 2B

CHMP2B is a subunit of the endosomal sorting complex required for transport (ESCRT-III). ESCRT-III is essential for the formation of multivesicular body and for the sorting of proteins at the endosomal membrane. Mutations in ESCRT-III block endosomal maturation into the lysosome and impede the following fusion of autophagosome with lysosome, leading to abnormal aggregation of autophagosomes. CHMP2B mutations (T104N, I29V, Q206H) have been found in ALS cohorts, with patients' brains exhibiting immature autophagosome accumulation and upregulated LC3-II and p62 levels, implicating autophagic flux impairment.

11.3.3.1.8 Optineurin

OPTN encodes a cytoplasmic protein containing 577 amino acids, involved in a diverse range of cellular processes, such as vesicle trafficking, autophagy, and nuclear factor kappa-light-chain-enhancer of activated B cells signal transduction. Genetic analyzes have uncovered OPTN mutations in cohorts of ALS patients, which account for approximately 3% of fALS and about 1% of sALS. OPTN contains an LIR that can bind to LC3-marked autophagosome, and a ubiquitin-binding domain in the C-terminal, facilitating binding with ubiquitinated proteins. As an autophagy receptor that binds with LC3, OPTN plays an important role in autophagy, especially in mitochondrial turnover. More studies will be needed to confirm the relationship between OPTN and autophagy in ALS.

11.3.3.1.9 TANK- binding kinase 1

TBK1 encodes a protein containing 729 amino acids, which belongs to the IKK-kinase family of kinases that play essential roles in innate immunity signaling pathways. In 2015, Cirulli et al. found that *TBK1* gene was closely associated with ALS by sequencing the whole exon in 2869 ALS patients and 6405 age- and sex-matched controls. Nonsense and frameshift mutations cause major disruption to TBK1 and decrease its expression, implying that TBK1 insufficiency contributes to the development of ALS. TBK1 controls the phosphorylation of autophagy adapters such as p62 and OPTN, playing a major role in autophagy. TBK1 phosphorylates the ubiquitin-associated domain (UBA) of OPTN on Ser-177, enhancing LC3 binding affinity and promoting autophagosome formation. It has also been reported that TBK1 coordinates with the autophagic machinery via phosphorylating Ser-403 on the UBA domain. TDP-43 and p62-positive inclusions were observed in the motor neurons of various brain regions of TBK1-mutated ALS/FTD patients. TBK1 was proven to function in autophagosome maturation, as in TBK1 knockdown cells, autophagosome maturation into autolysosomes was inhibited.

11.3.3.1.10 Factor-induced gene 4

FIG4 encodes the protein FIG4. Mutations in FIG4 have been identified in 1%–3% of ALS patients in Europe. FIG4 regulates the cellular level of PI(3,5)P2, and maintains endomembrane homeostasis and endosomal trafficking. PI(3,5)P2 is recognized by lysosomes and merges to gain access to the lysosomal machinery. Mice with FIG4 mutant show motor neuron degeneration and enlarged endosomal vesicles, suggesting FIG4 dysfunction inhibits the endo-lysosomal pathway. Further, FIG4 knockout has a deleterious impact on autophagy, as evidenced by the accumulation of p62, LC3-II, and autolysosomes in FIG4 knockout mice.

11.3.3.2 Autophagy modulation as therapy in amyotrophic lateral sclerosis

Current FDA-approved treatments for ALS offer limited benefit. Riluzole, the earliest approved drug, delays motor neuron degeneration by inhibiting glutamate excitotoxicity, resulting in a modest extension of patient survival. Edaravone, approved in 2017, has controversial efficacy, with some studies showing only a limited slowing of disease progression. The more recent 2023 approval of Tofersen targets ALS subtypes with SOD1 mutations by reducing the expression of the toxic mutant protein, though its efficacy requires further verification. Given the limitations of existing therapies, autophagy modulation represents an attractive alternative therapeutic avenue for ALS. However, since the disease involves defects in both autophagosome biogenesis and autophagic flux, autophagy inducers are unlikely to be a panacea for all ALS forms.

Evidence for the therapeutic potential of autophagy modulation in ALS is mixed. A recent phase 2 clinical trial (the RAP-ALS trial) found that rapamycin was well-tolerated and safe in ALS patients, though the trial's full completion was not reported, and only safety outcomes were confirmed at certain stages (Mandrioli et al., 2018, Mandrioli et al., 2023). This trial was prompted by promising animal studies in which rapamycin reduced TDP-43 inclusions and ameliorated motor deficits. Conversely, other preclinical studies have shown that rapamycin can exacerbate disease; it accelerated motor neuron degeneration and shortened lifespan in a transgenic mouse model expressing the mutant human G93A SOD1 gene. This detrimental effect is suggested to stem from the already impaired autophagic flux in these models, which may be worsened by rapamycin-induced autophagy initiation.

Results from mTOR-independent autophagy inducers are similarly inconsistent. Clinical trials demonstrated that lithium provided no benefit for disease progression or survival in ALS patients. In the G93A SOD1 transgenic mouse model, trehalose treatment improved early motor phenotypes and increased autophagy with reduced mSOD1 levels in the spinal cord, yet it failed to delay end-stage disease or extend overall survival (Li et al., 2015).

11.3.4 Autophagy and Huntington's disease

HD is a hereditary, progressive neurodegenerative disorder classified as a polyglutamine (PolyQ) disease. This category of diseases also includes dentatorubral-pallidoluysian atrophy, spinal and bulbar muscular atrophy, spinocerebellar ataxias (types 1, 2, 6, 7, and 17), and Machado-Joseph disease (SCA3). PolyQ diseases are caused by abnormal expansions of cytosine-adenine-guanine (CAG) repeats within a specific gene, which encodes an elongated PolyQ tract in the protein. The age of disease onset is inversely correlated with the length of the PolyQ tract, which promotes the formation of toxic oligomers and protein aggregates. HD, also known as hereditary chorea, is the most common PolyQ disease. It results from a CAG repeat expansion (>35 repeats) in the *huntingtin* (*Htt*) gene, encoding an elongated PolyQ tract at the N-terminus of the Htt protein. The clinical presentation of HD is characterized by widespread involuntary movements (chorea), cognitive decline, and psychiatric disturbances. The core neuropathology involves the premature death of GABAergic neurons in the striatum, leading to striatal atrophy. This is accompanied by the loss of pyramidal neurons in the motor cortex, cerebral cortical atrophy, and enlargement of the ventricular system. The mutant huntingtin protein (mHtt) with its expanded PolyQ tract forms perinuclear cytoplasmic aggregates and intranuclear inclusion bodies in neurons. HD is difficult to diagnose in its early stages, and current clinical management primarily focuses on symptomatic relief, as no curative therapies exist. Consequently, research is increasingly shifting towards disease-modifying strategies, with a major focus on reducing mHtt levels. The following sections detail the critical interplay between autophagy and the pathogenesis of HD.

11.3.4.1 Autophagy in the process of Huntington's disease

Substantial evidence indicates that autophagy activation promotes the degradation of Htt aggregates, while its inhibition leads to increased Htt aggregation. For instance, Rubinsztein et al. reported that mTOR is sequestered within polyQ aggregates of mHtt in cell cultures, transgenic mice, and HD patients. This sequestration inhibits mTOR signaling and consequently enhances autophagy activity. Conversely, treatment with autophagosome formation inhibitors like 3-MA or dimethyl adenosine, or the autophagosome-lysosome fusion inhibitor bafilomycin A1, leads to an increase insoluble mHtt protein levels.

Further evidence demonstrates that the autophagy process itself is impaired in HD, preventing the effective degradation of mHtt and exacerbating neurodegeneration. Defects have been identified at multiple stages of autophagy. For example, an increase in bilayer membrane structures in HD patients suggests an accumulation of autophagosomes and impaired autophagic flux. Furthermore, nonaggregated mHtt disrupts starvation-induced autophagosome biogenesis by interfering with wild-type ataxin-3, which normally deubiquitinates and stabilizes Beclin1. A specific autophagy defect in HD is manifested by the presence of "empty" cytosolic autophagosomes that fail to engulf cargo. This is likely due to the disruption of selective autophagy receptors like SQSTM1/p62 by mHtt, impairing the recognition and encapsulation of mHtt aggregates. The cargo recognition defects also inhibit organelle sequestration through autophagy, leading to abnormal storage of lipids, mitochondria, etc, in mHtt-expressing mice. Thus impaired autophagosome dynamics caused by mHtt expression result in the inefficient clearance of both dysfunctional mitochondria and mHtt aggregates, indicating that Htt plays a role in its own elimination.

The normal Htt protein also functions in regulating the axonal transport of autophagosomes via its interaction with huntingtin-associated protein-1 (HAP1). Loss of Htt or HAP1 disrupts the retrograde axonal transport of autophagosomes (Wong and Holzbaur, 2014). In addition, mHtt perturbs postGolgi trafficking to lysosomes by reducing interaction with the OPTN/Rab8 complex, which could contribute to dysfunctional autophagosome-lysosome dynamics (Toro et al., 2009). On a protective note, Rhes, a striatal-specific protein that binds mHtt, can activate autophagy by promoting the dissociation of BCL2 from Beclin1. Overexpression of Rhes has been shown to ameliorate HD-associated phenotypes in mice (Mealer et al., 2014).

11.3.4.2 Autophagy modulation as therapy in Huntington's disease

Autophagy deficiency is the core mechanism driving HD disease progression, leading to the toxic accumulation of mHtt and neuronal degeneration.

Consequently, therapeutic strategies aimed at activating or restoring the autophagy pathway hold significant promise for clearing mHtt and improving neurological function. These approaches include small-molecule drugs (Table 11.1), gene regulation, and comprehensive interventions. Despite existing challenges, the modulation of autophagy is poised to become a cornerstone of future disease-modifying therapies for HD.

Pharmacological activation of autophagy to enhance mHtt clearance has been shown to ameliorate HD pathology in various models. Beneficial effects have been demonstrated through mTOR-dependent pathways. For instance, the autophagy activator rapamycin increases the clearance of mHtt aggregates and reduces levels of soluble mHtt (Ravikumar et al., 2004, Ravikumar et al., 2006). Similarly, the rapamycin analog temsirolimus facilitated mHtt clearance via autophagy activation and rescued pathology in HD *Drosophila* and mouse models (Menzies et al., 2010).

mTOR-independent strategies are also effective. The autophagy inducers SMER10, SMER18, and SMER28 trigger the elimination of mHtt aggregates and reduce toxicity in cell culture and *Drosophila* HD models (Sarkar et al., 2007b). Notably, combining these SMER compounds with rapamycin produces enhanced protective effects (Sarkar et al., 2008). Furthermore, several natural bioactive compounds can stimulate autophagy and promote mHtt degradation. Genistein, for example, triggers mTOR-independent autophagy, enhances mHtt clearance, and improves neuronal function (Pierzynowska et al., 2024). Lithium, which can induce autophagy through both mTOR-dependent and -independent mechanisms, reduces both soluble and aggregated mHtt; a

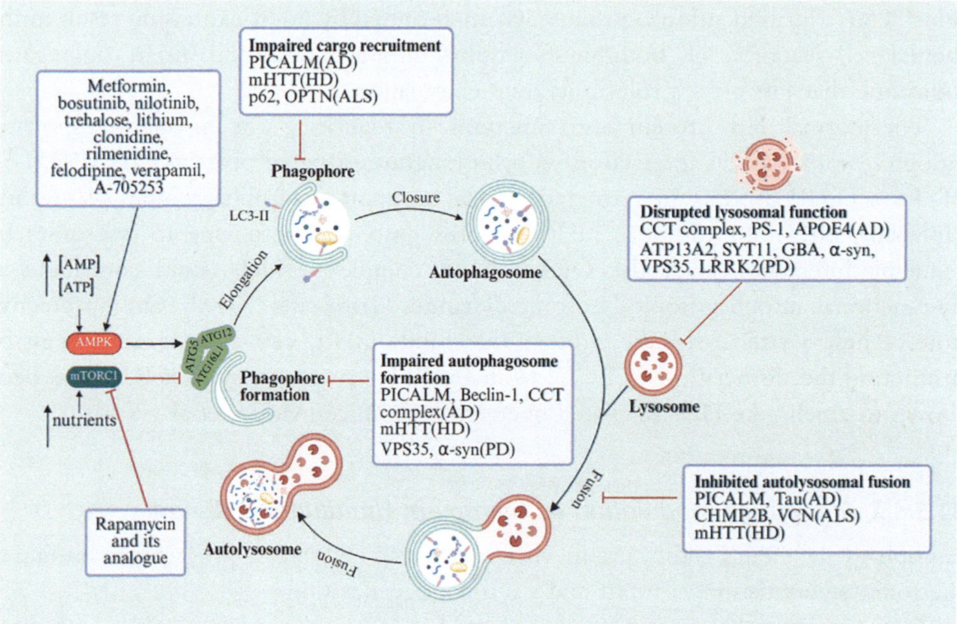

Figure 11.2 Disruption of autophagy machinery in neurodegenerative diseases and molecules that target autophagy as a therapeutic strategy for neurodegenerative diseases. Numerous genes implicated in neurodegenerative diseases have been associated with autophagy. These genes disrupt different steps of the autophagic flux, including autophagosome formation, cargo recruitment, lysosomal function, or autolysosomal fusion. Small molecules promote autophagy via inhibiting mTOR or activating the AMPK pathway (see Fig. 11.2).

combination therapy of lithium and rapamycin conferred protection against neurodegeneration in an HD *Drosophila* model (Sarkar et al., 2008). The Gi signaling activator clonidine, which induces mTOR-independent autophagy, enhances the removal of aberrant mHtt aggregates in HD models (Williams et al., 2008; Cansu Karabiyik et al., 2021). Additional inducers such as berberine, rilmenidine, and metformin have been shown to clear PolyQ aggregates and improve motor and behavioral deficits in PolyQ disease mice (see Fig. 11.2).

11.4 Conclusion

A significant decline in autophagic activity is a core hallmark of the aging process. As a central regulator of both aging and neurodegenerative diseases, the age-related waning of autophagy is directly linked to the loss of neuronal homeostasis. This dysfunction leads to the accumulation of toxic protein aggregates and impaired organelle renewal, ultimately accelerating neuronal degeneration. Currently, there are no curative treatments for most neurodegenerative diseases, such as AD, PD, and ALS, largely due to the incomplete understanding of their pathogenic mechanisms. Consequently, strategies aimed at clearing pathological protein aggregates have emerged as a major focus for developing disease-modifying therapies. Focusing on the development of safe and effective autophagy activation drugs, potentially in combination with multitarget intervention strategies (such as antioxidants to fight oxidative stress), represents a promising strategy to enhance the clearance of toxic aggregates and halt progressive neuronal death.

In some cases, especially in ALS, autophagy enhancers have been shown to be either beneficial or detrimental. This appears to depend on the stage of the disease, early detection and treatment with autopohagy inducers may be more productive.

Acknowledgement

This work was supported by National Natural Science Foundation of China (NSFC 32220103006, 82271524, 82401685), and Doctoral Scientific Research Foundation of Liaoning Science and Technology Department (2026-BS-0581, 2026-BS-0582).

References

Bharath, L.P., Nikolajczyk, B.S., 2020. Next steps in mechanisms of inflammaging. Autophagy 16 (12), 2285–2286. Available at https://doi.org/10.1080/15548627.2020.1822089.

Cansu Karabiyik, R.A.F., Park, S.J., Pavel, M., Rubinsztein, D.C., 2021. Autophagy in ageing and ageing-related neurodegenerative diseases [Review]. Ageing and Neurodegenerative Diseases 1 (2).

Chen, D., Li, P.W.L., Goldstein, B.A., Cai, W., Thomas, E.L., Chen, F., et al., 2013. Germline signaling mediates the synergistically prolonged longevity produced by double mutations in daf-2 and rsks-1 in C.elegans. Cell Reports 5 (6), 1600–1610. Available at https://doi.org/10.1016/j.celrep.2013.11.018.

Dulovic, M., Jovanovic, M., Xilouri, M., Stefanis, L., Harhaji-Trajkovic, L., Kravic-Stevovic, T., et al., 2014. The protective role of AMP-activated protein kinase in alpha-synuclein neurotoxicity in vitro. Neurobiology of Disease 63, 1–11. Available at https://doi.org/10.1016/j.nbd.2013.11.002.

Fernández, Á.F., Sebti, S., Wei, Y., Zou, Z., Shi, M., McMillan, K.L., et al., 2018b. Disruption of the beclin 1-BCL2 autophagy regulatory complex promotes longevity in mice. Nature 558 (7708), 136–140. Available at https://doi.org/10.1038/s41586-018-0162-7.

Fernández, Á.F., Sebti, S., Wei, Y., Zou, Z., Shi, M., McMillan, K.L., et al., 2018a. Author correction: disruption of the beclin 1–BCL2 autophagy regulatory complex promotes longevity in mice. Nature 561 (7723), E30. Available at https://doi.org/10.1038/s41586-018-0270-4.

Lapierre, L.R., De Magalhaes Filho, C.D., McQuary, P.R., Chu, C.C., Visvikis, O., Chang, J.T., et al., 2013. The TFEB orthologue HLH-30 regulates autophagy and modulates longevity in *Caenorhabditis elegans*. Nature Communications 4. Available at https://doi.org/10.1038/ncomms3267.

Li, Zhang, Yang, et al., 2009. Hypoxia increases Abeta generation by altering beta- and gamma-cleavage of APP. Neurobiology of Aging 30 (7), 1091–1098.

Li, Y., Guo, Y., Wang, X., Yu, X., Duan, W., Hong, K., et al., 2015. Trehalose decreases mutant SOD1 expression and alleviates motor deficiency in early but not end-stage amyotrophic lateral sclerosis in a SOD1-G93A mouse model. Neuroscience 298, 12–25. Available at https://doi.org/10.1016/j.neuroscience.2015.03.061.

Lieu, C.A., Dewey, C.M., Chinta, S.J., Rane, A., Rajagopalan, S., Batir, S., et al., 2014. Lithium prevents parkinsonian behavioral and striatal phenotypes in an aged parkin mutant transgenic mouse model. Brain Research 1591 (1), 111–117. Available at https://doi.org/10.1016/j.brainres.2014.10.032.

Mandrioli, J., D'Amico, R., Zucchi, E., Gessani, A., Fini, N., Fasano, A., et al., 2018. Rapamycin treatment for amyotrophic lateral sclerosis protocol for a phase II randomized, double-blind, placebo-controlled, multicenter, clinical trial (RAP-ALS trial). Medicine (United States) 97 (24). Available at https://doi.org/10.1097/MD.0000000000011119.

Mandrioli, J., D'Amico, R., Zucchi, E., De Biasi, S., Banchelli, F., Martinelli, I., et al., 2023. Randomized, double-blind, placebo-controlled trial of rapamycin in amyotrophic lateral sclerosis (RAP-ALS study). Research Square. Available at https://doi.org/10.21203/rs.3.rs-2789464/v1.

Mealer, R.G., Murray, A.J., Shahani, N., Subramaniam, S., Snyder, S.H., 2014. Rhes, a striatal-selective protein implicated in huntington disease, binds beclin-1 and activates autophagy. Journal of Biological Chemistry 289 (6), 3547–3554. Available at https://doi.org/10.1074/jbc.M113.536912.

Menzies, F.M., Hourez, R., Imarisio, S., Raspe, M., Sadiq, O., Chandraratna, D., et al., 2010. Puromycin-sensitive aminopeptidase protects against aggregation-prone proteins via autophagy. Human Molecular Genetics 19 (23), 4573–4586. Available at https://doi.org/10.1093/hmg/ddq385.

Mitchener, J.S., Shelburne, J.D., Bradford, W.D., Hawkins, H.K., 1976. Cellular autophagocytosis induced by deprivation of serum and amino acids in HeLa cells. American Journal of Pathology 83 (3), 485–498.

Mortimore, G.E., Hutson, N.J., Surmacz, C.A., 1983. Quantitative correlation between proteolysis and macro- and microautophagy in mouse hepatocytes during starvation and refeeding. Proceedings of the National Academy of Sciences of the United States of America 80 (8 I), 2179–2183. Available at https://doi.org/10.1073/pnas.80.8.2179.

Nilsson, P., Loganathan, K., Sekiguchi, M., Matsuba, Y., Hui, K., Tsubuki, S., et al., 2013. Aβ secretion and plaque formation depend on autophagy. Cell Reports 5 (1), 61–69. Available at https://doi.org/10.1016/j.celrep.2013.08.042.

Park, S.J., Ahmad, F., Philp, A., Baar, K., Williams, T., Luo, H., et al., 2012. Resveratrol ameliorates aging-related metabolic phenotypes by inhibiting cAMP phosphodiesterases. Cell 148 (3), 421–433. Available at https://doi.org/10.1016/j.cell.2012.01.017.

Pavel, M., Imarisio, S., Menzies, F.M., Jimenez-Sanchez, M., Siddiqi, F.H., Wu, X., et al., 2016. CCT complex restricts neuropathogenic protein aggregation via autophagy. Nature Communications 7. Available at https://doi.org/10.1038/ncomms13821.

Penniman, C.M., Bhardwaj, G., Nowers, C.J., Brown, C.U., Junck, T.L., Boyer, C.K., et al., 2023. Loss of FoxOs in muscle increases strength and mitochondrial function during aging. Journal of Cachexia, Sarcopenia and Muscle 14 (1), 243–259. Available at https://doi.org/10.1002/jcsm.13124.

Pierzynowska, K., Podlacha, M., Gaffke, L., Rintz, E., Wiśniewska, K., Cyske, Z., et al., 2024. Correction of symptoms of Huntington disease by genistein through FOXO3-mediated autophagy stimulation. Autophagy 20 (5), 1159–1182. Available at https://doi.org/10.1080/15548627.2023.2286116.

Pyo, J.O., Yoo, S.M., Ahn, H.H., Nah, J., Hong, S.H., Kam, T.I., et al., 2013. Overexpression of Atg5 in mice activates autophagy and extends lifespan. Nature Communications 4. Available at https://doi.org/10.1038/ncomms3300.

Ravikumar, B., Berger, Z., Vacher, C., O'Kane, C.J., Rubinsztein, D.C., 2006. Rapamycin pre-treatment protects against apoptosis. Human Molecular Genetics 15 (7), 1209–1216. Available at https://doi.org/10.1093/hmg/ddl036.

Ravikumar, B., Vacher, C., Berger, Z., Davies, J.E., Luo, S., Oroz, L.G., et al., 2004. Inhibition of mTOR induces autophagy and reduces toxicity of polyglutamine expansions in fly and mouse models of Huntington disease. Nature Genetics 36 (6), 585–595. Available at https://doi.org/10.1038/ng1362.

Rudnick, N.D., Griffey, C.J., Guarnieri, P., Gerbino, V., Wang, X., Piersaint, J.A., et al., 2017. Distinct roles for motor neuron autophagy early and late in the SOD1G93A mouse model of ALS. Proceedings of the National Academy of Sciences of the United States of America 114 (39). Available at https://doi.org/10.1073/pnas.1704294114 E8294.

Sarkar, S., Davies, J.E., Huang, Z., Tunnacliffe, A., Rubinsztein, D.C., 2007a. Trehalose, a novel mTOR-independent autophagy enhancer, accelerates the clearance of mutant huntingtin and α-synuclein. Journal of Biological Chemistry 282 (8), 5641–5652. Available at https://doi.org/10.1074/jbc.M609532200.

Sarkar, S., Perlstein, E.O., Imarisio, S., Pineau, S., Cordenier, A., Maglathlin, R.L., et al., 2007b. Small molecules enhance autophagy and reduce toxicity in Huntington's disease models. Nature Chemical Biology 3 (6), 331–338. Available at https://doi.org/10.1038/nchembio883.

Sarkar, S., Krishna, G., Imarisio, S., Saiki, S., O'Kane, C.J., Rubinsztein, D.C., 2008. A rational mechanism for combination treatment of Huntington's disease using lithium and rapamycin. Human Molecular Genetics 17 (2), 170–178. Available at https://doi.org/10.1093/hmg/ddm294.

Sebastián, D., Beltrà, M., Irazoki, A., Sala, D., Aparicio, P., Aris, C., et al., 2024. TP53INP2-dependent activation of muscle autophagy ameliorates sarcopenia and promotes healthy aging. Autophagy 20 (8), 1815–1824. Available at https://doi.org/10.1080/15548627.2024.2333717.

Siddiqi, F.H., Menzies, F.M., Lopez, A., Stamatakou, E., Karabiyik, C., Ureshino, R., et al., 2019. Author correction: felodipine induces autophagy in mouse brains with pharmacokinetics amenable to repurposing. Nature Communications 10 (1). Available at https://doi.org/10.1038/s41467-019-10536-y.

Toro, D.D., Alberch, J., Lazaro-Dieguez, F., Martin-Ibaflez, R., Xifro, X., Egea, G., et al., 2009. Mutant huntingtin impairs post-golgi trafficking to lysosomes by delocalizing optineurin/rab8 complex from the golgi apparatus. Molecular Biology of the Cell 20 (5), 1478–1492. Available at https://doi.org/10.1091/mbc.E08-07-0726.

Ulgherait, M., Midoun, A.M., Park, S.J., Gatto, J.A., Tener, S.J., Siewert, J., et al., 2021. Circadian autophagy drives iTRF-mediated longevity. Nature 598 (7880), 353–358. Available at https://doi.org/10.1038/s41586-021-03934-0.

Williams, A., Sarkar, S., Cuddon, P., Ttofi, E.K., Saiki, S., Siddiqi, F.H., et al., 2008. Novel targets for Huntington's disease in an mTOR-independent autophagy pathway. Nature Chemical Biology 4 (5), 295–305. Available at https://doi.org/10.1038/nchembio.79.

Wong, Y.C., Holzbaur, E.L.F., 2014. The regulation of autophagosome dynamics by huntingtin and HAP1 is disrupted by expression of mutant huntingtin, leading to defective cargo degradation. Journal of Neuroscience 34 (4), 1293–1305. Available at https://doi.org/10.1523/JNEUROSCI.1870-13.2014.

Xia, Q., Wang, H., Hao, Z., Fu, C., Hu, Q., Gao, F., et al., 2016. TDP-43 loss of function increases TFEB activity and blocks autophagosome-lysosome fusion. EMBO Journal 35 (2), 121–142. Available at https://doi.org/10.15252/embj.201591998.

Xiong, N., Jia, M., Chen, C., Xiong, J., Zhang, Z., Huang, J., et al., 2011. Potential autophagy enhancers attenuate rotenone-induced toxicity in SH-SY5Y. Neuroscience 199, 292–302. Available at https://doi.org/10.1016/j.neuroscience.2011.10.031.

PART 4

Contemporary biophysical methods and machine learning

CHAPTER 12

Intrinsically unfolded proteins and the role of protein condensation

Altijana Hromić-Jahjefendić[1], Jasmin Šutković[1], Abas Sezer[1], and Vladimir N. Uversky[2]
[1]Genetics and Bioengineering Study Program, Faculty of Engineering and Natural Sciences, International University of Sarajevo, Sarajevo, Bosnia and Herzegovina,
[2]Department of Molecular Medicine, University of South Florida, Tampa, FL, United States

12.1 Introduction

Proteins lacking a fixed or ordered 3D structure at physiological conditions, often in the absence of their interaction partners, are known as natively unfolded proteins (NUPs) or intrinsically disordered proteins (IDPs) (Ayyadevara et al., 2022). Usually, these proteins do not fold into stable secondary or tertiary structures like ordered globular proteins but exist instead as dynamic, heterogeneous ensembles of conformations, sometimes described as "protein clouds." In general, NUPs/IDPs can range from entirely unstructured to partially structured, presenting random coils, premolten globule, and molten globule-like states (Uversky and Dunker, 2010) or flexible linkers within larger proteins (Hsiao, 2024; Saurabh et al., 2023).

In regard to their structure, IDPs have a higher content of polar and charged amino acids than regular proteins and are deficient in bulky hydrophobic residues, which contribute to their inability to fold into an ordered state. As a result, IDPs/NUPs have high net charge and low hydrophobicity, disabling their folding into a stable 3D structure spontaneously (Seth et al., 2024; Gupta et al., 2025). Further, they consist of flexible, low-complexity, and repetitive sequences rich in amino acids like glycine, serine, glutamine, arginine, and proline. This unique composition reduces their hydrophobic content and increases net charge, preventing stable folding and enabling dynamic flexibility (Lotthammer et al., 2024). Due to their special structural features, IDPs engage in multivalent, transient, and weak interactions, such as electrostatic, π–π, cation–π, and hydrophobic contacts with various molecules (Már et al., 2023; Poudyal et al., 2023). These weak multivalent interactions are key drivers of the physicochemical phenomenon of liquid–liquid phase separation (LLPS) that leads to the condensation of proteins into membrane-less organelles (MLOs), including stress granules, P-bodies, nucleoli, Cajal bodies, and pathological condensates/aggregates of fused in sarcoma (FUS) and transactive response DNA-binding protein 43 (TDP-43) together with RNAs (Lang et al., 2024), found in several neurodegenerative diseases (Ibrahim et al., 2023; Scherer et al., 2024).

Molecular Insights into Proteinopathies
DOI: https://doi.org/10.1016/B978-0-443-34176-2.00005-5
© 2027 Elsevier Inc. All rights are reserved, including those for text and data mining, AI training, and similar technologies.

Despite this lack of stable structure, IDPs can undergo conformational changes and fold (at least partially) upon binding to specific partners, showing a disorder-to-order transition. Posttranslational modifications (PTMs) can further diversify the conformations these proteins can adopt (Liu et al., 2023; Uversky, 2025; Uversky, 2025b). Functionally, the cooperative and dynamic interactions of IDPs are caused by their intrinsic flexibility. Two examples of this flexibility are reflected in processes like DNA regulation and cell signaling (Wang et al., 2023). In addition, they can bind multiple partners, adopting distinct conformations with each, enabling versatile biological functions (Ghitti et al., 2025). Studies have shown that 30%–50% of all amino acid residues in eukaryotic proteins are found in disordered regions (Ghitti et al., 2025; Tunyasuvunakool et al., 2021).

IDPs' functional ability remains diverse. During cellular signaling, disordered proteins flexibly show transient interactions with multiple partners, often adopting a structure upon binding (Bondos et al., 2022). Due to their intrinsic flexibility, they can act as dynamic hubs in signaling networks, regulated by PTMs, which serve as molecular switches controlling signaling pathways and information flow (Trivedi and Nagarajaram, 2022). Further, IDPs are commonly found in the brain, where they play critical roles in preserving neuronal architecture, facilitating signaling pathways, and enabling adaptability in response to changes in cellular conditions (Norris et al., 2023).

Based on the above-described structural characteristics, the analytical detection of such proteins remains a highly challenging process. Techniques, such as nuclear magnetic resonance (NMR), single-molecule fluorescence, and small-angle scattering, have been instrumental in capturing the fluctuating behaviors of these proteins and their role in LLPS (Maiti et al., 2024). In addition, computational and bioinformatics tools now predict intrinsic disorder and phase separation tendencies directly from amino acid sequences (Lotthammer et al., 2024; Zhou, Zhou, et al., 2024). Both experimental and computational advanced methods enhance our complex understanding of IDPs and support deeper investigations into their biological functions and disease associations.

To summarize, the purpose of this chapter is to provide a comprehensive overview of NUPs/IDPs, exploring their unique structural and functional characteristics, especially their roles in cellular signaling, molecular recognition, and brain function. Further, it aims to clarify the molecular basis of protein condensation and its correlation to intrinsic disorder. The disease implications of dysregulated condensation, while also covering advanced methods and therapeutic strategies targeting IDPs, are revealed.

12.2 Structural and functional features of intrinsically disordered proteins

As is known, IDPs lack a stable tertiary structure, a feature rooted in their unique amino acid composition (Ayyadevara et al., 2022; Cascarina and Ross, 2025). In comparison with globular proteins that fold into well-defined three-dimensional shapes stabilized by hydrophobic cores, IDPs are typically enriched in polar and charged residues, such as arginine, serine, proline, and glycine, while lacking heavy hydrophobic amino acids like tryptophan, cysteine, and leucine (Hsiao, 2024; Sharma et al., 2025). This composition imparts a high net charge and low hydrophobicity, which together prevent the folding necessary for a rigid tertiary structure and promote strong interactions with the aqueous cellular environment (Firouzbakht et al., 2024). Therefore, IDPs behave as dynamic structures with rapidly interchanging conformations rather than static structures. These conformational changes can be compact, collapsed, and/or extended states, depending on environmental conditions and binding partners (Ghafouri et al., 2024).

Due to their intrinsic flexibility, IDPs are engaged in the phenomenon known as binding promiscuity (Uversky, 2025a, 2025b). Rather than binding to a single specific partner with a fixed structure, they interact transiently with multiple different molecules, adopting distinct conformations on each occasion, a process termed "induced folding" (Bondos et al., 2022; Morris et al., 2021; Liu et al., 2025). Therefore, IDPs have diverse biological functions, especially in signaling pathways where rapid, reversible interactions are essential (Bondos et al., 2022). In addition, phosphorylation, acetylation, glycosylation, and methylation, as important PTMs, play a crucial role in regulating the conformational ensembles of IDPs, thereby broadening their range of functional capabilities (Newcombe et al., 2022). Because IDPs can adapt structurally to their molecular environment, they are often found as hubs in signaling networks, where they coordinate complex cellular responses (Miao and Chong, 2025).

As an example, IDPs' structural plasticity allows them to serve as flexible scaffolds or hubs that facilitate the assembly of multiprotein complexes, regulate enzyme activities, and mediate transcriptional control (Shang et al., 2024; Sil et al., 2025). In this regard, IDPs can act as molecular switches, as they contain short linear motifs that support defined and selective protein interactions and associated activities (Cermakova and Hodges, 2023; Maiti and De, 2022). Further, they can act as sensors or recruiters for other proteins, binding to multiple partners in different conformations, enriching their role in diverse signaling pathways (Tripathi et al., 2025; Trivedi and Nagarajaram, 2022; Yorimitsu and Sato, 2025). Additionally, due to their adaptable conformations, IDPs are able to recognize diverse sequence motifs or structural features of other molecules, including proteins, nucleic acids, and membranes (Bhargava et al., 2024; Bondos et al., 2022; Chow et al., 2025). The IDP's flexible nature for multiple interactions also underlies the phenomenon of moonlighting, where a single IDP can perform multiple unrelated biological functions, depending on the cellular context or its binding partners (Gupta and Uversky, 2022). Moonlighting proteins are increasingly recognized as key players in human disease and microbial pathogenesis, making their identification crucial for understanding disease mechanisms and developing targeted therapies (Cedano et al., 2025).

12.3 Molecular mechanisms of protein condensation

Biomolecular condensation is a fundamental biophysical and cellular process whereby biomacromolecules, such as proteins and nucleic acids, dynamically organize into concentrated, membrane-less compartments through a process known as LLPS (Deng and Wan, 2024; Guan et al., 2025). LLPS is analogous to the separation of oil droplets from water, where molecules spontaneously demix and form distinct liquid phases, with multivalent weak interactions driving this process in biological systems. These biomolecular condensates serve as scaffolds for biochemical reactions, enabling efficient regulation and compartmentalization without the need for lipid membranes (Alberti et al., 2025). Due to the dynamic and reversible nature of biomolecular condensates, cells can rapidly respond to physiological changes, forming functional subcellular structures known as MLOs (Li et al., 2024).

At the core of LLPS are underlying thermodynamic principles and molecular driving forces. The process is governed by a balance of intermolecular interactions and entropy changes, with the system minimizing its free energy (Alberti et al., 2025; Wang and Zhang, 2024). Key forces driving protein condensation include electrostatic interactions between charged residues, hydrophobic effects arising from nonpolar amino acid side chains, and specialized interactions such as π–π stacking between aromatic rings and cation–π interactions (Netzer et al., 2024; Sanchez-Burgos et al., 2025; Zhou and Pang, 2018). The combined effect of these interactions promotes the formation of dense, dynamic liquid droplets in which biological molecules are concentrated to facilitate various cellular functions (Zwicker et al., 2025). One of the commonly known examples is the FUS protein, which has intrinsically disordered regions (IDRs), that is, low-complexity domains that facilitate multivalent interactions, such as electrostatic, π–π, and hydrophobic interactions (Song et al., 2025). Another well-studied example is nucleophosmin, a scaffold protein in the nucleolus that forms heterotypic condensates with ribosomal RNA (rRNA) via LLPS (Xu et al., 2025). More examples of proteins undergoing LLPS are shown in Table 12.1.

Condensates can explain how phase separation spatially and temporally controls biochemical reactions vital for cellular health. For example, stress granules are cytoplasmic condensates composed primarily of RNA and RNA-binding proteins (RBPs). They are formed under cellular stress conditions in order to stop translation and protect mRNAs (Baymiller and Moon, 2023). One of the largest nuclear condensates is the nucleolus, which organizes rRNA synthesis and ribosome assembly (Sheu-Gruttadauria et al., 2023). Synaptic condensates at neuronal synapses regulate signal transduction and synaptic plasticity by concentrating specific signaling molecules (Wu et al., 2025).

The combination of electrostatic and hydrophobic interactions in condensate formation is particularly critical. Electrostatic interactions between oppositely charged

Table 12.1 Examples of proteins that form heterotypic condensates.

Protein (scaffold) + partner molecule	Role of heterotypic condensate	References
PTBP1 (with RNA)	PTBP1 forms heterotypic condensates with specific RNAs; these cocondensates contribute to nuclear compartments and gene regulation by concentrating RNA and protein.	(Novakovic et al., 2025)
FUS PLD + ARID1A PLD (also other mSWI/SNF subunits)	The PLDs of FUS and mSWI/SNF subunits cocondense via heterotypic weak interactions, lowering the saturation concentration and forming mixed transcriptional-hub-like condensates.	(Davis et al., 2024)
Ydj1 (Hsp40) + Tau	Ydj1 (a chaperone) promotes heterotypic phase separation with Tau, influencing droplet formation and preventing aberrant solidification (linked to tauopathy).	(Rai et al., 2025)
Histone H1 + DNA (or chromatin)	Linker histone H1 undergoes liquid–liquid phase separation (LLPS) in the presence of DNA/chromatin, forming heterotypic condensates that may influence chromatin compaction and nuclear architecture.	(Bian et al., 2024)
hnRNPA1A + RNA	hnRNPA1A forms heterotypic condensates with RNA; RNA modulates condensation and influences liquid-to-amyloid transitions of the protein.	Li et al. (2024)
Caprin1 (with RNA/ATP)	Caprin1's intrinsically disordered region and its heterotypic interactions with RNA/ATP display salt- and ligand-dependent LLPS; these condensates can modulate cellular responses to stress.	Liu et al. (2025)
PTBP1 + structured RNA (U1-snRNA SL4)	PTBP1 with structured RNA undergoes cocondensation; specific protein–RNA contacts are preserved in the dense phase, highlighting heterotypic assembly beyond simple electrostatics.	(De Vries et al., 2024)

(Continued)

Table 12.1 (Continued)

Protein (scaffold) + partner molecule	Role of heterotypic condensate	References
Whi3 + RNA (various RNAs, e.g., CLN3, BNI1)	Heterotypic interactions between Whi3 protein and RNAs drive condensate formation; timing of RNA addition modulates composition.	(Lin et al., 2023)
FET proteins and mammalian SWI/SNF complex	Heterotypic PLD–PLD interactions, for example, mSWI/SNF subunits + FET family proteins (protein–protein heterotypic condensates).	(Davis et al., 2023)

Abbreviations: *PTBP1*, polypyrimidine tract binding protein 1; *FUS*, fused in sarcoma; *PLD*, prion-like domain; *ARID1A*, AT-rich interaction domain 1A; *mSWI/SNF*, mammalian SWItch/sucrose nonfermentable complex; *Hsp40*, heat shock protein 40; *DNA*, deoxyribonucleic acid; *hnRNPA1A*, heterogeneous nuclear ribonucleoprotein A1A; *U1-snRNA SL4*, U1 small nuclear RNA, stem loop 4; *FET*, FUS, EWS, and TAF15 protein family.

residues and RNA molecules drive the initial nucleation of condensates, while hydrophobic patches enhance the cohesive energy of the dense phase. π–π stacking (the attractive force is a combination of London dispersion forces and electrostatic interactions involving the quadrupole moments of the rings), especially among aromatic amino acids like phenylalanine and tyrosine, contributes to the structural organization within the condensates, stabilizing dynamic networks (Das et al., 2020; Netzer et al., 2024). In addition, the cation–π interactions, which are strong (typically contributing 2–5 kcal/mol to binding energy in aqueous or biological media), are noncovalent attractive forces between a positively charged cation and the face of an electron-rich aromatic system, like a π system control assembly by linking positively charged residues, such as arginine or lysine, to aromatic side chains (Hazra and Levy, 2023). Note that cation–π interactions are primarily driven by electrostatic attraction, as aromatic rings, while having no net dipole moment, possess a quadrupole moment, with a negative charge cloud above and below the plane of the ring and a positive charge around the edges (Dougherty, 1996).

In addition to the interplay of different interactions, the LLPS thermodynamic basis also involves enthalpic gains from multivalent interactions compensating for entropic costs due to the confinement of molecules within condensates (Mukherjee and Schäfer, 2023). Changes in environmental factors, such as temperature (Dignon et al., 2019), concentration (Chu and Wang, 2021), and molecular crowding (Cinar and Winter, 2020), affect the phase boundary, dictating whether proteins remain dispersed or condense.

Further, PTMs like phosphorylation can affect the interaction environment by changing charge distribution or exerting steric effects, thereby regulating condensate formation and control in response to cellular signals (Sridharan et al., 2022). This demonstrated adaptability helps cells to control condensate composition and functions accurately.

12.4 Interplay between intrinsically disordered proteins and posttranslational modifications

PTMs are essential biochemical modifications of proteins after their synthesis and serve as important regulators of protein function, stability, and interactions (Kitamura and Galligan, 2023). In the context of IDPs and LLPS, the PTMs regulate the behavior of these proteins by altering their charge, conformation, and interaction potential (Xiao and Xia, 2025). Phosphorylation, for instance, modifies the charge properties and interaction capabilities of IDPs, thereby regulating their phase-separation capacity (Usher et al., 2024). For example, TDP-43 hyperphosphorylation, often seen in amyotrophic lateral sclerosis (ALS) and frontotemporal dementia (FTD), suppresses the ability of this protein to form reversible liquid droplets and instead predisposes it to aggregate formation (Xiao and Xia, 2025).

Similarly, methylation and ubiquitination affect the condensate properties of proteins. For example, methylation of arginine residues modulates FUS's phase separation by weakening arginine-mediated interactions critical for droplet formation (Hofweber et al., 2018). Further, ubiquitination is known to label irregular phase-separated proteins for degradation, serving as a quality control checkpoint preventing irreversible aggregate formation (Müller and Hoppe, 2024; Pla-Prats and Thomä, 2022). Polyubiquitinated substrates can modulate the phase separation of ubiquilin-2, a disordered shuttle protein, thereby influencing condensation behavior and substrate degradation fate (Valentino et al., 2024). The combinatorial effect of these modifications determines the viscosity (Li et al., 2025), internal dynamics (Galvanetto et al., 2025), and maturation of condensates (Fedorov et al., 2025), forming an intricate regulatory network controlling IDP-driven phase separation. In an example study, a full-length tau protein has been shown to undergo LLPS that can be affected by interaction with α-synuclein or Cdk2-controlled phosphorylation. Tau protein undergoes LLPS in the presence of RNA and molecular crowding agents, forming liquid droplets that sequester tau in the dynamic environment (Siegert et al., 2021). However, abnormal phosphorylation patterns in tau enhance droplet maturation and conversion into fibrillar amyloid aggregates, correlated with neurofibrillary tangles. Similarly, α-synuclein undergoes phase separation driven by intrinsic disorder and electrostatic interactions, particularly involving its C-terminal acidic tail and interaction partners such as prion protein, as demonstrated by Rodríguez et al. (2024). These and similar studies emphasize that LLPS actually serves as an early event in pathological aggregation, pointing out the delicate balance between reversible assembly and irreversible aggregation dictated by intrinsic disorder and PTMs (Morris et al., 2024; Rodríguez et al., 2024; Siegert et al., 2021; Zhang et al., 2024).

The biophysical principles that regulate LLPS by IDPs are affected by cellular conditions, such as protein concentration, ionic strength, and the presence of RNA molecules (Keating and Pappu, 2021; Zaslavsky et al., 2023). It has been found that long noncoding RNAs like NEAT1 serve as molecular scaffolds that recruit RBPs to increase local concentration and drive phase separation, forming condensates such as paraspeckles (a type of MLOs), due to pathological aggregate formation under neurodegenerative-like conditions (Guo et al., 2021). In addition, He et al. (2025) demonstrated how numerous RNAs are involved in LLPS, thereby showing that many RNAs function as scaffolds through multivalent interactions with RNA-binding domains of proteins.

LLPS-mediated condensates exist in different forms, from highly dynamic liquids to viscous gels and ultimately solid fibrils (Wang et al., 2021a, 2021b). This change in the phase state of biological matter, supported by intrinsic disorder, happens along with increased intermolecular β-sheet structures and irreversible amyloid formation (Yamaguchi et al., 2025). The transition from reversible LLPS assemblies to irreversible pathological aggregates encapsulates a critical mechanistic link between cellular organization and neurodegeneration, as seen in several diseases (Parkinson's disease [PD], Huntington's disease, and Prion diseases) (Elbaum-Garfinkle, 2019; Xu et al., 2022; Zhou, Chen, et al., 2024).

12.5 Pathological consequences of dysregulated condensation

Pathological consequences of dysregulated biomolecular condensation are increasingly recognized as central to the development of neurodegenerative diseases from the class of proteinopathies and some other brain pathologies. Under physiological conditions, dynamic phase separation enables proteins and RNAs to form liquid-like biomolecular condensates that compartmentalize cellular processes without membranes (Hurtle et al., 2023). However, perturbations, such as mutations, stress, or aging, can dysregulate this dynamic process and convert liquid-like condensates into solid-like, irreversible aggregates, driving pathology. Mechanistically, this phase transition from reversible condensation to pathogenic aggregation is linked to the accumulation of protein deposits characteristic of neurodegenerative diseases, including ALS, FTD, Alzheimer's disease (AD), and PD (Hu et al., 2025).

In ALS and FTD, mutations in critical RBPs such as FUS and TDP-43 have been shown to drive aberrant phase separation that promotes the formation of solid-like inclusions, which disrupt neuronal function and survival. Both FUS and TDP-43 normally engage in LLPS that supports essential RNA processing and homeostasis functions (Rummens and Da Cruz, 2025). However, mutations associated with familial forms of ALS and FTD alter their phase separation propensity, favoring pathological transitions from dynamic, liquid-like condensates to more static, solid aggregates that accumulate in neurons. Structural domains rich in glycine, aromatic, and serine residues in TDP-43's C-terminal domain confer aggregation-prone properties, especially when disease-linked mutations destabilize normal conformations (Rhine et al., 2020). These aberrant inclusions sequester RNA and proteins and impair neuronal RNA metabolism,leading to widespread transcriptional dysregulation that affects genes critical for neuronal development and synaptic function. For example, mutated TDP-43 variants show compromised RNA-binding and splicing regulation, reducing cell viability and impeding neurite growth, which reflects profound functional deficits at the cellular level (Jiang and Ngo, 2022).

Similarly, in AD, the tau protein undergoes pathological phase separation that facilitates its assembly into neurofibrillary tangles, one of the key pathological hallmarks of AD. Normal tau proteins phase separate transiently to regulate microtubule stability and cellular trafficking, but under conditions of oxidative stress, hyperphosphorylation, or mutation, tau condensates become more viscous and mature into solid-like fibrillar aggregates. These tau aggregates disrupt cytoskeletal architecture and induce neurotoxicity through mechanisms including synaptic dysfunction and impaired intracellular transport (Jiang et al., 2025). Concurrently, α-synuclein, a presynaptic protein linked to PD, similarly undergoes a pathological phase transition from soluble monomers or liquid-like droplets into solid-like aggregates that form Lewy bodies. Dysregulated condensation of α-synuclein leads to amyloid fibril formation, disrupting synaptic vesicle release machinery and triggering dopaminergic neuron death. These pathological condensates in AD and PD lose fluidity, becoming sticky and prone to clumping, which seeds amyloid fibril networks and propagates aggregate spread, accelerating neurodegeneration (Mukherjee et al., 2023).

The mechanistic understanding of phase separation in these neurodegenerative contexts has offered pivotal insights into how disruption of protein homeostasis, via disturbed phase behavior, underlies disease pathogenesis. Aberrant phase transitions link early, potentially reversible dynamic condensates to pathological solid aggregates, providing a continuum of states that mediate toxicity (Zbinden et al., 2020). This conceptual framework highlights novel therapeutic opportunities aimed at modulating phase behavior, such as small molecules or chaperones that restore liquid-like properties or prevent pathological maturation into aggregates. For example, targeting PTMs or RNA interactions that govern phase separation properties holds promise for reducing toxic aggregation and rescuing neuronal function (Zhao et al., 2025).

Furthermore, recent studies emphasize that pathological phase separation is not restricted to neurons or classical proteinopathies but extends to other brain pathologies. In gliopathies, dysfunction of astrocytes and microglia involves altered biomolecular condensation, impacting glial support and neuroimmune signaling (Huang et al., 2025a), with alterations in phase separation dynamics potentially contributing to glial dysregulation and neuroinflammation (Franklin et al., 2021). Viral encephalopathies are another area where virus–host interactions can coopt or disrupt biomolecular condensates and phase separation machinery, potentially exacerbating neuroinflammatory responses or promoting viral replication complexes (Sonneville et al., 2025). Additionally, neuroinflammatory processes themselves are modulated by condensate dynamics in immune cells, affecting cytokine signaling and inflammasome assembly, which influences the progression of neurodegenerative diseases and other brain injuries. These findings imply a broad relevance of condensate dysregulation and phase separation pathology as a convergent mechanism in diverse neuropathologies (Rocamonde et al., 2023), not only in neuronal but also in glial and immune brain cell dysfunction (Adamu et al., 2024) (see Fig. 12.1).

At the molecular level, certain proteins involved in the condensates harbor IDRs that are crucial for their phase separation properties but also render them vulnerable to pathological aggregation (Adamu et al., 2024). Aging diminishes the regulation of these processes, as exemplified by decreased chaperone function that normally maintains condensate dynamics. For instance, in stem and progenitor cells, phase separation regulated by isomerases and chromatin modulators influences cellular stress responses and senescence, linking dysregulated condensation with cellular aging and potentially neurodegeneration (Wei et al., 2025). Thus, both genetic and environmental factors impacting phase separation kinetics and condensate material properties converge to drive pathological transitions. Pathological consequences of dysregulated condensation are drawn in Fig. 12.2.

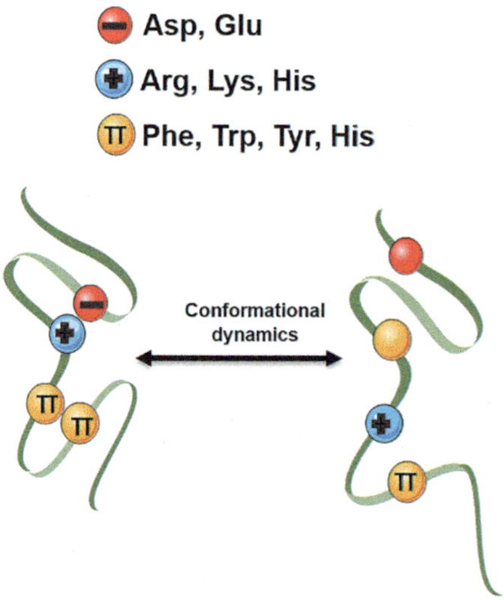

Figure 12.1 A schematic illustration of charge–π interplay in intrinsically disordered condensates. *The figure was generated using Servier Medical Art.*

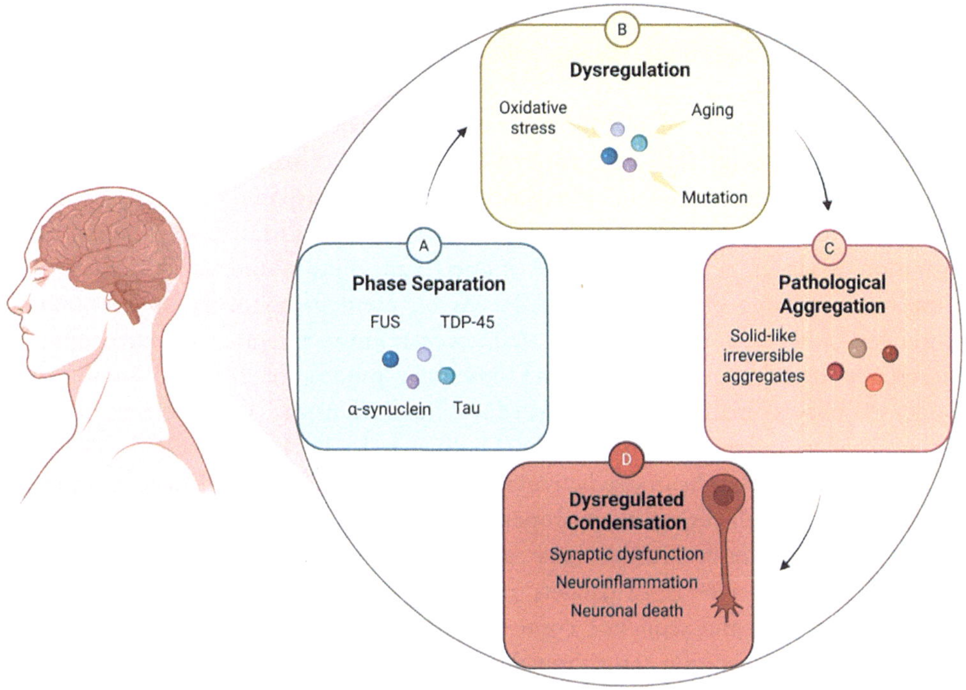

Figure 12.2 Consequences of dysregulated condensation. *The figure was generated using BioRender.*

12.6 Experimental and computational insights

Experimental and computational studies have profoundly advanced our understanding of IDPs and biomolecular condensates, which play critical roles in cellular organization and disease. Studying the structure and dynamics of IDPs and phase-separated condensates presents unique challenges due to their lack of stable tertiary structure and dynamic nature, but a variety of advanced experimental and computational methods have been developed and refined to elucidate their properties (Shi et al., 2025). Experimental approaches such as NMR spectroscopy, cryo-electron microscopy (cryo-EM), fluorescence microscopy, including fluorescence recovery after photobleaching (FRAP), and single-molecule imaging provide complementary insights into IDP conformations, intermolecular interactions, and condensate material properties (Naudi-Fabra et al., 2021). NMR, in particular, enables residue-specific characterization of IDP conformational ensembles and dynamics at atomic resolution even within condensed phases, illuminating molecular interactions that drive phase separation and maturation of condensates (Ahmed and Forman-Kay, 2022). Cryo-EM is increasingly applied to capture detailed architectures of phase-separated condensates and amyloid-like aggregates formed from IDPs, providing high-resolution structural snapshots that complement spectroscopic data (Li et al., 2025). Fluorescence microscopy and FRAP allow real-time visualization and quantification of condensate dynamics in living cells or in vitro, revealing their liquid-like or solid-like states, exchange kinetics, and response to perturbations (Wüstner, 2022).

Alongside experimental techniques, computational modeling has become indispensable for dissecting IDP and condensate behavior. Coarse-grained and all-atom molecular simulations complement experimental data by providing mechanistic insights into the molecular driving forces and structural transitions underpinning phase separation (Saar et al., 2023). These models can capture the effects of sequence composition, charge distribution, and PTMs on condensate formation and stability. Moreover, advanced computational frameworks now integrate spectroscopy and microscopy data to generate accurate ensemble models of IDPs in solution and condensed phases, enabling predictions of structural, dynamic, and interaction features that are challenging to measure directly (Mathur et al., 2024).

Recent breakthroughs in predictive algorithms have further enhanced the capacity to analyze IDP sequences and their propensity for phase separation. Traditional methods for predicting intrinsic disorder have evolved into sophisticated machine learning models that incorporate large datasets and diverse features such as residue physicochemical properties and sequence context (Chaurasiya and Lahiri, 2025). Particularly notable are next-generation algorithms designed specifically to predict phase separation propensity at residue-level resolution from primary sequences alone. For example, the phase separation transfer-learning prediction (PSTP) algorithm combines protein conformational embeddings with large language model embeddings to achieve state-of-the-art accuracy in identifying regions of proteins likely to drive phase separation (Feng et al., 2025). PSTP and similar models facilitate the rapid screening of thousands of natural and artificial protein variants to map phase separation domains, with important implications for understanding disease mutations and designing synthetic biomolecules. These algorithms run efficiently on typical Central Processing Units or Graphics Processing Units, are accessible via web servers and software packages, and represent a powerful resource for the biomedical community (Feng et al., 2025).

Notably, these predictive tools reveal that IDRs are central to phase separation but also highlight sequence features that modulate the balance between functional condensate formation and pathological aggregation. As the size of experimentally curated phase separation databases grows, predictive accuracy continues to improve, enabling more precise identification of molecular determinants. However, ongoing challenges remain in distinguishing genuine phase-separating proteins from those that can aggregate only under extreme or nonphysiological conditions, underscoring the need for careful training data curation and experimental validation (Ibrahim et al., 2023).

Together, the integration of complementary experimental methodologies and computational modeling forms a powerful framework for probing the complexity of

IDPs and biomolecular condensates. This synergy is crucial for elucidating how these dynamic assemblies regulate cellular processes and how their dysregulation contributes to diseases such as neurodegeneration and cancer (Zhu et al., 2025). The ongoing refinement of spectroscopic, imaging, and computational tools, coupled with emerging predictive algorithms, propels the field toward comprehensive models that connect sequence, structure, and function. Moreover, they facilitate the identification of novel therapeutic targets and small molecules aimed at modulating phase behavior to prevent pathological aggregation or engineer beneficial condensates (Zhang et al., 2025a, 2025b).

12.7 Therapeutic and translational perspectives

Therapeutic and translational perspectives targeting pathological protein condensation and aggregation have gained significant attention as accumulating evidence implicates dysregulated biomolecular condensation in neurodegenerative diseases and some other disorders. Strategies to modulate LLPS and prevent aberrant aggregation focus on multiple approaches, including small molecule modulators, peptides, and chaperone systems that target phase separation dynamics, as well as the identification of novel biomarkers and drug targets emerging from IDP research (Silva et al., 2023). These therapeutic avenues provide promising routes to intervene in disease cascades driven by condensate dysregulation.

12.7.1 Small molecules

Small molecules have emerged as a highly attractive and versatile class of modulators for biomolecular condensates, offering promising strategies to finely tune protein phase separation and aggregation implicated in a variety of diseases. These compounds are frequently designed either to restore normal phase behavior or to disrupt pathological condensate formation by altering intermolecular interactions, protein conformations, or the physicochemical environment within condensates (Usman et al., 2025). A classic example is 1,6-hexanediol, a small aliphatic alcohol widely used to dissolve liquid-like biomolecular condensates by disrupting weak hydrophobic interactions that drive phase separation. Despite its utility in experimental settings as a general probe of LLPS, the application of 1,6-hexanediol faces significant challenges: it lacks specificity, often interferes with diverse biological processes beyond condensates, and demonstrates notable cytotoxicity at commonly used concentrations (Düster et al., 2021). Recent studies recommend optimized conditions, such as a low concentration (1.5%) and brief exposure times (2 minutes) to minimize adverse cellular effects while maintaining effective condensate dissolution. However, even under these optimized conditions, 1,6-hexanediol affects chromatin dynamics and cell viability in sensitive cell types, limiting its direct therapeutic potential and underscoring the necessity for more selective condensate-targeting molecules (Liu et al., 2021).

More selective small molecules have been developed that modulate LLPS with improved specificity by targeting proteinsor other condensate components. Melatonin, a neurohormone with antioxidant properties, exemplifies such compounds capable of modulating phase separation behavior in disease-relevant proteins (Loh and Reiter, 2021). Recent work reveals that melatonin influences the phase behavior of aggregation-prone proteins involved in AD, including Aβ42 and tau (Zhang et al., 2025). Melatonin stabilizes these proteins in more soluble, monomeric forms, reducing their propensity to undergo pathological phase separation and aggregation. In vitro studies show that melatonin, at micromolar concentrations, inhibits amyloid fibril formation by Aβ peptides and produces soluble amorphous oligomers with diminished β-sheet content. Contrasting effects are observed with tau aggregates, where melatonin exhibits limited prevention of fibrillization but can fragment preformed fibrils, suggesting that its modulatory effects on phase separation are dependent on the specific protein and aggregation state (Loh and Reiter, 2023). Proposed mechanisms involve melatonin's capacity to regulate hydrophobic interactions and ATP-dependent molecular interactions that influence condensate dynamics and amyloid fibrillization. These findings position melatonin as an intriguing lead compound for modulating pathogenic phase transitions underlying neurodegeneration (Loh and Reiter, 2025).

A particularly innovative chemical strategy involves charge-driven covalent modification of folded domains within proteins containing IDRs, which are often central to phase separation. Proteins like G3BP1, a key stress granule nucleator, can be targeted by small molecules bearing carbonylacrylic amide functionality that covalently modify surface-exposed cysteines and introduce charged moieties (Chen et al., 2025). This modification shifts the conformational ensemble of G3BP1's IDRs towards more compact states with increased intra-domain interactions, thereby inhibiting aberrant stress granule assembly. Cellular experiments with mutated G3BP1 and structure–-activity relationship studies validate that these small molecules alter condensate formation by modulating IDR dynamics through a charge-driven mechanism (PMID: 36611004). Molecular dynamics simulations support a model where enhanced electrostatic interactions induce more compact IDR conformations, reducing the multivalent interactions that promote phase separation. Such charge-driven small molecules enable reversible and tunable control over condensate behavior with higher specificity and functional precision compared to general hydrophobic disruptors. This emerging chemical modality exemplifies targeted condensate modulation with promising applications in diseases involving stress granule pathology, including neurodegeneration and cancer (Vazquez et al., 2022).

Hydrophobicity is confirmed as a key driver for small molecule enrichment within condensates composed of hydrophobic low-complexity domains, while polar condensates recruit small molecules through more specific interactions, demonstrating that the physicochemical environment shapes molecular partitioning (Emelianova et al., 2025). These insights emphasize that successful condensate-targeting therapeutics must balance affinity and physicochemical compatibility with the condensate microenvironment to achieve selective modulation. Large-scale computational frameworks developed from these principles enable efficient prediction of small-molecule partitioning, facilitating rational drug design targeted toward distinct condensate types and disease contexts. Such approaches highlight the potential to engineer condensate-specific therapeutics with improved efficacy and minimized off-target effects by exploiting the interplay between compound properties and condensate composition (Ambadi Thody et al., 2024).

12.7.2 Peptide-based therapeutics

Peptide-based therapeutics have gained considerable attention as a highly specific and modular approach to modulate biomolecular condensates, which are critical cellular compartments formed via LLPS (Song et al., 2024). Inspired by naturally occurring protein interaction motifs, synthetic peptides can be rationally designed to mimic or competitively inhibit the multivalent interactions that drive phase separation, thus offering a targeted means to attenuate pathological aggregation while preserving physiological condensate functions. One of the key advantages of peptide-based therapeutics lies in their high specificity for condensate components, which emerges from the ability to tailor their amino acid sequences and physicochemical properties to interact with particular domains—especially IDRs or RNA-binding modules—critical for condensate formation and maintenance (Groß et al., 2015).

Recent studies demonstrate that peptides targeting RNA-binding domains of aggregation-prone proteins such as FUS and TDP-43 effectively prevent their aberrant phase transitions that underlie neurodegenerative diseases like ALS and FTD (Rummens and Da Cruz, 2025). These peptides can interfere with the multivalent interactions essential for nucleating and stabilizing pathological condensates, either by competitively binding to key interaction surfaces or by altering conformational dynamics within IDRs. This approach allows intervention at early disease stages before irreversible solid-like aggregates form, representing a promising therapeutic strategy to halt or delay neurodegeneration (Sanchez-Burgos et al., 2025). Additionally, peptides have been engineered to promote liquid-like behavior within condensates, stabilizing their dynamic and reversible states and thus preventing maturation into solid, neurotoxic aggregates. Such engineered peptides modulate the material properties of condensates to maintain cellular homeostasis (Kochman et al., 2025).

The modularity of peptide-based systems facilitates fine-tuning of their phase separation behavior and biological activity. Recent advances leverage a detailed understanding of molecular grammar—how sequence composition, net charge, hydrophobicity, and aromatic residue placement control interaction specificity and phase behavior—to guide peptide design (Shi et al., 2025). For instance, peptides can be constructed to localize preferentially at condensate interfaces, acting as "surfactants" that remodel phase boundaries. These interface-localizing peptides often feature amphiphilic architectures, with one segment enriched in aromatic residues that interact with condensate interiors and another segment with charged or uncharged residues that control exclusion from the dense phase, thereby enabling tailored modulation of condensate size, shape, and dynamics (Alfano et al., 2024).

Beyond direct inhibition of pathogenic condensates, peptide-based condensates themselves are being explored as synthetic biomimetic compartments that replicate the biophysical properties of intracellular phase-separated droplets. These peptide coacervates can partition and concentrate biologically relevant solutes, serving as vehicles for targeted drug delivery or as tools for tissue engineering by mimicking membrane-less organelles (Song et al., 2025). The diversity of amino acid residues and programmable sequence designs enables custom control over phase separation behavior, droplet viscoelasticity, responsiveness to environmental cues, and molecular partitioning. Such biomimicry expands the scope of peptide therapeutics beyond inhibition toward engineering beneficial condensates with novel biomedical applications, including localized delivery of therapeutic agents to pathological condensates (Calvário et al., 2025).

Peptide therapeutics also complement small molecule drugs by overcoming challenges such as limited specificity and off-target effects. Their ability to engage multiple weak interactions synergistically within condensates, modulate local physicochemical environments, and adapt dynamically to condensate changes offers distinct mechanistic advantages (Rossino et al., 2023). Importantly, peptides can be chemically modified to enhance stability, cellular uptake, and pharmacokinetics, addressing traditional limitations of peptide drugs. Moreover, combination therapies employing peptides alongside small molecules or chaperone modulators would seem to promise multipronged control over phase separation phenomena, broadening therapeutic windows and reducing toxicity (Zheng et al., 2025).

12.7.3 Molecular chaperone systems

Molecular chaperone systems constitute a critical endogenous mechanism for regulating protein folding and mitigating aggregation, including within biomolecular condensates that organize cellular biochemical processes (Bard and Drummond, 2024). Heat shock proteins (Hsp70, Hsp90) and cochaperones such as the carboxyl terminus of Hsc70-interacting protein (CHIP) play pivotal roles in controlling the assembly, maintenance, and clearance of pathogenic protein aggregates implicated in neurodegenerative diseases. These chaperones aid proteostasis by preventing aberrant protein–protein interactions, unfolding misfolded species, and targeting defective proteins for degradation via the ubiquitin–proteasome system or autophagy pathways (Ranek et al., 2018). In biomolecular condensates, which involve multivalent and often weak protein interactions among IDRs, chaperones regulate not only protein conformation but also condensate dynamics and material states, maintaining a balance between functional liquid-like assemblies and pathological solid aggregates (Vazquez et al., 2022).

For instance, CHIP has been shown to mediate the ubiquitination and subsequent proteasomal degradation of hyperphosphorylated tau and α-synuclein, key aggregation-prone proteins driving neurotoxicity in AD and PD, respectively (Ho and Wing, 2024). By promoting the clearance of these toxic species, CHIP reduces their accumulation within neuronal inclusions and attenuates downstream cellular dysfunction. Furthermore, enhancing the recruitment of molecular chaperones to aberrant condensates or pharmacologically activating their ATPase activity stabilizes aggregation-prone domains within droplets, preventing their maturation into solid, insoluble aggregates (VanPelt and Page, 2017). This stabilization of the dynamic, reversible droplet state is fundamental to preserving cellular homeostasis and preventing the seeding of pathogenic amyloids. Experimental models demonstrate that modulating chaperone activity influences the formation and dissolution kinetics of stress granules and other physiological condensates, highlighting chaperones as central regulators of condensate fate under both normal and stress conditions (Akaree et al., 2025).

Recent research efforts have focused on developing small molecules that can specifically modulate chaperone function or their interactions with cochaperones to harness their protective roles against proteinopathies linked to phase separation (Ho and Wing, 2024). Compounds that allosterically activate Hsp70 or enhance CHIP-mediated ubiquitination hold therapeutic promise by boosting the natural cellular machinery for quality control and aggregate clearance. Additionally, some small molecules facilitate the disaggregation of mature fibrils by chaperone complexes, offering avenues to reverse established pathology rather than solely prevent aggregation. Such strategies capitalize on the intricate network of chaperones and cochaperones that orchestrate protein quality control within condensates, extending therapeutic opportunities beyond traditional enzyme inhibition or receptor targeting (Hipp and Hartl, 2024). Importantly, the dual role of chaperones in regulating both physiological condensate assembly and pathological aggregate clearance underscores their candidacy for early disease intervention. By preventing aberrant phase transitions or promoting the resolubilization of nascent aggregates, chaperones may forestall irreversible cellular damage seen in neurodegenerative conditions (Akaree et al., 2025).

12.7.4 Other approaches

Beyond therapeutic modulation, the field is rapidly advancing in identifying novel biomarkers and drug targets arising from IDP and biomolecular condensate biology.

Systems biology approaches, including protein–protein interaction network analyses, have uncovered key regulatory hub genes and transcription factors that control proteins involved in pathological condensates. These regulatory molecules, along with their associated posttranslational modification enzymes, are emerging as attractive drug targets with the potential to modulate condensate formation and pathogenic aggregation (Qin et al., 2025). Advances in large-scale proteomics and transcriptomics now allow comprehensive mapping of disease-relevant condensates and their molecular composition, facilitating the discovery of mechanistically relevant biomarkers and signaling pathways that inform therapeutic development and clinical trial designs (Maiti et al., 2025).

For example, plasma proteomic studies have identified immune-related proteins such as interleukin receptors IL23R and IL7R, and thymidine phosphorylase, that are associated with neuroinflammatory and neurodegenerative states. These proteins hold promise as accessible biomarkers for disease risk, progression, and therapeutic response, potentially enabling early diagnostics and personalized treatment strategies (Huang et al., 2025b). Importantly, plasma proteomes reveal both disease-specific and transdiagnostic signatures across AD, PD, FTD, and ALS, reflecting common molecular pathways of neurodegeneration. The establishment of large harmonized datasets by consortia such as the Global Neurodegeneration Proteomics Consortium (GNPC) catalyzes biomarker discovery by integrating tens of thousands of biofluid samples with clinical and genetic data, accelerating translation to clinical practice (Imam et al., 2025).

Integration of experimental and computational technologies represents a crucial driver of translational progress in this domain. High-throughput screening platforms combined with advanced predictive algorithms enable efficient identification of candidate small molecules and peptides that favorably influence phase separation propensity and condensate stability. Molecular simulations of IDP dynamics and condensate material properties guide rational drug design by revealing relevant sequence–structure–function relationships underlying pathogenic aggregation. This integrated pipeline expedites the discovery of compounds capable of modulating condensate formation and dissolution under physiological conditions, thus improving the likelihood of clinical success (Zhu et al., 2025). Furthermore, distinguishing physiologically beneficial condensates from pathological assemblies requires deeper mechanistic understanding and robust biomarkers capable of discriminating between these states. Early phase clinical trials incorporating biomarkers of condensate activity are emerging, focusing on mechanism-driven interventions for neurodegenerative diseases such as ALS, AD, and PD. Trial designs leveraging umbrella and basket frameworks with biomarker-guided patient stratification show promise in improving therapeutic efficacy by matching treatments to molecular signatures linked to condensate dysfunction. The summary of therapeutic perspectives can be seen in Table 12.2.

Table 12.2 Therapeutic and translational strategies targeting pathological biomolecular condensation.

Therapeutic category	Mechanism of action	Representative compounds/systems	Key biological targets/effects	Advantages	Limitations/challenges	Representative references/PMIDs
Small molecule modulators	Alter intermolecular interactions within condensates to restore liquid-like states or prevent aggregation	1,6-Hexanediol (hydrophobic disruptor); melatonin (antioxidant and phase modulator); carbonylacrylic amide-based compounds (charge-driven covalent modifiers)	Disrupt weak hydrophobic forces (liquid–liquid phase separation [LLPS]); stabilize tau/Aβ monomers; modify cysteines in G3BP1 IDRs to inhibit stress granule formation	Rapid cell permeability; tunable activity; reversible modulation	Cytotoxicity (e.g., 1,6-hexanediol); lack of specificity; limited bioavailability	Emelianova et al. (2025), Gutiérrez-Garcia et al. (2023), Loh and Reiter (2023); Liu et al. (2021)
Peptide-based therapeutics	Compete with protein–-protein or protein–-RNA interactions; mimic molecular grammar governing LLPS	Peptides targeting TDP-43, fused in sarcoma, and RNA-binding domains; interface-localizing amphiphilic peptides; synthetic peptide coacervates	Inhibit aggregation-prone interactions; maintain liquid-like condensates; mimic membraneless organelles for drug delivery	High specificity; modular design; customizable sequence chemistry	Limited stability; delivery barriers; potential immunogenicity	Alfano et al. (2024), Calvário et al. (2025), Sanchez-Burgos et al. (2025)

(Continued)

Table 12.2 (Continued)

Therapeutic category	Mechanism of action	Representative compounds/ systems	Key biological targets/effects	Advantages	Limitations/ challenges	Representative references/ PMIDs
Molecular chaperone systems	Stabilize or refold aggregation-prone proteins; promote degradation of misfolded species	Hsp70, Hsp90, CHIP (cochaperone); Hsp70 activators	Regulate proteostasis; ubiquitinate tau and α-synuclein; prevent solid aggregate maturation	Endogenous regulation; reversible and ATP-dependent	Complex regulation; requires cochaperone coordination; ATP cost	Akaree et al. (2025), Ho and Wing (2024), Wang et al. (2021)
Systems biology and biomarker discovery	Identify key regulators, hub genes, and protein networks linked to condensate dysfunction	Proteomic profiling (IL23R, IL7R, thymidine phosphorylase); Global Neurodegeneration Proteomics Consortium datasets	Early detection of neurodegenerative and neuroinflammatory signatures; patient stratification for trials	Enables personalized medicine; mechanistic insight	Requires large cohort validation; interdisease variability	Huang et al. (2025), Imam et al. (2025)

Integrated experimental–computational pipelines	Combine in silico modeling, high-throughput screening, and experimental assays for target validation	AI-based phase separation predictors (PSTP, ML models) + HTS libraries	Predict sequence determinants of LLPS; identify small molecules and peptide modulators efficiently	Accelerate discovery; rational design of condensate modulators	Integration complexity; reliance on high-quality datasets	Zhang et al. (2025), Zhu et al. (2025)
Emerging clinical translation	Incorporate condensate biomarkers and phase behavior into trial design	Biomarker-driven umbrella and basket clinical trials for ALS, AD, and PD	Mechanism-driven therapeutic stratification	Improves efficacy and patient matching	Early stage framework; few completed trials	Ongoing (2024–2025) translational efforts

12.8 Conclusions and future directions

The growing understanding of IDPs has fundamentally reshaped our perception of brain proteinopathies by emphasizing their dynamic and regulatory roles that extend well beyond classical structural paradigms. Traditionally, proteins lacking a fixed three-dimensional structure were regarded as unstable or misfolded, whose presence was mostly associated with dysfunction or disease. However, a paradigm shift has occurred, recognizing IDPs as central orchestrators of cellular organization, particularly through their unique capacity to mediate LLPS. This biophysical process enables IDPs to drive the formation of biomolecular condensates, membrane-less compartments that dynamically concentrate specific proteins and nucleic acids in a reversible and highly regulated manner.

The ability of IDPs to be engaged in LLPS connects molecular disorder with precise spatiotemporal control of critical neuronal functions, including signaling pathways, synaptic plasticity, and cellular adaptation to stress. Unlike fixed complexes, these condensates provide a flexible scaffold that can rapidly assemble and disassemble in response to cellular cues, allowing the brain's intricate networks to maintain homeostasis and respond to changing environmental conditions. By integrating IDP biology into the broader context of brain proteinopathies, researchers have developed a unifying conceptual framework that links normal physiological biomolecular condensation processes with the pathological protein aggregation observed in neurodegenerative disorders. This continuum spans from healthy, reversible condensate formation that supports cellular compartmentalization to irreversible and toxic aggregation events underlying conditions such as AD, PD, and ALS.

In these diseases, dysregulated condensate dynamics often precede and potentiate the formation of insoluble protein aggregates, accompanied by disrupted RNA metabolism and maladaptive cellular stress responses. Understanding that IDPs act as drivers of both physiological organization and pathological misfolding marks a significant advancement in mechanistic insight. This shift also refines therapeutic strategies, moving away from targeting the static endpoints of protein aggregation toward earlier, reversible stages of condensate dysregulation. Such approaches hold promise for interventions that restore normal condensate behavior, thereby preventing or halting disease progression at its roots.

Looking ahead, the next frontier in IDP and condensate research lies in systems-level and single-cell investigations that capture the full dynamic complexity of these biomolecular assemblies within their native cellular and tissue contexts. Recent technological advances, such as multiomics integration, live-cell superresolution imaging, and high-throughput biophysical assays, now enable comprehensive mapping of condensate networks across diverse brain cell types, developmental stages, and pathological conditions. Single-cell proteomic and transcriptomic techniques allow detailed characterization of IDP expression patterns, PTMs, and condensate compositions at unprecedented resolution. These insights will reveal how stochastic cellular heterogeneity modulates vulnerability or resilience to neurodegeneration, uncovering cell type–specific condensate signatures linked to disease risk.

Furthermore, systems biology approaches that synergize structural modeling, network theory, and artificial intelligence are poised to elucidate emergent properties of condensate regulation at the organismal level. Such integrative frameworks can identify functional crosstalk between neuronal, glial, and immune cell condensates, revealing how intercellular communication influences brain health and disease. By constructing predictive, personalized models of brain proteinopathies grounded in condensate dynamics, researchers envision new therapeutic avenues that actively modulate these processes to restore cellular homeostasis and function.

In essence, the future of brain proteinopathy research lies at the intersection of IDP biology, multiscale experimental methods, and systems-level data modeling, transforming our understanding from a view of molecular chaos and aggregation to one of highly regulated cellular complexity and adaptability. This comprehensive perspective not only advances fundamental science but also drives innovation toward targeted, early-stage interventions that may ultimately revolutionize the treatment of devastating neurodegenerative disorders.

Last but not least, when a mutation of the gene is causing the proteinopathy, new avenues of how to change or edit the gene will also help in rare familial neurodegenerative disease cases.

Abbrevations

CPUs Central processing units
GPUs Graphics processing units

References

Adamu, A., Li, S., Gao, F., Xue, G., 2024. The role of neuroinflammation in neurodegenerative diseases: current understanding and future therapeutic targets. Frontiers in Aging Neuroscience 16, 1663–4365. Available at https://doi.org/10.3389/fnagi.2024.1347987.

Ahmed, R., Forman-Kay, J.D., 2022. NMR insights into dynamic, multivalent interactions of intrinsically disordered regions: from discrete complexes to condensates. Essays in Biochemistry 66 (7), 863–873. Available at https://doi.org/10.1042/EBC20220056.

Akaree, N., Secco, V., Levy-Adam, F., Younis, A., Carra, S., Shalgi, R., 2025. Regulation of physiological and pathological condensates by molecular chaperones. The FEBS Journal 292 (13), 3271–3297. Available at https://doi.org/10.1111/febs.17390.

Alberti, S., Arosio, P., Best, R.B., Boeynaems, S., Cai, D., Collepardo-Guevara, R., et al., 2025. Current practices in the study of biomolecular condensates: a community comment. Nature Communications 16 (1). Available at https://doi.org/10.1038/s41467-025-62055-8.

Alfano, C., Fichou, Y., Huber, K., Weiss, M., Spruijt, E., Ebbinghaus, S., et al., 2024. Molecular crowding: the history and development of a scientific paradigm. Chemical Reviews 124 (6), 3186–3219. Available at https://doi.org/10.1021/acs.chemrev.3c00615.

Ambadi Thody, S., Clements, H.D., Baniasadi, H., Lyon, A.S., Sigman, M.S., Rosen, M.K., 2024. Small-molecule properties define partitioning into biomolecular condensates. Nature Chemistry 16 (11), 1794–1802. Available at https://doi.org/10.1038/s41557-024-01630-w.

Ayyadevara, S., Ganne, A., Balasubramaniam, M., Shmookler Reis, R.J., 2022. Intrinsically disordered proteins identified in the aggregate proteome serve as biomarkers of neurodegeneration. Metabolic Brain Disease 37 (1), 147–152. Available at https://doi.org/10.1007/s11011-021-00791-8.

Bard, J.A.M., Drummond, D.A., 2024. Chaperone regulation of biomolecular condensates. Frontiers in Biophysics 2, 2813–7183. Available at https://doi.org/10.3389/frbis.2024.1342506.

Baymiller, M., Moon, S.L., 2023. Stress granules as causes and consequences of translation suppression. Antioxidants and Redox Signaling 39 (4–6), 390–409. Available at https://doi.org/10.1089/ars.2022.0164.

Bhargava, P., Yadav, P., Barik, A., 2024. Computational insights into intrinsically disordered regions in protein-nucleic acid complexes. International Journal of Biological Macromolecules 277 (Pt 1), 134021. Available at https://doi.org/10.1016/j.ijbiomac.2024.134021.

Bondos, S.E., Dunker, A.K., Uversky, V.N., 2022. Intrinsically disordered proteins play diverse roles in cell signaling. Cell Communication and Signaling 20 (1). Available at https://doi.org/10.1186/s12964-022-00821-7.

Calvário, J., Antunes, D., Cipriano, R., Kalafatovic, D., Mauša, G., Pina, A.S., 2025. Understanding biases in liquid–liquid phase separation: investigating amino acid enrichments in phase-separating proteins toward peptide design. Biomacromolecules 26 (11), 7247–7264. Available at https://doi.org/10.1021/acs.biomac.4c00224.

Cascarina, S.M., Ross, E.D., 2025. Protein activities driven by amino acid composition. Journal of Biological Chemistry 301 (10). Available at https://doi.org/10.1016/j.jbc.2025.110640.

Cedano, J., Huerta, M., Mozo-Villarias, A., Querol, E., 2025. Moonlighting proteins: some hypotheses on the structural origin of their multifunctionality. International Journal of Molecular Sciences 26 (21), 10375. Available at https://doi.org/10.3390/ijms262110375.

Cermakova, K., Hodges, H.C., 2023. Interaction modules that impart specificity to disordered protein. Trends in Biochemical Sciences 48 (5), 477–490. Available at https://doi.org/10.1016/j.tibs.2023.01.004.

Chaurasiya, D., Lahiri, T., 2025. Decoding disorder: advances in predicting intrinsically disordered proteins. Current Proteomics 22 (3), 100030. Available at https://doi.org/10.1016/j.curpro.2025.100030.

Chen, Z., Liu, L., Cattin, J., Knowles, T.P.J., Bernardes, G.J.L., 2025. A charge-driven strategy for covalent modification and modulation of biomolecular condensates. Journal of the American Chemical Society 147 (32), 28558–28563. Available at https://doi.org/10.1021/jacs.5c06625.

Chow, C.F.W., Lenz, S., Scheremetjew, M., Ghosh, S., Richter, D., Jegers, C., et al., 2025. SHARK-capture identifies functional motifs in intrinsically disordered protein regions. Protein Science 34 (4). Available at https://doi.org/10.1002/pro.70091.

Chu, W.-T., Wang, J., 2021. Thermodynamic and sequential characteristics of phase separation and droplet formation for an intrinsically disordered region/protein ensemble. PLoS Computational Biology 17 (3), e1008672. Available at https://doi.org/10.1371/journal.pcbi.1008672.

Cinar, H., Winter, R., 2020. The effects of cosolutes and crowding on the kinetics of protein condensate formation based on liquid–liquid phase separation: a pressure-jump relaxation study. Scientific Reports 10 (1), 2045–2322. Available at https://doi.org/10.1038/s41598-020-74271-x.

Das, S., Lin, Y.H., Vernon, R.M., Forman-Kay, J.D., Chan, H.S., 2020. Comparative roles of charge, π, and hydrophobic interactions in sequence-dependent phase separation of intrinsically disordered proteins. Proceedings of the National Academy of Sciences of the United States of America 117 (46), 28795–28805. Available at https://doi.org/10.1073/pnas.2008122117.

Deng, B., Wan, G., 2024. Technologies for studying phase-separated biomolecular condensates. Advanced Biotechnology 2, 10. Available at https://doi.org/10.1007/s44307-024-00020-0.

Dignon, G.L., Zheng, W., Kim, Y.C., Mittal, J., 2019. Temperature-controlled liquid-liquid phase separation of disordered proteins. ACS Central Science 5 (5), 821–830. Available at https://doi.org/10.1021/acscentsci.9b00102.

Dougherty, D.D.A., 1996. Cation-π interactions in chemistry and biology: a new view of benzene, Phe, Tyr, and Trp. Science (New York, N.Y.) 271 (5246), 163–168. Available at https://doi.org/10.1126/science.271.5246.163.

Düster, R., Kaltheuner, I.H., Schmitz, M., Geyer, M., 2021. 1,6-Hexanediol, commonly used to dissolve liquid–liquid phase separated condensates, directly impairs kinase and phosphatase activities. Journal of Biological Chemistry 296, 100260. Available at https://doi.org/10.1016/j.jbc.2021.100260.

Elbaum-Garfinkle, S., 2019. Matter over mind: liquid phase separation and neurodegeneration. Journal of Biological Chemistry 294 (18), 7160–7168. Available at https://doi.org/10.1074/jbc.rev118.001188.

Emelianova, A., Garcia, P.L., Tan, D., Joseph, J.A., 2025. Prediction of small-molecule partitioning into biomolecular condensates from simulation. JACS Au 5 (7), 3125–3139. Available at https://doi.org/10.1021/jacsau.5c00291.

Fedorov, D., Sammalisto, F.E., Harmat, A.L., Ahlberg, M., Koskela, S., Haataja, M.P., et al., 2025. Metastable liquid–liquid phase separation and aging lead to strong processing path dependence in mini-spidroin solutions. Advanced Functional Materials 35 (15). Available at https://doi.org/10.1002/adfm.202410421.

Feng, M., Liu, L., Xian, Z.N., Wei, X., Li, K., Yan, W., et al., 2025. PSTP: accurate residue-level phase separation prediction using protein conformational and language model embeddings. Briefings in Bioinformatics 26 (3). Available at https://doi.org/10.1093/bib/bbaf171.

Firouzbakht, A., Haider, A., Gaalswyk, K., Alaeen, S., Ghosh, K., Gruebele, M., 2024. HYPK: a marginally disordered protein sensitive to charge decoration. Proceedings of the National Academy of Sciences 121 (18), 0027–8424. Available at https://doi.org/10.1073/pnas.2316408121.

Franklin, H., Clarke, B.E., Patani, R., 2021. Astrocytes and microglia in neurodegenerative diseases: lessons from human in vitro models. Progress in Neurobiology 200, 101973. Available at https://doi.org/10.1016/j.pneurobio.2020.101973.

Galvanetto, N., Ivanović, M.T., Del Grosso, S.A., Chowdhury, A., Sottini, A., Nettels, D., et al., 2025. Material properties of biomolecular condensates emerge from nanoscale dynamics. Proceedings of the National Academy of Sciences of the United States of America 122 (23). Available at https://doi.org/10.1073/pnas.2424135122 e2424135122.

Ghafouri, H., Lazar, T., Del Conte, A., Tenorio Ku, L.G., Tompa, P., Tosatto, S.C.E., et al., 2024. PED in 2024: improving the community deposition of structural ensembles for intrinsically disordered proteins. Nucleic Acids Research 52 (1), D536. Available at https://doi.org/10.1093/nar/gkad947.

Ghitti, M., Colley, L.S., Mantonico, M.V., Musco, G., Bianchi, M.E., 2025. Intrinsic disorder and fuzzy interactions drive multiple functions of HMGB1. Trends in Biochemical Sciences 50 (10), 906–918. Available at https://doi.org/10.1016/j.tibs.2025.08.001.

Groß, A., Hashimoto, C., Sticht, H., Eichler, J., 2015. Synthetic peptides as protein mimics. Frontiers in Bioengineering and Biotechnology 3, 211. Available at https://doi.org/10.3389/fbioe.2015.00211.

Guan, H., Wang, H., Cai, X., Wang, J., Chai, Z., Wang, J., et al., 2025. Liquid-liquid phase separation of membrane-less condensates: from biogenesis to function. Frontiers in Cell and Developmental Biology 13. Available at https://doi.org/10.3389/fcell.2025.1600430.

Guo, Q., Shi, X., Wang, X., 2021. RNA and liquid-liquid phase separation. Non-coding RNA Research 6 (2), 92–99. Available at https://doi.org/10.1016/j.ncrna.2021.04.003.

Gupta, M.N., Uversky, V.N., 2022. Chapter 10 - Role of plasticity and disorder in protein moonlighting: blurring of lines between biocatalysts and other biologically active proteins. In: Gupta, M.N., Uversky, V.N. (Eds.), Structure and Intrinsic Disorder in Enzymology. Academic Press, pp. 279–301. Available at https://doi.org/10.1016/B978-0-323-99533-7.00003-0.

Gupta, V., Kumari, P., Sonowal, K., Sathe, A., Mehta, K., Salvi, P., 2025. Molecular intricacies of intrinsically disordered proteins and drought stress in plants. International Journal of Biological Macromolecules 292, 139314. Available at https://doi.org/10.1016/j.ijbiomac.2024.139314.

Gutiérrez-Garcia, R., Koyuncu, S., Hommen, F., Bilican, S., Lee, H.J., Fatima, A., et al., 2023. G3BP1-dependent mechanism suppressing protein aggregation in Huntington's models and its demise upon stress granule assembly. Human Molecular Genetics 32 (10), 1607–1621. Available at https://doi.org/10.1093/hmg/ddac304.

Hazra, M.K., Levy, Y., 2023. Cross-talk of cation−π interactions with electrostatic and aromatic interactions: a salt-dependent trade-off in biomolecular condensates. Journal of Physical Chemistry Letters 14 (38), 8460–8469. Available at https://doi.org/10.1021/acs.jpclett.3c01642.

He, Y., Bao, X., Chen, T., Jiang, Q., Zhang, L., He, L.N., et al., 2025. RPS 2.0: an updated database of RNAs involved in liquid–liquid phase separation. Nucleic Acids Research 53 (1), D299. Available at https://doi.org/10.1093/nar/gkae951.

Hipp, M.S., Hartl, F.U., 2024. Interplay of proteostasis capacity and protein aggregation: implications for cellular function and disease. Journal of Molecular Biology 436 (14). Available at https://doi.org/10.1016/j.jmb.2024.168615.

Ho, H.H., Wing, S.S., 2024. α-Synuclein ubiquitination – functions in proteostasis and development of Lewy bodies. Frontiers in Molecular Neuroscience 17. Available at https://doi.org/10.3389/fnmol.2024.1498459.

Hofweber, M., Hutten, S., Bourgeois, B., Spreitzer, E., Niedner-Boblenz, A., Schifferer, M., et al., 2018. Phase separation of FUS Is suppressed by its nuclear import receptor and arginine methylation. Cell 173 (3), 706. Available at https://doi.org/10.1016/j.cell.2018.03.004.

Hsiao, A.-S., 2024. Protein disorder in plant stress adaptation: from late embryogenesis abundant to other intrinsically disordered proteins. International Journal of Molecular Sciences 25 (2), 1178. Available at https://doi.org/10.3390/ijms25021178.

Hu, L., Huang, Z., Liu, Z., Zhang, Y., 2025. Biomolecular phase separation in tumorigenesis: from aberrant condensates to therapeutic vulnerabilities. Molecular Cancer 24 (1), 1476–4598. Available at https://doi.org/10.1186/s12943-025-02428-1.

Huang, M., Long, A., Hao, L., Shi, Z., Zhang, M., 2025a. Astrocyte in neurological disease: pathogenesis and therapy. MedComm 6 (8), e70299. Available at https://doi.org/10.1002/mco2.70299.

Huang, Y.N., Liu, S., Park, T., Chaudhuri, S., Kuchenbecker, L.A., Carrasquillo, M.M., et al., 2025b. Plasma proteomic analysis identifies proteins and pathways related to Alzheimer's risk. Alzheimer's and Dementia 21 (8). Available at https://doi.org/10.1002/alz.70579.

Hurtle, B.T., Xie, L., Donnelly, C.J., 2023. Disrupting pathologic phase transitions in neurodegeneration. Journal of Clinical Investigation 133 (13). Available at https://doi.org/10.1172/JCI168549.

Ibrahim, A.Y., Khaodeuanepheng, N.P., Amarasekara, D.L., Correia, J.J., Lewis, K.A., Fitzkee, N.C., et al., 2023. Intrinsically disordered regions that drive phase separation form a robustly distinct protein class. Journal of Biological Chemistry 299 (1). Available at https://doi.org/10.1016/j.jbc.2022.102801.

Imam, F., Saloner, R., Vogel, J.W., Krish, V., Abdel-Azim, G., Ali, M., et al., 2025. S. Lovestone, the global neurodegeneration proteomics consortium: biomarker and drug target discovery for common neurodegenerative diseases and aging. Nature Medicine 31 (8), 2556–2566. Available at https://doi.org/10.1038/s41591-025-03834-0.

Jiang, L., Ngo, S.T., 2022. Altered TDP-43 structure and function: key insights into aberrant RNA, mitochondrial, and cellular and systemic metabolism in amyotrophic lateral sclerosis. Metabolites 12 (8), 709. Available at https://doi.org/10.3390/metabo12080709.

Jiang, G., Xie, G., Li, X., Xiong, J., 2025. Cytoskeletal proteins and Alzheimer's disease pathogenesis: focusing on the interplay with tau pathology. Biomolecules 15 (6), 831. Available at https://doi.org/10.3390/biom15060831.

Keating, C.D., Pappu, R.V., 2021. Liquid-liquid phase separation: a widespread and versatile way to organize aqueous solutions. Journal of Physical Chemistry B 125 (45), 12399–12400. Available at https://doi.org/10.1021/acs.jpcb.1c08831.

Kitamura, N., Galligan, J.J., 2023. A global view of the human post-translational modification landscape. Biochemical Journal 480 (16), 1241–1265. Available at https://doi.org/10.1042/BCJ20220251.

Kochman, U., Sitka, H., Kuźniar, J., Czaja, M., Kozubek, P., Beszłej, J.A., et al., 2025. Neuronal death and biomolecular condensates: are there any new treatment options for Alzheimer's disease? Cells 14 (17), 1356. Available at https://doi.org/10.3390/cells14171356.

Lang, R., Hodgson, R.E., Shelkovnikova, T.A., 2024. TDP-43 in nuclear condensates: where, how, and why. Biochemical Society Transactions 52 (4), 1809–1825. Available at https://doi.org/10.1042/BST20231447.

Li, J.Q., Yan, Z.S., Ma, Y.Q., Ding, H.M., 2025. Sequence-dependent biomolecular phase separation driven by short-range interaction: from material properties to coarsening dynamics. Physical Review E 111 (5). Available at https://doi.org/10.1103/PhysRevE.111.054403.

Li, Y., Liu, Y., Yu, X.Y., Xu, Y., Pan, X., Sun, Y., et al., 2024. Membraneless organelles in health and disease: exploring the molecular basis, physiological roles and pathological implications. Signal Transduction and Targeted Therapy 9 (1). Available at https://doi.org/10.1038/s41392-024-02013-w.

Liu, W., Chen, L., Yin, D., Yang, Z., Feng, J., Sun, Q., et al., 2023. Visualizing single-molecule conformational transition and binding dynamics of intrinsically disordered proteins. Nature Communications 14 (1), 5203. Available at https://doi.org/10.1038/s41467-023-41018-x.

Liu, C., Wu, K., Choi, H., Han, H.L., Zhang, X., Watson, J.L., et al., 2025. Diffusing protein binders to intrinsically disordered proteins. Nature 644 (8077), 809–817. Available at https://doi.org/10.1038/s41586-025-09248-9.

Liu, X., Jiang, S., Ma, L., Qu, J., Zhao, L., Zhu, X., et al., 2021. Time-dependent effect of 1,6-hexanediol on biomolecular condensates and 3D chromatin organization. Genome Biology 22 (1), 230. Available at https://doi.org/10.1186/s13059-021-02455-3.

Loh, D., Reiter, R.J., 2021. Melatonin: regulation of biomolecular condensates in neurodegenerative disorders. Antioxidants 10 (9), 1483. Available at https://doi.org/10.3390/antiox10091483.

Loh, D., Reiter, R.J., 2023. Light, water, and melatonin: the synergistic regulation of phase separation in dementia. International Journal of Molecular Sciences 24 (6), 5835. Available at https://doi.org/10.3390/ijms24065835.

Loh, D., Reiter, R.J., 2025. Melatonin regulation of phase separation in Neuro-PASC: out-maneuvering Janus-faced amyloids. Exploration of Neuroscience 4. Available at https://doi.org/10.37349/en.2025.100678.

Lotthammer, J.M., Ginell, G.M., Griffith, D., Emenecker, R.J., Holehouse, A.S., 2024. Direct prediction of intrinsically disordered protein conformational properties from sequence. Nature Methods 21 (3), 465–476. Available at https://doi.org/10.1038/s41592-023-02159-5.

Maiti, S., De, S., 2022. Identification of potential short linear motifs (SLiMs) in intrinsically disordered sequences of proteins by fast time-scale backbone dynamics. Journal of Magnetic Resonance Open 10–11, 100029. Available at https://doi.org/10.1016/j.jmro.2021.100029.

Maiti, S., Singh, A., Maji, T., Saibo, N.V., De, S., 2024. Experimental methods to study the structure and dynamics of intrinsically disordered regions in proteins. Current Research in Structural Biology 7, 100138. Available at https://doi.org/10.1016/j.crstbi.2024.100138.

Maiti, S., Tripathi, S., Baggett, D.W., Phillips, A.H., Park, C.G., Wang, J., et al., 2025. Proteome-wide computational analyses reveal links between protein condensate formation and RNA biology. bioRxiv. Available at https://doi.org/10.1101/2025.03.03.640993.

Mathur, A.A., Ghosh, R., Nunes-Alves, A., 2024. Recent progress in modeling and simulation of biomolecular crowding and condensation inside cells. Journal of Chemical Information and Modeling 64 (24), 9063–9081. Available at https://doi.org/10.1021/acs.jcim.4c01520.

Miao, J., Chong, S., 2025. Roles of intrinsically disordered protein regions in transcriptional regulation and genome organization. Current Opinion in Genetics & Development 90, 102285. Available at https://doi.org/10.1016/j.gde.2024.102285.

Morris, O.M., Toprakcioglu, Z., Röntgen, A., Cali, M., Knowles, T.P.J., Vendruscolo, M., 2024. Aggregation of the amyloid-β peptide (Aβ40) within condensates generated through liquid–liquid phase separation. Scientific Reports 14 (1). Available at https://doi.org/10.1038/s41598-024-72265-7.

Morris, O.M., Torpey, J.H., Isaacson, R.L., 2021. Intrinsically disordered proteins: modes of binding with emphasis on disordered domains. Open Biology 11 (10). Available at https://doi.org/10.1098/rsob.210222.

Mukherjee, S., Sakunthala, A., Gadhe, L., Poudyal, M., Sawner, A.S., Kadu, P., et al., 2023. Liquid-liquid phase separation of α-synuclein: a new mechanistic insight for α-synuclein aggregation associated with Parkinson's disease pathogenesis. Journal of Molecular Biology 435 (1). Available at https://doi.org/10.1016/j.jmb.2022.167713.

Mukherjee, S., Schäfer, L.V., 2023. Thermodynamic forces from protein and water govern condensate formation of an intrinsically disordered protein domain. Nature Communications 14 (1). Available at https://doi.org/10.1038/s41467-023-41586-y.

Már, M., Nitsenko, K., Heidarsson, P.O., 2023. Multifunctional intrinsically disordered regions in transcription factors. Chemistry - A European Journal 29 (21). Available at https://doi.org/10.1002/chem.202203369.

Müller, L., Hoppe, T., 2024. UPS-dependent strategies of protein quality control degradation. Trends in Biochemical Sciences 49 (10), 859–874. Available at https://doi.org/10.1016/j.tibs.2024.06.006.

Naudi-Fabra, S., Blackledge, M., Milles, S., 2021. Synergies of single molecule fluorescence and nmr for the study of intrinsically disordered proteins. Biomolecules 12 (1), 27. Available at https://doi.org/10.3390/biom12010027.

Netzer, A., Baruch Leshem, A., Veretnik, S., Edelstein, I., Lampel, A., 2024. Regulation of peptide liquid–liquid phase separation by aromatic amino acid composition. Small (Weinheim an der Bergstrasse, Germany) 20 (38), 1613–6810. Available at https://doi.org/10.1002/smll.202401665.

Newcombe, E.A., Delaforge, E., Hartmann-Petersen, R., Skriver, K., Kragelund, B.B., 2022. How phosphorylation impacts intrinsically disordered proteins and their function. Essays in Biochemistry 66 (7), 901–913. Available at https://doi.org/10.1042/EBC20220060.

Norris, V., Oláh, J., Krylov, S.N., Uversky, V.N., Ovádi, J., 2023. The sherpa hypothesis: phenotype-preserving disordered proteins stabilize the phenotypes of neurons and oligodendrocytes. npj Systems Biology and Applications 9 (1). Available at https://doi.org/10.1038/s41540-023-00291-8.

Pla-Prats, C., Thomä, N.H., 2022. Quality control of protein complex assembly by the ubiquitin–proteasome system. Trends in Cell Biology 32 (8), 696–706. Available at https://doi.org/10.1016/j.tcb.2022.02.005.

Poudyal, M., Patel, K., Gadhe, L., Sawner, A.S., Kadu, P., Datta, D., et al., 2023. Intermolecular interactions underlie protein/peptide phase separation irrespective of sequence and structure at crowded milieu. Nature Communications 14 (1). Available at https://doi.org/10.1038/s41467-023-41864-9.

Qin, C., Wang, Y.L., Zheng, J., Wan, X.B., Fan, X.J., 2025. Current perspectives in drug targeting intrinsically disordered proteins and biomolecular condensates. BMC Biology 23 (1). Available at https://doi.org/10.1186/s12915-025-02214-x.

Ranek, M.J., Stachowski, M.J., Kirk, J.A., Willis, M.S., 2018. The role of heat shock proteins and co-chaperones in heart failure. Series B, Biological Sciences 373 (1738). Available at https://doi.org/10.1098/rstb.2016.0530.

Rhine, K., Makurath, M.A., Liu, J., Skanchy, S., Lopez, C., Catalan, K.F., et al., 2020. ALS/FTLD-linked mutations in FUS glycine residues cause accelerated gelation and reduced interactions with wild-type FUS. Molecular Cell 80 (4), 666. Available at https://doi.org/10.1016/j.molcel.2020.10.014.

Rocamonde, B., Hasan, U., Mathieu, C., Dutartre, H., 2023. Viral-induced neuroinflammation: different mechanisms converging to similar exacerbated glial responses. Frontiers in Neuroscience 17. Available at https://doi.org/10.3389/fnins.2023.1108212.

Rodríguez, L.C., Foressi, N.N., Celej, M.S., 2024. Liquid-liquid phase separation of tau and α-synuclein: a new pathway of overlapping neuropathologies. Biochemical and Biophysical Research Communications 741. Available at https://doi.org/10.1016/j.bbrc.2024.151053.

Rossino, G.G., Marchese, E., Galli, G., Verde, F., Finizio, M., Serra, M., et al., 2023. Peptides as therapeutic agents: challenges and opportunities in the green transition era. Molecules (Basel, Switzerland) 28 (20), 7165. Available at https://doi.org/10.3390/molecules28207165.

Rummens, J., Da Cruz, S., 2025. RNA-binding proteins in ALS and FTD: from pathogenic mechanisms to therapeutic insights. Molecular Neurodegeneration 20 (1). Available at https://doi.org/10.1186/s13024-025-00851-y.

Saar, K.L., Qian, D., Good, L.L., Morgunov, A.S., Collepardo-Guevara, R., Best, R.B., et al., 2023. Theoretical and data-driven approaches for biomolecular condensates. Chemical Reviews 123 (14), 8988–9009. Available at https://doi.org/10.1021/acs.chemrev.2c00586.

Sanchez-Burgos, I., Tejedor, A.R., Castro, A., Feito, A., Collepardo-Guevara, R., Espinosa, J.R., 2025. Charged peptides enriched in aromatic residues decelerate condensate ageing driven by cross-β-sheet formation. Nature Communications 16 (1). Available at https://doi.org/10.1038/s41467-025-62686-x.

Saurabh, S., Nadendla, K., Purohit, S.S., Sivakumar, P.M., Cetinel, S., 2023. Fuzzy Drug targets: disordered proteins in the drug-discovery realm. ACS Omega 8 (11), 9729–9747. Available at https://doi.org/10.1021/acsomega.2c07708.

Scherer, N.M., Maurel, C., Graus, M.S., McAlary, L., Richter, G., Radford, R.A.W., et al., 2024. RNA-binding properties orchestrate TDP-43 homeostasis through condensate formation in vivo. Nucleic Acids Research 52 (9), 5301–5319. Available at https://doi.org/10.1093/nar/gkae112.

Seth, S., Stine, B., Bhattacharya, A., 2024. Fine structures of intrinsically disordered proteins. The Journal of Chemical Physics 160 (1). Available at https://doi.org/10.1063/5.0176306.

Shang, B., Li, C., Zhang, X., 2024. How intrinsically disordered proteins order plant gene silencing. Trends in Genetics 40 (3), 260–275. Available at https://doi.org/10.1016/j.tig.2023.12.009.

Sharma, M., Gonzalez, D.T., Antonietti, M., Uversky, V., Djulbegovic, M., 2025. Intrinsic disorder in CYP1B1 and its implications in primary congenital glaucoma pathogenesis. Journal of Proteins and Proteomics 16 (3), 347–361. Available at https://doi.org/10.1007/s42485-025-00186-8.

Sheu-Gruttadauria, J., Yan, X., Stuurman, N., Vale, R.D., Floor, S.N., 2023. Nucleolar dynamics are determined by the ordered assembly of the ribosome. bioRxiv. Available at https://doi.org/10.1101/2023.09.26.559432.

Shi, M., Wu, Z., Zhang, Y., Li, T., 2025. Decoding intrinsically disordered regions in biomolecular condensates. Fundamental Research. Available at https://doi.org/10.1016/j.fmre.2025.01.013.

Siegert, A., Rankovic, M., Favretto, F., Ukmar-Godec, T., Strohäker, T., Becker, S., et al., 2021. Interplay between tau and α-synuclein liquid–liquid phase separation. Protein Science 30 (7), 1326–1336. Available at https://doi.org/10.1002/pro.4025.

Sil, S., Datta, I., Basu, S., 2025. Use of AI-methods over MD simulations in the sampling of conformational ensembles in IDPs. Frontiers in Molecular Biosciences 12. Available at https://doi.org/10.3389/fmolb.2025.1542267.

Silva, J.L., Foguel, D., Ferreira, V.F., Vieira, T.C.R.G., Marques, M.A., Ferretti, G.D.S., et al., 2023. Targeting biomolecular condensation and protein aggregation against cancer. Chemical Reviews 123 (14), 9094–9138. Available at https://doi.org/10.1021/acs.chemrev.3c00131.

Song, S., Ivanov, T., Doan-Nguyen, T.P., Da Silva, L.C., Xie, J., Landfester, K., et al., 2025. Synthetic biomolecular condensates: phase-separation control, cytomimetic modelling and emerging biomedical potential. Angewandte Chemie International Edition 64 (8), e202418431. Available at https://doi.org/10.1002/anie.202418431.

Song, S., Ivanov, T., Yuan, D., Wang, J., da Silva, L.C., Xie, J., et al., 2024. Peptide-based biomimetic condensates via liquid-liquid phase separation as biomedical delivery vehicles. Biomacromolecules 25 (9), 5468–5488. Available at https://doi.org/10.1021/acs.biomac.4c00814.

Song, S., Xie, H., Wang, Q., Sun, X., Xu, J., Chen, R., et al., 2025. Spatiotemporal deciphering of dynamic the FUS interactome during liquid-liquid phase separation in living cells. Nature Communications 16 (1), 4328. Available at https://doi.org/10.1038/s41467-025-59457-z.

Sonneville, R., Lannuzel, A., Mathieu, C., Gonzalez-Dunia, D., Gaudin, R., 2025. Triggers of viral encephalitis: a brain teaser to be solved!. Annals of Intensive Care 15 (1). Available at https://doi.org/10.1186/s13613-025-01451-1.

Sridharan, S., Hernandez-Armendariz, A., Kurzawa, N., Potel, C.M., Memon, D., Beltrao, P., et al., 2022. Systematic discovery of biomolecular condensate-specific protein phosphorylation. Nature Chemical Biology 18 (10), 1104–1114. Available at https://doi.org/10.1038/s41589-022-01062-y.

Tripathi, T., Uversky, V.N., Giuliani, A., 2025. 'Intelligent' proteins. Cellular and Molecular Life Sciences 82 (1). Available at https://doi.org/10.1007/s00018–025-05770-1.

Trivedi, R., Nagarajaram, H.A., 2022. Intrinsically disordered proteins: an overview. International Journal of Molecular Sciences 23 (22), 14050. Available at https://doi.org/10.3390/ijms232214050.

Tunyasuvunakool, K., Adler, J., Wu, Z., Green, T., Zielinski, M., Žídek, A., et al., 2021. Highly accurate protein structure prediction for the human proteome. Nature 596 (7873), 590–596. Available at https://doi.org/10.1038/s41586-021-03828-1.

Usher, E.T., Fossat, M.J., Holehouse, A.S., 2024. Phosphorylation of disordered proteins tunes local and global intramolecular interactions. Biophysical Journal 123 (23), 4082–4096. Available at https://doi.org/10.1016/j.bpj.2024.10.021.

Usman, A., Yusuf, A., Isah, M.B., Dang, M., Zhang, X., 2025. Small molecules as regulators of liquid-liquid phase separation: mechanisms and strategies for new drug discovery. FASEB Journal: Official Publication of the Federation of American Societies for Experimental Biology 39 (13). Available at https://doi.org/10.1096/fj.202501476R.

Uversky, V.N., 2025. Functional diversity of intrinsically disordered proteins and their structural heterogeneity: protein structure-function continuum. Progress in Molecular Biology and Translational Science 211, 1–15. Available at 10.1016/bs.pmbts.2024.11.006.

Uversky, V.N., 2025b. How to drug a cloud? Targeting intrinsically disordered proteins. Pharmacological Reviews 77 (2), 100016. Available at https://doi.org/10.1124/pharmrev.124.001113.

Uversky, V.N., Dunker, A.K., 2010. Understanding protein non-folding. Biochimica et Biophysica Acta - Proteins and Proteomics 1804 (6), 1231–1264. Available at https://doi.org/10.1016/j.bbapap.2010.01.017.

Valentino, I.M., Llivicota-Guaman, J.G., Dao, T.P., Mulvey, E.O., Lehman, A.M., Galagedera, S.K.K., et al., 2024. Phase separation of polyubiquitinated proteins in UBQLN2 condensates controls substrate fate. Proceedings of the National Academy of Sciences of the United States of America 121 (33). Available at https://doi.org/10.1073/pnas.2405964121.

VanPelt, J., Page, R.C., 2017. Unraveling the CHIP:Hsp70 complex as an information processor for protein quality control. Biochimica et Biophysica Acta - Proteins and Proteomics 1865 (2), 133–141. Available at https://doi.org/10.1016/j.bbapap.2016.11.005.

Vazquez, D.S., Toledo, P.L., Gianotti, A.R., Ermácora, M.R., 2022. Protein conformation and biomolecular condensates. Current Research in Structural Biology 4, 285–307. Available at https://doi.org/10.1016/j.crstbi.2022.09.004.

Wang, L., Bergkvist, L., Kumar, R., Winblad, B., Pavlov, P.F., 2021a. Targeting chaperone/co-chaperone interactions with small molecules: a novel approach to tackle neurodegenerative diseases. Cells 10 (10), 2596. Available at https://doi.org/10.3390/cells10102596.

Wang, B., Zhang, L., Dai, T., Qin, Z., Lu, H., Zhang, L., et al., 2021b. Liquid–liquid phase separation in human health and diseases. Signal Transduction and Targeted Therapy 6 (1), 290. Available at https://doi.org/10.1038/s41392-021-00678-1.

Wang, F., Zhang, Y., 2024. Physiology and pharmacological targeting of phase separation. Journal of Biomedical Science 31 (1). Available at https://doi.org/10.1186/s12929-024-00993-z.

Wang, X., Bigman, L.S., Greenblatt, H.M., Yu, B., Levy, Y., Iwahara, J., 2023. Negatively charged, intrinsically disordered regions can accelerate target search by DNA-binding proteins. Nucleic Acids Research 51 (10), 4701–4712. Available at https://doi.org/10.1093/nar/gkad045.

Wei, Y., Zhang, B., Bie, Q., He, B., 2025. Metabolic regulation in the senescence process of stem cells. Stem Cells Translational Medicine 14 (9). Available at https://doi.org/10.1093/stcltm/szaf041.

Wu, X., Shen, Z., Zhang, M., 2025. Phase separation–mediated compartmentalization underlies synapse formation and plasticity. Annual Review of Neuroscience 48 (1), 149–168. Available at https://doi.org/10.1146/annurev-neuro-112723-040159.

Wüstner, D., 2022. Dynamic mode decomposition of fluorescence loss in photobleaching microscopy data for model-free analysis of protein transport and aggregation in living cells. Sensors 22 (13), 4731. Available at https://doi.org/10.3390/s22134731.

Xiao, L., Xia, K., 2025. Functions of intrinsically disordered regions. Biology 14 (7), 810. Available at https://doi.org/10.3390/biology14070810.

Xu, X., van Haren, M.H.I., Smokers, I.B.A., Visser, B.S., White, P.B., Jansen, R.S., et al., 2025. Amino acids bind to phase-separating proteins and modulate biomolecular condensate stability and dynamics. Nature Communications 16 (1). Available at https://doi.org/10.1038/s41467-025-63755-x.

Xu, Y., Maya-Martinez, R., Radford, S.E., 2022. Controlling amyloid formation of intrinsically disordered proteins and peptides: slowing down or speeding up? Essays in Biochemistry 66 (7), 959–975. Available at https://doi.org/10.1042/EBC20220046.

Yamaguchi, K., Nakajima, K., Ogi, H., Goto, Y., 2025. Mechanism of amyloid fibril formation triggered by breakdown of supersaturation. npj Biosensing 2 (1), 3004–8656. Available at https://doi.org/10.1038/s44328-025-00028-z.

Yorimitsu, T., Sato, K., 2025. Potential ER tubular lumen sensing by intrinsically disordered regions. Journal of Cell Science 138 (5). Available at https://doi.org/10.1242/jcs.263696.

Zaslavsky, B.Y., Ferreira, L.A., Uversky, V.N., 2023. Chapter 1 - Biophysical principles of liquid–liquid phase separation. In: Uversky, V.N. (Ed.), Droplets of Life: Membrane-Less Organelles, Biomolecular Condensates, and Biological Liquid-Liquid Phase Separation. Academic Press, pp. 3–82. Available at https://doi.org/10.1016/B978-0-12-823967-4.00009-9.

Zbinden, A., Pérez-Berlanga, M., De Rossi, P., Polymenidou, M., 2020. Phase separation and neurodegenerative diseases: a disturbance in the force. Developmental Cell 55 (1), 45–68. Available at https://doi.org/10.1016/j.devcel.2020.09.014.

Zhang, Z., Xue, P., Bendlin, B.B., Zetterberg, H., De Felice, F., Tan, X., et al., 2025a. Melatonin: a potential nighttime guardian against Alzheimer's. Molecular Psychiatry 30 (1), 237–250. Available at https://doi.org/10.1038/s41380-024-02691-6.

Zhang, J., Che, Y., Liu, R., Wang, Z., Liu, W., 2025b. Deep learning-driven multi-omics analysis: enhancing cancer diagnostics and therapeutics. Briefings in Bioinformatics 26 (4), bbaf440. Available at https://doi.org/10.1093/bib/bbaf440.

Zhang, Y., Jin, C., Xu, X., Guo, J., Wang, L., 2024. The role of liquid-liquid phase separation in the disease pathogenesis and drug development. Biomedicine & Pharmacotherapy 180, 117448. Available at https://doi.org/10.1016/j.biopha.2024.117448.

Zhao, Y., Dai, E., Zhang, M., Wu, Y., Sun, D., Ding, J., 2025. Mechanisms and pathological significance of liquid-liquid phase separation in bacteria. FASEB Journal: Official Publication of the Federation of American Societies for Experimental Biology 39 (18). Available at https://doi.org/10.1096/fj.202500634RR.

Zheng, B., Wang, X., Guo, M., Tzeng, C.-M., 2025. Therapeutic peptides: recent advances in discovery, synthesis, and clinical translation. International Journal of Molecular Sciences 26 (11), 5131. Available at https://doi.org/10.3390/ijms26115131.

Zhou, H.X., Pang, X., 2018. Electrostatic interactions in protein structure, folding, binding, and condensation. Chemical Reviews 118 (4), 1691–1741. Available at https://doi.org/10.1021/acs.chemrev.7b00305.

Zhou, Y., Chen, X., Liu, C., Wang, Y., Xia, T., Gu, X., 2024. Liquid–liquid phase separation in neurodegenerative diseases: an updated understanding. Advanced Neurology 4 (1), 40. Available at https://doi.org/10.36922/an.4493.

Zhou, Y., Zhou, S., Bi, Y., Zou, Q., Jia, C., 2024. A two-task predictor for discovering phase separation proteins and their undergoing mechanism. Briefings in Bioinformatics 25 (6), 1467–5463. Available at https://doi.org/10.1093/bib/bbae528.

Zhu, Q., Raza, Z., Do-Ha, D., De Costa, E., Sasheva, P., McAlary, L., et al., 2025. Biomolecular condensates as emerging biomaterials: functional mechanisms and advances in computational and experimental approaches. Advanced Materials (Deerfield Beach, Fla.) 37 (36), e10115. Available at https://doi.org/10.1002/adma.202510115.

Zwicker, D., Paulin, O.W., Ter Burg, C., 2025. Physics of droplet regulation in biological cells. Reports on Progress in Physics. Physical Society (Great Britain). Available at https://doi.org/10.1088/1361-6633/ae12a7.

CHAPTER 13

Aβ peptides: structure and interaction with membranes as revealed by nuclear magnetic resonance

Lesia Kolomiiets[1,2] and Igor Zhukov[1]
[1]Institute of Biochemistry and Biophysics, Polish Academy of Sciences, Warsaw, Poland,
[2]Institute of Molecular Biology and Genetics, National Academy of Sciences of Ukraine, Kyiv, Ukraine

13.1 Introduction

Alzheimer's disease (AD) is a neurodegenerative disorder affecting humans and our closest cousins, chimpanzees (Edler et al., 2017). After decades of controversy and conflicting reports, aggregated amyloid-beta (*Aβ*) peptides have finally been established as an early and possibly causative factor in the cascade of events leading to neurodegeneration in AD.

Aβ peptides play a central role in AD pathogenesis, and their interactions with biological membranes are critical to understanding disease mechanisms. Aβ(1–40) dimers (hereafter Aβ40) bind directly to neuronal glutamate receptors, triggering a vicious cycle of neuronal hyperactivation that eventually leads to loss of synaptic activity, neuronal death, and irreversible disruption of brain circuits (Zott et al., 2019). At the same time, monomeric *Aβ* peptides are nontoxic and participate in memory-related processes (Sarkar et al., 2013).

Aβ is derived from the proteolytic cleavage of the transmembrane amyloid precursor protein (APP) by secretases and is processed along nonamyloidogenic and amyloidogenic pathways. The amyloidogenic pathway involves initial cleavage of APP by β-secretase, leading to the formation of *sAPPβ* and the 99-amino acid fragment, usually referred to as C99. In the last step, cleavage of C99 by γ-secretase is characterized by low fidelity and is initiated at the η-sites to produce the amyloid intracellular domain and Aβ(1−49) or Aβ(1−48) peptides. After additional cleavage, a variety of *Aβ* isoforms result, with the most prevalent Aβ40 and Aβ(1−42) (hereafter Aβ42) being minor but more amyloidogenic forms (Beel et al., 2008).

The "amyloid cascade hypothesis" (Hardy and Higgins, 1992) posits that Aβ aggregation is a central event in AD pathogenesis, leading to synaptic dysfunction, neuroinflammation, and ultimately neuronal death. Experimental data suggest that soluble oligomeric forms of Aβ, rather than mature fibrils or plaques, are the primary

Molecular Insights into Proteinopathies
DOI: https://doi.org/10.1016/B978-0-443-34176-2.00014-6
© 2027 Elsevier Inc. All rights are reserved, including those for text and data mining, AI training, and similar technologies.

toxic species responsible for synaptic dysfunction and cognitive impairment (Cheng et al., 2020). Understanding the structural properties of these oligomeric intermediates and their interactions with cellular membranes is therefore crucial for elucidating disease mechanisms and developing therapeutic interventions.

Nuclear magnetic resonance (NMR) spectroscopy has emerged as a powerful tool for elucidating the structural details of Aβ-membrane interactions at atomic resolution. This report comprehensively reviews the application of NMR techniques—including solution NMR, solid-state NMR (ssNMR), and dynamic nuclear polarization-enhanced methods—to characterize $A\beta$ peptide structure, orientation, and dynamics when bound to membrane systems. The data obtained reveal that membrane composition profoundly influences Aβ conformation, with lipids such as gangliosides and cholesterol directing peptide folding, oligomerization pathways, and membrane-disruption mechanisms. NMR studies have identified lipid contacts in a residue-specific manner and revealed that β-sheet architectures form the structural basis of Aβ toxicity. This chapter describes current knowledge of Aβ—membrane interactions derived from NMR investigations.

Biological membranes actively participate in $A\beta$ aggregation and toxicity. $A\beta$ peptides interact directly with neuronal membranes, and these interactions profoundly influence peptide folding, aggregation kinetics, and the formation of toxic species (Yagi-Utsumi et al., 2023; Weber et al., 2012). Lipid bilayers can act as two-dimensional templates that concentrate $A\beta$ peptides, accelerate nucleation, and promote specific aggregation pathways distinct from those occurring in bulk solution (Kenyaga et al., 2022).

The presence of specific lipids, such as gangliosides (particularly GM1), cholesterol, and anionic phospholipids, dramatically alters $A\beta$ binding affinity, conformation, and aggregation propensity (Yagi-Utsumi et al., 2023; Qiang et al., 2020). At the same time, $A\beta$ oligomers can perturb membrane integrity through multiple mechanisms, including lipid extraction, membrane thinning, domain reorganization, and the formation of ion-permeable pores, leading to calcium dysregulation and cellular dysfunction (Viles, 2023; Feuillie et al., 2020).

13.2 Application of nuclear magnetic resonance spectroscopy for analysis of amyloid-beta-membrane interactions

Solution-state NMR spectroscopy typically characterizes $A\beta$ structure in membrane-mimetic environments, including detergent micelles, lipid bicelles, and nanodiscs. These systems mimic biological membranes and support rapid molecular tumbling for high-resolution NMR. TOCSY experiments establish through-bond connectivities within amino acid spin systems for resonance assignment. Afterward, NOESY identifies through-space dipolar couplings between protons separated by less than ~5 Å, providing distance restraints for structure determination. Heteronuclear experiments further analyze chemical shift perturbations in ^{1}H–^{15}N or ^{1}H–^{13}C HSQC spectra to map Aβ-peptide–lipid interaction sites (Williamson, 2013). Additionally, NMR studies using spin-labeled lipids or peptide-attached paramagnetic probes provide long-range distance constraints (up to ~25 Å), revealing depth and orientation of peptide insertion relative to the membrane (Rajput et al., 2021).

In contrast, ssNMR spectroscopy is suited to studying Aβ peptides in lipid bilayers, where molecular tumbling is inhibited compared to solution NMR. ssNMR directly probes peptide structure, orientation, and dynamics in fully hydrated bilayers that closely mimic biological membranes (Tycko, 2011). Secondary structure propensities are determined from ^{13}Cα and ^{13}Cβ chemical shifts: residues in α-helical conformation show characteristic upfield ^{13}Cα and downfield ^{13}Cβ shifts, whereas residues in β-sheets show the opposite pattern (Wishart and Sykes, 1994).

Dipolar recoupling experiments, such as REDOR and TEDOR, reintroduce dipolar couplings and provide distance information between specific nuclei. By incorporating ^{13}C or ^{31}P labels into lipids and detecting dipolar couplings to peptide nuclei, these experiments measure peptide–lipid proximities (Gullion, 2008). Orientational restraints are obtained from experiments probing anisotropic interactions (e.g., ^{15}N chemical shift anisotropy, ^{1}H–^{15}N dipolar couplings) in aligned samples, providing information on peptide orientation and tilt angles relative to the membrane normal (Opella and Marassi, 2004).

Finally, ssNMR can characterize both peptide structure and lipid bilayer properties. Specifically, ^{31}P NMR of lipid phosphate headgroups and ^{2}H NMR of deuterated lipid acyl chains provide information on lipid order parameters, phase behavior, and dynamics (Weber et al., 2012; Davis, 1983). When Aβ binds, changes in lipid order reveal membrane perturbations that correlate with membrane disruption and toxicity.

13.3 Structural features of micelle-bound amyloid-beta peptides

Aβ peptides exhibit remarkable structural plasticity, adopting different conformations depending on environmental conditions.

In aqueous solution, the Aβ40 peptide adopts a random coil conformation, except for two α-helical regions: Gln15–Asp23 and Ile31–Met35 (Sticht et al., 1995). Moreover, further studies show that *Aβ*40 is structurally unstable; small changes in experimental conditions can significantly alter its average structure (Hou et al., 2004; Vivekanandan et al., 2011). Notably, ionic strength specifically influences Aβ40's conformational preferences and aggregation, shifting low molecular weight species from unstructured to α-helical intermediate and decreasing the β-sheet high molecular weight species in equilibrium with the monomer, depending on the anion's position in the Hofmeister series (Narayanan and Reif, 2005).

Sodium dodecyl sulfate (SDS) micelles are widely used for structural analysis of the Aβ40 and Aβ42 peptides in the micelle-bound state (Shao et al., 1999; Jarvet et al., 2007). There are two fragments, Gln15–Val24 and Gly29–Met35, in the *Aβ*40 peptide that adopt α-helical conformations in SDS micelles. The position of the α-helical fragments within the micelle was established by measuring the effect of paramagnetic resonance enhancement upon the addition of paramagnetic Mn^{2+} ions and 5-doxyl stearic acid to the sample (Jarvet et al., 2007). Experimental data demonstrate that the C-terminal α-helix is buried within the micelle's interior, suggesting a stable interaction with its hydrophobic core. Consequently, the central α-helix is positioned in the SDS headgroup region, with flexible regions that are mainly located outside the micelle (Jarvet et al., 2007) (see Fig. 13.1A). Comparison of ^{15}N R_1 relaxation rates evaluated for Aβ40 in SDS micelles with those in water reveals a propensity to adopt an α-helical

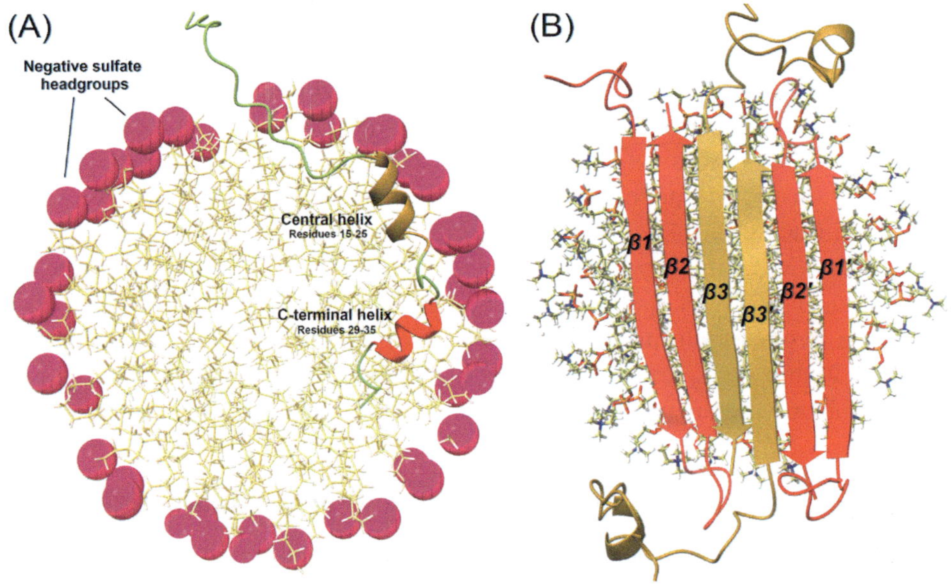

Figure 13.1 The 3D structures of amyloid-beta (Aβ) peptides in sodium dodecyl sulfate (SDS) and dodecylphosphocholine (DPC) micelles are presented. (A) The structure of the Aβ40 peptide in a negatively charged SDS micelle, as established on nuclear magnetic resonance (NMR) data (Jarvet et al., 2007), is presented. The negatively charged sulfate headgroups are shown as magenta balls. The two α-helices, the central residues 15–25, and the C-terminal one, residues 29–35, are highlighted in khaki and red, respectively. The positioning of α-helices was deduced from the NMR experiments with paramagnetic agents. (B) The Aβ42 peptide in neutral DPC micelles adopts a 3D structure with two subunits (shown in red and orange) that form a stable tetramer characterized by six β-strands (PDB 6RHY) (Ciudad et al., 2020). The DPC micelle is shown in the background as a stick. Both figures are prepared with the program ChimeraX (Meng et al., 2023). The SDS and DPC micelles were created with the CHARM-GUI server (Jo et al., 2008).

secondary structure, consistent with the peptide's dynamic properties in aqueous solution (Danielsson et al., 2006).

Under specific conditions, dodecylphosphocholine (DPC) micelles provide a platform for the formation of homogeneous and stable A*β* oligomers (Serra-Batiste et al., 2018). The A*β*42 tetramer structure embedded in a DPC micelle has been determined by solution NMR, providing the assignments of ^{1}H, ^{13}C, and ^{15}N resonances. Based on assigned chemical shifts and NOE constraints, the 3D structure consists of two monomers assembled into a *β*-barrel formed by six *β*-sheets (see Fig. 13.1B). The *β*-barrel assembly of A*β*42 oligomers is stabilized by hydrophobic interactions between residues in the barrel core. A*β*42 tetramers and octamers form pores, with hydrophilic residues on the *β*-sheet core edges mediating water and ion permeation. Lipid interactions further stabilize these pores, which are more permeable than those formed by fibrils (Ciudad et al., 2020).

13.4 Interaction of amyloid-beta with membranes—mechanism of pore formation

A key mechanism of A*β* neurotoxicity involves the formation of pores in cellular membranes, leading to ion dysregulation, calcium influx, and eventual neuronal damage. This process is mediated by A*β* oligomers, which are more toxic than mature fibrils because of their greater propensity to form ion-permeable pores (Nguyen et al., 2022). The critical aggregation concentration (CAC) represents the minimal concentration above which monomeric A*β* peptides start to aggregate into oligomers. The CAC limit for A*β*42 was determined by fluorescence correlation spectroscopy to be approximately 90 nM. In contrast, A*β*40 has a higher CAC of 0.5 ± 0.3 μM for the formation of metastable oligomers and 18.9 ± 2.2 μM for the formation of larger aggregates, indicating a lower propensity to aggregate than A*β*42 (Novo et al., 2018). These results suggest that *Aβ*42 has a high tendency to form oligomers that can insert into lipid bilayers and create pores. The pores are typically organized as *β*-barrels, cylindrical arrays of *β*-sheets that span the membrane (see Fig. 13.2). The formation of these *β*-barrel structures is critical for their pore-forming activity (Serra-Batiste et al., 2016; Sepehri and Lazaridis, 2023).

The sequence leading to A*β* pore (amyloid-pore) formation involves several key steps. First, A*β* peptides assemble into oligomers, which are stabilized by hydrogen bonding and hydrophobic interactions (Nguyen et al., 2022). Next, A*β* oligomers insert into the lipid bilayer, driven by the hydrophobic effect. The C-terminal region of A*β* plays a critical role in this process (Serra-Batiste et al., 2018; Sepehri and Lazaridis, 2023). Once inserted, these oligomers form *β*-barrel or edge pores, creating aqueous channels that disrupt membrane integrity. As a result, the pores permit the passage of ions (e.g., Ca^{2+}) and water, leading to cellular dysfunction (Ciudad et al., 2020; Serra-Batiste et al., 2016; Kandel et al., 2019).

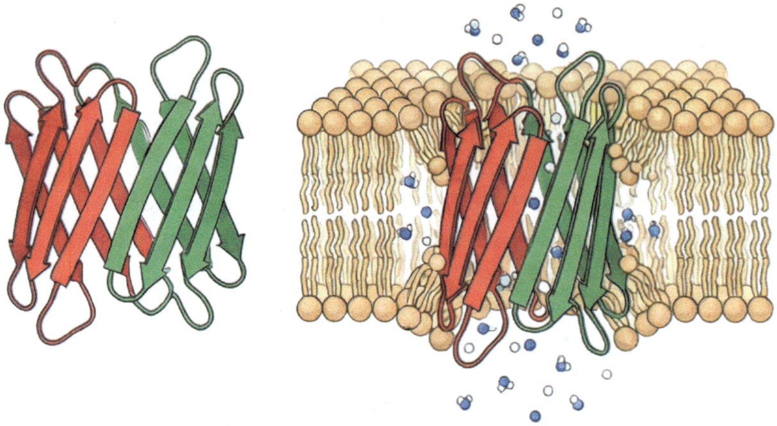

Figure 13.2 Insertion of amyloid-beta (Aβ) oligomers into lipid membranes. An increase in the concentration of the Aβ42 peptide triggers two tetramers to assemble into an asymmetric octameric form. Unlike traditional pore models, Aβ42 octamers adopt a β-sandwich topology similar to a β-barrel (*left panel*). The pore formed in the membrane by the β-sandwich characterized exposure of hydrophilic residues on both sides of the Aβ42 octamer outside the membrane (Ciudad et al., 2020) (*right panel*). As a result, the hydrophobic lipid tails facilitate rearrangements, creating lipid-stabilized "toroidal-like" pores (Serra-Batiste et al., 2016).

13.5 Impact of membrane composition on amyloid-beta structure

The composition of the lipid bilayer influences Aβ pore formation by affecting binding, structure, and aggregation. Biological membranes contain hundreds of lipid species; certain lipids play a critical roles in modulating Aβ–membrane interactions.

Cholesterol, a major lipid component of mammalian membranes, is particularly enriched in lipid rafts and profoundly influences membrane physical properties, such as fluidity, thickness, and phase behavior (Simons and Ikonen, 1997). High concentration of cholesterol stabilizes the membrane and inhibits pore formation, whereas low cholesterol level facilitates Aβ insertion and pore formation (Fernández-Pérez et al., 2018; Di Scala et al., 2016). Cholesterol can exert contradictory effects on Aβ-membrane interactions. It can either enhance or inhibit Aβ binding depending on concentration and membrane composition. High cholesterol levels can reduce Aβ insertion depth by increasing membrane order and thickness (Bera et al., 2016). At the lower concentration (15% w/w lipid), cholesterol promotes Aβ insertion and channel formation, while also accelerating primary nucleation and increasing the fibril formation rate by 20-fold (Habchi et al., 2018). Increased parallel β-sheet structures, associated with enhanced membrane disruption and cytotoxicity, have been detected in cardiolipin (CL), cholesterol-enriched membranes (Korshavn et al., 2017).

Despite limiting deep insertion, cholesterol-enriched membranes can promote Aβ aggregation at the membrane surface, potentially by concentrating peptides within cholesterol-rich domains (Di Scala et al., 2016). Furthermore, NMR studies show that cholesterol inclusion during Aβ aggregation increases parallel β-sheet content and enhances membrane disruption and cytotoxicity (Korshavn et al., 2016).

13.6 Ganglioside GM1 micelle affects the structure of amyloid-beta

Ganglioside GM1 has emerged as a key lipid in Aβ–membrane interactions. As a glycosphingolipid with a complex oligosaccharide headgroup, GM1 is enriched in lipid rafts and neuronal membranes (Yanagisawa et al., 1995). Gangliosides, especially GM1, interact with Aβ *to* promote pore formation, which is critical for Aβ oligomer neurotoxicity (Nguyen et al., 2022; Di Scala et al., 2016). The presence of anionic phospholipids, such as POPG, further enhances Aβ pore formation by stabilizing β-barrel structures (Kandel et al., 2019).

NMR and biophysical studies have demonstrated that GM1 acts as a "seed" that nucleates Aβ aggregation at membrane surfaces (Yagi-Utsumi et al., 2023; Kenyaga et al., 2022). Aβ binds to GM1-enriched membrane domains with high affinity and this binding dramatically accelerates its aggregation kinetics.

The unique double-layer antiparallel β-sheet architecture consists of two layers of Aβ molecules arranged antiparallely as demonstrated by ssNMR studies, revealing that Aβ assemblies formed on GM1-containing membranes adopt this architecture (Yagi-Utsumi et al., 2023; Kenyaga et al., 2022). A solvent-exposed hydrophobic surface generated by the antiparallel Aβ arrangement has demonstrated catalytic activity in converting additional monomeric Aβ into fibrils.

Antiparallel β-sheets assemblies formed on GM1 ganglioside-containing membranes adopt a distinctive double-layer antiparallel β-sheet architecture (Yagi-Utsumi et al., 2023; Kenyaga et al., 2022). This nonfibrillar assembly exposes a hydrophobic surface to the solvent and catalytically promotes the conversion of monomeric Aβ into fibrils. The antiparallel arrangement involves residues forming intermolecular hydrogen bonds with adjacent peptides in opposite orientations. Some membrane-associated oligomers display both parallel and antiparallel β-strand arrangements, contributing to structural heterogeneity (Tycko, 2015).

Importantly, the GM1-bound assembly does not itself convert into fibrils but instead serves as a catalytic platform for fibril formation. The exposed hydrophobic surface likely facilitates the recruitment of monomeric Aβ and promotes its conversion to a β-sheet structure (Yagi-Utsumi et al., 2023).

GM1–Aβ complexes have been detected in AD brains, and GM1 levels are altered in AD pathology (Ariga et al., 2008). The ability of GM1 to template specific Aβ structures at membrane surfaces suggests that GM1-rich lipid rafts may be critical sites for early Aβ aggregation in vivo.

13.7 Impact of cardiolipin lipids on the promotion of amyloid-beta aggregation

Cardiolipin (CL), a unique dimeric phospholipid enriched in mitochondrial membranes, has been shown to promote Aβ aggregation and increase the proportion of parallel β-sheet structures, which correlate with enhanced cytotoxicity (Korshavn et al., 2016). NMR studies reveal that CL induces specific conformational changes in Aβ, facilitating intermolecular contacts and accelerating fibril formation. As recently demonstrated, CL promotes membrane-inserted and pore-competent oligomer morphologies. The Aβ42 aggregates formed in the presence of CL had a parallel β-sheet content that correlated with higher toxicity (Zhaliazka et al., 2023).

CL, together with other anionic lipids, strongly accelerates the kinetics of Aβ oligomerization and fibrillization relative to zwitterionic lipids, which induce parallel β-sheet oligomers with increased toxicity (Zhaliazka et al., 2023). CL localizes Aβ interactions to mitochondria, facilitates membrane permeabilization and pore formation by oligomers, and is both a target and mediator of Aβ-driven mitochondrial dysfunction. Soluble prefibrillar Aβ42 species permeabilize mitochondrial-like vesicles and induce cytochrome c release in a manner that depends on the CL content of the membranes (Camilleri et al., 2013). Aβ exposure can increase mitochondrial reactive oxygen species (ROS), which oxidize and deplete CL, shifting transient permeability transition events to permanent modes and promoting apoptosis in cellular models (Oren et al., 2020). Loss or oxidation of CL is reported in AD model synaptic mitochondria, linking CL changes to early synaptic bioenergetic deficits (Ferguson, 2024).

13.8 Interaction of amyloid-beta peptides with phospholipids

Phosphatidylcholine (PC) is the most abundant phospholipid in mammalian cell membranes. The zwitterionic nature of the PC headgroup leads to relatively weak binding of Aβ to pure PC bilayers (McLaurin and Chakrabartty, 1996). Despite this, PC bilayers support Aβ aggregation, as evidenced by NMR studies that have characterized Aβ structures in PC-containing systems. In the presence of PC membranes, Aβ fibrils form with distinct morphologies compared to solution-grown fibrils and, as confirmed by cryo-EM and ssNMR, reveal lipid-bound fibril structures (Frieg et al., 2024). Lipid-fibril complexes can extract additional lipids from membranes, potentially contributing to membrane disruption.

Anionic lipids such as phosphatidylserine (PS), phosphatidylglycerol, and CL dramatically enhance Aβ binding through electrostatic interactions with the positively charged residues of Aβ (Lys16, Lys28, and the N-terminal amino group) (Zhao et al., 2004).

PS is enriched in the inner leaflet of plasma membranes but can be exposed on the outer leaflet during apoptosis or cellular stress. PS promotes Aβ binding and aggregation, as evidenced by NMR, which shows enhanced membrane perturbation in PS-containing bilayers (Terzi et al., 1997).

In addition to PS, phosphatidylethanolamine (PE) is a major component of bacterial and mitochondrial membranes and is present in significant amounts in mammalian membranes. Like the previously discussed lipids, PE has a smaller headgroup than phosphatidylcholine and, under certain conditions, can form nonlamellar phases such as the hexagonal HII phase, a nonbilayer structure with cylindrical lipid arrangements. NMR studies indicate that PE-containing bilayers support Aβ aggregation and may influence membrane curvature and defect formation (Tashima et al., 2004).

13.9 Binding amyloid-beta peptides perturb lipid order and dynamics

Recent studies have shown that ssNMR measurements of lipid order parameters (using ^{2}H NMR of deuterated lipids or ^{13}C NMR of labeled lipids) reveal that Aβ binding perturbs lipid organization (Weber et al., 2012; Qiang et al., 2020). As shown later, binding Aβ alters ^{31}P NMR chemical shifts and lineshapes, indicating changes in headgroup orientation and dynamics (Rajput et al., 2021). Moreover, Aβ can disorder lipid acyl chains, particularly in the upper portion near the headgroup–chain interface, leading to membrane thinning (Davis, 1983). Notably, different lipid species within the same bilayer show differential responses to Aβ binding, suggesting selective interactions (Qiang et al., 2020). Together, these results demonstrate the multifaceted effects of Aβ on membrane structure and dynamics.

13.10 Effect of Cu^{2+} and Zn^{2+} ions on amyloid-beta oligomerization

Zinc and copper ions are the most prevalent physiological transition-metal ions, playing a key role in modulating the aggregation and toxicity of amyloid peptides. Consequently, dysregulated copper homeostasis has been associated with AD development, as evidenced by increased copper concentration in amyloid plaques (Bush and Tanzi, 2008; Smith et al., 2007). The presence of three potential metal-binding histidine residues in Aβ peptides may point to toxic mechanisms mediated by metal ions (Liu et al., 2019). Furthermore, Cu^{2+}–Aβ complexes may facilitate the formation of

ROS via Fenton-like chemistry, resulting in lipid peroxidation, protein oxidation, and dityrosine formation (Guilloreau et al., 2007). The histidine side chains first anchor Cu^{2+} binding to the Aβ monomer (fast exchange rate), followed by deprotonation and/or severe line broadening of the backbone amide NH for Glu3–Val18 (intermediate exchange rate) (Hou and Zagorski, 2006).

Cu^{2+} modulates Aβ40 stability and conformation by inhibiting its fibrillation and altering its physical properties (Parthasarathy et al., 2011). Specifically, Cu^{2+} binding to Aβ produces a more compact complex, as shown by an increase in Aβ40's translational diffusion coefficient with rising Cu^{2+} concentration. This change reflects a decrease in the hydrodynamic radius from 16.8 to 15.8 Å. Based on ^{15}N CPMG relaxation dispersion and paramagnetic relaxation (PRE) measurements, two prevalent Cu^{2+} coordination modes characterized by three ligands, Asp1 (NH_2, and CO) and His6 (Hε) were identified (Fig. 13.3). The fourth ligand in the Cu^{2+}-Aβ complex was selected as the His13 (Nδ) or His14 (Nδ) forming a well-defined fold of the 23-residue N-terminal fragment of Aβ40 (Abelein et al., 2022).

Zinc concentration in the brain is estimated at 150 μM, 10 times higher than in the serum (Portbury and Adlard, 2017). Notably, zinc competes with copper for amyloid coordination, and supplementation has been reported to affect copper distribution in the brain (Alies et al., 2013). The presence of Zn^{2+} significantly reduces the elongation rate, thereby lowering the overall aggregation rate, though it does not completely inhibit amyloid formation (Abelein, 2023). NMR studies show that Zn^{2+} binding causes significant signal intensity attenuation in the N-terminal fragment (first 16–18 residues) of Aβ40. Moreover, Zn^{2+}-induced folding of Aβ40 mainly affects the N-terminus, which is normally assumed not to directly participate in amyloid formation

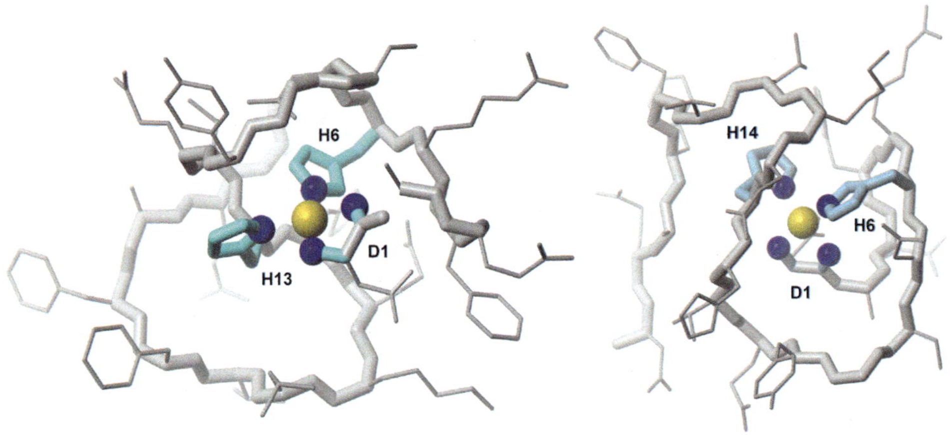

Figure 13.3 3D structures of the Cu^{2+}-Aβ complex were established based on NMR data combined with MD simulations. Shown are the coordination modes with the fourth binding ligands with His13 (left) and His14 (right).

and is not involved in the β-rich core of in vitro produced Aβ40 amyloid fibrils (Abelein et al., 2015). Modulation of the fibril ends by Zn^{2+} may cause perturbations of one or both steps, leading to retardation in the aggregation kinetics (Miller et al., 2010). This also explains why already very low relative concentrations of Zn^{2+} result in a significant reduction of the elongation rate. Zinc acts as a chaperone-mimicking agent, primarily by reducing the elongation rate of Aβ fibrils through transient binding and folding of the N-terminus, which modulates the fibril ends where elongation occurs (Abelein et al., 2015).

13.11 Conclusion

Aβ–membrane interactions represent a complex, multifaceted process involving dynamic conformational changes, lipid-specific binding, oligomerization, and membrane disruption. NMR spectroscopy is uniquely capable of providing atomic-resolution structural information in native-like lipid bilayers (membrane mimetics), allowing direct observation of Aβ's behavior in these environments. The insights gained from NMR studies have fundamentally shaped our understanding of Aβ oligomerization and aggregation mechanisms and contributed to understanding of the membrane-mediated toxicity.

References

Abelein, A., Gräslund, A., Danielsson, J., 2015. Zinc as chaperone-mimicking agent for retardation of amyloid β peptide fibril formation. Proceedings of the National Academy of Sciences 112 (17), 5407–5412. Available at https://doi.org/10.1073/pnas.1421961112.

Abelein, A., Ciofi-Baffoni, S., Mörman, C., Kumar, R., Giachetti, A., Piccioli, M., et al., 2022. Molecular structure of Cu(II)-bound amyloid-β monomer implicated in inhibition of peptide self-assembly in Alzheimer's disease. JACS Au 2 (11), 2571–2584. Available at https://doi.org/10.1021/jacsau.2c00438.

Abelein, A., 2023. Metal binding of Alzheimer's amyloid-β and its effect on peptide self-assembly. Accounts of Chemical Research 56 (19), 2653–2663. Available at https://doi.org/10.1021/acs.accounts.3c00370.

Alies, B., Sasaki, I., Proux, O., Sayen, S., Guillon, E., Faller, P., et al., 2013. Zn impacts Cu coordination to amyloid-β, the alzheimer's peptide, but not the ROS production and the associated cell toxicity. Chemical Communications 49 (12), 1214–1216. Available at https://doi.org/10.1039/c2cc38236a.

Ariga, T., McDonald, M.P., Yu, R.K., 2008. Role of ganglioside metabolism in the pathogenesis of Alzheimer's disease - a review. Journal of Lipid Research 49 (6), 1157–1175. Available at https://doi.org/10.1194/jlr.R800007-JLR200.

Beel, A.J., Mobley, C.K., Hak, J.K., Tian, F., Hadziselimovic, A., Jap, B., et al., 2008. Structural studies of the transmembrane C-terminal domain of the amyloid precursor protein (APP): does APP function as a cholesterol sensor? Biochemistry 47 (36), 9428–9446. Available at https://doi.org/10.1021/bi800993c.

Bush, A.I., Tanzi, R.E., 2008. Therapeutics for Alzheimer's disease based on the metal hypothesis. Neurotherapeutics: The Journal of the American Society for Experimental NeuroTherapeutics 5 (3), 421–432. Available at https://doi.org/10.1016/j.nurt.2008.05.001.

Bera, S., Korshavn, K.J., Kar, R.K., Lim, M.H., Ramamoorthy, A., Bhunia, A., 2016. Biophysical insights into the membrane interaction of the core amyloid-forming Aβ40 fragment K16-K28 and its role in the pathogenesis of Alzheimer's disease. Physical Chemistry Chemical Physics 18 (25), 16890–16901. Available at https://doi.org/10.1039/c6cp02023b.

Camilleri, A., Zarb, C., Caruana, M., Ostermeier, U., Ghio, S., Högen, T., et al., 2013. Mitochondrial membrane permeabilisation by amyloid aggregates and protection by polyphenols. Biochimica et Biophysica Acta (BBA) - Biomembranes 1828 (11), 2532–2543. Available at https://doi.org/10.1016/j.bbamem.2013.06.026.

Cheng, Q., Hu, Z.W., Tobin-miyaji, Y., Perkins, A.E., Deak, T., Qiang, W., 2020. Fibrillization of 40-residue β-amyloid peptides in membrane-like environments leads to different fibril structures and reduced molecular polymorphisms. Biomolecules 10 (6), 1–20. Available at https://doi.org/10.3390/biom10060881.

Ciudad, S., Puig, E., Botzanowski, T., Meigooni, M., Arango, A.S., Do, J., et al., 2020. Aβ(1-42) tetramer and octamer structures reveal edge conductivity pores as a mechanism for membrane damage. Nature Communications 11, 3014. Available at https://doi.org/10.1038/s41467-020-16566-1.

Danielsson, J., Andersson, A., Jarvet, J., Gräslund, A., 2006. 15N relaxation study of the amyloid β-peptide: structural propensities and persistence length. Magnetic Resonance in Chemistry 44 (S1), S114. Available at https://doi.org/10.1002/mrc.1814.

Davis, J.H., 1983. The description of membrane lipid conformation, order and dynamics by 2H-NMR. Biochimica et Biophysica Acta (BBA) - Reviews on Biomembranes 737 (1), 117–171. Available at https://doi.org/10.1016/0304-4157(83)90015-1.

Di Scala, C., Yahi, N., Boutemeur, S., Flores, A., Rodriguez, L., Chahinian, H., et al., 2016. Common molecular mechanism of amyloid pore formation by Alzheimer's β-amyloid peptide and α-synuclein. Scientific Reports 6, 28781. Available at https://doi.org/10.1038/srep28781.

Edler, M.K., Sherwood, C.C., Meindl, R.S., Hopkins, W.D., Ely, J.J., Erwin, J.M., et al., 2017. Aged chimpanzees exhibit pathologic hallmarks of Alzheimer's disease. Neurobiology of Aging 59, 107–120. Available at https://doi.org/10.1016/j.neurobiolaging.2017.07.006.

Ferguson, D.J., 2024. Uncovering cardiolipin's function in neurodegenerative diseases. Preprints 2024110154. Available at https://doi.org/10.20944/preprints202411.0154.v1.

Fernández-Pérez, E.J., Sepúlveda, F.J., Peters, C., Bascuñán, D., Riffo-Lepe, N.O., González-Sanmiguel, J., et al., 2018. Effect of cholesterol on membrane fluidity and association of Aβ oligomers and subsequent neuronal damage: a Double-Edged Sword. Frontiers in Aging Neuroscience 10. Available at https://doi.org/10.3389/fnagi.2018.00226.

Feuillie, C., Lambert, E., Ewald, M., Azouz, M., Henry, S., Marsaudon, S., et al., 2020. High speed AFM and nanoinfrared spectroscopy investigation of Aβ1–42 peptide variants and their interaction With POPC/SM/Chol/GM1 model membranes. Frontiers in Molecular Biosciences 7. Available at https://doi.org/10.3389/fmolb.2020.571696.

Frieg, B., Han, M., Giller, K., Dienemann, C., Riedel, D., Becker, S., et al., 2024. Cryo-EM structures of lipidic fibrils of amyloid-β (1-40). Nature Communications 15, 1297. Available at https://doi.org/10.1038/s41467-023-43822-x.

Guilloreau, L., Combalbert, S., Sournia-Saquet, A., Mazarguil, H., Faller, P., 2007. Redox chemistry of copper–amyloid-β: the generation of hydroxyl radical in the presence of ascorbate is linked to redox-potentials and aggregation state. Chembiochem: a European Journal of Chemical Biology 8 (11), 1317–1325. Available at https://doi.org/10.1002/cbic.200700111.

Gullion, T., 2008. Rotational-echo, double-resonance NMR. In: Webb, G.A. (Ed.), Modern Magnetic Resonance. Springer, Dordrecht, pp. 713–718. Available at 10.1007/1-4020-3910-7_89.

Habchi, J., Chia, S., Galvagnion, C., Michaels, T.C.T., Bellaiche, M.M.J., Ruggeri, F.S., et al., 2018. Cholesterol catalyses Aβ42 aggregation through a heterogeneous nucleation pathway in the presence of lipid membranes. Nature Chemistry 10 (6), 673–683. Available at https://doi.org/10.1038/s41557-018-0031-x.

Hardy, J.A., Higgins, G.A., 1992. Alzheimer's disease: the amyloid cascade hypothesis. Science (New York, N.Y.) 256 (5054), 184–185. Available at https://doi.org/10.1126/science.1566067.

Hou, L., Shao, H., Zhang, Y., Li, H., Menon, N.K., Neuhaus, E.B., et al., 2004. Solution NMR Studies of the Aβ(1-40) and Aβ(1-42) peptides establish that the Met35 oxidation state affects the mechanism of amyloid formation. Journal of the American Chemical Society 126 (7), 1992–2005. Available at https://doi.org/10.1021/ja036813f.

Hou, L., Zagorski, M.G., 2006. NMR reveals anomalous copper(II) binding to the amyloid Aβ peptide of Alzheimer's disease. Journal of the American Chemical Society 128 (29), 9260–9261. Available at https://doi.org/10.1021/ja046032u.

Jarvet, J., Danielsson, J., Damberg, P., Oleszczuk, M., Gräslund, A., 2007. Positioning of the Alzheimer Aβ(1–40) peptide in SDS micelles using NMR and paramagnetic probes. Journal of Biomolecular NMR 39 (1), 63–72. Available at https://doi.org/10.1007/s10858-007-9176-4.

Jo, S., Kim, T., Iyer, V.G., Im, W., 2008. CHARMM-GUI: A web-based graphical user interface for CHARMM. Journal of Computational Chemistry 29, 1859–1865. Available at https://doi.org/10.1002/jcc.20945.

Kandel, N., Matos, J.O., Tatulian, S.A., 2019. Structure of amyloid β 25–35 in lipid environment and cholesterol-dependent membrane pore formation. Scientific Reports 9 (1). Available at https://doi.org/10.1038/s41598-019-38749-7.

Kenyaga, J.M., Cheng, Q., Qiang, W., 2022. Early stage β-amyloid-membrane interactions modulate lipid dynamics and influence structural interfaces and fibrillation. Journal of Biological Chemistry 298 (10). Available at https://doi.org/10.1016/j.jbc.2022.102491.

Korshavn, K.J., Bhunia, A., Lim, M.H., Ramamoorthy, A., 2016. Amyloid-β adopts a conserved, partially folded structure upon binding to zwitterionic lipid bilayers prior to amyloid formation. Chemical Communications 52 (5), 882–885. Available at https://doi.org/10.1039/c5cc08634e.

Korshavn, K.J., Satriano, C., Lin, Y., Zhang, R., Dulchavsky, M., Bhunia, A., et al., 2017. Reduced lipid bilayer thickness regulates the aggregation and cytotoxicity of amyloid-β Journal of Biological Chemistry 292 (11), 4638–4650. Available at https://doi.org/10.1074/jbc.M116.764092.

Liu, Y., Nguyen, M., Robert, A., Meunier, B., 2019. Metal Ions in Alzheimer's disease: a key role or not? Accounts of Chemical Research 52 (7), 2026–2035. Available at https://doi.org/10.1021/acs.accounts.9b00248.

McLaurin, J.A., Chakrabartty, A., 1996. Membrane disruption by Alzheimer β-amyloid peptides mediated through specific binding to either phospholipids or gangliosides. Implications for neurotoxicity. Journal of Biological Chemistry 271 (43), 26482–26489. Available at https://doi.org/10.1074/jbc.271.43.26482.

Meng, E.C., Goddard, T.D., Pettersen, E.F., Couch, G.S., Pearson, Z.J., Morris, J.H., Ferrin, T.E., 2023. UCSF ChimeraX: Tools for structure building and analysis. Protein Science 32, e4792. Available at https://doi.org/10.1002/pro.4792.

Miller, Y., Ma, B., Nussinov, R., 2010. Zinc ions promote alzheimer aβ aggregation via population shift of polymorphic states. Proceedings of the National Academy of Sciences of the United States of America 107 (21), 9490–9495. Available at https://doi.org/10.1073/pnas.0913114107.

Narayanan, S., Reif, B., 2005. Characterization of chemical exchange between soluble and aggregated States of β-amyloid by solution-State NMR upon variation of salt conditions. Biochemistry 44 (5), 1444–1452. Available at https://doi.org/10.1021/bi048264b.

Nguyen, H.L., Linh, H.Q., Krupa, P., La Penna, G., Li, M.S., 2022. Amyloid β dodecamer disrupts the neuronal membrane more strongly than the mature fibril: understanding the role of oligomers in neurotoxicity. Journal of Physical Chemistry B 126 (20), 3659–3672. Available at https://doi.org/10.1021/acs.jpcb.2c01769.

Novo, M., Freire, S., Al-Soufi, W., 2018. Critical aggregation concentration for the formation of early Amyloid-β (1–42) oligomers. Scientific Reports 8, 1–8. Available at https://doi.org/10.1038/s41598-018-19961-3.

Opella, S.J., Marassi, F.M., 2004. Structure determination of membrane proteins by NMR spectroscopy. Chemical Reviews 104 (8), 3587–3606. Available at https://doi.org/10.1021/cr0304121.

Oren, O., Ben Zichri, S., Taube, R., Jelinek, R., Papo, N., 2020. Aβ42 double mutant inhibits Aβ42-induced plasma and mitochondrial membrane disruption in artificial membranes, isolated organs, and intact cells. ACS Chemical Neuroscience 11 (7), 1027–1037. Available at https://doi.org/10.1021/acschemneuro.9b00638.

Parthasarathy, S., Long, F., Miller, Y., Xiao, Y., McElheny, D., Thurber, K., et al., 2011. Molecular-level examination of Cu^{2+} binding structure for amyloid fibrils of 40-residue Alzheimer's β by solid-State NMR spectroscopy. Journal of the American Chemical Society 133 (10), 3390–3400. Available at https://doi.org/10.1021/ja1072178.

Portbury, S.D., Adlard, P.A., 2017. Zinc signal in brain diseases. International Journal of Molecular Sciences 18 (12). Available at https://doi.org/10.3390/ijms18122506.

Qiang, W., Doherty, K.E., Klees, L.M., Tobin-Miyaji, Y., 2020. Time-dependent lipid dynamics, organization and peptide-lipid interaction in phospholipid bilayers with incorporated β-amyloid oligomers. Journal of Physical Chemistry Letters 11 (19), 8329–8336. Available at https://doi.org/10.1021/acs.jpclett.0c01967.

Rajput, S., Sani, M.A., Keizer, D.W., Separovic, F., 2021. Utilizing magnetic resonance techniques to study membrane interactions of amyloid peptides. Biochemical Society Transactions 49 (3), 1457–1465. Available at https://doi.org/10.1042/BST20201244.

Sarkar, B., Das, A.K., Maiti, S., 2013. Thermodynamically stable amyloid-β monomers have much lower membrane affinity than the small oligomers. Frontiers in Physiology 4. Available at https://doi.org/10.3389/fphys.2013.00084.

Sepehri, A., Lazaridis, T., 2023. Putative structures of membrane-embedded amyloid β oligomers. ACS Chemical Neuroscience 14 (1), 99–110. Available at https://doi.org/10.1021/acschemneuro.2c00535.

Serra-Batiste, M., Ninot-Pedrosa, M., Bayoumi, M., Gairí, M., Maglia, G., Carulla, N., 2016. Aβ42 assembles into specific β-barrel pore-forming oligomers in membrane-mimicking environments. Proceedings of the National Academy of Sciences 113 (39), 10866–10871. Available at https://doi.org/10.1073/pnas.1605104113.

Serra-Batiste, M., Ninot-Pedrosa, M., Puig, E., Ciudad, S., Gairí, M., Carulla, N., 2018. Preparation of a well-defined and stable β-barrel pore-forming Aβ42 oligomer. In: Sigurdsson, E., Calero, M., Gasset, M. (Eds.), Amyloid Proteins. Methods in Molecular Biology, vol 1779. Humana Press, New York, NY, pp. 13–22. Available at https://doi.org/10.1007/978-1-4939-7816-8_2.

Shao, H., Jao, S.C., Ma, K., Zagorski, M.G., 1999. Solution structures of micelle-bound amyloid β-(1-40) and β-(1-42) peptides of Alzheimer's disease. Journal of Molecular Biology 285 (2), 755–773. Available at https://doi.org/10.1006/jmbi.1998.2348.

Simons, K., Ikonen, E., 1997. Functional rafts in cell membranes. Nature 387 (6633), 569–572. Available at https://doi.org/10.1038/42408.

Smith, D.G., Cappai, R., Barnham, K.J., 2007. The redox chemistry of the Alzheimer's disease amyloid β peptide. Biochimica et Biophysica Acta - Biomembranes 1768 (8), 1976–1990. Available at https://doi.org/10.1016/j.bbamem.2007.02.002.

Sticht, H., Bayer, P., Willbold, D., Dames, S., Hilbich, C., Beyreuther, K., et al., 1995. Structure of amyloid A4-(1–40)-Peptide of Alzheimer's Disease. European Journal of Biochemistry 233 (1), 293–298. Available at https://doi.org/10.1111/j.1432-1033.1995.293_1.x.

Tashima, Y., Oe, R., Lee, S., Sugihara, G., Chambers, E.J., Takahashi, M., et al., 2004. The effect of cholesterol and monosialoganglioside (GM1) on the release and aggregation of amyloid β-peptide from liposomes prepared from brain membrane-like lipids. Journal of Biological Chemistry 279 (17), 17587–17595. Available at https://doi.org/10.1074/jbc.m308622200.

Terzi, E., Hölzemann, G., Seelig, J., 1997. Interaction of Alzheimer β-amyloid peptide(1−40) with lipid membranes. Biochemistry 36 (48), 14845–14852. Available at https://doi.org/10.1021/bi971843e.

Tycko, R., 2011. Solid-State NMR studies of amyloid fibril structure. Annual Review of Physical Chemistry 62 (1), 279–299. Available at https://doi.org/10.1146/annurev-physchem-032210-103539.

Tycko, R., 2015. Amyloid polymorphism: structural basis and neurobiological relevance. Neuron 86 (3), 632–645. Available at https://doi.org/10.1016/j.neuron.2015.03.017.

Viles, J.H., 2023. Imaging amyloid-β membrane interactions: ion-channel pores and lipid-bilayer permeability in Alzheimer's disease. Angewandte Chemie - International Edition 62 (25). Available at https://doi.org/10.1002/anie.202215785.

Vivekanandan, S., Brender, J.R., Lee, S.Y., Ramamoorthy, A., 2011. A partially folded structure of amyloid-beta(1-40) in an aqueous environment. Biochemical and Biophysical Research Communications 411 (2), 312–316. Available at https://doi.org/10.1016/j.bbrc.2011.06.133.

Weber, D.K., Gehman, J.D., Separovic, F., Sani, M.A., 2012. Copper modulation of amyloid beta 42 interactions with model membranes. Australian Journal of Chemistry 65 (5), 472–479. Available at https://doi.org/10.1071/CH12015.

Williamson, M.P., 2013. Using chemical shift perturbation to characterise ligand binding. Progress in Nuclear Magnetic Resonance Spectroscopy 73, 1–16. Available at https://doi.org/10.1016/j.pnmrs.2013.02.001.

Wishart, D.S., Sykes, B.D., 1994. The 13C chemical-shift index: a simple method for the identification of protein secondary structure using 13C chemical-shift data. Journal of Biomolecular NMR 4 (2), 171–180. Available at https://doi.org/10.1007/BF00175245.

Yagi-Utsumi, M., Itoh, S.G., Okumura, H., Yanagisawa, K., Kato, K., Nishimura, K., 2023. The double-layered structure of amyloid-β assemblage on GM1-containing membranes catalytically promotes fibrillization. ACS Chemical Neuroscience 14 (15), 2648–2657. Available at https://doi.org/10.1021/acschemneuro.3c00192.

Yanagisawa, K., Odaka, A., Suzuki, N., Ihara, Y., 1995. GM1 ganglioside–bound amyloid β–protein (Aβ): a possible form of preamyloid in Alzheimer's disease. Nature Medicine 1 (10), 1062–1066. Available at https://doi.org/10.1038/nm1095-1062.

Zhaliazka, K., Matveyenka, M., Kurouski, D., 2023. Lipids uniquely alter the secondary structure and toxicity of amyloid beta 1–42 aggregates. FEBS Journal 290 (12), 3203–3220. Available at https://doi.org/10.1111/febs.16738.

Zhao, H., Tuominen, E.K.J., Kinnunen, P.K.J., 2004. Formation of amyloid fibers triggered by phosphatidylserine-containing membranes. Biochemistry 43 (32), 10302–10307. Available at https://doi.org/10.1021/bi049002c.

Zott, B., Simon, M.M., Hong, W., Unger, F., Chen-Engerer, H.J., Frosch, M.P., et al., 2019. A vicious cycle of β amyloid−dependent neuronal hyperactivation. Science (New York, N.Y.) 365 (6453), 559–565. Available at https://doi.org/10.1126/science.aay0198.

CHAPTER 14

Machine learning in prediction of protein aggregation and/or condensation states

Katja Venko
Theory Department, National institute of Chemistry, Ljubljana, Slovenia

14.1 Machine learning prediction of protein aggregation and/or condensation states

Protein aggregation in the complex cellular environment, including the presence of liquid–liquid phase-separated (LLPS) condensates, is currently attracting considerable interest. There is growing evidence that protein condensates are associated with neurodegenerative diseases that involve protein aggregation, such as Alzheimer's disease and amyotrophic lateral sclerosis. In cells, two types of condensate aggregation are present (Visser et al., 2024). In the first, idemic condensate aggregation, the aggregation-prone protein itself phase separates and undergoes a transition from liquid to solid, fibrillar state. In the second, host–guest condensate aggregation, an aggregation-prone protein interacts with a host condensate, whereby such condensates are predominantly composed of different biomolecules. From the perspective that in silico experiments involving predictions of the condensation and aggregation propensities/ regions of proteins are time- and cost-less consuming than "real" experiments, they can be used as an initial step in the process of understanding amyloid formation and can help in the development of new therapeutic strategies for the treatment of neurodegenerative diseases.

To date, more than 50 proteins are known to be associated with amyloid diseases. Of note, most of them are intrinsically disordered proteins (IDPs) that can undergo cross-beta sheet stacking to form insoluble, thermodynamically stable fibrillar aggregates (Sanchez de Groot et al., 2012). From a bioinformatic point of view, some insights into the amino acid code for the conversion of proteins between liquid- and solid-like condensates are available. In general, IDPs are thought to be more prone to undergo LLPS (Vendruscolo and Fuxreiter, 2022). Given that IDPs are often involved in both condensate and aggregate formation, the identification and prediction of protein condensate and aggregate prone regions is gaining increasing attention. After decades of development and improvement, the propensity of proteins to aggregate and/or condensate can be computationally determined based on the amino acid sequence.

Molecular Insights into Proteinopathies
DOI: https://doi.org/10.1016/B978-0-443-34176-2.00010-9
© 2027 Elsevier Inc. All rights are reserved, including those for text and data mining, AI training, and similar technologies.

Numerous computational methods are available to make sequence-based predictions about the protein's propensity to aggregate or condensate. As different parameters determining the tendency of proteins to form aggregates/condensates have been proposed, several computational tools based on various machine-learning strategies have been developed. Nowadays, single predictors and also so-called metapredictors are available. Metapredictors not only provide a single prediction, but their final prediction is based on multiple predictions, which are based on algorithms that combine results from several computational models, including predictions of condensable, amyloidogenic, or/and intrinsically disordered regions. In this respect, metapredictors produce more reliable and accurate predictions (Pintado-Grima et al., 2023; Venko and Žerovnik, 2023). However, the accuracy of aggregate/condensate prediction still remains a major challenge, and great improvements in prediction competence are expected in the future. In this chapter, we present the most commonly used machine-learning approaches for the prediction of protein condensates and aggregates of the amyloid and prion types (Fig. 14.1)

14.2 Predicting protein aggregation propensity

Over the last decades, several research groups have devoted much effort to understanding protein aggregation, and especially to the development of accurate sequence-based predictors of aggregation propensity and aggregation prone regions (ARPs). Anyhow, available computational methods still cannot reliably predict aggregation kinetics, since the majority of sequence-structural properties known to have mechanistic roles in protein aggregation do not correlate well with the aggregation kinetics. AggreRATE-Pred is the only tool that is feasible for both identifying molecular origins of aggregation (APR prediction) and assessing the change in aggregation kinetics (Rawat et al., 2020). If compared with previously developed aggregation prediction

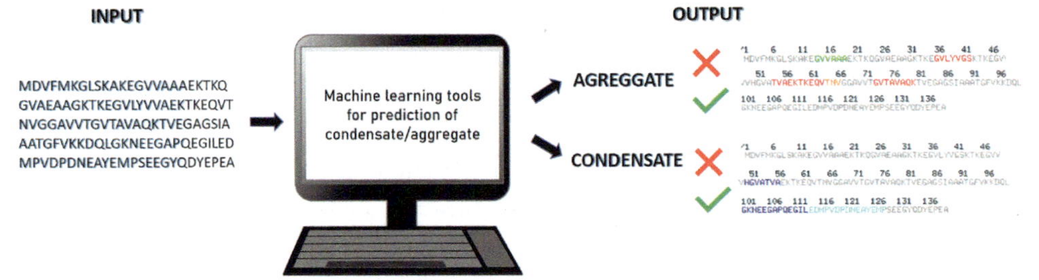

Figure 14.1 Scheme of an in silico experiment for prediction of aggregation/condensation propensity of α-synuclein; the amino acid sequence of α-synuclein serves as input data for machine-learning models, which calculate the propensity of the protein to form aggregates and/ or condensate; the output data are positive or negative prediction. In case of a positive prediction, some of the tools also give additional information about aggregation/condensation prone regions.

models, which have focused mostly on sequence-based properties, AggreRATE-Pred has advanced in considering both the structural properties and sequence-based properties to predict the change in aggregation rate upon point mutations. Other tools compute only the total aggregation propensity of a polypeptide chain and cannot provide information about how protein aggregates grow with time. The APRs are usually integral parts buried within the hydrophobic core of the native protein and consequently are often enriched with residues favoring β-strand formation, increased hydrophobicity, and low charge content (Louros et al., 2020). Considering these sequence propensities, various methods have been developed to predict aggregation potential from protein sequences (see Table 14.1). One of the first and up-to-date still very popular and frequently used machine-learning tools is AGGRESCAN (aggregation-prone segments in proteins), which is based on an aggregation-propensity scale for 20 amino acids that was revealed from *in vivo* experiments and on the assumption that short and specific sequence stretches, known as APRs or "hot spots," modulate protein aggregation (Conchillo-Solé et al., 2007). This algorithm is simple and fast; the implementation of online software was recently done (Pintado-Grima et al., 2023). TANGO is a statistical mechanics algorithm used to predict protein aggregation. It includes the physicochemical principles of β-sheet formation and the assumption that the core regions of an aggregate are fully buried. Authors have revealed that intermolecular β-sheet formation is a widespread underlying mechanism of protein aggregation (Fernandez-Escamilla et al., 2004). Generalized aggregation proneness is a method that can, with high accuracy, distinguish between amyloid fibril-forming and amorphous β-aggregating hexapeptides. Compared to novel methods, its shortcoming is a small dataset that was used in its development (Thangakani et al., 2014). Zyggregator is an algorithm for predicting protein aggregation propensities, which is based on the assumption that folding and aggregation are two closely related processes depending primarily on the fundamental physicochemical properties of polypeptide chains (Tartaglia and Vendruscolo, 2008). Prediction of amyloid fibril-forming segments (Pafig) is a support vector machine (SVM)-based model for the identification of hexapeptides associated with APRs. It is based on 41 physicochemical properties selected from the amino acid index database (Tian et al., 2009). Betascan is another phenomenological method that evaluates β-strand pairing propensity. The result is obtained from probabilities of residues to be H-bonded in amphiphilic β-sheets (Bryan et al., 2009). AggreProt estimates aggregation propensity based on the intrinsic amino acid properties and sequence patterns (Planas-Iglesias et al., 2024). While protein SOlubility based on Disorder and Aggregation (SODA) provides calculations of the propensity of a protein sequence to aggregate, intrinsic disorder, hydrophobicity, and secondary structure preferences. Moreover, a score to evaluate the difference in aggregation and solubility introduced by the mutations can be evaluated (Paladin et al., 2017). However, the majority of tools for APRs identification use amino acid

Table 14.1 Computational tools for prediction of aggregates.

Predictor	Description	References
TANGO	Prediction of protein aggregation	Fernandez-Escamilla et al. (2004)
AGGRESCAN	Prediction of "hot spots" of aggregation	Conchillo-Solé et al. (2007)
Zyggregator	Predicting protein aggregation propensities	Tartaglia and Vendruscolo (2008)
SAP	Spatial aggregation propensity	Chennamsetty et al. (2009)
Betascan	β-strand pairing propensity	Bryan et al. (2009)
Pafig	Prediction of amyloidogenic regions in protein	Garbuzynskiy et al. (2009)
GAP	Generalized aggregation proneness	Thangakani et al. (2014)
Aggrescan3D	Aggregation propensity in protein structures	Zambrano et al. (2015a, 2015b)
SolubiS	Identification of mutations that affects protein aggregation	Van Durme et al. (2016)
CamSol Structurally	Structure-based method of calculating APRs	Sormanni et al. (2017)
SODA	Protein SOlubility based on Disorder and Aggregation	Paladin et al. (2017)
AggScore	Visualizing of aggregation prone regions (APRs) distribution	Sankar et al. (2018)
SOLart	Structure-based RF model to predict protein solubility and aggregation	Hou et al. (2019)
Aggrescan3D 2.0	Prediction of aggregation and engineering of protein solubility	Kuriata et al. (2019)
AggreRATE-Pred	APR prediction and change in aggregation kinetics	Rawat et al. (2020)
ANuPP	Aggregation nucleation prediction in proteins	Prabakaran et al. (2021)
Aggrescan4D	pH-dependent protein aggregation propensity	Zalewski et al. (2024)
AggreProt	Per-residue aggregation profiles	Planas-Iglesias et al. (2024)
AggNet	Protein aggregation, deep learning, and protein language model	He et al. (2025)

residue composition and/or sequence pattern features (Rawat et al., 2020). In this regard, aggregation nucleation prediction in peptides and proteins (ANuPP), a web metaclassifier for ARPs identification, is a novelty (Prabakaran et al., 2021). It is a unique ensemble-classifier based on atom-level features and considers the diversity of aggregation mechanisms. The performance of ANuPP was evaluated using several different datasets, and results demonstrate that ANuPP is one of the best predictive methods for both amyloidogenic hexapeptide prediction and APR identification when compared with other currently available methods.

In general, the prediction of aggregation propensities of protein 3D structures still remains the bottleneck, due to the lack of known high-resolution structures. Although the algorithms for detecting aggregation-nucleating regions in proteins' primary sequences perform quite well, in the folded state, many of these regions are part of the inner core of the protein and probably do not contribute to aggregation unless the protein unfolds extensively (Graña-Montes and Ventura, 2015). Therefore the development of algorithms able to detect APRs at the protein surfaces is of high interest and has been in constant progress in recent years. Currently, the most well-known structure-based methods are spatial aggregation propensity (SAP), Aggrescan3D/4D, CamSol, and AggScore. To enhance predictive performance, they incorporate 3D structural information with atomic coordinates, surface hydrophobic patches, solvent accessibility, and the local microenvironment. SAP identifies the location and size of hot spots for aggregation based on the exposure of spatially adjacent hydrophobic amino acids. This commercially available tool was primarily developed for the prediction of antibody APRs, but it could also be used to identify the APRs of other proteins and peptides (Chennamsetty et al., 2009). In distinction to Aggrescan, the Aggrescan3D (A3D) (Zambrano et al., 2015b) and Aggrescan3D 2.0 (A3D2) (Kuriata et al., 2019) are tools for prediction of aggregation and engineering of protein solubility that are based on features associated with structural properties. With the A3D/A3D2 tools, the identified APR residues can be virtually mutated to test the impact on solubility or pathogenicity, and the dynamic fluctuations of protein structure in solution, which may influence aggregation propensity, can be studied. Building upon its predecessor, A3D, the Aggrescan4D is now the most advanced tool for assisting the improvement of protein solubility, since it is designed for predicting protein aggregation, leveraging structural information and the influence of pH (Zalewski et al., 2024). Further, several methods that are primarily developed for solubility profiling are also capable of APR evaluation. For example, SOLart is a structure-based random forest model to predict protein solubility and aggregation, which yields a scaled solubility score with values close to zero indicating APR proteins (Hou et al., 2019). SolubiS is a pipeline to identify mutations that reduce protein aggregation by reducing the intrinsic aggregation propensity of the sequence while taking care not to disrupt the thermodynamic stability of the native structure (using the TANGO and FoldX algorithms) (Van Durme et al., 2016). CamSol structurally corrected is a structure-based method of calculating APRs and mutated variants design, which accounts for the proximity of the amino acids in the structure and for their solvent exposure. The result is represented on the protein structure as colored spot patches of low solubility that may elicit self-assembly (Sormanni et al., 2017). AggScore is a commercially available method (Schrödinger's BioLuminate Suite). The algorithm is based on 3D structure and uses the hydrophobic and electrostatic patches distribution on the protein surface, factoring in the intensity and relative orientation of surface patches into an aggregation propensity function. It offers visualization of APR distribution on the surface of proteins and predicts the impact of residue mutation on aggregation behavior (Sankar et al., 2018). AggreProt is an ensemble of deep neural networks to predict APRs in protein sequences that was trained on experimentally evaluated hexapeptides. The method provides per-residue aggregation profiles along with information on solvent accessibility and transmembrane propensity with interactive sequence and structure viewers for comprehensive analysis. Since in all of the above mentioned predictors a series of structure-based and commonly used sequence features are integrated, they all predict with improved accuracy. However, based on benchmark comparisons, the newest approach, AggNet, outperforms existing methods and achieves state-of-the-art performance on protein aggregation prediction. AggNet is a deep-learning framework based on the protein language model ESM2 and AlphaFold2, which utilizes physicochemical, evolutionary, and structural information to discriminate amyloid and nonamyloid peptides and identify APRs (He et al., 2025).

14.2.1 Amyloid propensity

The transition of soluble proteins into insoluble amyloid fibrils is driven by specific self-propagating short sequence stretches that can be predicted on a genomic scale given input sequences. In regard to the propensities of different protein sequences to aggregate into amyloids mostly relying at the molecular level on the stability of the amyloid cross-beta structure, the amyloidogenicity prediction is also an interesting bioinformatics topic and can nowadays be performed with several computational tools. Anyhow, the accurate prediction of amyloid-forming determinants still remains challenging and in focus of further research (Louros et al., 2020). AmyLoad is a web server of amyloidogenic sequence fragments (over 1480 different entries, and the number is still growing). It allows own browsing sessions, add their own sequences into the database, and analyze query FASTA sequences with implemented amyloid predictors (Wozniak and Kotulska, 2015). Further, an updated and significantly expanded WALTZ-DB 2.0 database (Louros et al., 2019) is now the largest open-access repository for amyloid fibril formation determinants and provides information on experimentally determined amyloid-forming hexapeptide sequences. Due to constant updates and new entries of valuable experimental data in this database, the comprehensive enhancement of the development of new approaches for prediction of amyloid-forming regions is expected in the near future.

In Table 14.2 different prediction programs are gathered. One of the first-generation amyloid predictors is FoldAmyloid, which enables detection of amyloidogenic regions in the protein chain (Garbuzynskiy et al., 2009). The tool is based on two characteristics: the expected probability of hydrogen bond formation and the expected packing density of residues. Authors demonstrate that for the formation of amyloid fibrils, the regions with high expected probability of the formation of backbone–backbone hydrogen bonds and the regions with high expected packing density are responsible. On the other hand, Waltz is using a different classification method for predicting amyloidogenic regions in protein sequences. It uses a position-specific scoring matrix to identify amyloid-forming regions in functional amyloids and can distinguish between amyloid sequences and amorphous beta-sheet aggregates (Maurer-Stroh et al., 2010). SALSA also enables determination of the aggregation profile of sequences. The algorithm, which calculates an average β-strand propensity score of a peptide window using Chou and Fasman β-strand preference numbers, is integrated in the AMYPdb database (Pawlicki et al., 2008). PASTA 2.0 is an algorithm based on an energy function that evaluates the stability of putative cross-beta pairings between different sequence stretches. The aggregation data are also complemented and enriched with other information, such as intrinsic disorder and secondary structure predictions. The amyloid-forming regions can be correctly detected at high specificity after the rederivation on a larger dataset of globular protein domains (Walsh et al., 2014). Further, neural networks for calculating contact-dependent secondary structure propensity is a tool

Table 14.2 Computational tools for prediction of amyloids.

Predictor	Description	References
SALSA	Determining the aggregation profile of sequences in AMYPdb	Pawlicki et al. (2008)
FoldAmyloid	Detection of amyloidogenic regions in the protein chain	Garbuzynskiy et al. (2009)
NetCSSP	Neural networks for Contact-dependent Secondary Structure Propensity	Kim et al. (2009)
WALTZ	Predicting amylogenic regions in protein sequences	Maurer-Stroh et al. (2010)
MetAmyl	METapredictor for AMYLoid proteins	Emily et al. (2013)
AmylPred2	Amyloid propensity prediction	Tsolis et al. (2013)
PASTA 2.0	Prediction of amyloid-forming regions	Walsh et al. (2014)
FiSH Amyloid	Method for finding amyloidogenic segments in proteins	Gasior and Kotulska (2014)
AmyLoad	Server of amyloidogenic sequence fragments	Wozniak and Kotulska (2015)
APPNN	Amyloid Propensity Prediction Neural Network	Família et al. (2015)
ArchCandy	Prediction of 3D structures of amyloid fibrils	Ahmed et al. (2015)
AmyloGram	Amyloidogenic motifs revealed by n-gram analysis	Burdukiewicz et al. (2017)
WALTZ-DB 2.0	Repository for amyloid fibril formation determinants	Louros et al. (2020)
CORDAX	Clusters of amyloid sequence space	Louros et al. (2020)
BAP	Prediction of amyloidogenic hexapeptides	Keresztes et al. (2021)

based on neural networks that provides helical and beta propensity in a single profile with a dual mode of calculation. This method is well applicable to predicting nonnative secondary structures and amyloid fibril formation (Kim et al., 2009). FiSH Amyloid is another machine-learning method based on the cooccurrence of amino acid couples in short amyloidogenic fragments of proteins (Gasior and Kotulska, 2014). AmyloGram predicts amyloidogenic sequences using random forests and n-gram analysis (Burdukiewicz et al., 2017). Amyloid Propensity Prediction Neural Network (APPNN) is an amyloidogenicity propensity predictor based on the analysis of seven physicochemical and biochemical features (β-sheet frequency, atom-based hydrophobic moment, helix termination parameters, isoelectric point, etc.) (Família et al., 2015). It performs with the same accuracy as the Budapest amyloid predictor (BAP), which enables prediction of amyloidogenic hexapeptides. BAP is based on SVM algorithm, trained and tested on the WALTZ hexapeptide dataset (Keresztes et al., 2021).

In advance, MetAmyl and AmylPred2 are metapredictors based on a consensus approach, which combines the strength of different individual predictors into a single predictor. MetAmyl makes a metaprediction of sequences' amyloidogenicity based on the set of four individual predictors: Pafig, SALSA, WALTZ, and FoldAmyloid. The score for the metaprediction is obtained using a logistic regression model (Emily et al., 2013). AmylPred2 is an improved version of the previous method for amyloid propensity prediction (Frousios et al., 2009). It makes a consensus prediction based on 11 algorithms. The consensus score is defined as the hit overlap of at least n/2 out of n selected methods (Tsolis et al., 2013). The method is useful for understanding the properties of disease protein misfolding and for helping to control protein aggregation/solubility in biotechnology. Of interest, in the 3D structures of the majority of disease-related amyloid fibrils, it has been shown that the structures contain a β-strand-loop-β-strand motif called β-arch. Accordingly, based on the assumption that protein sequences that are able to form β-arcades are amyloidogenic, a novel bioinformatics approach was developed. ArchCandy allows one to distinguish between amyloidogenic/nonamyloidogenic sequences, to localize the amyloidogenic regions within the protein, to explain the effect of mutations on the amyloid-forming potential, and to predict the 3D structures of amyloid fibrils (Ahmed et al., 2015). With benchmark analysis, the high performance of this method was demonstrated. However, in general, the currently defined amyloid sequence space is strongly biased towards hydrophobic, beta-sheet-prone sequences that form the core of globular proteins and towards Q/N/Y-rich yeast prions. Therefore Louros et al. (2020) took advantage of the increasing amount of high-resolution structural information on amyloid cores, which are currently available in the Protein Data Bank, and implemented a machine-learning method, CORDAX. The method is based on clustering by t-distributed stochastic neighbour embedding and resulted in an expansion away from hydrophobic amyloid sequences towards clusters of lower aliphatic content and higher charge, or regions of helical and disordered propensities.

14.2.2 Prions and prion-like propensity

Since prions are a particular class of amyloids that can propagate their misfolded conformation and have unique compositional features, several bioinformatics tools capable of identifying novel pathological and functional polypeptides with prion-like properties are also available. Herein, we discuss the features of different databases and algorithms developed to study prions and prion-like proteins (see Table 14.3). ZipperDB is a database that contains predictions of fibril-forming segments within proteins identified by the 3D profile method. This method is a unique approach using the structural information to evaluate the likelihood of fibril formation for a particular sequence (Goldschmidt et al., 2010). Another interesting online database for the prediction of prion domains in complete proteomes is PrionScan (Espinosa Angarica et al., 2014). It has been used to understand prion/prionogenic protein functions and how their interaction networks have a substantial impact on gene regulation, to identify regions driving LLPS (Pancsa et al., 2021), or has been proposed as a predictor of prion-like proteins capable of LLPS (Maziuk et al., 2017). pWaltz is a prediction algorithm for determining proteins containing glutamine/asparagine rich prion-like domains (PrLD) and their amyloid cores. It was originally inspired by the Waltz amyloid prediction strategy, but employed a lower detection threshold to identify milder amyloids and used a larger sliding window corresponding to size of the minimum transmissible β-fold (Sabate et al., 2015). Further, PrionW is an advanced tool with the implementation of

Table 14.3 Computational tools for the prediction of prions.

Predictor	Description	References
ZipperDB	Database of fibril-forming segment predictions	Goldschmidt et al. (2010)
PLAAC	Prion-like amino acid composition	Lancaster et al. (2014)
PrionScan	Prediction of prion domains in complete proteomes	Espinosa Angarica et al. (2014)
pWaltz	Determining proteins of glutamine/asparagine rich prion-like domains	Sabate et al. (2015)
PrionW	Prion-like protein finder	Zambrano et al. (2015a)
AMYCO	Aggregation properties of prion-like proteins	Iglesias et al. (2019)
SGnn	Estimation of heat-induced stress granules	Iglesias et al. (2021)

the pWaltz algorithm, capable of working with complete protein sequences and identifying the compositional context and structural features needed for prion conversion (Zambrano et al., 2015a). Prion-like amino acid composition (PLAAC) is another web application that scans protein sequences for domains with PLAAC. The tool is based on a hidden-Markov model algorithm, which identifies probable prion subsequences (Lancaster et al., 2014). AMYCO is a server for fast, automated, and graphical evaluation of the impact of mutations on the aggregation properties of prion-like proteins (Iglesias et al., 2019). Its performance is better than that of previous state-of-the-art predictors. Moreover, the AMYCO implementation has been used to gain insights into prion evolution. While SGnn enables recruitment of PrLDs to heat-induced stress granules (SGs) across complete proteomes. The SGnn algorithm is based on three parameters that are relevant to SG localization (i.e., PrLD aggregation propensity, electrostatic interactions, and free cysteine content) (Iglesias et al., 2021).

14.3 Predicting protein condensation propensity

The relatively small number of experimentally validated phase-separating proteins and the difficulty in discovering them remains a bottleneck in the development of accurate predictors for condensates (see Table 14.4). Despite these limitations, in the last decade, novel approaches have been constantly developed. In general, most machine-learning classifiers for predicting LLPS propensity are based on the amino acid sequence input. For example, phase-separating protein (PSPer) highlights the regions that are likely to phase-separate through a particular mechanism involving RNA interactions. It can prioritize PSPer among proteins containing similar RNA-binding domains, intrinsically disordered regions, and prions (Orlando et al., 2019). PSPredictor is a sequence-based tool for the query and prediction of potential phase separation proteins by integrating the database LLPSDB (Li et al., 2019) and the machine-learning methods. PSPredictor enables users to determine the most similar proteins in LLPSDB and their experimentally validated phase-separation conditions (Chu et al., 2022). PSAP is a random forest classifier used to predict the probability of proteins mediating phase separation. This classifier is trained on a set of 90 high-confidence human proteins (van Mierlo et al., 2021). A recent survey by Pancsa et al. (2021) is a useful showcase of advancements in LLPS bioinformatics, including comprehensive discussions of limitations of the LLPS methods and benchmarking analyses on a set of proteins.

Moreover, PhaSePred is a novel metapredictor to discover candidate PSPers (Chen et al., 2022). Its advantage is the integration of multiple machine-learning models for the prediction of PSPers. Moreover, this predictor can sufficiently predict self-assembling or partner-dependent protein categories. This is an advantage of PhaSePred, since most of the current computational tools preferentially predict only self-assembling proteins and perform poorly in screening partner-dependent proteins. PhaSePred prediction is based on the

Table 14.4 Computational tools for prediction of condensates.

Predictor	Description	References
catGRANULE	Liquid phase separation	Bolognesi et al. (2016)
PScore	Pi–pi contacts of phase separation	Vernon et al. (2018)
LARKS	Low-complexity aromatic-rich kinked segments	Hughes et al. (2018)
PSPer	Phase-separating protein	Orlando et al. (2019)
FuzDrop	Identification of droplet-promoting regions	Hardenberg et al. (2020)
PSAP	Probability of proteins to mediate phase separation	van Mierlo et al. (2021)
DeePhase	Distinguishing liquid–liquid phase-separated-prone proteins	Saar et al. (2021)
PSPredictor	Prediction of liquid–liquid PSP	Chu et al. (2022)
PhaSePred	Metapredictor for PSPers	Chen et al. (2022)
PSPire	Screen of PSPers lacking IDRs	Hou et al. (2024)

integration of scores from representative single predictors, including catGRANULE, classification of intrinsically disordered ensemble regions (CIDER), DeePhase, FuzDrop, low-complexity aromatic-rich kinked segments (LARKS), PLAAC, PScore, and ZipperDB. PLAAC and ZipperDB predict PrLDs and fibril-forming segments, respectively. CIDER is a web-server for calculating various sequence parameters associated with disordered protein sequences (Holehouse et al., 2017). Further, PScore, catGRANULE, and LARKS are first-generation LLPS propensity predictors. Since these predictors were generated on different training samples, the differences in their predictive behavior can be determined. CatGRANULE was trained on granule participants (Bolognesi et al., 2016). On the other hand, the PScore predictive tool is based on pi interaction frequency, highlighting proteins involved in biomaterials and RNA processing (Vernon et al., 2018). LARKS is an algorithm for the identification of protein interaction motifs that provide adhesion of low-complexity domains in protein gels and membrane-less assemblies (Hughes et al., 2018). Compared with the first-generation predictors, FuzDrop and DeePhase were developed on a larger number of training datasets from different databases (LLPSDB) (Li et al., 2019), PhaSepDB (You et al., 2020), and PhaSePro (Mészáros et al., 2020), and consequently can provide a broader perspective for screening LLPS proteins. The predictor DeePhase is a neural network-based model and has high performance for distinguishing LLPS-prone proteins from structured proteins and for identifying them within the human proteome. Authors of this tool highlighted that LLPS-prone proteins are more disordered, less hydrophobic, and of lower Shannon entropy (Saar et al., 2021). FuzDrop is a method based on the concept that the droplet state is stabilized by the large conformational entropy associated with nonspecific side-chain interactions. This method performs a sequence-based identification of droplet-promoting regions and aggregation-promoting regions within proteins, which can spontaneously phase separate (Hardenberg et al., 2020). Most of the tools are often biased towards PSPs with a high content of IDRs and frequently undervalue potential PSPs without IDRs. Therefore PSPire, a machine-learning predictor, is designed based on the XGBoost classifier and improved availability of high-accuracy atomic coordinates of proteins in the full human proteome. PSPire incorporates both residue-level and structure-level features for the precise prediction of PSPs. Compared to the best current predictors, PSAP and PhaSePred, PSPire shows a notable improvement in identifying PSPs without IDRs. Evaluations demonstrate that PSPire (Hou et al., 2024) is a state-of-the-art condensate predictor and significantly outperforms current predictors, highlighting the significant value of protein structural information in phase separation.

14.4 Future perspective

Worth highlighting is how a recent avenue of AlphaFold (Jumper et al., 2021) and related publicly accessible high-accuracy structure predictions in the AlphaFoldDB database (https://alphafold.ebi.ac.uk) has massively broken down the limitations in determinationof 3D protein structures. Consequently, it is likely that in the next years the appearance of numerous highly reliable tools able to predict the aggregation/ condensation prone regions embedded in protein 3D structures (high-resolution experimental structures or AlphaFold structures - generated by artificial intelligence algorithms) will rise. These acievements will open new ways for application in the field of biological protein design, such as formulation of antibodies, and other therapeutic peptides/ proteins with an aim to prevent and even cure diseases caused by protein aggregation (Pintado-Grima et al., 2023). However, in the last decade, impressive innovations in prediction of aggregation and condensation of proteins have been made. The currently available algorithms fairly well enable an automated, sequence and structure -based predictions of aggregation/condensation properties for any protein of scientific or industrial interest.

In parallel, another comprehensive bioinformatics discipline focused on protein–protein interactions is evolving. These discoveries are especially relevant to protein condensates, so-called membrane-less organelles, since some of the condensing proteins are scaffolds, whereas others are partners (Visser et al., 2024). The introduction of AlphaFold2 was a revolution in modeling the structure of proteins and their interactions. Moreover, AlphaFold3 with an even more updated diffusion-based architecture is capable of predicting the joint structure of complexes including proteins, nucleic acids, small molecules, ions and modified residues (Abramson et al., 2024). It is foreseen that in the near future the integration of knowledge from many previous specialized tools of aggregation/condensation and protein–protein interaction predictions will enable high-accuracy modeling across biomolecular space within a single unified deep-learning framework.

Acknowledgement

The author thanks collaborators and mentors, especially prof. Marjana Novič (NIC, Ljubljana) and prof. Eva Žerovnik (IJS, Ljubljana) for useful discussions on the subject. This research was financially supported by Slovenian Research and Innovation Agency (ARIS), grant number P1-0017.

References

Abramson, J., Adler, J., Dunger, J., Evans, R., Green, T., Pritzel, A., et al., 2024. Accurate structure prediction of biomolecular interactions with AlphaFold 3. Nature 630 (8016), 493–500. Available at https://doi.org/10.1038/s41586-024-07487-w.

Ahmed, A.B., Znassi, N., Château, M.-T., Kajava, A.V., 2015. A structure-based approach to predict predisposition to amyloidosis. Alzheimer's & Dementia 11 (6), 681–690. Available at https://doi.org/10.1016/j.jalz.2014.06.007.

Bolognesi, B., Lorenzo Gotor, N., Dhar, R., Cirillo, D., Baldrighi, M., Tartaglia, G.G., et al., 2016. A concentration-dependent liquid phase separation can cause toxicity upon increased protein expression. Cell Reports 16 (1), 222–231. Available at https://doi.org/10.1016/j.celrep.2016.05.076.

Bryan, A.W., Menke, M., Cowen, L.J., Lindquist, S.L., Berger, B., 2009. BETASCAN: probable beta-amyloids identified by pairwise probabilistic analysis. PLoS Computational Biology 5 (3), e1000333. Available at https://doi.org/10.1371/journal.pcbi.1000333.

Burdukiewicz, M., Sobczyk, P., Rödiger, S., Duda-Madej, A., MacKiewicz, P., Kotulska, M., 2017. Amyloidogenic motifs revealed by n-gram analysis. Scientific Reports 7 (1). Available at https://doi.org/10.1038/s41598-017-13210-9.

Chen, Z., Hou, C., Wang, L., Yu, C., Chen, T., Shen, B., et al., 2022. Screening membraneless organelle participants with machine-learning models that integrate multimodal features. Proceedings of the National Academy of Sciences of the United States of America 119 (24). Available at https://doi.org/10.1073/pnas.2115369119.

Chennamsetty, N., Voynov, V., Kayser, V., Helk, B., Trout, B.L., 2009. Design of therapeutic proteins with enhanced stability. Proceedings of the National Academy of Sciences of the United States of America 106 (29), 11937–11942. Available at https://doi.org/10.1073/pnas.0904191106.

Chu, X., Sun, T., Li, Q., Xu, Y., Zhang, Z., Lai, L., et al., 2022. Prediction of liquid–liquid phase separating proteins using machine learning. BMC Bioinformatics 23, 72. Available at https://doi.org/10.1186/s12859-022-04599-w.

Conchillo-Solé, O., de Groot, N.S., Avilés, F.X., Vendrell, J., Daura, X., Ventura, S., 2007. AGGRESCAN: a server for the prediction and evaluation of "hot spots" of aggregation in polypeptides. BMC Bioinformatics 8. Available at https://doi.org/10.1186/1471-2105-8-65.

Emily, M., Talvas, A., Delamarche, C., Tosatto, S.C.E., 2013. MetAmyl: a METa-predictor for AMYLoid proteins. PLoS One 8 (11), e79722. Available at https://doi.org/10.1371/journal.pone.0079722.

Espinosa Angarica, V., Angulo, A., Giner, A., Losilla, G., Ventura, S., Sancho, J., 2014. PrionScan: an online database of predicted prion domains in complete proteomes. BMC Genomics 15 (1). Available at https://doi.org/10.1186/1471-2164-15-102.

Família, C., Dennison, S.R., Quintas, A., Phoenix, D.A., 2015. Prediction of peptide and protein propensity for amyloid formation. Plos One 10, e0134679.

Fernandez-Escamilla, A.M., Rousseau, F., Schymkowitz, J., Serrano, L., 2004. Prediction of sequence-dependent and mutational effects on the aggregation of peptides and proteins. Nature Biotechnology 22 (10), 1302–1306. Available at https://doi.org/10.1038/nbt1012.

Frousios, K.K., Iconomidou, V.A., Karletidi, C.M., Hamodrakas, S.J., 2009. Amyloidogenic determinants are usually not buried. BMC Structural Biology 9, 44. Available at https://doi.org/10.1186/1472-6807-9-44.

Garbuzynskiy, S.O., Lobanov, M.Y., Galzitskaya, O.V., 2009. FoldAmyloid: a method of prediction of amyloidogenic regions from protein sequence. Bioinformatics (Oxford, England) 26 (3), 326–332. Available at https://doi.org/10.1093/bioinformatics/btp691.

Gasior, P., Kotulska, M., 2014. FISH Amyloid - a new method for finding amyloidogenic segments in proteins based on site specific co-occurence of aminoacids. BMC Bioinformatics 15 (1), 54. Available at https://doi.org/10.1186/1471-2105-15-54.

Goldschmidt, L., Teng, P.K., Riek, R., Eisenberg, D., 2010. Identifying the amylome, proteins capable of forming amyloid-like fibrils. Proceedings of the National Academy of Sciences of the United States of America 107 (8), 3487–3492. Available at https://doi.org/10.1073/pnas.0915166107.

Graña-Montes, R., Ventura, S., 2015. Protein aggregation and its prediction. In: Scapin, G., Patel, D., Arnold, E. (Eds.), Multifaceted Roles of Crystallography in Modern Drug Discovery. NATO Science for Peace and Security Series A: Chemistry and Biology. Springer, Dordrecht, pp. 115–127. Available at https://doi.org/10.1007/978-94-017-9719-1_10.

Hardenberg, M., Horvath, A., Ambrus, V., Fuxreiter, M., Vendruscolo, M., 2020. Widespread occurrence of the droplet state of proteins in the human proteome. Proceedings of the National Academy of Sciences of the United States of America 117 (52), 33254–33262. Available at https://doi.org/10.1073/PNAS.2007670117.

He, W., Xu, X., Li, H., Zhou, J., Gao, X., 2025. AggNet: advancing protein aggregation analysis through deep learning and protein language model. Protein Science 34 (2). Available at https://doi.org/10.1002/pro.70031.

Holehouse, A.S., Das, R.K., Ahad, J.N., Richardson, M.O.G., Pappu, R.V., 2017. CIDER: resources to analyze sequence-ensemble relationships of intrinsically disordered proteins. Biophysical Journal 112 (1), 16–21. Available at https://doi.org/10.1016/j.bpj.2016.11.3200.

Hou, Q., Kwasigroch, J.M., Rooman, M., Pucci, F., 2019. SOLart: a structure-based method to predict protein solubility and aggregation. Bioinformatics 36, 1445–1452.

Hou, S., Hu, J., Yu, Z., Li, D., Liu, C., Zhang, Y., 2024. Machine learning predictor PSPire screens for phase-separating proteins lacking intrinsically disordered regions. Nature Communications 15 (1). Available at https://doi.org/10.1038/s41467-024-46445-y.

Hughes, M.P., Sawaya, M.R., Boyer, D.R., Goldschmidt, L., Rodriguez, J.A., Cascio, D., et al., 2018. Atomic structures of low-complexity protein segments reveal kinked b sheets that assemble networks. Science (New York, N.Y.) 359 (6376), 698–701. Available at https://doi.org/10.1126/science.aan6398.

Iglesias, V., Conchillo-Sole, O., Batlle, C., Ventura, S., 2019. AMYCO: evaluation of mutational impact on prion-like proteins aggregation propensity. BMC Bioinformatics 20 (1). Available at https://doi.org/10.1186/s12859-019-2601-3.

Iglesias, V., Santos, J., Santos-Suárez, J., Pintado-Grima, C., Ventura, S., 2021. SGnn: A web server for the prediction of prion-like domains recruitment to stress granules upon heat stress. Frontiers in Molecular Biosciences 8. Available at https://doi.org/10.3389/fmolb.2021.718301.

Jumper, J., Evans, R., Pritzel, A., Green, T., Figurnov, M., Ronneberger, O., et al., 2021. Highly accurate protein structure prediction with AlphaFold. Nature 596 (7873), 583–589. Available at https://doi.org/10.1038/s41586-021-03819-2.

Keresztes, L., Szögi, E., Varga, B., Farkas, V., Perczel, A., Grolmusz, V., 2021. The budapest amyloid predictor and its applications. Biomolecules 11 (4), 500. Available at https://doi.org/10.3390/biom11040500.

Kim, C., Choi, J., Lee, S.J., Welsh, W.J., Yoon, S., 2009. NetCSSP: web application for predicting chameleon sequences and amyloid fibril formation. Nucleic Acids Research 37, W469. Available at https://doi.org/10.1093/nar/gkp351.

Kuriata, A., Iglesias, V., Pujols, J., Kurcinski, M., Kmiecik, S., Ventura, S., 2019. Aggrescan3D (A3D) 2.0: prediction and engineering of protein solubility. Nucleic Acids Research 47 (1), W300. Available at https://doi.org/10.1093/nar/gkz321.

Lancaster, A.K., Nutter-Upham, A., Lindquist, S., King, O.D., 2014. PLAAC: a web and command-line application to identify proteins with prion-like amino acid composition. Bioinformatics (Oxford, England) 30 (17), 2501–2502. Available at https://doi.org/10.1093/bioinformatics/btu310.

Li, Q., Peng, X., Li, Y., Tang, W., Zhu, J.A., Huang, J., et al., 2019. LLPSDB: a database of proteins undergoing liquid–liquid phase separation in vitro. Nucleic Acids Research 48, D320–D327.

Louros, N., Konstantoulea, K., De Vleeschouwer, M., Ramakers, M., Schymkowitz, J., Rousseau, F., 2019. WALTZ-DB 2.0: an updated database containing structural information of experimentally determined amyloid-forming peptides. Nucleic Acids Research 48 (D1), D389–D393.

Louros, N., Orlando, G., De Vleeschouwer, M., Rousseau, F., Schymkowitz, J., 2020. Structure-based machine-guided mapping of amyloid sequence space reveals uncharted sequence clusters with higher solubilities. Nature Communications 11 (1). Available at https://doi.org/10.1038/s41467-020-17207-3.

Maurer-Stroh, S., Debulpaep, M., Kuemmerer, N., De La Paz, M.L., Martins, I.C., Reumers, J., et al., 2010. Exploring the sequence determinants of amyloid structure using position-specific scoring matrices. Nature Methods 7 (3), 237–242. Available at https://doi.org/10.1038/nmeth.1432.

Maziuk, B., Ballance, H.I., Wolozin, B., 2017. Dysregulation of RNA binding protein aggregation in neurodegenerative disorders. Frontiers in Molecular Neuroscience 10. Available at https://doi.org/10.3389/FNMOL.2017.00089.

Mészáros, B., Erdos, G., Szabó, B., Schád, É., Tantos, Á., Abukhairan, R., et al., 2020. PhaSePro: the database of proteins driving liquid-liquid phase separation. Nucleic Acids Research 48 (1), D360. Available at https://doi.org/10.1093/nar/gkz848.

Orlando, G., Raimondi, D., Tabaro, F., Codicè, F., Moreau, Y., Vranken, W.F., 2019. Computational identification of prion-like RNA-binding proteins that form liquid phase-separated condensates. Bioinformatics (Oxford, England) 35 (22), 4617–4623. Available at https://doi.org/10.1093/bioinformatics/btz274.

Paladin, L., Piovesan, D., Tosatto, S.C.E., Pancsa, R., Vranken, W., Mészáros, B., 2017. Computational resources for identifying and describing proteins driving liquid-liquid phase separation. Briefings in Bioinformatics 45.

Pancsa, R., Vranken, W., Mészáros, B., 2021. Computational resources for identifying and describing proteins driving liquid–liquid phase separation. Briefings in Bioinformatics 22 (5), bbaa408.

Pawlicki, S., Le Béchec, A., Delamarche, C., 2008. AMYPdb: a database dedicated to amyloid precursor proteins. BMC Bioinformatics 9. Available at https://doi.org/10.1186/1471-2105-9-273.

Pintado-Grima, C., Bárcenas, O., Bartolomé-Nafría, A., Fornt-Suñé, M., Iglesias, V., Garcia-Pardo, J., et al., 2023. A review of fifteen years developing computational tools to study protein aggregation. Biophysica 3 (1), 1–20. Available at https://doi.org/10.3390/biophysica3010001.

Planas-Iglesias, J., Borko, S., Swiatkowski, J., Elias, M., Havlasek, M., Salamon, O., et al., 2024. AggreProt: a web server for predicting and engineering aggregation prone regions in proteins. Nucleic Acids Research 52 (W1), W159. Available at https://doi.org/10.1093/nar/gkae420.

Prabakaran, R., Rawat, P., Kumar, S., Gromiha, M.M., 2021. Evaluation of in silico tools for the prediction of protein and peptide aggregation on diverse datasets. Briefings in Bioinformatics 22 (6). Available at https://doi.org/10.1093/bib/bbab240.

Rawat, P., Prabakaran, R., Kumar, S., Gromiha, M.M., 2020. AggreRATE-Pred: a mathematical model for the prediction of change in aggregation rate upon point mutation. Bioinformatics (Oxford, England) 36 (5), 1439–1444. Available at https://doi.org/10.1093/bioinformatics/btz764.

Saar, K.L., Morgunov, A.S., Qi, R., Arter, W.E., Krainer, G., Lee, A.A., et al., 2021. Learning the molecular grammar of protein condensates from sequence determinants and embeddings. Proceedings of the National Academy of Sciences of the United States of America 118 (15). Available at https://doi.org/10.1073/pnas.2019053118.

Sabate, R., Rousseau, F., Schymkowitz, J., Ventura, S., Fetrow, J.S., 2015. What makes a protein sequence a prion? PLoS Computational Biology 11 (1), e1004013. Available at https://doi.org/10.1371/journal.pcbi.1004013.

Sanchez de Groot, N., Torrent, M., Villar-Piqué, A., Lang, B., Ventura, S., Gsponer, J., et al. (2012) Evolutionary selection for protein aggregation, Biochemical Society Transactions40, 1032-1037.

Sankar, K., Krystek, S.R., Carl, S.M., Day, T., Maier, J.K.X., 2018. AggScore: prediction of aggregation-prone regions in proteins based on the distribution of surface patches. Proteins: Structure, Function and Bioinformatics 86 (11), 1147–1156. Available at https://doi.org/10.1002/prot.25594.

Sormanni, P., Amery, L., Ekizoglou, S., Vendruscolo, M., Popovic, B., 2017. Rapid and accurate in silico solubility screening of a monoclonal antibody library. Scientific Reports 7 (1). Available at https://doi.org/10.1038/s41598-017-07800-w.

Tartaglia, G.G., Vendruscolo, M., 2008. The Zyggregator method for predicting protein aggregation propensities. Chemical Society Reviews 37 (7), 1395–1401. Available at https://doi.org/10.1039/b706784b.

Thangakani, A.M., Kumar, S., Nagarajan, R., Velmurugan, D., Gromiha, M.M., 2014. GAP: towards almost 100 percent prediction for β-strand-mediated aggregating peptides with distinct morphologies. Bioinformatics 30, 1983–1990.

Tian, J., Wu, N., Guo, J., Fan, Y., 2009. Prediction of amyloid fibril-forming segments based on a support vector machine. BMC Bioinformatics 10 (1). Available at https://doi.org/10.1186/1471-2105-10-S1-S45.

Tsolis, A.C., Papandreou, N.C., Iconomidou, V.A., Hamodrakas, S.J., 2013. A consensus method for the prediction of 'aggregation-prone' peptides in globular proteins. PLoS One 8 (1). Available at https://doi.org/10.1371/journal.pone.0054175.

Van Durme, J., De Baets, G., Van Der Kant, R., Ramakers, M., Ganesan, A., Wilkinson, H., et al., 2016. Solubis: a webserver to reduce protein aggregation through mutation. Protein Engineering, Design and Selection 29 (8), 285–289. Available at https://doi.org/10.1093/protein/gzw019, http://peds.oxfordjournals.org/.

van Mierlo, G., Jansen, J.R.G., Wang, J., Poser, I., van Heeringen, S.J., Vermeulen, M., 2021. Predicting protein condensate formation using machine learning. Cell Reports 34 (5). Available at https://doi.org/10.1016/j.celrep.2021.108705.

Vendruscolo, M., Fuxreiter, M., 2022. Sequence determinants of the aggregation of proteins within condensates generated by liquid-liquid phase separation. Journal of Molecular Biology 434, 167201.

Venko, K., Žerovnik, E., 2023. Protein condensates and protein aggregates: in vitro, in the cell, and in silico. Frontiers in Bioscience-Landmark 28 (8). Available at https://doi.org/10.31083/j.fbl2808183.

Vernon, R.M.C., Chong, P.A., Tsang, B., Kim, T.H., Bah, A., Farber, P., et al., 2018. Pi-Pi contacts are an overlooked protein feature relevant to phase separation. eLife 7. Available at https://doi.org/10.7554/eLife.31486.

Visser, B.S., Lipiński, W.P., Spruijt, E., 2024. The role of biomolecular condensates in protein aggregation. Nature Reviews Chemistry 8, 686–700.

Walsh, I., Seno, F., Tosatto, S.C.E., Trovato, A., 2014. PASTA 2.0: an improved server for protein aggregation prediction. Nucleic Acids Research 42 (W1), W301. Available at https://doi.org/10.1093/nar/gku399.

Wozniak, P.P., Kotulska, M., 2015. AmyLoad: website dedicated to amyloidogenic protein fragments. Bioinformatics (Oxford, England) 31 (20), 3395–3397. Available at https://doi.org/10.1093/bioinformatics/btv375.

You, K., Huang, Q., Yu, C., Shen, B., Sevilla, C., Shi, M., et al., 2020. PhaSepDB: a database of liquid-liquid phase separation related proteins. Nucleic Acids Research 48 (1), D354. Available at https://doi.org/10.1093/nar/gkz847.

Zalewski, M., Iglesias, V., Bárcenas, O., Ventura, S., Kmiecik, S., 2024. Aggrescan4D: a comprehensive tool for pH-dependent analysis and engineering of protein aggregation propensity. Protein Science 33 (10), e5180. Available at https://doi.org/10.1002/pro.5180.

Zambrano, R., Conchillo-Sole, O., Iglesias, V., Illa, R., Rousseau, F., Schymkowitz, J., et al., 2015a. PrionW: a server to identify proteins containing glutamine/asparagine rich prion-like domains and their amyloid cores. Nucleic Acids Research 43, W331–W337.

Zambrano, R., Jamroz, M., Szczasiuk, A., Pujols, J., Kmiecik, S., Ventura, S., 2015b. AGGRESCAN3D (A3D): server for prediction of aggregation properties of protein structures. Nucleic Acids Research 43, W306–W313.

Index

Note: Page numbers followed by "*f*," "*t*," and "*b*" refer to figures, tables, and boxes, respectively.

D

E

K

L

M

N

O

P

Q

R

S

T

U

V

W

X

Z

9780443341762